HARCOURT
Math

Harcourt School Publishers

Orlando • Boston • Dallas • Chicago • San Diego

www.harcourtschool.com

ISBN 0-15-320750-7

3 4 5 6 7 8 9 10 032 10 09 08 07 06 05 04 03 02

Senior Author

Evan M. Maletsky
Professor of Mathematics
Montclair State University
Upper Montclair, New Jersey

Mathematics Advisors

Jerome Dancis
*Associate Professor of
Mathematics*
University of Maryland
College Park, Maryland

Tom Roby
*Assistant Professor of
Mathematics*
California State University
Hayward, California

Authors

Angela Giglio Andrews
Math Teacher, Scott School
Naperville District #203
Naperville, Illinois

Jennie M. Bennett
Instructional Mathematics Supervisor
Houston Independent School District
Houston, Texas

Grace M. Burton
*Chair, Department of Curricular Studies
Professor, School of Education*
University of North Carolina
  at Wilmington
Wilmington, North Carolina

Howard C. Johnson
*Dean of the Graduate School
Associate Vice Chancellor
  for Academic Affairs
Professor, Mathematics and
  Mathematics Education*
Syracuse University
Syracuse, New York

Lynda A. Luckie
Administrator/Math Specialist
Gwinnett County Public Schools
Lawrenceville, Georgia

Joyce C. McLeod
Visiting Professor
Rollins College
Winter Park, Florida

Vicki Newman
Classroom Teacher
McGaugh Elementary School
Los Alamitos Unified School District
Seal Beach, California

Janet K. Scheer
Executive Director
Create A Vision
Foster City, California

Karen A. Schultz
College of Education
Georgia State University
Atlanta, Georgia

Program Consultants and Specialists

Janet S. Abbott
Mathematics Consultant
California

Lois Harrison-Jones
*Education and Management
  Consultant*
Dallas, Texas

Elsie Babcock
*Director, Mathematics and Science
  Center; Mathematics Consultant*
Wayne Regional Educational
  Service Agency
Wayne, Michigan

Arax Miller
*Curriculum Coordinator and English
  Department Chairperson*
Chamlian School
Glendale, California

William J. Driscoll
Professor of Mathematics
Department of Mathematical Sciences
Central Connecticut State University
New Britain, Connecticut

Rebecca Valbuena
Language Development Specialist
Stanton Elementary School
Glendora, California

iii

UNIT 1
CHAPTERS 1–4

Number Sense and Operations

TECHNOLOGY LINK

**Harcourt Math
Newsroom Video:**
Chapter 1, p. 18

E-Lab:
Chapter 2, p. 43
Chapter 4, p. 75

**Mighty Math Calculating
Crew:**
Chapter 1, p. 24
Chapter 3, p. 54
Chapter 4, p. 67

Multimedia Glossary:
*The Learning Site at
www.harcourtschool.com/
mathglossary*

UNIT 2
CHAPTERS 5–6
Statistics and Graphing

Chapter 5

TECHNOLOGY LINK

**Harcourt Math
Newsroom Video:**
Chapter 5, p. 95

E-Lab:
Chapter 6, p. 129

Data Toolkit:
Chapter 5, p. 102
Chapter 6, p. 122

Multimedia Glossary:
The Learning Site at
www.harcourtschool.com/
mathglossary

Chapter 6

UNIT 3
CHAPTERS 7–10

Fraction Concepts and Operations

TECHNOLOGY LINK

UNIT 4
CHAPTERS 11–12
Algebra: Integers

TECHNOLOGY LINK

Harcourt Math Newsroom Video:
Chapter 11, p. 234

E-Lab:
Chapter 12, p. 249

Mighty Math Astro Algebra:
Chapter 12, pp. 253

Multimedia Glossary:
The Learning Site at
www.harcourtschool.com/
mathglossary

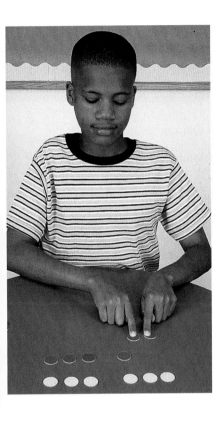

UNIT 5
CHAPTERS 13–15

Algebra: Expressions and Equations

TECHNOLOGY LINK

Harcourt Math Newsroom Video:
Chapter 13, p. 273

E-Lab:
Chapter 13, p. 277
Chapter 14, p. 286
Chapter 15, pp. 296, 305

Mighty Math Astro Algebra:
Chapter 15, p. 298

Multimedia Glossary:
The Learning Site at
www.harcourtschool.com/
mathglossary

xii

Chapter 15
ALGEBRA MULTIPLICATION AND DIVISION EQUATIONS **294**

UNIT WRAPUP

Geometry and Spatial Reasoning

TECHNOLOGY LINK

**Harcourt Math
Newsroom Video:**
Chapter 17, p. 338

E-Lab:
Chapter 18, p. 353
Chapter 19, p. 369

**Mighty Math Cosmic
Geometry:**
Chapter 16, pp. 321, 323
Chapter 17, pp. 334, 345

Multimedia Glossary:
The Learning Site at
www.harcourtschool.com/
mathglossary

UNIT 7
CHAPTERS 20–23
Ratio, Proportion, Percent, and Probability

TECHNOLOGY LINK

Harcourt Math Newsroom Video:
Chapter 20, p. 391

E-Lab:
Chapter 20, pp. 387, 398
Chapter 21, p. 417
Chapter 22, p. 435

Mighty Math Cosmic Geometry:
Chapter 20, p. 397

Mighty Math Number Heroes:
Chapter 23, p. 448

Multimedia Glossary:
The Learning Site at
www.harcourtschool.com/
mathglossary

Chapter 22

Chapter 23

UNIT 8
CHAPTERS 24–27

Measurement

TECHNOLOGY LINK

**Harcourt Math
Newsroom Video:**
Chapter 25, p. 485

E-Lab:
*Chapter 26, p. 501
Chapter 27, p. 520*

Multimedia Glossary:
The Learning Site at
**www.harcourtschool.com/
mathglossary**

UNIT 9
CHAPTERS 28–30

Algebra: Patterns and Relationships

TECHNOLOGY LINK

Harcourt Math Newsroom Video:
Chapter 28, p. 539

E-Lab:
Chapter 28, p. 541
Chapter 30, p. 581

Mighty Math Cosmic Geometry:
Chapter 29, p. 554

Mighty Math Astro Algebra:
Chapter 30, p. 569

Multimedia Glossary:
The Learning Site at
www.harcourtschool.com/mathglossary

STUDENT HANDBOOK

WELCOME!

The authors of *Harcourt Math* want you to be a good mathematician, but most of all we want you to enjoy learning math and feel confident that you can do it. We invite you to share your math book with family members. Take them on a guided tour through your book!

THE GUIDED TOUR

Choose a chapter you are interested in. Show your family some of these things in the chapter that will help you learn.

✓ CHECK WHAT YOU KNOW

Do you need to review any skills before you begin the next chapter? If you do, you will find help in the Handbook in the back of your book.

THE MATH LESSONS

☑ **Quick Review** to check the skills you need for the lesson.

☑ **Learn section** to help you study problems, models, examples, and questions that give you different ways to learn.

☑ **Check for Understanding** to make sure you understood the lesson.

☑ **Practice and Problem Solving** to help you practice what you have just learned.

☑ **Mixed Review and Test Prep** to keep your skills sharp and prepare you for important tests. Look back at the pages shown by each problem to get help if you need it.

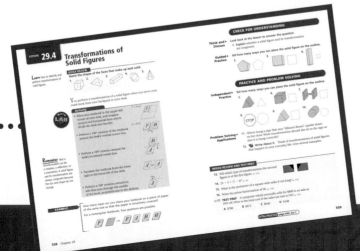

STUDENT HANDBOOK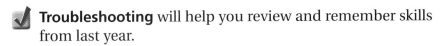

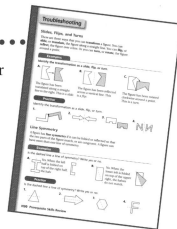

Now show your family the **Student Handbook** in the back of your book. The sections will help you in many different ways.

☑ **Troubleshooting** will help you review and remember skills from last year.

☑ **Extra Practice** will help you make sure that you are ready to move on to the next lesson.

☑ **Sharpen Your Test-Taking Skills** will help you feel confident that you can do well on a test.

☑ **Skills Review** will help you review addition, subtraction, multiplication, and division of whole numbers and decimals to improve your skills.

☑ **Answers to Selected Exercises** will help you check your answers to part of an assignment.

Invite your family members to:

▶ talk with you about what you are learning.

▶ help you correct errors you have made on completed work.

▶ help you set a time and find a quiet place to do math homework.

▶ solve problems with you as you play, shop, and do household chores together.

▶ visit **The Learning Site** at **www.harcourtschool.com**

▶ have *FUN WITH MATH!*

Have a great year!

The Authors

FOCUS ON
PROBLEM SOLVING

Analyze
Choose
Solve
Check

Good problem solvers need to be good thinkers. They also need to know the strategies listed on page 1. This plan can help you learn how to think through a problem.

Analyze the problem.

What are you asked to find?	Restate the question in your own words.
What information is given?	Look for numbers. Find how they are related.
Is there information you will not use? If so, what?	Decide whether you need all the information you are given.

Choose a strategy to solve.

What strategy will you use?	Think about some problem solving strategies you can use. Then choose one.

Solve the problem.

How can you use the strategy to solve the problem?	Follow your plan. Show your solution.

Check your answer.

Look back at the problem. Does the answer make sense? Explain.	Be sure you answered the question that is asked.
What other strategy could you use?	Solving the problem by another method is a good way to check your work.

Try It

Here's how you can use the problem-solving steps to solve a problem.

Draw a Diagram

During summer vacation, Kayla will visit her cousin and then her grandmother. She will be gone for 5 weeks and 2 days and will spend 9 more days with her cousin than with her grandmother. How long will she stay with each?

PROBLEM SOLVING STRATEGIES

▶ **Draw a Diagram or Picture**

Make a Model

Predict and Test

Work Backward

Make an Organized List

Find a Pattern

Make a Table or Graph

Solve a Simpler Problem

Write an Equation

Use Logical Reasoning

Analyze

Kayla will visit her cousin and grandmother for 5 weeks and 2 days this summer. I need to find out how long she will stay with each if she spends 9 more days with her cousin than with her grandmother.

Choose

I can *draw a diagram* to show how long she will stay. I can use boxes for the length of each stay. The length of the boxes will represent the lengths of the stays.

Solve

5 weeks and 2 days is 37 days in all. So, my diagram looks like this:

Step 1

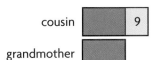

Sum of all parts (days) = 37.
37 − 9 = 28
28 ÷ 2 = 14

Step 2

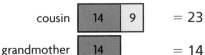

cousin $14 \quad 9$ = 23

grandmother 14 = 14   So, each part = 14.

So, Kayla will stay with her cousin for 23 days and with her grandmother for 14 days.

Check

23 days is 9 days longer than 14 days. The total of the two stays is 23 + 14, or 37 days. This solution fits the description of Kayla's vacation trip.

Draw a Diagram

Eight basketball teams will play in a tournament. A team is out after it loses once. How many games does a team have to win in order to win the tournament?

Analyze

There are 8 teams. If a team loses, it is out of the tournament. If a team wins, it keeps playing.

Choose

You can *draw a diagram* to show which teams play in each game of the tournament. Use the diagram to show each pair of teams playing against each other.

Solve

Use the diagram to pair up the 8 teams. Make up a winner for each game. Then show the next games until the tournament has a winner.

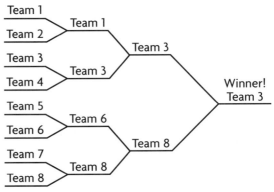

So, a team must win 3 games to win the tournament.

Check

Follow Team 3 through the tournament listing. Count the number of times it wins.

Problem Solving Practice

1. **What if** half as many teams play in the tournament? How many games will have to be played to determine the winner?

2. After the game, the teams have sandwiches at their coaches' houses. You can build your own sandwich with lettuce, tomato, and ham. How many different ways can you stack the 3 fillings to make a sandwich?

Make a Model

Alice has three pieces of ribbon. Their lengths are 7 in., 10 in., and 12 in. How can Alice use these ribbons to measure a length of 15 in.?

Analyze

Each of the three ribbons is a different length. No ribbon is 15 in. long, but you can combine the ribbon lengths to measure 15 in.

Choose

You can *make a model* of each piece of ribbon. Then you can organize the ribbons to show new lengths.

Solve

When you put two pieces together, you can form lengths of 17, 19, and 22 in. All of these are too long.

One Way
Alice can place the 7-in. piece above the 10-in. piece to show that the 10-in. piece extends 3 in. beyond it. Then that 3 in. piece and the 12-in. piece together will make 15 in.

Another Way
Alice can place the 7-in. piece above the 12-in. piece to show that the 12-in. piece extends 5 in. beyond it. Then that 5 in. and the 10-in. piece together will make 15 in.

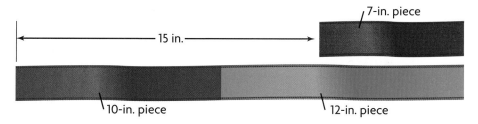

Check

Check that you used the correct lengths for the ribbon pieces. Then add the lengths of the two longer pieces and subtract the length of the short one. The result is 15 in.

Problem Solving Practice

1. Draw pictures of models you could make to measure other lengths with the three pieces of ribbon.

2. **What if** Andy is building a stand by using four cubes? He stacks the cubes, one on top of the other, and paints the outside of each cube (not the bottom). How many faces of the cubes are painted?

Predict and Test

Margaret sells tickets to a pizza dinner at her school. An adult ticket costs $3.50. A student ticket costs $2.50. Margaret sells 10 more adult tickets than student tickets. She collects $155 in ticket sales. A total of 450 people attend the pizza dinner. How many student tickets has Margaret sold?

Analyze
Student tickets cost $2.50 and adult tickets cost $3.50. Margaret sells 10 more adult tickets than student tickets. Her sales are $155. The number of people at the dinner is given, but this information is not needed.

Choose
You can use *predict and test* to solve. Use number sense and the needed information to predict how many student tickets Margaret has sold. Then test your prediction and revise it if needed.

Solve
Make a table to show each prediction and its result. Be sure you have 10 more adult tickets than student tickets each time.

PREDICTION		TEST	
Adult	Student	Sales	
20	10	(20 × $3.50) + (10 × $2.50) = $95	too low, revise
40	30	(40 × $3.50) + (30 × $2.50) = $215	too high, revise
30	20	(30 × $3.50) + (20 × $2.50) = $155	✓ correct

So, Margaret has sold 20 student tickets.

Check
Check that each multiplication is correct: 30 × $3.50 = $105, and 20 × $2.50 = $50; $105 + $50 = $155. The answer checks.

Problem Solving Practice

1. The sum of Margaret's and her younger brother's ages is 38. The difference between their ages is 8. How old is Margaret's brother?

2. Tables for the pizza dinner are set up in 3 rooms. One room has 11 tables to seat 98 people. Some tables are for 8 people, and others are for 10 people. How many tables for 8 people are set up in this room?

Work Backward

Joe and his brother Tim go shopping. At the toy store, they use half of their money to buy a video game. Then they go to the pizza parlor and spend half of the money they have left on a pizza. Then they spend half of the remaining money to rent a video. After these stops, they have $4.50 left. How much money did they have at the start?

Analyze

You need to find how much money they had at the start. You know that they had $4.50 left and that they spent half of their money at each of three stops.

Choose

Start with the amount you know they have left—$4.50—and *work backward*. Knowing how much they have left will help you calculate first how much they spent to rent the video, then how much the pizza cost, then the price of the video game, and finally how much money they had at the start.

Solve

amount they had left → $4.50

$4.50 → twice that amount before a video → 2 × $4.50 = $9

$9 → twice that amount before a pizza → 2 × $9 = $18

$18 → twice that amount before a video game → 2 × $18 = $36

So, the amount they had at the start was $36.

Check

Put $36 in the original problem to check the amount they spent at each stop.

$36 ÷ 2 = $18, $18 ÷ 2 = $9, $9 ÷ 2 = $4.50

The amount they had left matches the amount given in the problem.

Problem Solving Practice

1. The Lauber family has 4 children. Joe is 5 years younger than his brother Mark. Tim is half as old as his brother Joe. Mary, who is 10, is 3 years older than Tim. How old is Mark?

2. If you divide this mystery number by 4, add 8, and multiply by 3, you get 42. What is the mystery number?

Make an Organized List

At Fun City Amusement Park you can throw 3 darts at a target to score points and win prizes. If each dart lands within the target area, how many different total scores are possible?

2
5
10

Analyze

If all 3 darts hit the center circle, you get 30 points. This is the highest score. If all 3 darts hit the outside circle, you get 6 points. This is the lowest score. If the 3 darts hit a different combination, other scores are possible.

Choose

Make an organized list to determine all possible hits and score totals. List the value of each dart and the total for all three darts.

Solve

First consider all 3 darts hitting the center circle. List the value of each dart and the total score. Then consider 2 darts hitting the center circle and the third dart hitting a different circle. List the value of each dart and the total scores. Do the same for 1 dart hitting the center circle and no darts hitting the center circle.

3 Darts Hit Center	2 Darts Hit Center	1 Dart Hits Center	0 Darts Hit Center
10 + 10 + 10 = 30	10 + 10 + 5 = 25	10 + 5 + 5 = 20	5 + 5 + 5 = 15
	10 + 10 + 2 = 22	10 + 5 + 2 = 17	5 + 5 + 2 = 12
		10 + 2 + 2 = 14	5 + 2 + 2 = 9
			2 + 2 + 2 = 6

So, there are 10 possible scores.

Check

Make sure that all possible combinations of scores are listed and that each set of scores in the list is different.

Problem Solving Practice

1. The Yogurt Store at Fun City sells 3 flavors of yogurt: chocolate, vanilla, and strawberry. You want to get a scoop of each flavor in a waffle cone. How many different ways can the scoops be arranged?

2. How many ways can you make change for a quarter by using dimes, nickels, and pennies?

Find a Pattern

A contractor can build stairways to a deck or patio with any number of steps. She uses the pattern at the right to build them. How many blocks are needed to build a stairway with 7 steps?

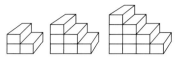

Analyze

As the number of steps increases, so does the number of blocks. You must find the number of blocks needed for a stairway with 7 steps.

Choose

You can *find a pattern* for the number of blocks needed. For each stairway, count the number of blocks to make each step and then find the total number of blocks. Use the pattern to find the number of blocks for 7 steps.

Solve

The pattern shows that the number of the step in the stairway is the same as the number of blocks needed to make it. The 2nd step is made with 2 blocks, the 3rd step is made with 3 blocks, and so on.

Number of Step	Side View	Number of Blocks	Cumulative Total
2		2	$2 + 1 = 3$
3		3	$3 + 2 + 1 = 6$
4		4	$4 + 3 + 2 + 1 = 10$

So, a stairway with 7 steps has 7 blocks in the 7th step, 6 blocks in the 6th step, and so on. The total number of blocks is $7 + 6 + 5 + 4 + 3 + 2 + 1 = 28$ blocks.

Check

Sketch a stairway with 7 steps. Check the number of blocks needed. The number in the sketch matches the answer.

$7 + 6 + 5 + 4 + 3 + 2 + 1 = 28$

Problem Solving Practice

1. A cereal company adds baseball cards to certain boxes of cereal. Cards are added to the 3rd box, the 6th box, the 11th box, and the 18th box of cereal. If this pattern continues, how many boxes will have baseball cards in them when a case of 40 is ready to be shipped? Explain.

Describe the pattern and find the missing numbers.

2. 1, 4, 16, 64, 256, ▪, ▪, 16,384

Make a Table

In math, Mrs. Laurence gave her class a 100-point test. Students who score 80 or above are eligible to join the Math Club. The test scores are in the box below. How many students scored 80 or above? How many students scored below 80?

90	83	80	77	78	91	92	73
62	83	79	88	72	85	93	84
75	68	82	75	94	70	98	82

Analyze

You have the test scores for the class. You need to find how many students scored below 80 and how many students scored 80 or above.

Choose

You can *make a table* and tally the test scores. This organizes the data and makes it easier to answer questions about the scores.

Solve

Make a two-column table for the range of scores. As you read each test score, place a tally mark across from the appropriate range. Be sure you end up with 24 tallies, one for each student.

SCORES	TALLIES
60–69	||
70–79	||||| |||
80–89	||||| |||
90–99	|||||	

Now use the table to answer the questions.

SCORES OF 80 OR ABOVE	SCORES BELOW 80
8 students scored 80–89.	8 students scored 70–79.
6 students scored 90–99.	2 students scored 60–69.

So, 14 students scored 80 or above, and 10 students scored below 80.

Check

Recount the tallies in each row and check your addition.

Problem Solving Practice

1. Starting at 3:00, the director will give each student 5 minutes to try out for the talent show. The students, in order, are Ben, Tarek, Jan, Ed, and Frank. At what time does Frank start?

2. Kelly reads stories to children at the library. There are three sessions, each lasting 45 minutes, with 30 minutes between sessions. If Kelly starts reading at 10:00 A.M., at what time does she finish?

Solve a Simpler Problem

Make a table of the first 2 odd numbers and their sum, the first 3 odd numbers and their sum, and so on. How does the table show a pattern? What is the sum of the first 20 odd numbers?

Analyze

You know how to begin finding sums for the first 2 odd numbers and the first 3 odd numbers. You need to find a pattern in the table to help you find the sum of the first 20 odd numbers.

Choose

You can begin by *solving a simpler problem*. Start with sums for the first 2 odd numbers, followed by sums for 3 odd numbers, 4 odd numbers, and 5 odd numbers.

Solve

Organize the data in a table and look for a pattern in the sums.

ODD NUMBERS	SUM	PATTERN
1 + 3	4	2^2
1 + 3 + 5	9	3^2
1 + 3 + 5 + 7	16	4^2
1 + 3 + 5 + 7 + 9	25	5^2

The table shows a pattern in the sums that relates to square numbers. Extend the pattern in the table to 20 odd numbers.

6 odd numbers → 36

7 odd numbers → 49

8 odd numbers → 64

20 odd numbers → 400

So, the sum of the first 20 odd numbers is 20^2, or 400.

Check

Extend the table a few more rows to check that there really is a pattern of square numbers. For example, find sums for the first 9, 10, 11, and 12 odd numbers. The answers check.

Problem Solving Practice

1. Martha has 5 pairs of slacks and 4 blouses for school. How many different outfits can she make with these items?

2. What is the least 5-digit number that can be divided by 50 with a remainder of 17?

Write an Equation

Aschool district can afford to build a gym with three basketball courts with an area of 18,700 square feet. The gym needs to be 110 feet wide to leave walking room around the courts. How long will the gym be?

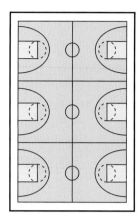

Analyze

The gym is rectangular. To find the area a rectangle covers, you multiply the length by the width. You know the area and the width of the gym and need to find its length.

Choose

You can *write an equation* for the area of the rectangle. Begin with the formula for the area of a rectangle. Use the numbers you know to find the missing number.

Solve

You find the area of a rectangle by multiplying its length by its width.

$$A = l \times w$$ Write the formula.
$$18,700 = l \times 110$$ Use the numbers you know.
$$l = 18,700 \div 110$$ Write a related division equation.
$$l = 170$$

So, the length of the gym should be 170 feet.

Check

Place your answer in the original equation.

$$A = l \times w$$
$$A = 170 \times 110$$
$$A = 18,700$$

The product matches the information in the problem.

Problem Solving Practice

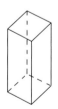

1. A rectangular box of crackers has a volume of 128 cubic inches. The area of its base is 16 square inches. What is the height of the box?

2. The perimeter of an isosceles triangle is 70 cm. The sides of equal length each measure 28 cm. What is the length of the unknown side?

Use Logical Reasoning

Christine, Elizabeth, and Sharon like different sports. One girl likes to snow ski, one likes to run track, and the other likes to swim. Elizabeth is a good friend of the girl who likes track. Sharon is shorter than the girl who likes to ski. Elizabeth does not like cold weather. Which girl likes which sport?

Analyze

You know that there are three girls and that each one likes a different sport. Clues will help you determine which girl likes which sport.

Choose

You can *use logical reasoning* to solve this problem.

Solve

Make a table. List the sports and the names of the friends. Work with the clues one at a time. Write "yes" in a box if the clue applies to that girl. Write "no" in a box if the clue does not apply. Only one box in each row and column can have a "yes" in it.

a. Elizabeth is a good friend of the girl who likes track, so Elizabeth does not run track. Write "no" in the appropriate box.

	ski	track	swim
Christine	yes	no	no
Elizabeth	no	no	yes
Sharon	no	yes	no

b. Sharon is shorter than the girl who likes to ski, so Sharon does not ski. Write "no" in the appropriate box.

c. Elizabeth does not like cold weather. Write "no" in the appropriate box. With the other sports eliminated, Elizabeth must be the person who likes to swim. Write "yes" in the appropriate box. For Christine and Sharon, write "no" in each box under *swim*.

So, Christine likes to snow ski, Sharon likes track, and Elizabeth likes to swim.

Check

Compare your answers to the clues in the problem. Make sure none of your conclusions conflict with the clues.

Problem Solving Practice

- Brent, Michael, and Tim always order their favorite pizzas: pepperoni, three-cheese, and sausage. Brent is allergic to pepperoni. Tim is going to see a movie with the boy who loves sausage. Mike does not like meat on his pizza. Which kind of pizza does each boy order?

Compare Strategies

Mrs. Hagen is eager to plant her spring garden. She wants to plant 24 tomato plants in a rectangular array. For each plant, there will be a 1-foot square of space. How many different arrays can she make?

Analyze   There are 24 plants to be put into a rectangular array. You need to find how many arrays are possible.

Choose   Lexi and Jake use different strategies to solve this problem. Lexi chooses to *draw a diagram*, and Jake chooses to *write an equation*.

Solve

Lexi's Way: Draw a Diagram

Lexi uses graph paper to draw diagrams of rectangles to show all possible arrays. Each rectangle represents a garden that covers 24 square feet. The rectangles are 1 × 24, 2 × 12, 3 × 8, and 4 × 6.

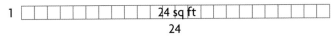

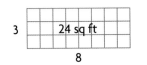 

Jake's Way: Write an Equation

Jake uses the equation $A = l \times w$ for finding the area of a rectangle to identify the different rectangles that could be formed. Each length and width needs to be a whole number.

$A = l \times w$ Think: What

$24 = 1 \times 24$ are the factors

$24 = 2 \times 12$ of 24?

$24 = 3 \times 8$

$24 = 4 \times 6$

So, Lexi and Jake both find that Mrs. Hagen can choose from four rectangular arrays.

Check   Lexi can do the problem Jake's way, and Jake can try it Lexi's way.

Problem Solving Practice

1. Mr. Sargent is buying math games for his class. One game costs $2.95, one costs $3.75, and one costs $6.00. He wants 6 of each game for his class. What is the total cost of these games?

2. How many triangles are in the figure at the right?

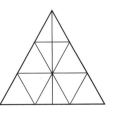

Multistep Problems

The Enrichment Program Committee at the Franklin School orders 500 T-shirts to sell at the school Field Day. Each T-shirt costs the school $4.20 and sells for $13.50. All 500 T-shirts are sold. After paying for the T-shirts, the school has reached its goal of raising $4,000. How much money has the school made?

Look at the facts in the problem. T-shirts cost $4.20 each and sell for $13.50. The committee orders and sells 500 T-shirts. You need to find how much money has been made after the costs are deducted. You know your answer should be at least $4,000.

One Way

First subtract to find how much the school has made on each T-shirt.

money earned on 1 T-shirt: $13.50 − $4.20 = $9.30

Then multiply to find the total earned.

money earned on 500 T-shirts: $9.30 × 500 = $4,650

Another Way

Find the total sales and subtract the total cost.

Total sales: 500 × $13.50 = $6,750

Total cost: 500 × $4.20 = $2,100

Total earned = $6,750 − $2,100 = $4,650

So, the school has made $4,650 on the sale of the T-shirts.

Problem Solving Practice

1. At the school Field Day, frozen yogurt cones sell for $2.25. Lin buys one for herself and one for each of her 4 friends. How much change does she get back from $20?

2. Fran is making a cabinet for her shell collection. She buys 3 boards at $2.75 each and 4 hinges at $0.99 each. What is the total cost of the supplies?

3. Mrs. Smith bought bread for $0.98, eggs for $1.17, ground turkey for $3.18, and 3 bunches of carrots for $0.65 each. She gave the clerk $10.03. How much change should Mrs. Smith receive?

CHAPTER 1

Whole Number Applications

The first Space Shuttle mission, STS-1, went up in 1981 with two astronauts. Since then, NASA has flown more than 100 Shuttle missions.

PROBLEM SOLVING To the nearest hour, the first mission, STS-1, lasted 30 hours. The STS-78 mission lasted 406 hours. About how many hours longer did the STS-78 mission last than the STS-1 mission?

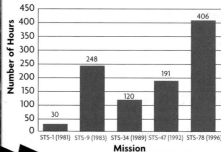

SELECTED SPACE SHUTTLE MISSIONS

Number of Hours

30	248	120	191	406
STS-1 (1981)	STS-9 (1983)	STS-34 (1989)	STS-47 (1992)	STS-78 (1996)

Mission

Check What You Know

Use this page to help you review and remember important skills needed for Chapter 1.

✔ Vocabulary

Choose the best term from the box.

	factor
	difference
	dividend
	divisor
	product
	sum

1. In $8 \times 3 = 24$, the __?__ is 24.

2. In $48 \div 12 = 4$, the __?__ is 12.

3. In $40 - 30 = 10$, the __?__ is 10.

4. In $4\overline{)412}$, the __?__ is 412.

5. In $3 + 8 + 5 = 16$, the __?__ is 16.

✔ Place Value of Whole Numbers (For Intervention, see p. H3.)

Name the place value of the digit 8.

6. 56,485,013	7. 2,403,815	8. 4,568,137
9. 423,859,100	10. 207,108,629	11. 813,593,005
12. 375,178,955,276	13. 351,480,994,133	14. 748,630,123,944

Give the value of the blue digit.

15. 321,497,580	16. 321,497,580	17. 321,497,580
18. 628,705,814,956	19. 628,705,814,956	20. 628,705,814,956

✔ Round Whole Numbers (For Intervention, see p. H4.)

Round to the nearest thousand.

21. 2,467	22. 5,609	23. 28,500	24. 299
25. 34,831	26. 19,089	27. 6,136	28. 43,712
29. 134,612	30. 217,501	31. 59,832	32. 539,513

Round to the nearest ten thousand.

33. 51,677	34. 228,260
35. 12,435	36. 78,562
37. 639,108	38. 40,500
39. 90,499	40. 381,810

LOOK AHEAD

In Chapter 1 you will

- estimate and compute with whole numbers
- evaluate expressions
- use mental math to compute and to solve equations

Estimate with Whole Numbers

Learn how to estimate sums, differences, products, and quotients of whole numbers.

Remember that when rounding, you look at the digit to the right of the place to which you are rounding.

- If that digit is 5 or greater, round up.
- If that digit is less than 5, round down.

QUICK REVIEW

1. 30 + 40
2. 25 + 60
3. 70 − 20
4. 100 − 45
5. 1,200 + 1,200 + 1,200

The longest throw of a flying disc was 656 ft 2 in., made by Scott Stokely in 1995. Carmen, David, and Leona each threw a disc. Is the total distance of the three throws close to the record distance?

You don't need an exact sum to answer the question, so estimate.

134	→	130
148	→	150
+183	→	+180
		460

Round each number to the nearest ten.

NAME	DISTANCE
David	134 ft
Leona	148 ft
Carmen	183 ft

The estimate, 460, is not close to the record distance of 656 ft, so the total distance is not close either.

Use **clustering** to estimate a sum when the addends are about the same.

EXAMPLE 1

Estimate 1,802 + 2,182 + 1,999.

1,802
2,182
+1,999

The three addends all cluster around 2,000, so use 2,000 for each number.

3 × 2,000 = 6,000   *Multiply.*

So, the sum is about 6,000.

You can use rounding to estimate a difference.

EXAMPLE 2

Estimate 31,928 − 20,915.

Round to the nearest ten thousand.	*Round to the nearest thousand.*
30,000	32,000
−20,000	−21,000
10,000	11,000

So, both 10,000 and 11,000 are reasonable estimates.

You can show an estimate by using the "approximately equal to" symbol, ≈.

$5{,}125 - 1{,}920 \approx 3{,}000$ *Read: 5,125 − 1,920 is approximately equal to 3,000.*

An estimate that is less than the exact answer is an **underestimate**. An estimate that is greater than the exact answer is an **overestimate**.

$$\begin{array}{r} 366 \\ +198 \\ \hline 564 \end{array}$$ *Exact answer* $$\begin{array}{r} 370 \\ +200 \\ \hline 570 \end{array}$$ *Round up.*
Overestimate

$$\begin{array}{r} 144 \\ \times 123 \\ \hline 17{,}712 \end{array}$$ *Exact answer* $$\begin{array}{r} 100 \\ \times 100 \\ \hline 10{,}000 \end{array}$$ *Round down.*
Underestimate

EXAMPLE 3

Students set up 28 rows of 36 seats each for a talent show in the school cafeteria. About how many programs should the school print for the show?

Estimate 28×36. To make sure there are enough programs, find an overestimate.

$$\begin{array}{r} 28 \\ \times 36 \\ \end{array} \rightarrow \begin{array}{r} 30 \\ \times 40 \\ \end{array}$$ *Round each factor up to the nearest ten.*

$$\begin{array}{r} 30 \\ \times 40 \\ \hline 1{,}200 \end{array}$$ *Multiply. Because each factor is rounded up, the product is an overestimate.*

So, the school should print about 1,200 programs.

Remember that compatible numbers are numbers that divide without a remainder, are close to the actual numbers, and are easy to compute with mentally.

To estimate a quotient, use rounding or compatible numbers.

EXAMPLE 4

The employees of the Briar Creek office building have collected 1,545 lb of paper to be recycled. The building has 36 offices. About how many pounds of paper, on average, did the employees in each office collect?

Estimate $1{,}545 \div 36$.

$1{,}600 \div 40$ *4 is compatible with 16, so use 1,600 ÷ 40.*
$1{,}600 \div 40 = 40$ *Divide.*

So, the employees in each office collected an average of about 40 lb.

- Is $1{,}488 \div 36$ easier to estimate using rounding or compatible numbers? Explain.

Math **I**dea ▶ Some of the strategies you can use to estimate are rounding, clustering, and compatible numbers.

Think and Discuss ▶ Look back at the lesson to answer each question.

1. **Explain** why $6,000 + 1,500$ is an overestimate or an underestimate of $6,108 + 1,524$.

2. **Describe** how to estimate a quotient by using compatible numbers. Give an example to illustrate your answer.

Guided Practice ▶ Estimate the sum or difference.

3. 723 +819	4. 2,940 3,140 +2,834	5. 4,480 4,100 +3,967	6. 5,449 4,869 +4,834
7. 667 −133	8. 8,855 −2,268	9. 34,855 −11,268	10. 67,184 −49,650

Estimate the product or quotient.

11. 36 × 9	12. 59 ×33	13. 48 ×29	14. 490 × 66

15. $321 \div 4$ 16. $1,544 \div 28$ 17. $4,156 \div 64$ 18. $8,429 \div 39$

Independent Practice ▶ Estimate the sum or difference.

19. 1,700 2,008 +2,324	20. 293 348 +343	21. 5,765 5,948 +6,324	22. 43,643 +84,211
23. 389 − 43	24. 3,556 −3,339	25. 44,123 −29,512	26. 667,184 −249,650

27. $17,809 − 2,145$ 28. $321,059 + 42,950$

TECHNOLOGY LINK

To learn more about estimation with whole numbers, watch the **Harcourt Math Newsroom Video** *Weather Stacked Stadium.*

Estimate the product or quotient.

29. 364 × 12	30. 53 ×41	31. 482 ×299	32. 1,874 × 582

33. $1,844 \div 22$ 34. $3,575 \div 56$ 35. $6,435 \div 529$ 36. $21,416 \div 521$

37. $4,135 \times 784$ 38. $62,217 \div 4,889$

Tell whether the estimate is an *overestimate* or an *underestimate.* Then show how the estimate was determined.

39. $352 + 675 \approx 1,100$ 40. $4,134 + 47 \approx 4,100$ 41. $96 \times 19 \approx 2,000$

42. $709 + 151 \approx 850$ 43. $291 \times 28 \approx 9,000$ 44. $25 \times 29 \approx 900$

Use estimation to compare. Write < or > for ●.

45. 614×41 ● $21{,}119 + 1{,}899$ **46.** $18{,}391 \div 19$ ● 59×21

47. $4{,}012 - 3{,}508$ ● $3{,}624 \div 6$ **48.** $12{,}283 + 19{,}971$ ● $209{,}910 \div 7$

49. 711×63 ● $28{,}520 + 16{,}990$ **50.** 513×52 ● $29{,}190 - 1{,}986$

Problem Solving ▶ Applications

Use Data   For 51–52, use the table.

51. About how many more library books does Harvard have than the University of Illinois at Urbana?

52. Estimate the total number of library books at Yale and the University of California at Berkeley.

| LARGEST UNIVERSITY LIBRARIES IN THE UNITED STATES (1999) ||
Library	Books
Harvard University	12,877,360
Yale University	9,485,823
University of Illinois–Urbana	8,474,737
University of California–Berkeley	8,078,685
University of Texas	7,019,508

53. The theater of the Natural Science and History Museum was filled to capacity for 276 shows. The theater holds 36 people. How many people attended the shows?

54. ✏ **Write About It** Is it easier to use rounding or compatible numbers to estimate $756 \div 74$? Explain.

55. Two numbers, each rounded to the nearest hundred, have a product of 60,000. What are two possible numbers?

MIXED REVIEW AND TEST PREP

For 56–58, find the perimeter and area.

56.

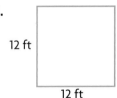

12 ft   12 ft

57.

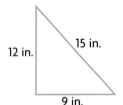

12 in.   15 in.   9 in.

58.

10 cm   3 cm

⭐ **59. TEST PREP** Ms. Cannon baked a total of 60 apple and blueberry pies. She baked a dozen more apple pies than blueberry pies. How many apple pies did she bake?

A 18   **B** 24   **C** 36   **D** 48

⭐ **60. TEST PREP** Allen and Mei began working at the same time. It took Allen 50 minutes to mow the lawn, while Mei took $1\frac{1}{4}$ hr to paint the fence. How much longer did Mei work than Allen?

F 85 min   **G** 65 min   **H** 35 min   **J** 25 min

EXTRA PRACTICE page H32, Set A

Use Addition and Subtraction

Learn how to add and subtract whole numbers.

Remember that you add when joining groups of different sizes and subtract when taking away or comparing groups.

PROGRAM	YEARS	TOTAL HOURS IN SPACE
Mercury	1961–1963	54
Gemini	1965–1966	1,940
Apollo	1968–1972	7,506
Skylab	1973–1974	12,352

During the early years of space flight, the total time United States astronauts spent in space increased with each program.

EXAMPLE 1

How many hours did U.S. astronauts spend in space?

Find 54 + 1,940 + 7,506 + 12,352.

Estimate to check for reasonableness.

Round to the nearest thousand. | Find the sum.

```
    54            0           54
 1,940        2,000        1,940    Compare your estimate.
 7,506   →    8,000        7,506    21,852 is close to 22,000, so
+12,352     +12,000      +12,352    the sum is reasonable.
             22,000        21,852
```

So, U.S. astronauts spent 21,852 hours in space.

EXAMPLE 2

How many more hours did U.S. astronauts spend in Skylab than on the Gemini missions?

Find 12,352 − 1,940.

Estimate to check for reasonableness. | Find the difference.

```
12,352       12,000       12,352    Compare your estimate.
- 1,940   → - 2,000       - 1,940   10,412 is close to 10,000, so
             10,000        10,412   the difference is reasonable.
```

So, U.S. astronauts spent 10,412 more hours in Skylab than on the Gemini missions.

CHECK FOR UNDERSTANDING

Think and Discuss ▶ Look back at the lesson to answer each question.

1. **Explain** how you know whether to add or subtract when solving a word problem.

2. **Explain** why it is a good idea to find an estimate before or after you find the exact answer.

Guided Practice ▶ Find the sum or difference. Estimate to check.

3. $835 + 604$

4. $6,901 + 342 + 67$

5. $40,190 - 13,982$

PRACTICE AND PROBLEM SOLVING

Independent Practice ▶ Find the sum or difference. Estimate to check.

6. $9,500 - 289$

7. $21,670 + 14,704$

8. $31,227 + 56,995$

9. $\begin{array}{r} 999,999 \\ +111,385 \end{array}$

10. $\begin{array}{r} 987,654 \\ -456,789 \end{array}$

11. $\begin{array}{r} 50,000,000 \\ -\ 3,604,381 \end{array}$

Solve.

12. $1,485 + 2,019 + 1,310 + 3,665 + 798$

13. $43,875 + 81,420 - 38,288 + 12,108 - 23,990$

Problem Solving Applications ▶ **Use Data** For 14–15, use the graph.

14. How many more calories are in 1 cup of almonds than in 1 cup of carrots mixed with 1 cup of raisins? Estimate to check.

15. **Write a problem** that uses data from the graph and that can be solved by adding or subtracting.

16. When the international space station is completed, it will be 290 ft long, which is 206 ft longer than Skylab. Skylab was 41 ft longer than Mir, the Russian space station. How long is Mir?

International space station

MIXED REVIEW AND TEST PREP

17. $6,785 + 4,521$

18. Complete. 36 in. = ■ ft

19. List the factors of 24.

20. List the first six multiples of 9.

21. **TEST PREP** Which is a prime number?

    A 15   **B** 27   **C** 31   **D** 50

EXTRA PRACTICE page H32, Set B

Use Multiplication and Division

Learn how to multiply and divide whole numbers.

1. 7×3 **2.** $72 \div 9$ **3.** 12×4 **4.** $300 \div 25$
5. 800×40

Sometimes you can multiply to solve a word problem.

EXAMPLE 1

Sixth-grade students sold 132 books of carnival ride coupons. How many ride coupons did they sell if there were 18 in each book?

Multiply. 132×18 Estimate. $130 \times 20 = 2,600$

$$
\begin{array}{r}
132 \\
\times\ 18 \\
\hline
1\ 056 \\
+1\ 32 \\
\hline
2,376
\end{array}
$$

Compare the exact product to your estimate. Since 2,376 is close to the estimate of 2,600, the exact product is reasonable.

Remember that you multiply when joining equal-sized groups, and you divide when separating into equal-sized groups or when finding how many in each group.

So, the students sold 2,376 coupons.

You can omit the zero placeholders when you multiply. Just be careful to line up the products correctly.

Correct	Incorrect
132	132
$\times\ 24$	$\times\ 24$
528	528
$+264$	$+264$

EXAMPLE 2

Season tickets to an amusement park are on sale for $125 each. On the first day of the sale, the amusement park sold 12,383 tickets. How much money did the amusement park receive for season tickets that day?

Multiply. $12,383 \times 125$

Estimate. $12,000 \times 130 = 1,560,000$

$$
\begin{array}{r}
12,383 \\
\times\ \ \ \ 125 \\
\hline
61\ 915 \\
247\ 66 \\
+1\ 238\ 3 \\
\hline
1,547,875
\end{array}
$$

Compare the exact product to your estimate. Since 1,547,875 is close to the estimate of 1,560,000, the exact product is reasonable.

So, the amusement park received $1,547,875.

Sometimes you have to use division to solve word problems.

EXAMPLE 3

Remember that the procedure for dividing is divide, multiply, subtract, compare, and bring down.

Mrs. Lopez is redesigning the company cafeteria to seat 540 employees. Each table in her design seats 12 employees. How many tables will she need?

Divide. 540 ÷ 12

Estimate. 480 ÷ 12 = 40

$$
\begin{array}{r}
45 \\
12\overline{)540} \\
-48 \\
\hline
60 \\
-60 \\
\hline
0
\end{array}
$$

Compare the exact quotient to your estimate. Since 45 is close to the estimate of 40, the exact quotient is reasonable.

So, Mrs. Lopez will need 45 tables in her design.

• What if each table seats 10 employees? How many tables will Mrs. Lopez need?

Sometimes a division problem has a zero in the quotient.

EXAMPLE 4

A school collected 2,568 newspapers. The newspapers were bundled in packages of 25. How many packages of newspapers did the school bundle?

Divide. 2,568 ÷ 25

Estimate. 2,500 ÷ 25 = 100

$$
\begin{array}{r}
102\ r18 \\
25\overline{)2,568} \\
-25 \\
\hline
06 \\
-0 \\
\hline
68 \\
-50 \\
\hline
18
\end{array}
$$

Compare the exact answer to your estimate. Since 102 r18 is close to the estimate of 100, the exact quotient is reasonable.

So, the school bundled 102 packages of newspapers.

In Example 4 there is a remainder. Some calculators allow you to show a whole-number remainder.

2,568 ⟦÷R⟧ 25 ⟦=⟧ ⟦102   R18⟧

You can express a remainder with an *r*, or you can express it as a fractional part of the divisor or as a decimal. The quotient and remainder in Example 4 can also be expressed as $102\frac{18}{25}$ or as 102.72.

CHECK FOR UNDERSTANDING

Think and Discuss ▶ Look back at the lesson to answer each question.

1. **Explain** why the school bundled 102 packages of newspapers instead of 103 packages of newspapers in Example 4.

2. **Tell** the different ways to express the remainder for the division problem 153 ÷ 6.

Guided Practice ▶ Multiply or divide. Estimate to check.

3. $1,113 \times 712$ 4. $2,115 \div 72$ 5. $16,225 \times 219$

Multiply or divide.

6. $\begin{array}{r} 13 \\ \times 14 \\ \hline \end{array}$ 7. $8\overline{)432}$ 8. $12\overline{)144}$ 9. $\begin{array}{r} 962 \\ \times 40 \\ \hline \end{array}$

10. 159×340 11. $7,658 \times 111$ 12. $7,044 \div 14$ 13. $1,068 \div 19$

PRACTICE AND PROBLEM SOLVING

Independent Practice ▶ Multiply or divide. Estimate to check.

14. $2,250 \div 18$ 15. $4,904 \times 196$ 16. $193,200 \div 46$

17. $7,021 \times 498$ 18. $249,900 \div 49$ 19. $24,587 \times 71$

Multiply or divide.

TECHNOLOGY LINK

More Practice: Use **Mighty Math Calculating Crew,** *Captain Nick Knack,* Level V.

20. $16\overline{)1,664}$ 21. $\begin{array}{r} 298 \\ \times 89 \\ \hline \end{array}$ 22. $\begin{array}{r} 5,233 \\ \times 238 \\ \hline \end{array}$ 23. $52\overline{)728}$

24. $4\overline{)412}$ 25. $\begin{array}{r} 380 \\ \times 55 \\ \hline \end{array}$ 26. $\begin{array}{r} 2,382 \\ \times 12 \\ \hline \end{array}$ 27. $24\overline{)626}$

28. $\begin{array}{r} 327 \\ \times 123 \\ \hline \end{array}$ 29. $26\overline{)2,314}$ 30. $68\overline{)24,820}$ 31. $\begin{array}{r} 5,470 \\ \times 240 \\ \hline \end{array}$

32. $29\overline{)13,253}$ 33. $\begin{array}{r} 6,378 \\ \times 291 \\ \hline \end{array}$ 34. $\begin{array}{r} 2,009 \\ \times 562 \\ \hline \end{array}$ 35. $120\overline{)10,080}$

Divide. Write the remainder as a fraction.

36. $5\overline{)49}$ 37. $7,349 \div 20$ 38. $386 \div 15$ 39. $4\overline{)3,385}$

40. **ALGEBRA** What is the least whole number, n, for which it is true that $n \div 8 > 542 + 258$?

41. **ALGEBRA** What is the least whole number, n, for which it is true that $70 \times n > 29,000$?

Problem Solving ▶ Applications

42. Yuji's parents bought an entertainment center for $1,176. They plan to pay for it with 14 equal monthly payments. How much will each payment be if a $3.50 service charge is added every month?

43. Lincoln Middle School had a car wash to raise money. The students charged $3.00 for every car and $5.00 for every van. If they washed 23 cars and 18 vans, how much did they earn?

44. ❓ **What's the Error?** Describe the error. Then solve the problem correctly.

$$\begin{array}{r} 62 \text{ r}15 \\ 49\overline{)30,395} \end{array}$$

45. There are 5 children in Brenda's family. Brenda is 16 years old. Brenda's twin sisters are 6 years old, and her brothers are 10 and 12 years old. What is the mean of the 5 children's ages?

MIXED REVIEW AND TEST PREP

46. How much more than 96,784 is 142,981? (p. 20)

47. Complete. 4 lb = ■ oz

48. Order 804, 824, 818, and 803 from least to greatest.

49. TEST PREP How many meters are in 1,800 millimeters?

   A 0.18 m   **B** 1.8 m   **C** 18 m   **D** 180 m

50. TEST PREP Teresa is buying perfume for her mother. The prices are $16.19, $15.89, $15.99, and $17.00. How much will she save by buying the least expensive instead of the most expensive?

   F $0.01   **G** $0.11   **H** $0.81   **J** $1.11

PROBLEM SOLVING LiNKUP to Reading

Strategy • Use Context   Many word problems contain words such as *more than, fewer than, twice as many,* and *total*. Be sure to interpret these words within the context of the problem before you choose an operation.

Use Data   **For each problem, write the words that help you choose the operation. Then solve the problem.**

1. How many more calories will you burn in 1 hr by skiing than by hiking?

2. On Saturday, Connie spent an hour in gymnastics class and then walked for 1 hr. How many total calories did she burn during these two activities?

3. Clara burned half as many calories while raking the lawn for 1 hr as she did while jogging for 1 hr. How many calories did she burn while raking the lawn?

CALORIES BURNED PER HOUR	
Activity	**Calories**
Walking (at 2 mi per hr)	112
Gymnastics	128
Hiking	191
Jogging	224
Cross-country skiing	326

PROBLEM SOLVING STRATEGY
Predict and Test

Learn how to use the strategy *predict and test* to solve problems with whole numbers.

You can solve some problems by using your number sense to predict a possible answer. You should then test your answer and revise your prediction if necessary.

There were 25 problems on a test. For each correct answer, 4 points were given. For each incorrect answer, 1 point was subtracted. Tania answered all 25 problems. Her score was 85. How many correct answers did she have?

Analyze

What are you asked to find?

What facts are given?

Is there any numerical information you will not use? If so, what?

Choose

What strategy will you use?

You can use the strategy *predict and test.* Use the given information and your number sense to predict about how many correct answers Tania had. Then test your prediction, and revise it if necessary.

Solve

How will you solve the problem?

Make a table to show your prediction, your test of it, and any revisions you need. Be sure that the total of correct and incorrect problems is 25.

PREDICTION		TEST	
Correct	Incorrect	SCORE	
20	5	$(20 \times 4) - 5 = 75$	*too low, so revise*
21	4	$(21 \times 4) - 4 = 80$	*too low, so revise*
22	3	$(22 \times 4) - 3 = 85$	← *correct*

So, Tania had 22 correct answers.

Check

How can you check your answer?

What if Tania's score were 65? How many incorrect answers would she have?

PROBLEM SOLVING PRACTICE

PROBLEM SOLVING STRATEGIES

Draw a Diagram or Picture

Make a Model

▶ **Predict and Test**

Work Backward

Make an Organized List

Find a Pattern

Make a Table or Graph

Solve a Simpler Problem

Write an Equation

Use Logical Reasoning

Solve by predicting and testing.

1. Rodney bought a total of 40 oranges and apples. He bought 14 fewer apples than oranges. How many of each fruit did he buy?

2. The Mighty Tigers soccer team played a total of 25 games. They won 9 more games than they lost, and 2 games ended in ties. How many games did they win?

3. The perimeter of a rectangular garden is 40 ft. If the length is 6 ft more than the width, what are the length and width of the garden?

   A $l = 12$ ft, $w = 6$ ft

   B $l = 13$ ft, $w = 7$ ft

   C $l = 6$ ft, $w = 12$ ft

   D $l = 7$ ft, $w = 13$ ft

4. The perimeter of a rectangular lawn is 28 yd. If the length is 4 yd more than the width, what are the length and width of the lawn?

   F $l = 10$ yd, $w = 4$ yd

   G $l = 9$ yd, $w = 5$ yd

   H $l = 5$ yd, $w = 9$ yd

   J $l = 4$ yd, $w = 10$ yd

MIXED STRATEGY PRACTICE

5. Rosalia waters her tomato plants every other day. She waters her pepper plants every 3 days. If she waters both on April 20, what are the next three dates on which she will water both?

6. Stacy spent a total of $28.45. She bought a ticket for a basketball game for $8.50, food for $7.95, and some T-shirts for $6.00 each. How many T-shirts did she buy?

7. Sam has 98 baseball cards. This is 2 more than twice as many as Paul has. How many cards does Paul have?

8. Use the table below. If the pattern continues, how many miles in all will four runners run on the fifth day?

9. Melina and her two sisters collect stamps. Melina has twice as many as her older sister, who has 33 stamps. Melina has three times as many as her younger sister. How many stamps do they have in all?

10. ⓘ **What's the Question?** The sum of the ages of Jeff, Elijah, and Stefan is 41. Jeff is 14 years old. Stefan is 3 years older than Elijah. The answer is 12 years old.

11. The train that leaves at 11:45 A.M. usually arrives in New York City 34 minutes after that. Today it arrived at 12:24 P.M. How late was the train?

ALGEBRA
Use Expressions

Learn how to identify, write, and evaluate expressions involving whole numbers.

In a basketball game, a team scored 27 points in the first half and 38 points in the second half. To represent the total points, you could use a numerical expression. A **numerical expression** is a mathematical phrase that includes only numbers and operation symbols.

$$27 + 38 \leftarrow \text{total points}$$

Here are other examples of numerical expressions.

$$60 + 25 \qquad 42 \div 7 \qquad 16 - 3 \qquad 51 \times 36 \qquad 30 + 12 + 41$$

If you didn't know how many points the team scored in the second half, you could use a variable to represent the points. A **variable** is a letter or symbol that can stand for one or more numbers. An expression that includes a variable is called an **algebraic expression**.

$27 + p \leftarrow$ *Use p to represent points scored in second half.*

Here are other examples of algebraic expressions.

$$5 + n \qquad 7 \times a \qquad k - 3 \qquad y \div 2 \qquad 6 \times 5 \times b$$

In an expression, there are several ways to show multiplication.

$7 \times a$ can be written as $7a$, $7(a)$, or $7 \cdot a$.

Word expressions can be translated into numerical or algebraic expressions.

Vocabulary
numerical expression
variable
algebraic expression
evaluate

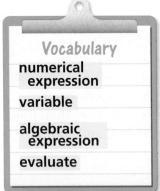

EXAMPLE 1

Write a numerical or algebraic expression for the word **expression.**

A. three dollars less than five dollars $\qquad 5 - 3$

B. two times a distance, *d* $\qquad 2 \times d$, 2(*d*), 2*d*, or 2 · *d*

To **evaluate** a numerical expression, you find its value. To evaluate an algebraic expression, replace the variable with a number and then find the value.

EXAMPLE 2

Evaluate each expression.

A. $a + 150$, for $a = 18$

$a + 150$ *Replace a*
$18 + 150$ *with 18.*
168 *Add.*

B. $b \div 10 \times 3$, for $b = 120$

$b \div 10 \times 3$ *Replace b with*
$120 \div 10 \times 3$ *120.*
12×3 *Divide and*
36 *then multiply.*

Think and ▶
Discuss

Look back at the lesson to answer each question.

1. **Explain** the difference between a numerical expression and an algebraic expression. Give some examples of each.

2. **Show** four different ways to write an algebraic expression for the product of the number 10 and the variable g.

Guided ▶
Practice

Write a numerical or algebraic expression for the word expression.

3. forty-six less than one hundred twenty-five

4. one hundred seven more than y

5. y divided by fifteen

Evaluate each expression.

6. 21×15

7. $100 - g$, for $g = 54$

8. $s \div 8$, for $s = 720$

Independent ▶
Practice

Write a numerical or algebraic expression for the word expression.

9. twenty-five $20 bills

10. q more than two hundred fifteen

11. seventy-six decreased by k

12. x divided by fourteen

Evaluate each expression.

13. 15×31

14. $3,021 + 915$

15. $10,340 - 1,340$

16. $k - 65$, for $k = 95$

17. $\frac{d}{7} \times 2$, for $d = 490$

18. $100b$, for $b = 54$

19. $m \div n$, for $m = 1,230$ and $n = 410$

20. cd, for $c = 5$ and $d = 200$

Problem Solving ▶
Applications

21. Let n represent the number of free throws Nathan scored. Bryan scored 12 more free throws than Nathan. Write an algebraic expression to show how many free throws Bryan scored.

22. ✎ **Write About It** Explain how to evaluate an algebraic expression when you know the value of each variable. Give an example.

23. **REASONING** Tiffany, Deidre, Luisa, Kendall, and Ann Marie are runners. Kendall can outrun Luisa and Tiffany, but Deidre can outrun Kendall. Luisa can outrun Ann Marie, but Deidre can outrun Luisa. Which one of the girls is the fastest runner?

MIXED REVIEW AND TEST PREP

24. 530×42 (p. 22)

25. $3,870 \div 18$ (p. 22)

26. $1,234 + 453$ (p. 20)

27. $8,000 - 357$ (p. 20)

★**28. TEST PREP** Christina bought 3 pens and 1 notebook for $5.85. A pen cost $1.20. How much did the notebook cost?

A $1.20   **B** $1.25   **C** $1.85   **D** $2.25

EXTRA PRACTICE page H32, Set D

ALGEBRA
Mental Math and Equations

Learn how to use mental math to solve equations.

QUICK REVIEW

Evaluate each expression.

1. 25×8
2. $80 + d$, for $d = 37$
3. $g \div 8$, for $g = 72$
4. $450 - 225$
5. Find the missing factor. $\blacksquare \times 8 = 24$

Vocabulary

equation

solution

An **equation** is a statement showing that two quantities are equal. These are equations:

$$6 + 7 = 13 \qquad 24 \div 3 = 8 \qquad k - 3 = 1 \qquad 2d = 18 \qquad a + b = 11$$

If an equation contains a variable, you can solve the equation by finding the value of the variable that makes the equation true. That value is the **solution**.

EXAMPLE 1

Which of the numbers 8, 9, and 10 is a solution of the equation $12p = 108$?

Remember that a variable is a letter or symbol that stands for one or more numbers.

Replace *p* with 8.	Replace *p* with 9.	Replace *p* with 10.
$12(8) \overset{?}{=} 108$	$12(9) \overset{?}{=} 108$	$12(10) \overset{?}{=} 108$
$96 = 108$ *false*	$108 = 108$ *true*	$120 = 108$ *false*

The solution is 9 because $12(9) = 108$.

• Which of the numbers 4, 5, and 6 is a solution of the equation $222 \div n = 37$?

Some equations with variables can be solved by using mental math. Think of the value of the variable that makes the equation true. Then check your answer.

EXAMPLE 2

The Statue of Liberty's hand is about 16 ft long. The index finger is 8 ft long. What is the length of the palm of her hand? Solve the equation $16 = c + 8$ by using mental math.

| $16 = c + 8$ | *What number added to 8 gives 16?* |
| $8 = c$ | *The solution is 8.* |

Check:

| $16 = 8 + 8$ | *Replace c with 8.* |
| $16 = 16$ | *8 + 8 is equal to 16.* |

• Solve the equation $m \times 7 = 56$ by using mental math.

Think and ▶
Discuss

Look back at the lesson to answer each question.

1. **Tell** whether 4 is a solution of the equation $x + 3 = 9$. If it is not, find the solution.

2. **Give an example** of an equation with a solution of 5.

Guided ▶
Practice

Determine which of the given values is the solution of the equation.

3. $f \div 7 = 3$; $f = 19, 20,$ or 21 4. $t + 9 = 20$; $t = 10, 11,$ or 12

Solve each equation by using mental math.

5. $7 = x + 3$ 6. $\dfrac{h}{9} = 3$ 7. $4 \times k = 16$

PRACTICE AND PROBLEM SOLVING

Independent ▶
Practice

Determine which of the given values is the solution of the equation.

8. $3h = 39$; $h = 11, 12,$ or 13 9. $17 - x = 12$; $x = 5, 6,$ or 7

10. $48 + s = 57$; $s = 8, 9,$ or 10 11. $3 = 54 \div k$; $k = 16, 17,$ or 18

Solve each equation by using mental math.

12. $p - 7 = 7$ 13. $9m = 81$ 14. $13 + r = 30$

15. $x - 16 = 4$ 16. $h \div 8 = 7$ 17. $14 = k - 15$

18. $87 = e \div 10$ 19. $12 \times v = 240$ 20. $t \div 6 = 125$

21. $12 + 4 + d = 25$ 22. $3 \times 4 = c - 8$ 23. $p + 14 = 32 - 12$

Problem Solving ▶
Applications

24. The equation $w + 12 = 40$ describes the number of men and women riding the bus to a convention. If w is the number of women riding the bus, how many men are riding the bus?

25. Mr. Murakami teaches 5 classes of 25 students each. One hundred of his students are sixth graders. How many are not sixth graders?

26. **(?) What's the Question?** A roller coaster has 7 cars. Fifty-six people can ride the roller coaster at one time. The answer is 8.

MIXED REVIEW AND TEST PREP

27. Evaluate $a + 14$ for $a = 27$. (p. 28)   28. $525 \div 25$ (p. 22)

29. Find $4,310 - 1,900 + 3,450 - 870$. (p. 20)   30. Find the greatest common factor of 15 and 35.

⭐ 31. **TEST PREP** Andre left the house at 8:45 A.M. He arrived home $4\frac{1}{2}$ hours later. At what time did Andre arrive home?

A 11:45 A.M.   **B** 12:45 A.M.   **C** 12:45 P.M.   **D** 1:15 P.M.

EXTRA PRACTICE page H32, Set E

1. **VOCABULARY** A way to estimate a sum when all of the addends are about the same is ___?___ . (p. 16)

2. **VOCABULARY** A letter or symbol that can stand for one or more numbers is a(n) ___?___ . (p. 28)

3. **VOCABULARY** A statement showing that two quantities are equal is a(n) ___?___ . (p. 30)

Estimate. (pp. 16–19)

4. 593
    $+724$

5. $1,420$
    $+5,791$

6. 935
    -549

7. $2,371$
    $-1,456$

8. $43,816$
    $-39,972$

9. 48×6

10. 308×67

11. $374 \div 7$

12. $276 \div 42$

13. $3,764 \div 591$

Find the sum or difference. (pp. 20–21)

14. $4,762$
     $+39,038$

15. $9,724 - 286$

16. $50,031$
     $- \ 9,352$

17. 737
     $4,650$
     $+11,821$

18. $678,040$
     $-329,193$

Multiply or divide. (pp. 22–25)

19. 526
     $\times \ 42$

20. 123×12

21. $2,250 \div 18$

22. 189
     $\times 108$

23. $40\overline{)3,206}$

Evaluate each expression for $a = 63$, $b = 150$, and $c = 7$. (pp. 28–29)

24. $a + 305$

25. $b - 36$

26. $300 \div b$

27. $3c$

28. $a \div 9$

29. $215 - b$

30. $112 \div c$

31. $a \times 5$

32. $2a + 4$

33. $a + b$

Solve each equation by using mental math. (pp. 30–31)

34. $3m = 27$

35. $14 = q + 6$

36. $20 = y - 9$

37. $w \div 50 = 5$

38. $74 + a = 85$

Solve. (pp. 26–27)

39. At school during spirit week, Ming sold a total of 36 red and blue ribbons. She sold 6 more red ribbons than blue ribbons. How many of each color did she sell?

40. Colton worked two days on a project for school. He worked a total of 195 minutes. If he worked 45 minutes longer on the first day, how long did Colton work each day?

TIP!

Understand the problem.

See item **6**.

The words describe an algebraic expression. *Less* describes the relationship of the given number to the unknown number.

Also see problem **1**, p. H62.

Choose the best answer.

1. Which is the best estimate for this sum?
   6,204 + 5,893 + 6,028 + 5,991

   A 22,000   C 28,000

   B 24,000   D 30,000

2. A new jogging track at a recreation center is 580 feet longer than the old track. The new track is 1,320 feet long. How long was the old track?

   F 740 ft   H 1,900 ft

   G 860 ft   J Not here

3. There are 9,800 beads to put in boxes. Each box holds 48 beads. How many boxes are needed to hold all of the beads?

   A 48 boxes   C 205 boxes

   B 204 boxes   D Not here

4. Which is the best estimate for this product?
   81 × 409

   F 32,000

   G 28,000

   H 3,200

   J 2,800

5. What is the value of $35 + b$ for $b = 5$?

   A 7   C 40

   B 30   D 175

6. Which algebraic expression represents the expression "28 less than a number, q"?

   F $28 - q$

   G $q + 28$

   H $q \div 28$

   J $q - 28$

7. A bus can carry 16 people. There are 9 people on board and there is room for x more people. Which equation shows this relationship?

   A $16 + x = 9$ C $9 + x = 16$

   B $x - 9 = 16$ D $x - 16 = 9$

8. Which algebraic expression represents the expression "7 more than a number, t"?

   F $7 \times t$

   G $t - 7$

   H $7 - t$

   J $t + 7$

9. $1,225 \times 37$

   A 12,250   C 45,325

   B 34,225   D 48,390

10. $115,309 - 67,899$

    F 47,410   H 58,410

    G 57,410   J 58,510

Write What You Know

11. Describe two ways to estimate $9,184 \div 26$.

12. Explain how you would use mental math to solve the equation $8 \times p = 24$.

2 Operation Sense

A bird can reach almost any part of its body with its beak because it has an extremely flexible neck. The bones in the neck are called cervical vertebrae. Humans have 7 cervical vertebrae. Most birds have more cervical vertebrae than humans or other mammals. Flamingos have $4^2 + 8 \times 2 - 13$ cervical vertebrae.

PROBLEM SOLVING About how many times as many cervical vertebrae does a flamingo have as a human?

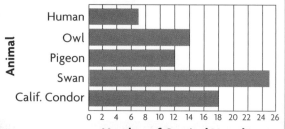

CERVICAL VERTEBRAE

Animal: Human, Owl, Pigeon, Swan, Calif. Condor

Number of Cervical Vertebrae
0 2 4 6 8 10 12 14 16 18 20 22 24 26

DATA LINK

Check What You Know

Use this page to help you review and remember important skills needed for Chapter 2.

 Repeated Multiplication (For Intervention, see p. H15.)

Find the product.

1. $3 \times 3 \times 3$

2. $2 \times 2 \times 2 \times 2$

3. $4 \times 4 \times 4$

4. $5 \times 5 \times 5 \times 5$

5. $10 \times 10 \times 10 \times 10$

6. $9 \times 9 \times 9$

7. $8 \times 8 \times 8$

8. $6 \times 6 \times 6$

9. $7 \times 7 \times 7$

10. $10 \times 10 \times 10$

11. $4 \times 4 \times 4 \times 4$

12. $5 \times 5 \times 5$

 Properties (For Intervention, see p. H2.)

Name the property illustrated.

13. $48 + 13 + 5 = 13 + 48 + 5$

14. $8 \times (3 + 1) = (8 \times 3) + (8 \times 1)$

15. $0 \times 999 = 0$

16. $15 \times 1 = 15$

17. $(9 + 5) + 10 = 9 + (5 + 10)$

18. $27 + 36 = 36 + 27$

19. $(4 + 2) \times 7 = (4 \times 7) + (2 \times 7)$

20. $7 \times 9 \times 2 = 2 \times 9 \times 7$

21. $1 \times 148 = 148$

22. $(2 \times 9) \times 5 = 2 \times (9 \times 5)$

23. $8 \times (20 + 8) = (8 \times 20) + (8 \times 8)$

24. $6 \times 3 = 3 \times 6$

 Use of Parentheses (For Intervention, see p. H15.)

Solve the equation.

25. $5 \times (4 + 3) = (a \times 4) + (a \times 3)$

26. $7 + (3 + 9) = (7 + m) + 9$

27. $4 + 6 + 3 = 4 + (6 + r)$

28. $(9 \times 4) + (9 \times 2) = h \times (4 + 2)$

29. $4 \times (8 - 4) = (4 \times t) - (4 \times 4)$

30. $(5 + 6) + (4 + 7) = (5 + 4) + (r + 7)$

Evaluate the expression.

31. $(4 + 7) + 9$

32. $5 \times (6 + 2)$

33. $(1 + 7) \times 4$

34. $(7 + 8) + 3$

35. $4 + (7 - 3)$

36. $6 \times (9 - 3)$

37. $2 \times (3 + 5 + 8)$

38. $(8 + 5) + (3 + 1)$

39. $4 + (5 + 7) + 2$

40. $2 \times (3 \times 1)$

> ### LOOK AHEAD
>
> **In Chapter 2 you will**
> - use properties and mental math to find sums, differences, products, and quotients
> - use exponents
> - use order of operations

MENTAL MATH
Use the Properties

Learn how to use properties and mental math to find sums, differences, products, and quotients.

Vocabulary

compensation

Remember these properties:

Commutative Property
$4 + 8 = 8 + 4$
$6 \times 7 = 7 \times 6$

Associative Property
$(6 + 8) + 5 = 6 + (8 + 5)$
$(2 \times 9) \times 5 = 2 \times (9 \times 5)$

Distributive Property
$12 \times 32 = 12 \times (30 + 2) = (12 \times 30) + (12 \times 2)$

QUICK REVIEW

Solve by using mental math.

1. $d + 13 = 33$ **2.** $25 - g = 17$
3. $6m = 54$ **4.** $9 = t \div 11$
5. $340 + 34 + 3$

Use the table to find how many bones the spine, chest, and shoulders have in all. Mentally find the sum by first reordering using the Commutative Property.

$26 + 25 + 4 = 26 + 4 + 25$ *Commutative Property*
$\qquad\qquad\quad = 30 + 25$ *Use mental math.*
$\qquad\qquad\quad = 55$

Mentally find the sum by regrouping using the Associative Property.

$(26 + 25) + 4 = 26 + (25 + 4)$ *Associative Property*
$\qquad\qquad\quad = 26 + 29$ *Use mental math.*
$\qquad\qquad\quad = 55$

BONES IN THE HUMAN BODY	
Part	**Number of Bones**
Head	28
Spine	26
Throat	1
Chest	25
Shoulders	4
Arms	6
Hands	54
Legs	10
Feet	52

So, the spine, chest, and shoulders have a total of 55 bones.

You can use the Distributive Property to mentally solve a problem.

EXAMPLE 1

How many bones are in 5 models of the human spine?
$5 \times 26 = 5 \times (20 + 6)$ *Break 26 into parts.*
$\qquad\quad = (5 \times 20) + (5 \times 6)$ *Use the Distributive Property. Multiply mentally.*
$\qquad\quad = 100 + 30$ *Add the products.*
$\qquad\quad = 130$

So, there are 130 bones in 5 models.

You can also use the Commutative and Associative Properties. Try to make partial products that end in 0.

EXAMPLE 2

James has 8 storage boxes on each of 6 shelves. Each box contains 5 items. How many items are there altogether?

Commutative Property
$(8 \times 6) \times 5 = (8 \times 5) \times 6$
$\qquad\qquad\quad = 40 \times 6$
$\qquad\qquad\quad = 240$

Associative Property
$(8 \times 6) \times 5 = 8 \times (6 \times 5)$
$\qquad\qquad\quad = 8 \times 30$
$\qquad\qquad\quad = 240$

A strategy you can use for some addition and subtraction problems is **compensation**. For addition, change one number to a multiple of 10 and then adjust the other number to keep the balance.

EXAMPLE 3

Mr. Forge and his friends play basketball for an hour on Fridays and Saturdays. On Friday they scored a total of 44 points, and on Saturday they scored 57 points. Use compensation to find the total points scored for both days.

$44 + 57 = (44 + 6) + (57 - 6)$ *Add 6 to 44 and subtract 6 from 57.*

     $= 50 + 51$ *Use mental math to add.*

     $= 101$

So, the total points scored is 101.

When you use compensation to subtract, you have to do the same thing to each number. Since it's easy to subtract numbers ending in zero, try to make the second number a multiple of 10.

EXAMPLE 4

Use compensation to find $128 - 56$.

$128 - 56 = (128 + 4) - (56 + 4)$ *Add 4 to 128 and to 56 before subtracting.*

     $= 132 - 60$

     $= 72$

So, the difference is 72.

You can sometimes divide mentally by breaking a number into smaller parts that are each divisible by the divisor.

EXAMPLE 5

Use mental math to find $396 \div 4$.

$396 = 360 + 36$ *Break 396 into parts.*

$360 \div 4 = 90$ and $36 \div 4 = 9$ *Divide each part by 4 mentally.*

$90 + 9 = 99$ *Add the parts of the quotient.*

So, $396 \div 4 = 99$.

• Tell another way to break 396 into two parts to divide by 4.

Math Idea ▶ Using the number properties and other mental math strategies will help you add, subtract, multiply, and divide mentally.

CHECK FOR UNDERSTANDING

Think and ▶ **Discuss**

Look back at the lesson to answer each question.

1. **Tell** how using the Associative Property in Example 2 made the problem easier to solve.

2. **Explain** two ways to use compensation to find $349 + 138$ mentally.

37

Use mental math to find the value.

3. 12×17 **4.** $45 + 9 + 15$ **5.** $124 + 17 + 16$ **6.** 9×36

7. $(6 + 37) + 13$ **8.** $2 \times 9 \times 50$ **9.** 5×29 **10.** 11×43

11. $39 + 16$ **12.** $83 + 38$ **13.** $426 \div 3$ **14.** $16 + 35$

15. $279 \div 3$ **16.** $137 - 51$ **17.** $65 - 22$ **18.** $567 \div 7$

PRACTICE AND PROBLEM SOLVING

Use mental math to find the value.

19. 24×7 **20.** $73 - 27$ **21.** 45×11 **22.** 12×35

23. $87 + 98$ **24.** $(12 + 23) + 8$ **25.** 4×27 **26.** $18 + 26$

27. 4×53 **28.** $64 - 29$ **29.** $24 + 32 + 16$ **30.** 19×14

31. $126 + 118$ **32.** $293 - 137$ **33.** $765 \div 9$

34. $32 + 36$ **35.** $19 + 26$ **36.** $4 \times 6 \times 50$

37. $25 \times 30 \times 2$ **38.** $172 \div 4$ **39.** $1{,}526 - 498$

40. $40 \times 15 \times 2$ **41.** $(4 \times 33) + (4 \times 7)$ **42.** $(6 \times 24) + (6 \times 6)$

43. $192 \div n$ for $n = 3$ **44.** $c \times 9 \times 5$ for $c = 8$

45. $p \div 12$ for $p = 624$ **46.** $a + 19 + 32$ for $a = 18$

Name each missing reason.

47. $80 \times 3 = (8 \times 10) \times 3$ 80 means 8×10.

$\quad\quad\quad = 8 \times (10 \times 3)$ Associative Property of Multiplication

$\quad\quad\quad = 8 \times (3 \times 10)$ $\underline{\quad ? \quad}$

$\quad\quad\quad = (8 \times 3) \times 10$ $\underline{\quad ? \quad}$

$\quad\quad\quad = 24 \times 10$ $\underline{\quad ? \quad}$

$\quad\quad\quad = 240$ $\underline{\quad ? \quad}$

48. What if the product of three whole numbers is 210? Without using 1 as a factor, what are the possible choices for the numbers?

Use Data   **For 49–51, use the data below.**

49. Use mental math to find how many CDs were bought in all. Explain how you got your answer.

50. If Nick and Selena each gave 12 CDs to Brenda, how many would Brenda have then?

51. How many CDs would Brenda, Selena, Ricardo, and Nick each have if they shared their CDs equally?

CDs Bought
Brenda 12
Selena 17
Nick 25
Ricardo 18

52. A toy store has 7 boxes on each of 4 shelves. Each box has 25 items in it. How many items altogether are on the 4 shelves?

53. Ann needs 250 signatures on a petition. On Monday she got 23 signatures, on Tuesday she got 3 times as many as on Monday, and on Wednesday and Thursday she got 45 each. How many more signatures does she need?

54. Jocelyn has $253.47. Her aunt gives her $87.95 more. Jocelyn buys a pair of shoes for $39.99, three T-shirts for $7.77 each, and two pairs of jeans for $60.22. Use estimation to find about how much money Jocelyn has left.

55. **Write About It**   Explain how to use compensation to add two numbers. Give an appropriate example to support your explanation.

MIXED REVIEW AND TEST PREP

56. Use mental math to solve. $a \div 7 = 21$ (p. 30)

57. Multiply. 732×46 (p. 22)   **58.** Divide. $64{,}270 \div 35$ (p. 22)

59. TEST PREP   Rob changes 4 quarts of oil in his car every 3,000 miles. How many quarts of oil will Rob have used after driving 9,000 miles? (p. 22)

    A 3   **B** 9   **C** 12   **D** 36

60. TEST PREP   Joe bought 3 basketballs for $22.99 each and a net for $5.99. Which number sentence can be used to find the total cost of the basketballs and net? (p. 28)

    F $3 \times (22.99 + 5.99)$ **H** $(3 \times 5.99) + 22.99$

    G $(3 \times 22.99) + 5.99$ **J** $(3 + 22.99) + 5.99$

PROBLEM SOLVING ☀ THINKER'S CORNER

MATH FUN   Practice using mental math strategies to solve this puzzle.

1. Copy the diagram. Place the values of the expressions below in the circles so that every sum of three numbers in a line is the same.

$84 \div 2$
$36 + 8$
$28 + 16 + 2$
$3 \times 8 \times 2$ $8 + 14 + 32$
$28 + 22$ $448 \div 8$
4×13 $4 + 38 + 16$

2. Use mental math. What is the sum of each row of three numbers?

LESSON 2.2

ALGEBRA
Exponents

Learn how to represent numbers by using exponents.

Vocabulary

exponent

base

Remember that when you multiply two or more numbers to get a product, the numbers multiplied are called factors.

$8 \times 3 \times 4 = 96$

The numbers 8, 3, and 4 are factors of 96.

QUICK REVIEW

1. 3×3

2. 6×6

3. $4 \times 4 \times 4$

4. $2 \times 2 \times 2 \times 2$

5. $9 \times 9 \times 9$

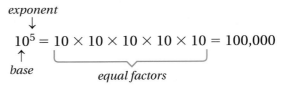

Some football stadiums can seat over 100,000 people. Large numbers can be hard to understand. On the right are four ways to write 100,000 using smaller numbers.

$10 \times 10,000$

$10 \times 10 \times 1,000$

$10 \times 10 \times 10 \times 100$

$10 \times 10 \times 10 \times 10 \times 10$

Another way to write 100,000 is by using exponents. An **exponent** shows how many times a number called the **base** is used as a factor.

$$\overset{\textit{exponent}}{\underset{\textit{base}}{10}}{}^{5} = \underbrace{10 \times 10 \times 10 \times 10 \times 10}_{\textit{equal factors}} = 100,000$$

EXPONENT FORM	READ	VALUE
10^1	The first power of ten	10
$10^2 = 10 \times 10$	Ten squared, or the second power of ten	100
$10^3 = 10 \times 10 \times 10$	Ten cubed, or the third power of ten	1,000

EXAMPLE 1

Find the values of 2^4, 4^2, and 6^3.

$2^4 = 2 \times 2 \times 2 \times 2$	$4^2 = 4 \times 4$	$6^3 = 6 \times 6 \times 6$
$= 16$	$= 16$	$= 216$
2 is a factor four times.	4 is a factor two times.	6 is a factor three times.

Note: The first power of any number equals that number.

$$6^1 = 6 \qquad 9^1 = 9 \qquad 10^1 = 10$$

The zero power of any number, except zero, is defined to be 1.

$$6^0 = 1 \qquad 19^0 = 1 \qquad 10^0 = 1$$

EXAMPLE 2

Write 125 using an exponent and the base 5.

$125 = 5 \times 25 = 5 \times 5 \times 5$ *Find the equal factors.*

$\quad\;\; = 5^3$ *Write the base and the exponent.*

So, $125 = 5^3$.

Think and Discuss ▶ Look back at the lesson to answer each question.

1. **Tell** how many zeros there are in the standard form of 10^7.

2. **Explain** how to write $6 \times 6 \times 6 \times 6$ using exponents.

Guided Practice ▶ Write the equal factors. Then find the value.

3. 2^3 4. 5^2 5. 3^4 6. 9^3 7. 1^4

PRACTICE AND PROBLEM SOLVING

Independent Practice ▶ Write the equal factors. Then find the value.

8. 4^5 9. 7^3 10. 1^{12} 11. 5^3 12. 2^5

13. 34^2 14. 13^2 15. 10^8 16. 20^2 17. 2^{10}

18. 10^4 19. 3^0 20. 15^2 21. 25^1 22. 90^2

Write in exponent form.

23. $12 \times 12 \times 12$ 24. $1 \times 1 \times 1 \times 1 \times 1$ 25. $4 \times 4 \times 4 \times 4$

26. $2 \times 2 \times 2 \times 2 \times 2$ 27. $n \times n$ 28. $y \times y \times y \times y$

Express with an exponent and the given base.

29. 64, base 8   30. 216, base 6   31. 10,000; base 10

32. Write 64 using a base of 8, a base of 4, and a base of 2.

Problem Solving Applications ▶

33. **Use Data** When did Oregon first have a population greater than 10^6? Explain.

34. Ben saves movie ticket stubs. He puts them in albums with 2^4 pages. Each page holds 3^2 stubs. How many albums does he need for 720 stubs?

35. **?** **What's the Question?** Scott has 3^2 video games and Aaron has 2^3 games. The answer is 1.

MIXED REVIEW AND TEST PREP

36. Use mental math to find the value of $279 \div 3$. (p. 30)

37. Round 45,621 to the nearest thousand. (p. 16)

38. $943,012 - 57,806$ (p. 20)   39. $32,047 \div 43$ (p. 22)

⭐ 40. **TEST PREP** Which expression is equivalent to $(3 + 5) \times 2$? (p. 28)

     A $(7 - 3) + 10$ **B** $(4 \times 2) + 8$ **C** $(2 \times 3) + 5$ **D** $10 + (7 \times 10)$

EXTRA PRACTICE page H33, Set B

Explore Order of Operations

Explore how to evaluate expressions by using order of operations.

You need a calculator.

Vocabulary

order of operations

algebraic operating system (AOS)

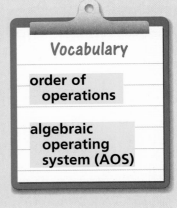

Erin's older brother, Todd, was working on his homework. At the top of the page, he wrote, "**P**lease **E**xcuse **M**y **D**ear **A**unt **S**ally." Erin didn't understand. She said, "We don't have an aunt named Sally."

Todd said, "You'll see why I wrote this on my paper."

Activity 1

• Use paper and pencil to find the value of $23 + 12 \times (6 - 2)^2$.

• How does your answer compare with the answer of a classmate?

Math Idea ▶ When you find the value of an expression with more than one operation, you need to use the **order of operations**.

1. Perform operations in parentheses.
2. Clear exponents.
3. Multiply and divide from left to right.
4. Add and subtract from left to right.

• Use the order of operations to evaluate $(41 + 31) \div 2^2 - 8$.

Think and Discuss
• What do the underlined letters in "**P**lease **E**xcuse **M**y **D**ear **A**unt **S**ally" represent?

• List the order of operations you would use to find the value of $3^3 + 5 - 3 \times (21 \div 7)$. Explain why the order is important.

Practice
Tell the order in which you would perform the operations in each expression. Then find the value of the expression.

1. $(120 - 14) + 4^2 \times 3$

2. $3 + 5^2 \times 2 \div 10 - 4$

3. $9 \times 1 + 12 \times 2 \div 8 + 4$

4. $5 + (7 - 4)^2 - 8 \div 2$

5. $16 - 2^3 - (9 - 7) \times 4$

You can use a calculator to evaluate expressions with more than one operation. Some calculators use an **algebraic operating system (AOS)** that automatically follows the order of operations.

Activity 2

- Use your calculator to find the value of $8 \div 2 + 6 \times 3 - 4$.

- Following the order of operations, use paper and pencil to find the value of $8 \div 2 + 6 \times 3 - 4$.

- Exchange papers with a classmate, and check each other's work.

Think and Discuss

- How does the calculator value for $8 \div 2 + 6 \times 3 - 4$ compare with the value you got by using paper and pencil? Does your calculator use an AOS?

To find the value of an expression with a calculator that does not use an AOS, follow the order of operations or use the memory keys.

Follow the order of operations to find the value of the expression $2 + 6 \times 3^2 - 4$.

3 2 6 2 4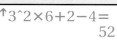

Use the memory keys to find the value of the expression $9^2 + 6 \div 2 \times 4$.

9 2   6 2 4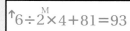

- When you enter values into a calculator that does not have an AOS, how do you know which values to enter first?

Practice

1. A calculator shows the display 9 as the value of $12 + 15 \div 3$. Does the calculator use an AOS? Explain.

2. How could you use memory keys to evaluate the expression $12 + 15 \div 3$?

3. How could you use the order of operations to evaluate the expression $12 + 15 \div 3$?

TECHNOLOGY LINK

More Practice: Use E-Lab, *Order of Operations.* www.harcourtschool.com/elab2002

Use a calculator to find the value.

4. $12 + 8 \times 4^2$ **5.** $9 + (6 - 2) \times 5$ **6.** $18 \div (6 - 4) + 5$

MIXED REVIEW AND TEST PREP

7. Find the value for 6^3. (p. 40)

8. $34{,}056 + 2{,}207$ (p. 20)

9. Evaluate $6p - q$ for $p = 7$ and $q = 11$. (p. 28)

10. $807 \div 45$ (p. 22)

⭐ **11. TEST PREP** A machine makes 2^5 bolts per second. How many minutes does it take to make 9,600 bolts? (p. 40)

A 5   **B** 12   **C** 300   **D** 1,920

ALGEBRA
Order of Operations

Learn how to use the order of operations.

To evaluate an expression that contains more than one operation, you use rules called the order of operations. The first letters of these words help you remember the order of operations.

Please **E**xcuse **M**y **D**ear **A**unt **S**ally

Parentheses
Exponents
Multiplication or **D**ivision
Addition or **S**ubtraction

EXAMPLE 1

Tell the operations used to evaluate the expression.

$35 \div 7 + 5 \times 3^2$	*Clear exponents.*
$35 \div 7 + 5 \times 9$	*Divide.*
$5 + 5 \times 9$	*Multiply.*
$5 + 45$	*Add.*
50	

EXAMPLE 2

Find the value of the expression
$285 + 93 \div (3 - 2) \times 3 \times 4^2$.

$285 + 93 \div (3 - 2) \times 3 \times 4^2$	*Operate inside parentheses.*
$285 + 93 \div 1 \times 3 \times 4^2$	*Clear exponents.*
$285 + 93 \div 1 \times 3 \times 16$	*Divide.*
$285 + 93 \times 3 \times 16$	*Multiply twice.*
$285 + 4,464$	*Add.*
$4,749$	

CHECK FOR UNDERSTANDING

Think and Discuss ▶ Look back at the lesson to answer each question.

1. Show where to insert parentheses to make this equation true.
$420 - 100 \div 40 = 8$

2. Tell which operation you would do last to evaluate the expression $7 + 8 - 2^3 \div 4$.

Guided Practice ▶ Evaluate the expression.

3. $30 - 15 \div 3$ **4.** $5^2 - (40 \div 4) \div 2$ **5.** $5^2 + 10^2 \div 25 - 1$

PRACTICE AND PROBLEM SOLVING

Independent ▶ Practice

Evaluate the expression.

6. $45 \div 15 + 2 \times 3$

7. $7 \times 2^2 + 6 - 9$

8. $3 + 4 \times 250$

9. $12 + (36 \div 4)^2 - 25$

10. $2^6 - (27 - 8) + (5^2 - 21)$

11. $(43 + 57) \times (9 - 6)^0$

12. $4^4 - (5^3 - 7^2) + (3^3 - 25)^3$

13. $(6^2 + 3^2)^2 \div 5 \times 3 + 3$

14. $(24 + 1^8) \times (7 - 5)^3$

15. $(7 \times 4)^2 - (34 + 1^8) \times 2^3$

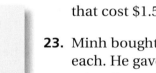 **Evaluate the expression for $a = 4$ and $b = 7$.**

16. $21 \div b + 8$

17. $a \times 31 - 8^2$

18. $(8 - a) \div 2 + 7$

19. $b^2 \div 7 \times (6 + 5)$

20. $(a^2 + b^2) \times 4$

21. $(b^2 - a^2) \div 3$

Problem Solving ▶ Applications

For 22–23, write and evaluate an expression to solve.

22. Heather mailed 3 packages that cost $2.50 each and 2 packages that cost $1.50 each. How much did she spend on postage?

23. Minh bought a watermelon for $3.25 and 3 cantaloupes for $1.29 each. He gave the clerk $20. How much change did he get in return?

24. **?** **What's the Error?**   Joe and Brett found the value of $2 + 6 \times 3^2 - 4$. Joe said the answer is 68 and Brett said the answer is 52. Decide who made the error and describe what the error is.

25. The Island Theater holds 236 people. It was filled to capacity for each of its 43 shows last week. This week 8,299 people attended shows at the theater. How many people attended shows at the Island Theater during these two weeks?

MIXED REVIEW AND TEST PREP

26. What are the equal factors and the value for 7^3? (p. 40)

27. Use mental math to find the value. 34×6 (p. 36)

28. Subtract. $2,500 - 1,646$ (p. 20)

29. Divide. $163 \div 5$ (p. 22)

30. **TEST PREP**   Ricky's math class collected a total of 1,364 pennies and quarters. They collected 234 more pennies than quarters. How much money did Ricky's class collect? (p. 26)

A $149.24

B $158.34

C $229.40

D $237.12

EXTRA PRACTICE page H33, Set C

PROBLEM SOLVING SKILL
Sequence and Prioritize Information

Analyze
Choose
Solve
Check

Learn how to solve problems by sequencing and prioritizing information.

Mrs. Rucki gets paid on the first workday of each month. She deposits her check in the bank and uses the money to pay her bills. Of the money that's left she uses part for spending money and puts the rest into savings.

On March 1, the following occurred:

- Mrs. Rucki had bills of $740, $85, $102, $33, and $52 to pay.
- She received her monthly paycheck for $1,570.
- She wanted to save at least $300 from her check.
- She needed at least $250 spending money.

The sequence, or order, in which Mrs. Rucki does these things is important. For example, she cannot use the money until the bank gives her credit for her deposit. That may take three days.

Mrs. Rucki made the list at the left. The list shows the things to do and the order she planned to do them.

- Why does it make sense for Mrs. Rucki to write "Put the rest into my savings account" last?

Suppose Mrs. Rucki follows the sequence above. How much will she have left to put into her savings account?

$740 + $85 + $102 + $33 + $52 = $1,012 *Find the total of her bills.*

$1,570 - $1,012 = $558 *Find the amount left after she pays her bills.*

$558 - $250 = $308 *Find what is left after she withdraws spending money.*

So, Mrs. Rucki could put $308 into her savings account.

Math Idea ▶ The order in which parts of an activity are carried out is often important to success. Some parts of an activity may be more important than others.

For 1–3, use the schedule at the right. Each show is 50 minutes long.

1. Jennifer and her family will visit Ocean World Park from 9:30 to 4:00. They want to see the Water Skiing show at 10:00. Name the order in which they can see all the other shows.

2. The Jackson family wants to see the Underwater Acrobats at 12:00 and then take 45 minutes to eat lunch. What other show could they see before leaving the park at 3:00?

   A Aquarium Tour
   B Animal Acts
   C Whale Acts
   D Water Skiing

Ocean World Park Show Times

9:00, 12:00	Underwater Acrobats
9:00, 3:00	Whale Acts
10:00, 2:00	Animal Acts
10:00, 1:00	Water Skiing
11:00, 4:00	Aquarium Tour

3. Michele wants to go on the Aquarium Tour at 11:00. Before she leaves the park, she wants to see the Animal Acts show at 2:00. At which other time could she see a show after the Aquarium Tour?

   F 9:00   **H** 1:00
   G 10:00   **J** 2:00

MIXED APPLICATIONS

4. Ed wants to buy 3 boxes of cereal at $2.99 each, 2 melons at $1.29 each, and 5 cans of juice at $0.79 each. Ed has $12.80. How much more money does he need to buy all of the items?

5. At the Discount Book Barn, books cost $5.00 and magazines cost $1.50. José bought some books and magazines for $21. How many of each did he buy?

6. **Use Data**   Use the graph below. The highest waterfall in the world is Angel Falls in Venezuela. About how much higher is Angel Falls than Ribbon Falls?

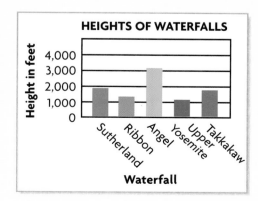

HEIGHTS OF WATERFALLS

7. Tim has saved $1,475 to make a down payment on a new car. He wants to save a total of three times that amount plus $275. How much money does Tim want to save for a down payment?

8. Ron and his two cousins own videotapes. Ron has twice as many as his older cousin, who has 27 videos. Ron has three times as many as his younger cousin. How many videotapes do they have in all?

9. Harry hammered nails into a board to make a circular pegboard. The nails were the same distance apart, and the sixth nail was directly opposite the eighteenth nail. How many nails formed the circle?

10. **Write About It**   Make a list of the steps you would follow to make a scrambled egg.

1. **VOCABULARY** In the expression 2^3, the number 2 is the __?__. (p. 40)

2. **VOCABULARY** To find the value of an expression that has more than one operation, you need to use the __?__. (p. 42)

3. **VOCABULARY** In the expression 8^4, the number 4 is the __?__. (p. 40)

Use mental math to find the value. (pp. 36–39)

4. $19 + 43$
5. $76 - 37$
6. $32 + (48 + 83)$

7. 26×12
8. $5,986 \times 1$
9. $4 \times 6 \times 25$

Write the equal factors. Then find the value. (pp. 40–41)

10. 6^2
11. 9^2
12. 3^5
13. 5^4
14. 2^0

15. 10^6
16. 7^5
17. 25^1
18. 4^3
19. 12^2

Evaluate the expression. (pp. 44–45)

20. $4 \times 5 - 6 \times 3$
21. $(12 \times 7) + 9^2$

22. $13 + 4 \times (20 + 35)$
23. $4^3 - 4 \times 12$

24. $36 \div (24 - 18) + 9$
25. $45 - (12 \times 3) \div 6$

26. $16 \times 9 \div 2^3$
27. $(100 - 28) \div 3^2$

28. $43 + (6^2 - 3^3)^2$
29. $(19 + 9^2)^2 \div 5^2 + 94$

Use Data For 30–31, use the table at the right.

30. Michael wants to surprise his family by having dinner ready when they get home. The table shows the length of time each of the foods needs to cook. If Michael wants to have everything ready to eat at the same time, in what order should he start cooking the food? (pp. 46–47)

31. Suppose Michael decides to cook baked potatoes instead of rice. The potatoes take 75 minutes to bake. In what order should he start cooking the food? (pp. 46–47)

32. Doreen has $40.00. She wants to buy 3 pairs of socks for $2.75 each, gloves for $9.99, and 2 T-shirts for $8.79 each. How much money will she have left? (pp. 42–43)

33. Each of three pyramids of Egypt is made up of about 2.5 million large stones. Is the total number of stones greater than or less than 10^8? (pp. 40–41)

Cooking Times for Food

Type of Food	Cooking Time
Chicken	55 Minutes
Peas	5 Minutes
Rice	20 Minutes
Stuffing	15 Minutes
Dinner Rolls	12 Minutes

TIP!

Look for important words.
See item **5**.

An important word is **not**. The words **not** true mean you need to find the one expression among the answer choices that is **not** an equality.

Also see problem **2**, p. H62.

Choose the best answer.

1. Which can be expressed as 3^5?

   A $3 \times 3 \times 3 \times 3 \times 3$ **C** 5×3

   B $5 \times 5 \times 5$ **D** 3×3

2. Lori has 79 stickers. This is three more than twice as many as Beth has. How many stickers does Beth have?

   F 34   **H** 38

   G 36   **J** 40

3. Calvin spent $43 on a video game and a CD. The video game cost $13 more than the CD. How much did the video game cost?

   A $13   **C** $30

   B $28   **D** $56

4. Kim's class has been saving pennies to go on a field trip to the Gem and Mineral Museum. The class has collected 10,000 pennies. How is 10,000 written using an exponent with the base 10?

   F 1^4 **H** 10^4

   G 10^2 **J** 10^5

5. Which number sentence is **not** true?

   A $30 \times 12 \times 8 = 12 \times 30 \times 8$

   B $30 + 8 + 12 = 30 + 12 + 8$

   C $30 - 12 + 8 = 30 + 8 - 12$

   D $30 + 8 \times 12 = 30 \times 12 + 8$

6. Which expression is equivalent to $(51 + 32) + 71$?

   F $(51 + 71) + (32 + 71)$

   G $51 + (32 + 71)$

   H $93 + 51$

   J $19 + 71$

7. What is the value of $42 - d$ for $d = 6$?

   A 7   **C** 48

   B 36   **D** Not here

8. Which algebraic expression represents the expression "13 less than a number, t"?

   F $13 \times t$ **H** $13 - t$

   G $t + 13$ **J** $t - 13$

9. Which is the best estimate for this sum?
   $$15{,}192 + 3{,}751 + 1{,}551$$

   A 18,000   **C** 20,000

   B 19,000   **D** 21,000

10. Which is another way to write $y \times y \times y$?

    F $3 \times y$ **H** y^3

    G $y \times 3y$ **J** y^4

Write What You Know

11. Explain how you can use compensation to find the sum $48 + 57$ mentally.

12. Evaluate $(3^2 + 7) \div (4 - 2)$. Explain how you used the order of operations to find your answer.

CHAPTER 3

Decimal Concepts

Wheat and corn are very important crops to farmers in the United States. These crops are grown across vast fields in Kansas and many other states. We eat wheat and corn at almost every meal in our cereals, breads, and pastas. We even eat them at parties.

PROBLEM SOLVING Which ingredients of the party mix occur in almost equal amounts?

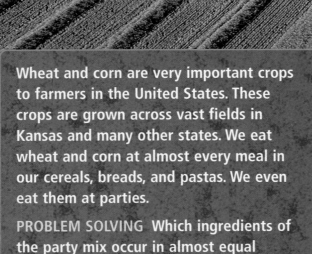

Party Mix Ingredients

Corn Chips 0.42

Puffed Corn 0.18

Wheat Chips 0.16

Puffed Wheat 0.24

DATA LINK

Check What You Know

Use this page to help you review and remember important skills needed for Chapter 3.

Represent Decimals (For Intervention, see p. H2.)

Write the decimal that is modeled.

1.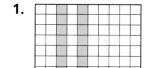

2.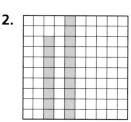

3.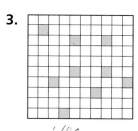

4.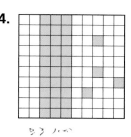

Write and Read Decimals (For Intervention, see p. H3.)

Write the decimal.

5. 836 and 23 hundredths

6. 364 and 2 tenths

7. 93 thousand, 450 and 38 hundredths

8. 595 thousand, 821 and 9 tenths

9. 306 thousand, 7 and 6 hundredths

Complete to show how to read the numbers.

10. 1,463.05 1 thousand, 463 and __?__

11. 204.7 204 and __?__ tenths

12. 32,617.45 32 __?__, 617 and 45 __?__

13. 4,382.1 4 thousand, 382 and __?__

Compare Whole Numbers (For Intervention, see p. H3.)

Compare the numbers. Write <, >, or = for ●.

14. 143 ● 140

15. 808 ● 880

16. 716 ● 716

17. 691 ● 961

18. 94 ● 49

19. 405 ● 305

20. 383 ● 383

21. 5,937 ● 397

22. 4,062 ● 4,206

23. 689 ● 648

24. 6,098 ● 6,908

25. 4,801 ● 4,108

Round Decimals (For Intervention, see p. H4.)

Round to the nearest whole number.

26. 3.64

27. 1.49

28. 6.938

29. 41.8

30. 18.70

31. 72.06

Round to the nearest tenth.

32. 69.64

33. 26.37

34. 52.489

35. 8.630

36. 9.479

37. 14.507

38. 90.63

39. 55.58

40. 26.397

LOOK AHEAD

In Chapter 3 you will

- represent, compare, and order decimals
- estimate with decimals
- use decimals and percents

Represent, Compare, and Order Decimals

Learn how to use place value to express, compare, and order decimals.

Remember that when reading a number with a decimal point, read the decimal point as "and." Read 8.2 as "eight and two tenths."

Genna read that it costs $0.03 to use her hair dryer for 30 minutes, $0.08 to use her clock for a month, and $0.53 to wash and dry a load of laundry.

Place value helps you understand numbers. The digits and the position of each digit determine a number's value. Read each number on the place-value chart. These numbers are part of the decimal system. Notice that 3 has a value of 3 thousandths, 3 tens, or 3 ten-thousandths, depending on its position in the number.

PLACE VALUE								
	Ten Thousands	Thousands	Hundreds	Tens	Ones	Tenths	Hundredths	Thousandths	Ten Thousandths
0.053					0	0	5	3	
32.4				3	2	4			
8.0023					8	0	0	2	3

When you read and write numbers, you are using place value.

EXAMPLE 1

A. Standard form: 0.053
Expanded form: 0.05 + 0.003
Word form: *fifty-three thousandths*

B. Standard form: 32.4
Expanded form: 30 + 2 + 0.4
Word form: *thirty-two and four tenths*

C. Standard form: 8.0023
Expanded form: 8 + 0.002 + 0.0003
Word form: *eight and twenty-three ten-thousandths*

Math Idea ▶ Knowing the place value of digits will help you read, write, and calculate numbers correctly, including decimal numbers.

Mark notices that one jar of cinnamon contains 2.6 oz and another contains 2.3 oz. He wants to buy the jar with the greater amount of cinnamon.

You can use a number line to compare 2.6 and 2.3.

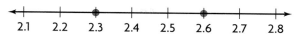

Since 2.6 is to the right of 2.3 on the number line, 2.6 is greater than 2.3.

$$2.6 > 2.3$$
↓
greater than

Since 2.3 is to the left of 2.6 on the number line, 2.3 is less than 2.6.

$$2.3 < 2.6$$
↓
less than

So, the 2.6-oz jar has more cinnamon.

You can also use place value to compare decimal numbers.

TECHNOLOGY LINK

More Practice: Use **Mighty Math Calculating Crew**, *Nautical Number Line*, Levels O and P.

EXAMPLE 2

Compare 7.28 and 7.2. Use < or >.

Start at the left.

7.28   7.2   *Compare the ones digits. They are the same.*

7.28   7.2   *Compare the tenths digits. They are the same.*

7.28   7.20   *Add a zero so both numbers have the same number of places. Compare the hundredths digits. 8 is greater than 0.*

So, 7.28 > 7.2, and 7.2 < 7.28.

• Which is greater, 7.2 or 7.08? Explain.

Remember that you can add a zero to the right of a decimal without changing its value.
7.2 = 7.20
2 tenths is the same as 20 hundredths.

You can use place value to order two or more decimal numbers.

EXAMPLE 3

The prices for the same kind of CD player in four different stores are $132.95, $132.50, $130.25, and $135.25. Order the prices of the CD players from least to greatest.

Compare every possible pair of numbers.

$132.95 > $132.50 $132.95 > $130.25 $132.95 < $135.25

$132.50 > $130.25 $132.50 < $135.25 $130.25 < $135.25

So, the list of numbers in order from least to greatest is $130.25, $132.50, $132.95, $135.25.

• List 9.365, 9.271, 9.356, and 9.065 in order from greatest to least.

**Think and ▶
Discuss**

Look back at the lesson to answer each question.

1. **Name** the number that is 4 hundredths greater than 2.0369.

2. **Tell** the place immediately to the right of hundred-thousandths.

**Guided ▶
Practice**

Read the number. Write the value of the blue digit.

3. 15,425.007   4. 2,654,000.25   5. 550.76

Write the number in expanded form.

6. 0.605   7. 0.00103   8. 12.0089   9. 342.046

Compare the numbers. Write <, >, or = for ●.

10. 1.15 ● 1.14   11. 92.3 ● 92.30   12. 0.82 ● 0.84

Write the numbers in order from least to greatest.

13. 1.361, 1.351, 1.363   14. 125.3, 124.32, 125.33

PRACTICE AND PROBLEM SOLVING

**Independent ▶
Practice**

Read the number. Write the value of the blue digit.

15. 5.0547   16. 827.142   17. 345.79456

Write the number in expanded form.

18. 46.00105   19. 0.0362   20. 1,500.1   21. 2.456

Compare the numbers. Write <, >, or = for ●.

22. 99.06 ● 99.6   23. 133.2 ● 133.32   24. 707.07 ● 707.07

25. 32.630 ● 32.63   26. 457.3685 ● 457.5683   27. 49.302 ● 49.203

28. $1 + 0.1 + 0.05$ ● $1 + 0.1 + 0.04$

29. $5 + 0.2 + 0.003$ ● $5 + 0.3 + 0.02$

Write the numbers in order from least to greatest.

30. 1.41, 1.21, 1.412, 1.12   31. 1.45, 1.05, 0.405, 1.25, 1.125

32. 35.2, 35.72, 35.171, 35.7   33. 9.82, 9.082, 8.91, 9.285, 9.85

Write the numbers in order from greatest to least.

34. 5.004, 5.040, 5.4   35. 125.33, 125.3, 125.35, 125.4

36. $3\frac{1}{10}$, 3.001, $3\frac{1}{100}$, 3   37. 14.01, $14\frac{1}{10}$, 41.01, $14\frac{3}{100}$

Problem Solving ▶
Applications

38. The world's smallest cut diamond is 0.0009 inch in diameter and weighs 0.0012 carat. Write both numbers in words.

39. A movie studio announced that the box office sales for its new release reached nine million, four hundred fifty-six thousand, three hundred two dollars in the first week. Write ten times that amount in standard and word forms.

40. **What's the Error?** Jon says 8.01 and 8.10 are equal because each number has the same digits. Explain his error.

Use Data For 41–42, use the graph at the right.

41. Estimate the total amount of rainfall during one year in Seattle.

42. Gene would like to visit Seattle to do some outdoor sightseeing. He would like to visit when the normal amount of rainfall is below 2 inches. When might be the best time for Gene to visit?

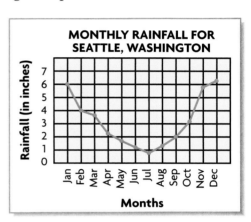

MONTHLY RAINFALL FOR SEATTLE, WASHINGTON

MIXED REVIEW AND TEST PREP

43. $12 + 3^3 \times 8$ (p. 42)   **44.** 234×38 (p. 22)   **45.** Evaluate $406 \div c$ for $c = 14$. (p. 28)

⭐ **46. TEST PREP** Which shows the value of 4^4? (p. 40)

A 16   **B** 124   **C** 256   **D** 2,414

⭐ **47. TEST PREP** If you do homework for 35 to 45 minutes a day, which is a reasonable estimate of the number of hours you do homework for 8 days? (p. 16)

F less than 2 hours   **H** between 6 and 8 hours
G between 4 and 6 hours   **J** more than 8 hours

PROBLEM SOLVING ThiNker's CorNer

Spin a Decimal Practice comparing and ordering decimals as you play this game.

Materials: a spinner numbered 0–9, a place-value chart

- In a small group, decide whether the player with the greatest decimal or least decimal will win. Taking turns, spin the pointer six times. After each spin, write the digit on your place-value chart. Once you have written it on your chart, the digit cannot be moved or erased.

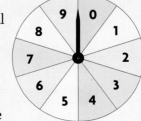

- Take turns reading your six-digit numbers aloud. The player with the greatest or the least decimal wins the round and receives a point.

- Continue playing until a player has five points.

EXTRA PRACTICE page H34, Set A

55

PROBLEM SOLVING STRATEGY
Make a Table

Learn how to solve problems by organizing data in a table.

Andrew chooses new in-line skates from the models below.

RC-204 $99.95; PRX-100 $78.50; D-500 $99.99; OP-1000 $78.99; ZZ-2 $91.50; ZA-45 $99.25

Andrew wants to buy the second most expensive model. His parents want him to buy the second least expensive model. Which models are these? What is the difference in price of the two models?

Analyze   What are you asked to find?

What information is given?

Choose   What strategy will you use?

You can use the strategy *make a table* to show the prices of the in-line skates in order from least to greatest.

Solve   How will you solve the problem?

Compare the prices and order them in a table. Then find the second most expensive model, the second least expensive model, and the difference in their prices.

$78.50 < $78.99 < $91.50 < $99.25 < $99.95 < $99.99

MODEL	PRICE	
PRX-100	$78.50	
OP-1000	$78.99	← *second least expensive*
ZZ-2	$91.50	
ZA-45	$99.25	
RC-204	$99.95	← *second most expensive*
D-500	$99.99	

Subtract: $99.95 − $78.99 = $20.96

So, the difference in price of models RC-204 and OP-1000 is $20.96.

Check   How can you check your answer?

What if Andrew chose model ZZ-2? How much more would he spend than if he bought model OP-1000?

Solve the problem by making a table.

Below are the fractions of games won by 8 baseball teams.

Hawks	0.650	Bulldogs	0.725
Tigers	0.750	Lions	0.490
Angels	0.675	Flames	0.700
Dolphins	0.550	Giants	0.695

1. Which team is in second place?

A Tigers   **B** Flames   **C** Bulldogs   **D** Angels

2. How many teams are behind the Giants?

F 2   **G** 3   **H** 4   **J** 5

Use Data For 3–4, reorder the data in the table at the right from greatest to least.

3. Which appliance uses the greatest amount of electricity?

4. Which appliance uses the least amount of electricity?

Electricity Used By Appliances

Appliance	Electricity (In Kilowatts)
Refrigerator	0.6
Air conditioner	1.5
Color TV	0.33
Iron	1.2
Coffeepot	0.9
Toaster	1.2

MIXED STRATEGY PRACTICE

5. A calculator, pen, and notebook cost $14.00 altogether. The calculator costs $9.00 more than the pen and $8.50 more than the notebook. How much does each item cost?

6. Kelly left the house with $16.00. She had $4.50 left after buying a movie ticket for $6.75, buying two snacks for $1.75 each, and paying for a bus ride. How much did she pay for the bus ride?

7. Carlene makes greeting cards. It costs $0.20 to make each card. She then sells them for $0.75 each. How many cards does she need to sell in order to make a profit of $33.00?

8. Frank walks 5 blocks to school for every 3 blocks Robert walks. They walk a total of 24 blocks. How many blocks from school does Robert live?

9. In a survey, teens prefer the Boomer portable stereo over the Blaster, but not as much as the Soundmaster. The Tekesound was preferred above all the others. Which portable stereo was least preferred?

10. (?) **What's the Question?** Glen read 69 pages each day for 6 days. He then read 23 pages each day for 4 days. The answer is about 500.

Estimate with Decimals

Learn how to estimate decimal sums, differences, products, and quotients.

You can estimate sums, differences, products, and quotients of decimals. To estimate with decimal numbers, use the same methods you used to estimate with whole numbers.

EXAMPLE 1

A long-distance phone company charges the rates shown at the right for calls from the United States. DeAnn made one-minute calls to India, Jordan, and Pakistan. About how much did the three calls cost?

Estimate $1.79 + $1.87 + $2.17.

$1.79   *The three addends*
$1.87   *cluster around $2.00.*
+$2.17   *So, multiply $2.00 by 3.*

3 × $2.00 = $6.00 *Multiply.*

So, the three calls cost about $6.00.

WIRED WORLD PHONE COMPANY	
Country	**Rate per Minute**
Argentina	$0.39
China	$0.49
France	$0.13
Germany	$0.09
India	$1.79
Ireland	$0.15
Jordan	$1.87
Pakistan	$2.17

EXAMPLE 2

Remember that the symbol ≈ means "is approximately equal to."

Estimate.

A. 36.4 × 18.25

Round to the nearest ten.

$$\begin{array}{c} 36.4 \\ \times 18.25 \end{array} \rightarrow \begin{array}{c} 40 \\ \times 20 \\ \hline 800 \end{array}$$

So, 36.4 × 18.25 ≈ 800.

B. 162.8 ÷ 8.16

Use compatible numbers.

$$8.16\overline{)162.8} \rightarrow 8\overline{)160} \; (20)$$

So, 162.8 ÷ 8.16 ≈ 20.

CHECK FOR UNDERSTANDING

Think and ▶ Discuss

Look back at the lesson to answer each question.

1. Tell how you could estimate the sum of 4.79, 18.99, and 3.09.

2. Explain how to use compatible numbers to estimate 423.2 ÷ 2.7.

Estimate.

3. $18.7 + 23.1$ **4.** $123.76 \div 9$ **5.** $185.32 - 101.99$

6. 39.83×36 **7.** $67.8 + 66.1 + 71.7$ **8.** 817.3×11

PRACTICE AND PROBLEM SOLVING

Estimate.

9. $6.7 + 9.4 + 15.82$ **10.** 12.2×8.3 **11.** $82.5 \div 9.3$

12. $\$266.08 - \$97.30$ **13.** $9.8 + 38.2$ **14.** 31.5×2.8

15. 6.8×18.2 **16.** $103.08 - 45.32$ **17.** 56.20×30.7

18. 689.89 **19.** $1,038.54$ **20.** $234.91 \div 5.79$
$\; -\;\; 98.5$ $\; \times\;\; 26.12$

21. $\$7,805.90$ **22.** $81.5 \times 23.1 \div 3.9$ **23.** $18.2 \times (7.2 - 4.9)$
$\; +\;\; 9,158.43$

Estimate to compare. Write $<$ or $>$ for each ⬤.

24. 4.32×8.56 ⬤ 40 **25.** 25 ⬤ $81.27 \div 4.1$ **26.** $34.6 - 12.4$ ⬤ 14

Use Data   For 27–29, the table shows the types of waste that make up a typical 100 pounds of garbage in the United States.

27. About how many pounds of newspapers and other paper are included in every 300 pounds of garbage thrown away?

What We Throw Away	
Type of Garbage	**Weight (in lb)**
Food Waste	8.6
Yard Waste	17.4
Newspapers	7.9
Other Paper	30.8
Metals	9.2
Glass	5.2
Other	20.9

28. About how many more pounds of food waste than glass are thrown away for every 500 pounds of garbage?

29. 📓 **Write a problem** involving estimation. Use the data about yard waste, other paper, and metals shown in the table.

MIXED REVIEW AND TEST PREP

30. Write the value of the digit 8 in the number 342.285. (p. 52)

31. Write 2.523, 2.325, 2.532, and 2.235 in order from least to greatest. (p. 52)

32. $20,817 - 19,805$ (p. 20)

33. $25,801 \div 23$ (p. 22)

⭐ **34. TEST PREP**   Evaluate $a \times 32$ for $a = 426$. (p. 28)

A 472 **B** 2,130 **C** 13,522 **D** 13,632

EXTRA PRACTICE page H34, Set B

Decimals and Percents

Learn how to write a decimal as a percent and a percent as a decimal.

QUICK REVIEW

1. 2,457 + 4,541　　　2. 3,470 − 350　　　3. 6 × 5

4. 50 ÷ 5　　　5. Write the decimal for $\frac{2}{100}$.

Vocabulary

percent

The graph at the right shows the responses to a question about what people in the United States like to eat for breakfast.

Percent means "per hundred" or "hundredths." The symbol used to write a percent is %.

40 percent: $40\% = \frac{40}{100}$

So, 40 out of 100 have toast or a roll.

26 percent: $26\% = \frac{26}{100}$

So, 26 out of 100 have cold cereal.

BREAKFAST FOODS

26% cold cereal

40% toast, roll

34% eggs, meat

Activity

You need: two 10 × 10 grids (decimal squares)

• On one grid, shade complete squares to make the first letter of your first name. On the other grid, shade complete squares to make the first letter of your last name. Make each letter as large as possible. Some examples are shown below.

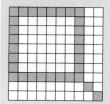 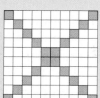 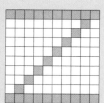

• Since 30 out of 100 squares are shaded for the letter *Q*, you can write 0.30 or 30%. What decimals and percents can be written for the squares shaded for *X* and *Z*?

• Count the number of complete squares you shaded on each of your grids. What percent of the squares are shaded?

Remember that when you read a decimal, you name the place with the least value. For 0.93, the place with the least value is hundredths. The number is read as "93 hundredths."

You can think about place value when you change decimals to percents or percents to decimals.

EXAMPLES

A. Write 0.08 as a percent.
0.08 is 8 hundredths.
So, 0.08 = 8%.

B. Write 32% as a decimal.
32% is 32 hundredths.
So, 32% = 0.32.

Think and Discuss ▶ Look back at the lesson to answer the question.

1. **Discuss** how you know that Sharon has 18 red cars if 18% of her 100 model cars are red.

Guided Practice ▶ Write the decimal and percent for the shaded part.

2. 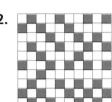   3. 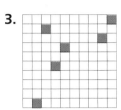   4.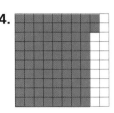

Write the corresponding decimal or percent.

5. 70%   6. 0.20   7. 0.03   8. 84%   9. 50%

PRACTICE AND PROBLEM SOLVING

Independent Practice ▶ Write the decimal and percent for the shaded part.

10. 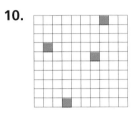   11. 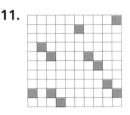   12.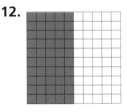

Favorite Types of Music

Alternative	21
Rhythm and Blues	14
Rap	21
Rock	20
Other or none	24

Write the corresponding decimal or percent.

13. 62%   14. 0.05   15. 28%   16. 45%   17. 53%

18. 0.63   19. 0.85   20. 33%   21. 0.4   22. 7%

Problem Solving Applications ▶

23. **Use Data**   The table shows how 100 teens responded to a survey. Write a decimal and a percent to show the number of teens who did not choose Alternative music.

24. What percent shows how many more students chose Rap and Rock music than Rhythm and Blues?

25. **Write About It**   Explain how to write 0.6 as a percent.

MIXED REVIEW AND TEST PREP

26. Order 27.8, 27.5, 27.82 from least to greatest. (p. 52)

27. Evaluate. $35 + 17 \times 3^2 - 16$ (p. 44)   28. 168×92 (p. 22)   29. $3,470 \div 42$ (p. 22)

30. **TEST PREP**   Which shows the sum $34,904 + 15,456 + 6,943$? (p. 20)

   A 55,920   **B** 57,303   **C** 68,870   **D** 72,780

EXTRA PRACTICE page H34, Set C

1. VOCABULARY A word that means "per hundred" is __?__ . (p. 60)

Write the value of the blue digit. (pp. 52–55)

2. 3.2497   **3.** 14.5805   **4.** 0.09003

5. 628.0402   **6.** 1.81738   **7.** 78.05124

Write the numbers in order from least to greatest. (pp. 52–55)

8. 2.365, 2.305, 2.3, 2.35, 2.035   **9.** 125.3, 124.32, 125.33, 12.245, 120.4

Estimate. (pp. 58–59)

10. 27.6 + 135.2   **11.** 4.8 × 2.3   **12.** 30.7 − 6.25   **13.** 89.75 ÷ 8

14. 8.45 + 8.99 + 9.2   **15.** 219.48 − 107.43   **16.** 416.2 × 31   **17.** 40.02 ÷ 6.3

Write the decimal and percent for the shaded part. (pp. 60–61)

18. 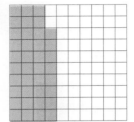   **19.** 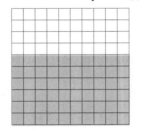   **20.**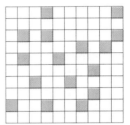

Write the corresponding decimal or percent. (pp. 60–61)

21. 74%   **22.** 0.07   **23.** 39%   **24.** 0.61

25. 0.6   **26.** 3%   **27.** 0.04   **28.** 84%

Solve.

29. The county library charges a fine of $0.10 a day for overdue books. The university library charges a fine of $0.50 for the first day and $0.05 for each additional day. On what day would overdue books have the same fine at both libraries? (pp. 56–57)

30. Planes leave Washington, D.C., for New York City every 45 min. The first plane leaves at 5:45 A.M. What is the departure time closest to 4:30 P.M.? (pp. 56–57)

31. Donna is on the decoration committee. She spent $15.90 on streamers, $12.15 on balloons, $6.84 on tape, $19.98 on banner paper, and $13.22 on banner paint. What is a reasonable estimate of the amount she spent? (pp. 58–59)

32. Frank earns $6.25 per hour. One week he worked 16 hours. About how much did Frank earn that week? (pp. 58–59)

33. Kirk has to list 125.3, 124.32, 125.33, 12.345, 120.4 in order from greatest to least. Which number should he list third? (pp. 52–55)

Get the information you need.

See item **7**.

Recall that a percent is the ratio of a number to 100. Write the ratio as a decimal.

Also see problem **3**, p. H63.

Choose the best answer.

1. Which of these numbers rounds to 450 when rounded to the nearest ten and to 500 when rounded to the nearest hundred?

   A 415   C 452

   B 428   D 478

2. Earth's orbit is more than 100,000,000 kilometers from the sun. How is this number written in exponential notation?

   F 10^7 km   H 10^9 km

   G 10^8 km   J 10^{10} km

3. What is the value of $2 + 3 \times 4$?

   A 11   C 20

   B 14   D Not here

4. Each of the 42 members of the band contributed $3.75 toward a gift for the band director. Which is a reasonable estimate of the total amount collected?

   F Between $80 and $90

   G Between $90 and $110

   H About $120

   J About $160

5. Kate has 4 bags of birdseed that have masses of 2.5 kilograms, 1.25 kilograms, 1.9 kilograms, and 2.15 kilograms. Which shows the bags in order from least mass to greatest?

   A 1.9 kg, 2.15 kg, 2.5 kg, 1.25 kg

   B 2.5 kg, 2.15 kg, 1.9 kg, 1.25 kg

   C 2.15 kg, 1.25 kg, 2.5 kg, 1.9 kg

   D 1.25 kg, 1.9 kg, 2.15 kg, 2.5 kg

6. A restaurant manager bought a total of 73 apples and oranges. She bought 11 more oranges than apples. How many of each kind of fruit did she buy?

   F 31 apples and 42 oranges

   G 27 apples and 46 oranges

   H 36 apples and 37 oranges

   J 42 apples and 31 oranges

7. How is 6% written as a decimal?

   A 6.0   B 0.6   C 0.06   D 0.006

8. Which is greater than 16.30?

   F 16.45   H 16.03

   G 16.23   J 1.730

9. Which is the value of $(3 - 2) \times 5 + 6^2$?

   A 29   B 40   C 41   D 122

10. Which of these is in order from least to greatest?

    F 0.0310, 0.301, 0.310

    G 0.310, 0.0310, 0.301

    H 0.301, 0.0310, 0.310

    J 0.0310, 0.310, 0.301

Write What You Know

11. With tax, a CD player costs $77.75. Ken saves $4.85 each week. Find a reasonable estimate of the number of weeks he must save for the new CD player. Explain your method.

12. Explain how you can write a decimal as a percent. Use your method to write 0.36 as a percent.

CHAPTER 4

Decimal Operations

Today's computers use optical laser writing technology to store billions of bytes of information. A Jaz® disk holds 1 gigabyte of information. A gigabyte is about 1,000,000,000 (10^9) bytes. Sometimes capacity is given in megabytes. A megabyte is about 1,000,000 (10^6) bytes.

PROBLEM SOLVING A 3.5-inch floppy disk holds just 1.44 megabytes of information. About how many bytes is that?

STORAGE CAPACITY IN GIGABYTES

DATA LINK

Bar chart titled "Storage Capacity in Gigabytes" with vertical axis "Type of Disk" listing Zip®, CD-ROM, Jaz®, DVD-RAM and horizontal axis "Number of Gigabytes" from 0 to 6.

Check What You Know

Use this page to help you review and remember important skills needed for Chapter 4.

✔ Whole-Number Operations (For Intervention, see p. H4.)

Add or subtract.

1. $7 + 28 + 12$
2. $45 - 15$
3. $63 - 19$
4. $19 + 41 + 27 + 23$
5. $34 - 17 - 7$
6. $27 + 56 + 100$
7. $143 + 79$
8. $213 - 88$

Multiply.

9. 63×4
10. 49×9
11. 19×76
12. 88×32

13. 80×50
14. 75×11
15. 200×15
16. 340×20

Divide.

17. $4\overline{)96}$
18. $5\overline{)127}$
19. $9\overline{)423}$
20. $7\overline{)760}$

21. $32\overline{)448}$
22. $20\overline{)3,660}$
23. $37\overline{)1,073}$
24. $23\overline{)4,715}$

✔ Multiply Decimals by 10, 100, and 1,000 (For Intervention, see p. H13.)

Multiply.

25. 4.3×10
26. $9.61 \times 1,000$
27. 8.4×100
28. $25.397 \times 1,000$
29. 194.05×100
30. 408.08×10

✔ Remainders (For Intervention, see p. H5.)

Divide. Write the remainder as a decimal.

31. $4\overline{)35}$
32. $5\overline{)56}$
33. $8\overline{)100}$

34. $12\overline{)243}$
35. $15\overline{)2,412}$

Divide. Write the remainder as a fraction.

36. $6\overline{)45}$
37. $8\overline{)77}$
38. $12\overline{)134}$

39. $14\overline{)550}$
40. $18\overline{)459}$

> **LOOK AHEAD**
>
> **In Chapter 4 you will**
> - add, subtract, multiply, and divide decimals
> - evaluate decimal expressions and solve equations

Add and Subtract Decimals

Learn how to add and subtract decimals.

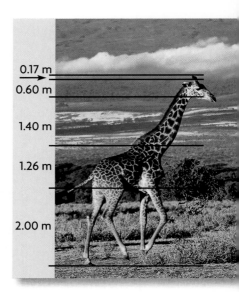

The tallest known mammal was a 6.1-m giraffe named George. Born in Kenya, George spent most of his life in the Chester Zoo in England.

To find the height of the giraffe shown, you must add five partial heights.

Math Idea ▶ When you add or subtract decimals, align the decimal points first and then add or subtact the digits, one place at a time.

      2.00   *Align the decimal points.*
      1.26
      1.40
      0.60
    +0.17   *Place the decimal point.*
      5.43   *Then add.*

So, the total height is 5.43 m.

You can use estimation to check for reasonableness.

EXAMPLE 1

Jamie has $85.75. Running shoes cost $68.45. How much money will Jamie have left after buying the running shoes?

Estimate.
        $85.75   → $86   *Round to the nearest dollar.*
      − 68.45   − 68
                           $18
Find the answer.

        $85.75
      − 68.45   *Align the decimal points.*
        $17.30   *Place the decimal point. Then subtract.*

Use your estimate to check the reasonableness of the answer. Compare the two. Since $17.30 is close to the estimate of $18, the answer is reasonable.

So, Jamie will have $17.30 left.

Some decimal numbers that you add or subtract do not have the same number of decimal places.

The samples below show how Courtney and Jamie found 17.06 + 5.493. Think about which sum is reasonable.

TECHNOLOGY LINK

More Practice: Use
**Mighty Math
Calculating Crew,**
Nautical Number Line,
Level R.

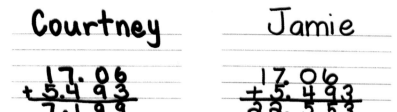

Courtney

17.06
+5.4 93
7.1 9 9

Jamie

17.06
+5.4 93
22.553

Use estimation to find which sum is reasonable.

17.06 + 5.493 → 17 + 5 = 22   Jamie's sum is reasonable.

Remember that you can add zeros to the right of a decimal without changing its value. Use zeros to make all decimals have the same number of decimal places.

EXAMPLE 2

Add. 3.45 + 7 + 0.835
Estimate.

3.45		3	*Round to the nearest whole number.*
7	→	7	
+0.835		+1	
		11	

Find the answer.

3.45		3.450	*Align the decimal points.*
7	or	7.000	*Use zeros as placeholders.*
+0.835		+0.835	
		11.285	*Place the decimal point. Add.*

Compare the answer to your estimate. 11.285 is close to the estimate of 11. The answer is reasonable.

So, 3.45 + 7 + 0.835 = 11.285.

EXAMPLE 3

Find the difference. 351.4 − 65.25
Estimate.

351.4	→	350	
− 65.25		− 70	*Round to the nearest ten.*
		280	

Find the answer.

351.4		351.40	*Align the decimal points.*
− 65.25	or	− 65.25	*Use a zero as a placeholder.*
		286.15	*Place the decimal point. Subtract.*

Compare the answer to your estimate. 286.15 is close to the estimate of 280. The answer is reasonable.

So, 351.4 − 65.25 = 286.15.

Think and ▶
Discuss

Look back at the lesson to answer each question.

1. **Tell** why it is important to align the decimal points when you add or subtract.

2. **Explain** the steps you would use to find $67 - 34.58$.

Guided ▶
Practice

Add or subtract. Estimate to check.

3. $\$6.18$
  $-\ 5.55$

4. 0.45
  0.5
 $+1.349$

5. 6
  5.43
  1.4
 $+5.755$

6. 10.72
 $-\ 1.3$

7. $3.2 + 2.68 + 15.043$

8. $142.108 - 63.8$

Copy the problem. Place the decimal point correctly in the answer.

9. $37.5 - 0.19 = 3731$

10. $0.431 + 1.549 + 2.017 = 3997$

11. $6 + 118.59 + 0.35 = 12494$

12. $9.7 - 3.01 = 669$

Independent ▶
Practice

Add or subtract. Estimate to check.

13. 50.28
 $+37.52$

14. 153.95
 $+434.16$

15. 805.41
 $+633.25$

16. 31.62
 $-\ 5.8$

17. $3.2 - 2.6$

18. $735.1 + 37 + 105.73$

19. $370.92 - 83.247$

20. $275.2 - 86.05$

21. $123.1 + 140 + 225.45$

22. $\$8 + \$215.49 + \$0.75$

23. $620.87 - 91.386$

24. $56.60 - 8.476$

Copy the problem. Place the decimal point correctly in the answer.

25. $23.64 + 233.5 = 25714$

26. $\$25.67 + \$7.16 + \$0.35 = \$3318$

27. $11.2 - 1.78 = 942$

28. $4.98 - 3.235 = 1745$

Estimate to determine if the given sum is reasonable.
Write *yes* or *no*.

29. $14.78 + 122.4 = 137.18$

30. $\$32.76 + \$8.09 + \$0.49 = \$41.34$

31. $58.02 - 9.473 = 3.671$

32. $427.7 - 39.27 = 388.43$

ALGEBRA Evaluate each expression for $d = 4.3$.

33. $d - 3.05$

34. $1 + d + 0.7$

35. $8 + d$

36. $37.60 - d$

37. $d - 2.084$

38. $(d + 16.05) - 4.5$

Problem Solving ▶
Applications

39. Jake's batting average is 0.325. Last year it was 0.235. What is the difference between his average last year and this year?

40. Karen has saved $15.75, $18.36, $9.07, and $20.37 to buy a camera. How much more does she need to save to buy a camera that costs $165.45?

41. Estimate $30.53 + 95.7 + 75.12$. Is the sum more than or less than your estimate? Explain how you know.

42.  **Write About It** Why is it important to estimate the answer when you add and subtract decimals?

43. Four clubs collected 3,905; 3,950; 3,590; and 3,509 lbs of paper each. Order the four weights of paper. Then find the difference between the least and the greatest weights of paper collected.

MIXED REVIEW AND TEST PREP

44. Write 0.05 as a percent. (p. 60)

45. Write the decimal for 46%. (p. 60)

46. Evaluate $a \div c$ for $a = 4,602$ and $c = 37$. (p. 28)

47. TEST PREP Which shows the value of 7^4? (p. 40)

    A 283   **B** 343   **C** 2,381   **D** 2,401

48. TEST PREP Tanya bought milk for $2.09, two loaves of bread for $1.05 each, cheese for $4.50, and three bottles of juice for $3.00 each. How much did Tanya pay for the items? (p. 20)

    F $12.04   **G** $12.64   **H** $17.69   **J** $23.04

PROBLEM SOLVING LiNKÜP to Science

Microbiologist Microbiologists are scientists who specialize in the study of the tiny cells that make up every living thing. While cells vary widely in size, most plant and animal cells are so small that they can be seen only with a microscope. Microbiologists often must use decimals when measuring the sizes of cells and their structures, as in these examples:

Average plant cell	0.000035 meter
Small bacterium	0.0000002 meter
Cell wall or membrane	0.0000000075 meter

- Use your school library to find the sizes of blood cells, skin cells, and nerve cells in the human body. How much larger or smaller is each of these cell types than the average plant cell?

EXTRA PRACTICE page H35, Set A

Multiply Decimals

Learn how to multiply decimals.

You need colored pencils and decimal squares.

QUICK REVIEW

1.	42	2.	83	3.	52	4.	93	5. 36×4
	$\times\ 7$		$\times\ 4$		$\times\ 6$		$\times\ 5$	

You can use a model to find the product of a decimal number and a whole number.

Activity

- To find 3×0.14, shade 0.14, or 14 small squares, three times. Use a different color and shade a different group of 14 small squares each time.

- Count the number of shaded squares. What is 3×0.14?

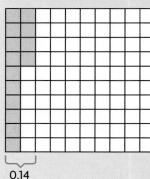

0.14

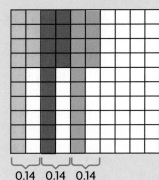

0.14 0.14 0.14

- Use a decimal square to find 5×0.18.

- Describe how you shaded your decimal square.

- What is 5×0.18?

Sometimes, when the factors are greater, as in 8×1.52, it is easier to compute the product by not using decimal squares.

EXAMPLE 1

Ed buys 9 sports cards at $1.12 each. How much does he spend?

Multiply. $\$1.12 \times 9$

Estimate. $\$1.12 \times 9 \rightarrow \$1 \times 9 = \$9$

Find the answer.

$$\begin{array}{r} \$1.12 \\ \times\quad 9 \\ \hline \$10.08 \end{array}$$ *Multiply as with whole numbers. Since the estimate is $9, place the decimal point after the 10.*

Since the estimate is $9, the answer is reasonable.

So, Ed pays $10.08 for sports cards.

Multiply a Decimal by a Decimal

You can use a decimal square or paper and pencil to find the product of two decimals.

EXAMPLE 2

Multiply. 0.2 × 0.6

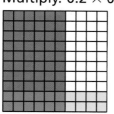

Shade 6 columns blue for 0.6.

Shade 2 rows yellow for 0.2.

The area in which the shading overlaps shows the product, or 0.2 of 0.6.

So, 0.2 × 0.6 = 0.12.

Place the decimal point in a product by estimating or by adding the number of decimal places in the factors.

$$
\begin{array}{r}
0.2 \leftarrow 1 \text{ decimal place} \\
\times 0.6 \leftarrow 1 \text{ decimal place} \\
\hline
0.12 \leftarrow 1 + 1, \text{ or } 2, \text{ decimal places}
\end{array}
$$

EXAMPLE 3

Mr. Ponti works 37.5 hr per week. He earns $8.70 an hour. How much does he earn in a week?

Multiply. $8.70 × 37.5

Estimate. $8.70 × 37.5 → $9 × 38 = $342

Find the answer.

$$
\begin{array}{r}
\$8.70 \leftarrow 2 \text{ decimal places} \\
\times\ 37.5 \leftarrow 1 \text{ decimal place} \\
\hline
4350 \\
6090 \\
+2610 \\
\hline
326.250 \leftarrow 3 \text{ decimal places}
\end{array}
$$

Multiply as with whole numbers.
Place the decimal point in the product.

Since the estimate is $342, the answer is reasonable.

So, Mr. Ponti earns $326.25.

When you multiply decimals, you sometimes have to insert zeros in the answer.

EXAMPLE 4

Multiply. 0.037 × 0.062

$$
\begin{array}{r}
0.037 \leftarrow 3 \text{ decimal places} \\
\times 0.062 \leftarrow 3 \text{ decimal places} \\
\hline
74 \\
+222 \\
\hline
0.002294
\end{array}
$$

Multiply as with whole numbers.
Place the decimal point in the product.
The answer must have 6 decimal places, so place 2 zeros to the left of 2.

So, 0.037 × 0.062 = 0.002294.

You can also use the Distributive Property to multiply with decimals.

EXAMPLE 5

Multiply. 6 × 5.9

Use estimation to place the decimal point in the product.
Estimate. 6 × 5.9 → 6 × 6 = 36

Find the answer.
6 × 5.9 = (6 × 5) + (6 × 0.9)   *Use the Distributive Property.*

$\quad\quad$ = 30 + 5.4   *Since the estimate is 36, place the*
$\quad\quad$ = 35.4   *decimal point after the 35.*

Since the estimate is 36, the answer is reasonable.

So, 6 × 5.9 = 35.4.

CHECK FOR UNDERSTANDING

**Think and ▶
Discuss**

Look back at the lesson to answer the question.

1. **Explain** how you would place the decimal point in the product 0.27 × 0.476.

**Guided ▶
Practice**

Use the decimal square shown to help you multiply.

2.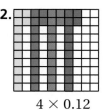
$\quad$ 4 × 0.12

3.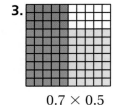
$\quad$ 0.7 × 0.5

4.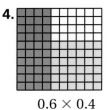
$\quad$ 0.6 × 0.4

Tell the number of decimal places there will be in the product.

5. 3.62 × 7

6. 2.15 × 8.18

7. 4.04 × 5.2

Copy the problem. Place the decimal point in the product.

8. 9 × 5.4 = 486

9. 0.7 × 4.1 = 287

10. 2.2 × 0.55 = 1210

Multiply. Estimate to check.

11. 0.42 × 2.9

12. 1.25 × 0.4

13. 3.23 × 8

PRACTICE AND PROBLEM SOLVING

**Independent ▶
Practice**

Use the decimal square shown to help you multiply.

14.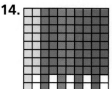
$\quad$ 5 × 0.18

15.
$\quad$ 0.3 × 0.8

16.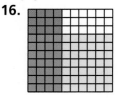
$\quad$ 0.7 × 0.4

Tell the number of decimal places there will be in the product.

17. 3.79×8.2 **18.** 0.876×0.2 **19.** 1.3842×0.91

Copy the problem. Place the decimal point in the product.

20. $6.37 \times 2.91 = 185367$ **21.** $20.4 \times 9.52 = 194208$

22. $7.32 \times 3 = 2196$ **23.** $0.82 \times 0.5 = 410$ **24.** $32.5 \times 0.06 = 1950$

Multiply. Estimate to check.

25. 3×4.6	**26.** 9×2.5	**27.** 7.3×5
28. 8.2×5	**29.** 1.2×4.1	**30.** 0.9×6.3
31. 0.2×0.4	**32.** 6.3×0.9	**33.** 0.21×2.1
34. 6.15×2.4	**35.** 4.08×1.35	**36.** 6.21×0.95
37. 24.63×1.09	**38.** 29.147×5.61	**39.** 0.189×2.09
40. 118.001×0.37	**41.** 148.9×0.006	**42.** $1,200.5 \times 8.2$

Problem Solving ▶ Applications

43. The recipe for a pie calls for 2.25 lb of apples and 0.75 lb of walnuts. Apples are $1.60 per pound and walnuts are $4.95 per pound. How much will the apples and walnuts cost in all?

44. Keith has a wall that is 5.2 m wide. He has 3 bookcases that are each 1.9 m wide. Is there enough room for all of the bookcases to be placed against the wall? Explain.

45. ✐ **Write a problem** that uses multiplication of two decimals to find the answer. The product must have four decimal places.

46. **Use Data** Look at the graph at the right. Rochelle had to earn 134 points in the first four rounds to advance in a competition. Rochelle says that she did advance. Is this a reasonable statement? Explain.

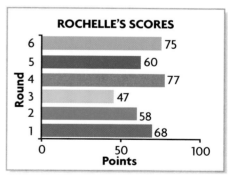

ROCHELLE'S SCORES

Round 6: 75
Round 5: 60
Round 4: 77
Round 3: 47
Round 2: 58
Round 1: 68

(Points: 0 to 100)

MIXED REVIEW AND TEST PREP

47. Add. $46.2 + 3.45 + 16$ (p. 66)   **48.** Subtract. $604.5 - 76.38$ (p. 66)

49. Evaluate the expression $a \div c$ for $a = 14,067$ and $c = 47$. (p. 28)

⭐**50.** **TEST PREP** Which shows the value of m when $8 = m \div 9$? (p. 36)

  A 70   **B** 72   **C** 720   **D** 7,200

⭐**51.** **TEST PREP** Which is the best estimate for $21,563 \div 43$? (p. 16)

  F 500   **G** 700   **H** 5,000   **J** 7,000

EXTRA PRACTICE page H35, Set B

73

Explore Division of Decimals

Explore how to use a model to divide decimals.

You need decimal squares, colored pencils, scissors.

Remember

1 tenth (0.1) = 1 column

1 hundredth (0.01) = 1 small square

You can shade and cut apart decimal squares to divide a decimal by a whole number.

Activity 1

Find $3.66 \div 3$.

• Shade 3.66 decimal squares.

 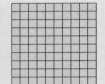

• Divide the shaded wholes into 3 equal groups. Divide the 66 hundredths into 3 equal groups.

 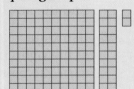 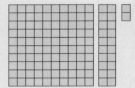

• What decimal names each group? What is the quotient?

• Use decimal squares to find $1.32 \div 4$.

Think and Discuss
• Find $366 \div 3$. How is the quotient the same as for $3.66 \div 3$? How is it different?

• Find $132 \div 4$. How is the quotient the same as for $1.32 \div 4$? How is it different?

Practice
Use decimal squares to find the quotient.

1. $4.04 \div 4$

2. $3.25 \div 5$

3. $1.35 \div 3$

You can shade and cut apart decimal squares to divide a decimal by a decimal.

TECHNOLOGY LINK

More Practice: Use E-Lab, *Exploring Division of Decimals*. www.harcourtschool.com/elab2002

Activity 2

Find 3.6 ÷ 1.2.

- Shade 3.6 decimal squares.

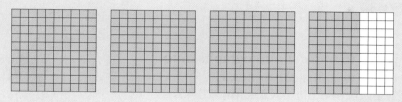

- Cut apart the 6 tenths.

- Divide the shaded squares and shaded tenths into equal groups of 1.2. How many groups of 1.2 are in 3.6? What is the quotient?

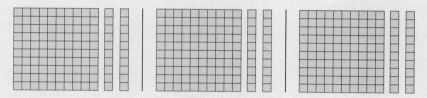

- Use decimal squares to find 3.2 ÷ 1.6.

- How many groups of 1.6 are in 3.2?

Think and Discuss

- Find 36 ÷ 12. How is the quotient the same as for 3.6 ÷ 1.2? How is the problem different from 3.6 ÷ 1.2?

- You know that 3.6 ÷ 12 = 0.3 and 3.6 ÷ 1.2 = 3. What do you think 3.6 ÷ 0.12 equals?

Practice

Use decimal squares to find the quotient.

1. 7.8 ÷ 1.3   **2.** 5.6 ÷ 0.8   **3.** 1.56 ÷ 0.52

4. 5.5 ÷ 1.1   **5.** 3.6 ÷ 0.9   **6.** 1.8 ÷ 0.45

7. 0.42 ÷ 0.14   **8.** 10.8 ÷ 2.7   **9.** 0.64 ÷ 0.08

MIXED REVIEW AND TEST PREP

10. Multiply. 64.7 × 3.6 (p. 70)

11. Write the decimal for 57%. (p. 60)

12. Write the value of 3^4. (p. 40)

13. Divide. 3,759 ÷ 42 (p. 22)

14. TEST PREP Evaluate the expression. 12 × 25 ÷ (12 + 18) (p. 44)

     A 8   **B** 10   **C** 43   **D** 300

Divide with Decimals

Learn how to divide a decimal by a whole number and how to divide a decimal by a decimal.

Remember that compatible numbers are numbers that divide without a remainder, are close to the actual numbers, and are easy to compute mentally.

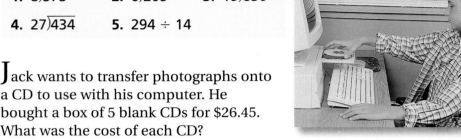

Jack wants to transfer photographs onto a CD to use with his computer. He bought a box of 5 blank CDs for $26.45. What was the cost of each CD?

Dividing a decimal by a whole number is like dividing whole numbers.

Divide. $26.45 \div 5$

Use compatible numbers to estimate.

$26.45 \div 5 \quad \rightarrow \quad \$25 \div 5 = \$5$

Find the answer.

$$\begin{array}{r} 5.29 \\ 5\overline{)26.45} \\ -25 \\ \hline 14 \\ -10 \\ \hline 45 \\ -45 \\ \hline 0 \end{array}$$

Place a decimal point above the decimal point in the dividend.

Divide.

Since the estimate is $5, the answer is reasonable.
So, each CD cost $5.29.

Sometimes you have to place a zero in the quotient when the dividend is less than the divisor.

EXAMPLE 1

Divide. $0.285 \div 3$

$$\begin{array}{r} 0.095 \\ 3\overline{)0.285} \\ -0 \\ \hline 28 \\ -27 \\ \hline 15 \\ -15 \\ \hline 0 \end{array}$$

Place a decimal point above the decimal point in the dividend.

Divide. Since 2 tenths is less than 3 ones, place a zero in the tenths place in the quotient.

So, $0.285 \div 3 = 0.095$.

Activity

- Use a calculator to find the first three quotients in each set.

- Look for a pattern. Try to predict the last quotient in each set.

Set A	Set B
$0.48 \div 0.03 = \blacksquare$	$0.621 \div 0.023 = \blacksquare$
$4.8 \div 0.3 = \blacksquare$	$6.21 \div 0.23 = \blacksquare$
$48 \div 3 = \blacksquare$	$62.1 \div 2.3 = \blacksquare$
$480 \div 30 = \blacksquare$	$621 \div 23 = \blacksquare$

- Describe the pattern that helped you predict the last quotient in each set.

- Look at $4.8 \div 0.3$. Multiply both numbers by 10. How do the quotients for $4.8 \div 0.3$ and $48 \div 3$ compare? How does multiplying the divisor and the dividend by 10 affect the quotient?

To divide a decimal by a decimal, first multiply the divisor and the dividend by a power of 10 to change the divisor to a whole number.

$$0.7\overline{)62.44} \quad \rightarrow \quad 7\overline{)624.4}$$

THINK: $0.7 \times 10 = 7$

$62.44 \times 10 = 624.4$

EXAMPLE 2

Divide. $22.8 \div 0.8$

$0.8\overline{)22.8}$ *Make the divisor a whole number by multiplying the divisor and dividend by 10.*

$0.8 \times 10 = 8 \qquad 22.8 \times 10 = 228$

$$\begin{array}{r} 28.5 \\ 8\overline{)228.0} \\ -16 \\ \hline 68 \\ -64 \\ \hline 40 \\ -40 \\ \hline 0 \end{array}$$

Place the decimal point in the quotient. Divide.

Since there is a remainder, place a zero in the tenths place in the dividend, and continue to divide.

So, $22.8 \div 0.8 = 28.5$.

- Think about $63.7 \div 0.24$. To change the divisor to a whole number, you multiply by 100. What does the dividend become?

77

Sometimes there aren't enough places in the dividend to move the decimal to the right.

EXAMPLE 3

Divide. 158.4 ÷ 0.12

$0.12\overline{)158.40}$ *Multiply the divisor and dividend by 100.*
0.12 × 100 = 12   158.4 × 100 = 15,840
Write a zero in the dividend.

```
        1320
   12)15840      Divide.
    −12
      38
     −36
       24          Since the remainder is zero, the quotient is a
      −24          whole number. You do not need to put the
        00         decimal point in the answer.
```

So, 158.4 ÷ 0.12 = 1,320.

CHECK FOR UNDERSTANDING

Think and Discuss ▶ Look back at the lesson to answer each question.

1. **Explain** how to change the divisor and the dividend before you solve the problem 55.8 ÷ 0.18.

2. **Compare** the quotients 4.5 ÷ 1.5 and 45 ÷ 15.

Guided Practice ▶ Rewrite the problem so that the divisor is a whole number.

3. 9.6 ÷ 1.6   4. 73.6 ÷ 0.5   5. 48.24 ÷ 2.4

Copy the problem. Place the decimal point in the quotient.

6. 28.50 ÷ 2.50 = 1140   7. 34.178 ÷ 2.3 = 1486   8. 62.44 ÷ 7 = 892

Divide. Estimate to check.

9. 7.88 ÷ 4   10. 33.66 ÷ 11   11. $0.55\overline{)2.42}$

PRACTICE AND PROBLEM SOLVING

Independent Practice ▶ Rewrite the problem so that the divisor is a whole number.

12. 48.4 ÷ 0.4   13. 8.19 ÷ 0.09   14. 3.7 ÷ 2.1

15. 2.39 ÷ 0.05   16. 45.218 ÷ 0.23   17. 233.58 ÷ 10.2

Copy the problem. Place the decimal point in the quotient.

18. 325.8 ÷ 3 = 1086   19. 53.07 ÷ 8.7 = 61   20. 10.2 ÷ 2 = 51

21. 84.87 ÷ 12.3 = 69   22. 274.89 ÷ 1.5 = 18326

Divide. Estimate to check.

23. $11\overline{)109.01}$ **24.** $90\overline{)10.8}$ **25.** $60\overline{)12.6}$

26. $0.38\overline{)13.3}$ **27.** $6.41\overline{)135.892}$ **28.** $38.2\overline{)469.86}$

29. $44.28 \div 5.4$ **30.** $80.1 \div 9$ **31.** $90.3 \div 6$

32. $1.26 \div 0.2$ **33.** $13.2 \div 0.06$ **34.** $42.5 \div 0.05$

Evaluate each expression.

35. $3 - 1.6 \times 0.4$ **36.** $5 - 2.4 \div 8$ **37.** $0.09 \div 3 + 3$

Problem Solving ▶
Applications

38. Emily rents 5 DVD movies for $28.75. Is the price for one movie closer to $5 or to $6? Explain.

39. Jonelle saves $4.95 every week to buy a video that costs $29.70. She has already saved $9. For how many more weeks does she need to save money to have enough to buy the video?

40. ❓ **What's the Error?** Michael divided 4.25 by 0.25 and got a quotient of 0.17. Explain the error. What is the correct quotient?

41. Andrew reads 22 pages the first day. He plans to increase the number of pages he reads by 4 a day until he finishes the book. How many pages will he be reading at the end of 3 days?

MIXED REVIEW AND TEST PREP

42. Multiply. 2.051×8.6 (p. 70)

43. Is 208.605 greater than or less than 206.605? (p. 52)

44. Use mental math to find $210 \div 42$. (p. 36)

45. TEST PREP Which shows the value of r when $18 + r = 52$? (p. 30)

  A 34   **B** 30   **C** 24   **D** 14

46. TEST PREP Jocelyn works 12 hours per week. She earns $7.25 an hour. How much does she earn in 4 weeks? (p. 70)

  F $87   **G** $174   **H** $290   **J** $348

PROBLEM SOLVING   THiNKer's CorNer

REASONING The ancient Greeks discovered that there is one particular number that can be used to relate the dimensions of any circle. They called this particular number π (pi) and found that it is the same number regardless of the size of the circle.

• Determine the approximate value of π by dividing the distance around each circle described below (the circumference) by the distance across that circle (the diameter). Is the answer the same for each?

	Circumference	Diameter
1. Circle 1	6.28 in.	2 in.
2. Circle 2	9.42 in.	3 in.

EXTRA PRACTICE page H35, Set C

PROBLEM SOLVING SKILL
Interpret the Remainder

Analyze
Choose
Solve
Check

Learn how to interpret the remainder in a division problem.

The sixth-grade class at Hightown Middle School has an annual class picnic. With the help of some students, Ms. Gordon is planning for the picnic this year.

Ms. Gordon needs 163 boxed drinks for lunch. A package holds 6 boxes. How many packages of drinks will Ms. Gordon need to buy?	$\begin{array}{r} 27\ r1 \\ 6\overline{)163} \\ -12 \\ \hline 43 \\ -42 \\ \hline 1 \end{array}$	Since 27 packages hold only 162 total boxes, increase the quotient by 1. So, Ms. Gordon needs to buy 28 packages of drinks.

Ms. Gordon has 158 ft of ribbon to use for games. She will cut the ribbon into 8-ft lengths. How many 8-ft lengths of ribbon will Ms. Gordon have?	$\begin{array}{r} 19\ r6 \\ 8\overline{)158} \\ -8 \\ \hline 78 \\ -72 \\ \hline 6 \end{array}$	The remainder is not enough for another 8-ft length of ribbon. Drop the remainder. So, Ms. Gordon will have 19 pieces of ribbon.

Students collected 158 prizes. The prizes were put in packages of 5 each. The remaining prizes were given to another class in the school. How many prizes were given to the other class?	$\begin{array}{r} 31\ r3 \\ 5\overline{)158} \\ -15 \\ \hline 8 \\ -5 \\ \hline 3 \end{array}$	The remainder is your answer. So, 3 prizes were given to the other class.

Talk About It ▶ • What does the remainder in the first problem mean?

• **What if** Ms. Gordon buys 28 packages of boxed drinks? How many extra boxed drinks will Ms. Gordon buy?

• **What if** Ms. Gordon has a total of 165 ft of ribbon? How many 8-ft lengths of ribbon will she have? How long is the ribbon that's too short?

PROBLEM SOLVING PRACTICE

Solve the problem by interpreting the remainder.

A total of 39 students and adults tour the science museum to see an exhibit about the shaping of the Earth's surface. The tour director can take groups of up to 5 on each tour. All 39 students and adults need to see the exhibit.

1. How many complete groups of 5 people can the tour director take?

   A 5 groups   **B** 6 groups   **C** 7 groups   **D** 8 groups

2. What is the least number of tours the director will have to give to accommodate all 39 students?

   F 6 tours   **G** 7 tours   **H** 8 tours   **J** 9 tours

3. James has $4.39 to buy magnets at the science museum. Each magnet costs $0.95. He wants to buy as many magnets as he can. How many magnets can James buy?

4. Sharon bought a package of 15 postcards at the museum. She gave the same number of postcards to each of her 4 teachers and kept the ones left over. How many postcards did Sharon keep?

MIXED APPLICATIONS

5. A train that is scheduled to arrive at 5:15 P.M. arrives 20 minutes late. If the train left at 9:30 A.M., how long was the trip?

6. A total of 51 students and teachers are using cars to go on a field trip. Six people ride in each car. How many cars are needed for the field trip?

7. Mark estimates that he needs 1 minute to solve a short homework problem and 5 minutes for each long problem. If he has 25 short and 8 long problems, how long should his homework take?

8. Tina makes bracelets for her friends. She uses 3 red beads for every 7 yellow beads to make a pattern. For one bracelet, she uses a total of 50 beads. How many of each color does she use?

9. A board game has 3 times as many red playing pieces as blue. It has 5 times as many green pieces as blue. There are 12 blue pieces. How many playing pieces are there in all?

10. Edgar has twice as many library books as his brother. If Edgar has 10 books in all, how many library books must he return to have the same number as his brother?

11. ❓ **What's the Question?**   Use the table at the right. Each Presidential term in office is 4 years. The answer is that he served more than one term but fewer than two.

United States Presidents	
President	**Years in Office**
Richard Nixon	1969-1974
Gerald Ford	1974-1977
Jimmy Carter	1977-1981
Ronald Reagan	1981-1989
George Bush	1989-1993

ALGEBRA
Decimal Expressions and Equations

Learn how to evaluate expressions and solve equations with decimals.

QUICK REVIEW

Solve by using mental math.

1. $s + 14 = 36$ **2.** $18 \div p = 9$

3. $57 - d = 42$ **4.** $w \div 7 = 7$

5. $a \times 8 = 24$

Different groups of friends eat lunch together at a cafe. The cost of the lunch is $6.45 per person. Write an expression to find the total cost of a lunch for a group of friends.

Let *a* be the number of friends buying lunch.

$a \times 6.45$ or $6.45a$ *Write the expression.*

The number of friends varies. What is the total cost for 7 friends?

$a \times 6.45$ *Write the expression.*

7×6.45 *Replace a with 7.*

45.15 *Multiply.*

So, the total cost is $45.15.

• Evaluate the expression $w \div 3 + 9.3$ for $w = 4.8$.

You solved equations with whole numbers by using mental math. You can use the same methods to solve some equations with decimals.

EXAMPLE

Solve the equation $h \div 6 = 0.6$ by using mental math.

$h \div 6 = 0.6$ *What number divided by 6 = 0.6?*

$h = 3.6$ THINK: $6 \times 0.6 = 3.6$.

$h \div 6 = 0.6$ *Check your answer. Replace h with 3.6.*

$3.6 \div 6 = 0.6$

$0.6 = 0.6$

So, $h = 3.6$.

• Solve. $t + 4.24 = 9.48$

CHECK FOR UNDERSTANDING

Think and ▶ Discuss

Look back at the lesson to answer each question.

1. **What if** 12 friends went to lunch at the cafe? Show how you would find the total cost of the lunch.

2. **Explain** how you would solve the equation $3.2 = a \div 2$.

Guided ▶ Practice

Evaluate each expression.

3. $a + 3.4$
for $a = 8.3$

4. $1.6 \div b$
for $b = 0.4$

5. $9.16 - a$
for $a = 4.08$

Solve each equation by using mental math.

6. $\frac{4.8}{k} = 8$

7. $m - 12.7 = 6.3$

8. $3t = 21.9$

PRACTICE AND PROBLEM SOLVING

Independent ▶ Practice

Evaluate each expression.

9. $2h$
for $h = 2.3$

10. $9.6 \div a$
for $a = 3$

11. $j + 7.1$
for $j = 6.9$

12. $4.17 - c$
for $c = 1.09$

13. $m \div 6 + 3.6$
for $m = 1.8$

14. $g + h - 3.2$
for $g = 4.1$ and
$h = 2.3$

Solve each equation by using mental math.

15. $r + 8.1 = 15.8$

16. $1.7 = \frac{d}{4}$

17. $4a = 32.8$

18. $x - 2.4 = 8.6$

19. $p + 11.1 = 28.7$

20. $3.2r = 7.1 + 5.7$

Problem Solving ▶ Applications

21. Let n represent the number of miles Jeremy rides his bicycle to attend 6 baseball practices. Write an expression to show how far he travels for one practice. Evaluate the expression for $n = 28.8$ mi.

22. **❓ What's the Error?** Explain the error at the right. Give the correct solution.

$$24.8 + x = 30$$
$$x = 54.8$$

23. David jogs 3 mi each weekday and 7 mi each Saturday. Ray jogs 18 mi each week. How much farther does David jog in a week?

MIXED REVIEW AND TEST PREP

24. $4.38 \div 7.5$ (p. 76)

25. $18 \div 6 + (16 \times 3) - 14$ (p. 44)

26. $6,045 - 973$ (p. 20)

27. Order from least to greatest. 3.58, 3.08, 3.85, 3.508 (p. 52)

⭐ 28. **TEST PREP** The sum of two numbers is 35. Their difference is less than 10. Which is not a possible pair? (p. 26)

A 15, 20   **B** 17, 18   **C** 10, 25   **D** 22, 13

EXTRA PRACTICE page H35, Set D

Add or subtract. Estimate to check. (pp. 66–69)

1.
$$\begin{array}{r} 3.9 \\ 4 \\ +5.91 \\ \hline \end{array}$$

2.
$$\begin{array}{r} 7.6 \\ -0.95 \\ \hline \end{array}$$

3.
$$\begin{array}{r} 3.02 \\ 0.17 \\ +4.338 \\ \hline \end{array}$$

4. $19.3 - 2.56$

5. $0.126 + 5.3 + 3.04$

6. $245 - 39.05$

Tell the number of decimal places there will be in the product. Then multiply. (pp. 70–73)

7.
$$\begin{array}{r} 8.3 \\ \times 12.9 \\ \hline \end{array}$$

8.
$$\begin{array}{r} 7.82 \\ \times\ 4.5 \\ \hline \end{array}$$

9.
$$\begin{array}{r} 0.13 \\ \times 2.07 \\ \hline \end{array}$$

10.
$$\begin{array}{r} 53.6 \\ \times 1.23 \\ \hline \end{array}$$

11. 20.01×8.2

12. 6.9×17.4

13. 4.91×6.2

14. 7.02×5.5

Divide. Estimate to check. (pp. 76–79)

15. $1.4\overline{)35}$

16. $3.7\overline{)5.92}$

17. $0.45\overline{)1.08}$

18. $0.25\overline{)85}$

19. $22.8 \div 3$

20. $9.72 \div 2.7$

21. $33.33 \div 1.1$

22. $6.9 \div 0.3$

Evaluate each expression. (pp. 82–83)

23. $(2.3 + c) + 1.7$ for $c = 8$

24. $3 \times d + b$ for $d = 1.7$ and $b = 5.4$

25. $c \div 2 - a$ for $c = 8$ and $a = 2.3$

26. $(5.4 - a) + d$ for $a = 2.3$ and $d = 1.7$

27. $4 \times a - c$ for $a = 2.3$ and $c = 8$

28. $(c - 5.4) \times c$ for $c = 8$

29. $(d + c) - 5$ for $d = 1.7$ and $c = 8$

30. $(a + d) \div 4$ for $a = 2.3$ and $d = 1.7$

Solve each equation by using mental math. (pp. 82–83)

31. $c + 14.07 = 32.97$

32. $6a = 24.78$

33. $8d = 6.4$

34. $7.14 - g = 3.24$

35. $2.4 + f = 4.76$

36. $4y = 29.6$

37. $1.86 \div r = 6.2$

38. $t + 3.6 = 4.5 + 3.3$

Solve. (pp. 80–81)

39. There are 9 golf balls in a box. How many boxes does Hector need if he wants to give away 500 golf balls as souvenirs?

40. On Tuesday, Sabrina gets an invitation to a party that is in 20 days. She can go to the party if it does not fall on the weekend. Can she go to the party? Explain.

Choose the answer.
See item **4**.

If your answer doesn't match one of the choices, check your computation and the placement of the decimal point. If your computation is correct, mark the letter for Not here.

Also see problem **6**, p. H64.

Choose the best answer.

1. Elise bought 4 balloons. Each balloon cost $0.98, including tax. Which is the total cost of the 4 balloons?

   A $3.62   C $3.96

   B $3.92   D $4.00

2. At the end of a trip, the odometer of a car read 4,572.1 miles. The odometer read 2,998.7 miles at the beginning of the trip. How far did the car go?

   F 7,570.8 mi   H 2,684.4 mi

   G 7,460.8 mi   J 1,573.4 mi

3. $0.27 + 1.098$

   A 1.368   C 1.125

   B 1.255   D Not here

4. $1.65 \div 0.5$

   F 0.33   H 33

   G 3.3   J Not here

5. Stu wants to buy a poster that costs $5.99, a desk lamp that costs $22.18, and bookends that cost $18.98. What is the cost of the 3 items he wants to buy, before tax is added?

   A $46.15   C $47.15

   B $47.00   D $48.15

6. What is the value of $x + 16.714$ for $x = 20.3$?

   F 37.014   H 3.614

   G 36.014   J 0.3614

7. A sweatshirt costs $9 more than a T-shirt. Together, the shirts cost $23. What is the cost of each type of shirt?

   A T-shirt: $4, sweatshirt: $19

   B T-shirt: $16, sweatshirt: $7

   C T-shirt: $12, sweatshirt: $11

   D T-shirt: $7, sweatshirt: $16

8. In which pair of numbers is there a 5 in the hundredths place of both numbers?

   F 553.621; 8,516.2

   G 87.35; 62,531.9

   H 36.157; 4,062.058

   J 513.26; 792.56

9. 7.8×0.3

   A 2.34   C 8.1

   B 3.24   D Not here

Write What You Know

10. Baseballs are sold in boxes of 12. During a typical game, 5 baseballs are used. A team has 20 games to play this season. Tell the steps you would use to find the number of baseballs the coach should buy.

11. What does the equation below mean? Explain the steps you can use to solve it. Then solve.

    $$s + 1.1 = 5.9$$

MATH DETECTIVE

Who Am I?

REASONING   Use the given information plus your knowledge of division to find the mystery number. Be prepared to explain how you solved the mystery.

Mystery Number 1

I'm the least whole number that gives a remainder of 4 when divided by 5 and when divided by 9. Who am I?

Mystery Number 2

I'm the first whole number greater than 240 that gives a remainder of 5 when divided by 9. Who am I?

Mystery Number 3

I'm the first whole number greater than 100 that gives a remainder of 3 when divided by 8 and when divided by 7. Who am I?

Mystery Number 4

I'm the least whole number that when divided by 4, by 5, and by 7 gives a remainder of 3. Who am I?

Think It Over!

- **Write About It**   Explain how you found each mystery number.

- **Stretch Your Thinking**   I'm the least number that when divided by 5 and when divided by 7 produces a remainder of 3. How much do you have to add to me to get a remainder of 0 when you divide by 5 and when you divide by 7?

Scientific Notation

820,000 pounds lifting off!

The weight of one type of passenger jet is about 820,000 lb. This number, 820,000, is written using *standard notation*.

You can also write the weight using **scientific notation**. Scientific notation is a shorthand method for writing large numbers.

A number written in scientific notation has two parts separated by a multiplication symbol.

$$8.2 \times 10^5$$

The first part is a number that is at least 1 but less than 10.

The second part is a power of 10.

To write the number 820,000 in scientific notation:

Count the number of places the decimal point must be moved to the left to form a number that is at least 1 but less than 10.

$$820,000 \;\rightarrow\; 8.2$$

5 places

Since the decimal point moved 5 places, the power of 10 is 5.

$$820,000 = 8.2 \times 10^5$$

The airplane's weight written in scientific notation is 8.2×10^5 lb.

EXAMPLE

About 32,500,000 passengers used Newark International Airport in 1998. Write the number of passengers in scientific notation.

$$32,500,000 \;\rightarrow\; 3.25 \times 10^7$$

7 places

The number of passengers was about 3.25×10^7.

TALK ABOUT IT

• Is 1.2×10^9 greater than 9.98×10^8 ? Explain.

• Show how to correctly write 782.5×10^8 in scientific notation.

TRY IT

Write the number in scientific notation.

1. 602,000   **2.** 199   **3.** 3,400,000

4. 8,540   **5.** 5,010,000   **6.** 113,000

7. 72   **8.** 48,900   **9.** 26,200,000

VOCABULARY

1. An expression that includes a variable is a(n) __?__ expression. (p. 28)

2. When you __?__ a decimal by a power of 10, the decimal point moves one place to the right for each factor of 10. (p. 87)

EXAMPLES

EXERCISES

Chapter 1

- **Add and subtract whole numbers.**
  (pp. 20–21)

  $$3,921 \\ 68 \\ +\ 205 \\ \overline{4,194}$$

  $$3,000 \\ -1,650 \\ \overline{1,350}$$

Find the sum or difference.

3. $756 + 902$ **4.** $4,293 + 256 + 19$

5. $3,511 - 1,345$ **6.** $729 + 8 + 3,996$

7. $16,092 - 5,618$ **8.** $25,080 - 19,387$

- **Multiply and divide whole numbers.**
  (pp. 22–25)

  $$273 \\ \times\ 86 \\ \overline{1\ 638} \\ +21\ 84 \\ \overline{23,478}$$

  $$\begin{array}{r} 203\ r16 \\ 41\overline{)8,339} \\ -8\ 2 \\ \overline{13} \\ -\ 0 \\ \overline{139} \\ -123 \\ \overline{16} \end{array}$$

Find the product or quotient.

9. $\begin{array}{r} 57 \\ \times 38 \end{array}$ **10.** $\begin{array}{r} 430 \\ \times 17 \end{array}$ **11.** $\begin{array}{r} 276 \\ \times 809 \end{array}$

12. $489 \div 26$

13. $9,671 \div 42$

14. $9,538 \div 19$

- **Use mental math to solve equations.**
  (pp. 30–31)

  Solve. $w \div 6 = 7$ THINK: *What number divided by 6 equals 7?*

  $w = 42$ *42 ÷ 6 = 7*

Solve each equation by using mental math.

15. $b + 5 = 12$ **16.** $a - 3 = 5$

17. $3y = 18$ **18.** $k \div 5 = 4$

Chapter 2

- **Use the order of operations to evaluate expressions.** (pp. 44–45)

  Evaluate. $25 \div 5 + (8 - 4)^2 \times 3$

  $25 \div 5 + (4)^2 \times 3$ *Operate in parentheses.*
  $25 \div 5 + 16 \times 3$ *Clear exponents.*
  $5 + 16 \times 3$ *Divide.*
  $5 + 48$ *Multiply.*
  53 *Add.*

Evaluate the expression.

19. $3 \times 6 + 7^2 - 9$

20. $(8 + 7) \div 3 + (5 - 3)^3$

21. $(15 - 3 \times 4) + 8 \div 2$

22. $9^2 - 20 \times 2 + 5$

23. $32 + (8^2 - 50) \times 2$

24. $16 \div 2^3 + 4 \times 3$

Chapter 3

- **Compare and order decimals.** (pp. 52–55)

  Compare 2.8 and 2.83.

  2.8   2.83   *Start at the left.*
  2.8   2.83   ← *same number of ones*
  2.8   2.83   ← *same number of tenths*
  2.80 2.83   *Add a zero to compare.*
  2.8 < 2.83, or 2.83 > 2.8

Compare the numbers. Write <, > or = for ●.

25. 3.72 ● 3.7   **26.** 5.02 ● 5.021

Write the numbers in order from least to greatest. Use <.

27. 1.67, 1.76, 1.607, 1.706, 1.076

28. 0.0014, 0.4001, 0.0401, 0.0041, 0.014

- **Write decimals as percents and percents as decimals.** (pp. 60–61)

  Write 0.05 as a percent.
  0.05 is 5 hundredths.
  So, 0.05 is 5%.

Write the percent or decimal.

29. 0.28   **30.** 7%

31. 47%   **32.** 0.6

Chapter 4

- **Add and subtract decimals.** (pp. 66–69)

  Find the difference. 8.2 − 6.391

    8.200   *Align the decimal points.*
  −6.391   *Use zeros as placeholders.*
    1.809   *Place the decimal point. Subtract.*

Find the sum or difference.

33. 29.6 + 0.935   **34.** 26.53 + 5.238

35. 76.03 − 58.94   **36.** 347.31 − 48.896

- **Multiply and divide decimals.** (pp. 70–79)

  Find the quotient. 2.46 ÷ 0.6

  $$\begin{array}{r} 4.1 \\ 0.6\overline{)2.46} \\ -24 \\ \hline 06 \\ -6 \\ \hline 0 \end{array}$$

  Make the divisor a whole number.
  0.6 × 10 = 6
  2.46 × 10 = 24.6
  Divide.

Find the product or quotient.

37. 75.8 × 6   **38.** 4.83 × 0.9

39. 3.92 × 0.58   **40.** 68.49 × 3.6

41. 36.48 ÷ 12   **42.** 43.5 ÷ 0.3

43. 59.04 ÷ 4.8   **44.** 8.094 ÷ 0.95

- **Evaluate expressions with decimals.** (pp. 82–83)

  Evaluate $r + s + 3$ for $r = 6.9$ and $s = 4.8$.
  $r + s + 3$ *Replace r with 6.9 and s with 4.8.*
  $6.9 + 4.8 + 3 = 14.7$

Evaluate each expression for $c = 2.5$, $d = 3.2$, and $f = 4.1$.

45. $(c \times 3) + d$ **46.** $(10 - d) + f$

47. $\dfrac{d}{8}$ **48.** $f - c + d$

PROBLEM SOLVING APPLICATIONS

49. Jake has half as many pennies as his sister has. Together their pennies are worth 72 cents. How many pennies does Jake have? (p. 26)

50. Movie tickets cost $3.75 each for the matinee. How many tickets can Ms. Hamil buy with $20? How much money will she have left over? (p. 76)

Performance Assessment

TASK A • Programming Problem

As part of a larger computer program you are building, you need to make a module that will evaluate any numerical expression the user puts into it.

- Write down the steps the computer must go through to evaluate any expression that is put into it.

- Make up your own numerical expression that includes at least three different operations and has part of the expression in parentheses. Explain the steps the computer will go through to evaluate it.

- Tell how the computer will evaluate the following expression:
  $3 \times 0.2 + 9 \div 0.3 - (4^2 - 2^3)$

TASK B • Pizza Party

Peggy is giving a pizza party to celebrate her birthday. To plan it she needs some information about the pizzas she can order.

Pizza Diameter	Number of Slices	Cost per Pizza
12 inches	4	$7
14 inches	6	$9
16 inches	8	$11

- Suppose Peggy has $30 to spend on the pizzas. What are some different ways to spend the money? Make a list. Which way gets her closest to spending all of the $30? What combination of pizzas should she buy to get the greatest number of slices?

Next, she'll have to decide how many friends she can invite.

- How many slices do you think a typical friend will eat? Find a way for Peggy to order exactly enough pizzas so that each friend gets the same number of slices and none are left over. Describe what you would do. Remember to stay within the budget.

Data ToolKit
Use Spreadsheet Formulas to Find Sums and Differences

The table below shows money Dave earned by helping in his parents' store last year. Dave uses a spreadsheet to find the total amount he earned.

Month	Amount	Month	Amount	Month	Amount
January	$25.15	May	$45.50	September	$47.77
February	$20.95	June	$56.85	October	$32.95
March	$31.25	July	$54.60	November	$68.25
April	$38.75	August	$59.00	December	$75.35

- Enter the months into Column A and the amounts into Column B.
- Click in cell B13. Click f_x.
- Highlight *SUM* and click *OK*.
- Type B1:B12. This tells the spreadsheet to add the cells from B1 to B12.
- Press *Enter*.

	Dave's Earnings		
All	**A**	**B**	**C**
1	January	$25.15	
2	February	$20.95	
3	March	$31.25	
4	April	$38.75	
5	May	$45.50	
6	June	$56.85	
7	July	$54.60	
8	August	$59.00	
9	September	$47.77	
10	October	$32.95	
11	November	$68.25	
12	December	$75.35	
13		$556.37	

So, Dave earned a total of $556.37 last year.

How much more did Dave earn in December than in January?

- Click in an empty cell. Type =.
- Click in cell B12. Type −. Click in cell B1. Type *Enter*.

So, Dave earned $50.20 more in December.

Practice and Problem Solving

Make a spreadsheet to find the sum and difference.

1. Kim earned $38.19 in January, $24.35 in February, $68.90 in March, and $47.60 in April. What is the total amount she earned? How much more did she earn in March than in April?

2. Chris earned $125.14 in May, $108.75 in June, $98.50 in July, $116.90 in August, and $130.25 in September. What is the total amount he earned? How much more did he earn in September than in July?

3. **REASONING** Explain the steps you follow to add and subtract numbers on a spreadsheet.

Multimedia Math Glossary www.harcourtschool.com/mathglossary

4. **Vocabulary** Look up *clustering* in the Multimedia Math Glossary. Write a problem that can be answered using the example shown in the glossary.

PROBLEM SOLVING ON LOCATION
With Energy

Wind Energy

Wind energy has been used for centuries to pump water, grind grain, and power sailboats. Today, many communities use windmills called wind turbines to generate electricity. Several important wind turbine projects are located in Minnesota.

Use Data **For 1–8, use the table.**

Wind Turbines in Minnesota	
Location	**Number of Turbines**
Lakota Ridge	15
Lake Benton I	143
Lake Benton II	138

1. Order the locations from greatest number of turbines to least.

2. How many more turbines are there at Lake Benton I than at Lake Benton II?

3. There are about 50 times as many wind turbines in California as there are at the three sites in Minnesota combined. About how many wind turbines are there in California?

4. How many times as many turbines are there at Lake Benton II as there are at Lakota Ridge? Write your answer as a fraction in simplest form and as a mixed number.

Electricity is measured in kilowatt-hours (kWh). The turbines at Lakota Ridge generate 30 million kWh per year. The turbines at Lake Benton II generate 355 million kWh per year.

5. About how many kWh are generated each day at Lake Benton II?

6. About how many kWh are generated each year by each turbine at Lake Benton II?

7. About how many kWh are generated each year by each turbine at Lakota Ridge?

8. About how many kWh are generated each day at Lakota Ridge?

Solar Energy

Each day, huge amounts of free energy strike the Earth in the form of radiation from the sun. This *solar* energy can be converted to heat and electricity for human use. Solar energy, like wind power, is a *renewable* resource.

The Reilly Elementary School is one of several schools in Chicago that use solar energy. These schools are part of the Million Solar Roofs program. By using the sun's energy, the schools can provide for some or all of their heating needs. This reduces their dependence on oil, coal, and other *nonrenewable* resources.

1. The solar project at Reilly Elementary School installed 120 solar power units at a total cost of $500,000. What was the average cost of a solar power unit, rounded to the nearest hundred dollars?

2. Out of every dollar spent for the Reilly solar units, the Chicago school system paid 10 cents. If the total cost was $500,000, how much money did the Chicago school system provide?

Use Data   For 3–5, use the table.

3. A quadrillion is a million billion. How is the amount for fossil fuels written in standard form?

U.S. Energy Consumption by Energy Source	
Source	**Amount (in quadrillion btu)**
Fossil Fuels	80
Nuclear Power	7
Renewable Sources	7

4. Write an equation you can use to find how much more energy is consumed from fossil fuels than from renewable sources. Then solve the equation.

5. Suppose the United States set a goal of using renewable energy sources for half of its energy use. By how much would the nation's current renewable sources have to increase?

These wind turbines are located at Lake Benton, Minnesota. When the blades spin, they drive generators, which in turn produce electricity.

CHAPTER 5

Collect and Organize Data

Hubbard Glacier in Alaska

If you could look at Earth from space, you would see that most of its surface is covered with water. However, most of that water is salt water, which is unusable for drinking, watering crops, or manufacturing.

PROBLEM SOLVING What information about Earth's water supply can you find in the graph?

DATA LINK

WATER ON THE EARTH

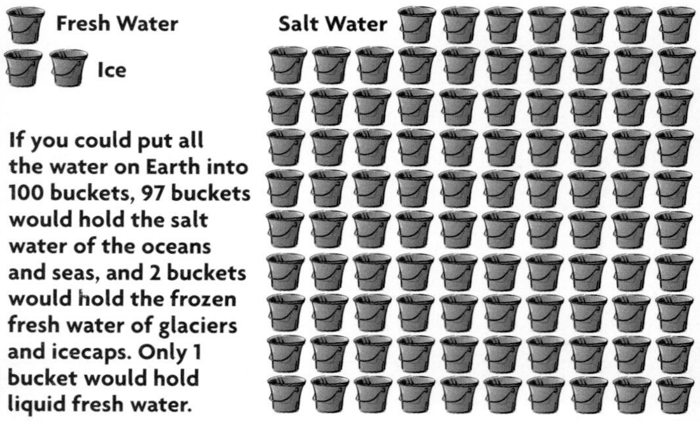

Fresh Water

Ice

Salt Water

If you could put all the water on Earth into 100 buckets, 97 buckets would hold the salt water of the oceans and seas, and 2 buckets would hold the frozen fresh water of glaciers and icecaps. Only 1 bucket would hold liquid fresh water.

Check What You Know

Use this page to help you review and remember important skills needed for Chapter 5.

✓ Vocabulary

Choose the best term from the box.

| mean |
| median |
| mode |
| range |

1. The middle number in a group of numbers arranged in numerical order is the __?__ .

2. The sum of a group of numbers divided by the number of addends is the __?__ .

3. The difference between the greatest number and the least number in a set of data is the __?__ .

✓ Reading a Table (For Intervention, see p. H5.)

STUDENTS' FAVORITE SPORTS				
	Gymnastics	**Basketball**	**Baseball**	**Football**
Boys	4	29	16	13
Girls	17	14	9	12

Use the data in the table above to answer the questions.

4. How many boys like baseball the best?

5. How many more girls than boys like gymnastics the best?

6. Which sport was selected by more boys than any other?

7. How many girls like gymnastics or football the best?

8. How many students were surveyed?

✓ Mean, Median, and Mode (For Intervention, see p. H6.)

Find the mean for each set of data.

9. 7, 6, 11, 7, 9

10. 84, 73, 92, 77, 90

11. 8.9, 8.8, 7.9, 8.3

12. 4.6, 5.1, 6.2, 5.4, 4.8, 5.7

Find the median and the mode for each set of data.

13. 16, 32, 24, 10, 48, 32, 28

14. 10, 15, 7, 8, 12, 13, 13, 19, 21

15. 12.1, 10.8, 11.6, 10.8, 11

16. 1.4, 0.9, 3.6, 2.5, 0.9, 6.2, 2.4

✓ Range (For Intervention, see p. H6.)

Find the range for each set of data.

17. 6, 3, 7, 5, 9

18. 105, 102, 116, 103, 96

LOOK AHEAD

In Chapter 5 you will
- identify different kinds of samples and determine if surveys are biased
- organize data in frequency tables and line plots
- work with and analyze measures of central tendency

Samples

Learn how to identify different types of samples and to determine if a sample is representative of the population.

Vocabulary

- survey
- population
- sample
- convenience sample
- random sample
- systematic sample

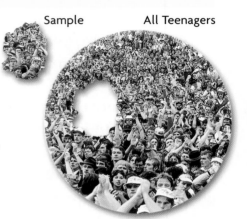

Sample   All Teenagers

A **survey** is a method of gathering information about a group. Surveys are usually made up of questions or other items that require responses.

You can survey the **population**, the entire group of individuals or objects, such as all teenagers. Or, if the population is large, you can survey a part of the group, called a **sample**.

EXAMPLE 1

Suppose Jenna wants to find out the favorite game among students in her math class. Describe the population. Should Jenna survey the population or use a sample? Explain.

The population consists of all the students in Jenna's math class. Jenna should survey the population since it is small.

- **What if** the population is changed to all of the 1,800 students in Jenna's school?

Surveying the population is not always possible, so samples are used.

EXAMPLE 2

Renee asked students at Roosevelt Middle School to indicate their favorite sport. She surveyed two samples and then the entire population of 600 students. How do the results from the two samples compare with the results from the population?

SAMPLE SIZE	NUMBER CHOOSING BASKETBALL	PERCENT OF SAMPLE
50	18	36%
100	33	33%

POPULATION	NUMBER CHOOSING BASKETBALL	PERCENT OF POPULATION
600	192	32%

The results from the samples are different from, but close to, the results from the population.

Types of Samples

There are many types of sampling methods. The table below shows three different sampling methods.

TECHNOLOGY LINK

To learn more about samples, watch the **Harcourt Math Newsroom Video** *What is in a Poll?*

TYPE	DEFINITION	EXAMPLE
convenience sample	The most available individuals or objects in the population are selected to obtain results quickly.	Choose a specific location, such as the cafeteria or the library, and survey students as they walk by you.
random sample	Every individual or object in the population has an equal chance of being selected. This produces the sample that is most representative of the population.	Assign a number to each student and then choose students by randomly selecting numbers with a computer.
systematic sample	An individual or object is randomly selected and then others are selected using a pattern.	Randomly choose a student from a list of students, and then choose every fourth student.

EXAMPLE 3

Identify each type of sample.

A. A bicycle shop owner randomly surveys a customer and then surveys every tenth customer after that to determine when customers expect to buy their next new bike.

This is a systematic sample since individuals are selected according to a pattern.

B. A sixth-grade middle school math teacher randomly selects 50 students in her math classes and asks them if they have a home computer.

This is a random sample since everyone has an equal chance of being selected.

C. The owner of a movie rental shop asks customers in the shop on a given day to fill out a card naming their favorite movie.

This is a convenience sample since the most available individuals are selected.

Math Idea ▶ The results from a sample can depend on how the sample is chosen. It is important to select a sample that is representative of the population. For example, if the population includes men and women, then the sample must include both men and women.

Think and ▶
Discuss

Look back at the lesson to answer each question.

1. **Explain** why a sample is often used rather than a population when conducting a survey.

2. **Tell** when you might survey a population rather than a sample.

3. **Discuss** why a random sample is more likely to represent a population than a convenience sample.

Guided ▶
Practice

Describe the population.

4. Suppose a random sample of 1,250 women in a city was taken in order to determine the favorite type of music of all women in the city.

Tell whether you would survey the population or use a sample. Explain.

5. You want to find out where the students in your class want to go for a field trip.

6. You want to know the favorite mountain bike of teenagers in your state.

For 7–9 determine the type of sample. Write *convenience*, *random*, or *systematic*.

7. A camp director was asked to conduct a survey to find which team game the campers liked best. She randomly selected one camper and then surveyed every fifth person on her list of campers.

8. The names of one hundred high school students are drawn from a box containing all student names.

9. The owners of a sports store ask shoppers in the store a question about the kind of running shoes they like.

Independent ▶
Practice

Describe the population.

10. To find out the average salary of female architects, a researcher randomly selected 100 female architects.

Tell whether you would survey the population or use a sample. Explain.

11. You want to find out the daily exercise habits of the varsity basketball team.

12. You want to know the favorite kinds of food of all teens in a large city.

For 13–15, determine the type of sample. Write *convenience*, *random*, or *systematic*.

13. A cafeteria worker randomly selected a student and then asked every tenth student who entered the cafeteria a question.

14. When conducting a survey, Rachel selected 50 people by drawing names out of a hat without looking.

15. To find out the favorite movie of students at her school, a teacher asked students in her math class to complete a survey.

Problem Solving ▶ Applications

16. Jim asked Lake Middle School students to indicate their favorite pet. Some results are in the tables at the right. He surveyed two samples and then the population of 320 students. How do the results from the two samples compare with the results from the population?

SAMPLE SIZE	NUMBER CHOOSING DOGS	PERCENT OF SAMPLE
10	2	20%
50	24	48%

POPULATION	NUMBER CHOOSING DOGS	PERCENT OF POPULATION
320	160	50%

Dalton wants to conduct a survey to find out which computer game is most popular among the students in his school.

17. How can Dalton get a random sample?

18. How can Dalton get a convenience sample?

19. 📖 **Write About It** Choose a topic you could find out about by conducting a survey. Then tell how you would choose a random sample.

20. A target has three circles, one inside of the other. The bull's-eye is 10 points, the middle circle is 6 points, and the outer circle is 4 points. If 3 darts are thrown and only 2 hit the target, what are the possible scores?

MIXED REVIEW AND TEST PREP

Evaluate for *a* = 3.1 and *b* = 6.2. (p. 82)

21. $a + b$ **22.** $a \times b$ **23.** $b \div a$

⭐**24. TEST PREP** Which is the product of 5.25 and 1.9? (p. 70)

A 9.975   **B** 99.75   **C** 997.5   **D** 9,975

⭐**25. TEST PREP** Martha saved $0.50 one week. Then she saved $0.50 more each week than she had the week before. How many weeks did it take to save a total of $10.50? (p. 66)

F 5 weeks   **G** 6 weeks   **H** 7 weeks   **J** 8 weeks

EXTRA PRACTICE page H36, Set A

Bias in Surveys

Learn how to determine whether a sample or question in a survey is unbiased.

Vocabulary

unbiased sample

biased sample

biased question

Does the earth revolve around the sun or does the sun revolve around the earth? In a recent survey of over 1,000 adults, 79% knew the correct answer.

When you collect data from a survey, your sample should represent the whole population. Every individual in that population should have an equal chance of being selected. Then the sample is an **unbiased sample**.

If individuals or groups from the population are not represented in the sample, then the sample is a **biased sample**. If the sample for the survey above included males only, it would be biased since the survey was for all adults.

EXAMPLE 1

Latosha wants to find out how many hours Polk Middle School students spend on the Internet. If she surveys students from this school, which samples below will be biased? Explain.

A. 200 girls

B. 200 athletes

C. 200 randomly selected students

D. Students who ride their bikes to school

Choices A, B, and D are biased. Choice A excludes boys, choice B only includes athletes, and choice D excludes students who don't ride their bikes to school.

Sometimes, questions are biased. A **biased question** leads to a specific response or excludes a certain group.

EXAMPLE 2

Is the following question biased?

Do you agree with a well-known movie critic that movies longer than two hours are boring?

This question is biased since it leads you to agree with the movie critic.

Think and ▶
Discuss

Look back at the lesson to answer the question.

1. **Write** the question in Example 2 so that it is not biased.

Guided ▶
Practice

Tell whether the sample is *biased* or *unbiased*. Explain.

The local gym wants to find out how its members feel about the new exercise equipment.

2. Female members

3. Members under age 20

4. Members who used the gym in August

5. 50 randomly selected members

PRACTICE AND PROBLEM SOLVING

Independent ▶
Practice

Tell whether the sample is *biased* or *unbiased*. Explain.

The Middletown Mall is conducting a survey of its shoppers to find out which days they prefer to shop.

6. 400 teenage shoppers

7. 400 randomly selected shoppers

8. 400 female shoppers

9. Shoppers who enter the record store

Determine whether the question is biased. Write *biased* or *unbiased*.

10. Is basketball your favorite sport?

11. Do you agree with the fruit industry president that green apples taste better than red apples?

Problem Solving ▶
Applications

12. **REASONING** Students at Memorial Middle School are being surveyed to find out their choice of a new school mascot. If 300 students are randomly selected, do equal numbers of boys and girls have to be chosen? Explain.

13. 📓 **Write About It** Is this question biased? Explain.
I think pizza is the best choice for lunch, don't you?

MIXED REVIEW AND TEST PREP

14. A store owner randomly selects a customer and then surveys every tenth customer. What type of sample is this? (p. 94)

Find the quotient. (p. 76)

15. $3.6 \div 0.9$

16. $1.44 \div 0.12$

17. $270 \div 0.03$

⭐ 18. **TEST PREP** Replace ● with the missing operation $3^2 \times (6 ● 4) = 90$. (p. 44)

　　A ÷　　　B ×　　　C −　　　D +

PROBLEM SOLVING STRATEGY
Make a Table

Learn how to solve problems by displaying information in a table.

QUICK REVIEW

Find the value.

1. 6^2 **2.** 3^3 **3.** 2^3 **4.** 2^4 **5.** 2^5

Mrs. Donovan asked the students in her class to tell the number of movies they had seen since the start of the school year. The results are shown at the right.

Number of Movies Seen

1 5 3 2 1 6 1 2 3 3 4 5
1 1 6 3 4 2 2 2 4 3 2 1
2 2 4 5 1 3 2 1

How many students saw at least 3 movies? How many students saw fewer than 3 movies?

Analyze

What are you asked to find?

What information is given?

Is there information you will not use? If so, what?

Choose

What strategy will you use?

You can make a tally table to organize the data. Then you can use the table to answer the questions.

Solve

How can you organize the tally table?

Make a row for each of the numbers of movies the students have seen. Then read the data and make a tally mark for each piece of information next to the appropriate number.

Next, use the table to answer the questions.

There are $6 + 4 + 3 + 2$, or 15, students who saw at least 3 movies and $8 + 9$, or 17, students who saw fewer than 3 movies.

NUMBER OF MOVIES SEEN	
1	卌 |||
2	卌 ||||
3	卌	
4					
5				
6			

Check

How can you check your answer?

What if two students were absent that day and they each had seen 5 movies? How many students would have seen at least 3 movies? fewer than 3 movies?

PROBLEM SOLVING PRACTICE

Solve the problem by displaying the data in a table.

Members of the math club took a survey to find out how each person gets to school. These are the results.

skateboard	bike	in-line skates	bike
walk	bus	bike	walk
skateboard	bike	bike	bus
walk	skateboard	in-line skates	bike
bus	walk	bus	walk
skateboard	bike	bus	walk

PROBLEM SOLVING STRATEGIES

Draw a Diagram or Picture
Make a Model
Predict and Test
Work Backward
Make an Organized List
Find a Pattern
▶ **Make a Table or Graph**
Solve a Simpler Problem
Write an Equation
Use Logical Reasoning

1. Make a table using the data above. How many categories of data are in your table?

 A 2   **B** 3   **C** 4   **D** 5

2. How many of the math club members walk or ride a skateboard to school?

 F 8   **G** 9   **H** 10   **J** 11

3. How many of the math club members ride a bike or the bus to school?

4. How many more math club members walk to school than ride in-line skates to school?

MIXED STRATEGY PRACTICE

5. Julia receives a $50.00 gift certificate to a music store. If she wants to buy three $13.99 CDs and two $7.99 cassettes, how much of her own money will Julia have to add to the $50.00 gift certificate?

6. A basketball team scores 23 points, including six 1-point foul shots. How many 2-point and 3-point baskets could they have made? Make a list of all the possibilities.

7. An automobile club sells books of 10 movie tickets for $45.00. If tickets usually cost $8.75 each, how much do you save on 10 tickets by buying the book?

8. A fence separating two gardens is 24 ft long. If there is a post in the ground every 3 ft, how many posts are in the fence?

9. On a wildlife outing, Enrique spots 3 more seagulls than egrets and 5 more egrets than geese. If he spots 8 geese, how many birds does he spot in all?

10. If today is Tuesday, what day of the week will it be 200 days from today?

11. 📓 **Write a problem** using the data from the math club survey at the top of the page.

101

Frequency Tables and Line Plots

Learn how to record and organize data collected in a survey.

Vocabulary

frequency table
cumulative frequency

QUICK REVIEW

1. 70
   +20
2. 35
   +40
3. 67
   −30
4. 50
   −15
5. 240 ÷ 6

In 2000, the largest iceberg on record contained enough ice to provide everyone on earth with 4 gallons of water per day for 40 years! Most of an iceberg is below the surface of the water. The data in the table below show the number of icebergs a scientist observed every day for 20 days.

NUMBER OF ICEBERGS OBSERVED				
12	15	11	16	15
16	16	15	13	16
16	14	12	14	16
15	13	15	14	15

You can use a line plot to record data.

EXAMPLE 1

Use the data above to make a line plot.

Step 1: Draw a horizontal line.

Step 2: On your line, write the numerical values for the number of icebergs, using vertical tick marks.

Step 3: Plot the data.

Each X represents the number of icebergs the scientist observed during one day.

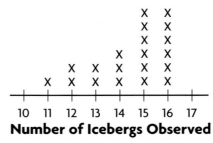

Number of Icebergs Observed

TECHNOLOGY LINK

More Practice: Use *Data ToolKit* to make frequency tables and line plots.

A line plot helps you see if there are any gaps or extremes in the data. You can also see where data cluster. The line plot in Example 1 shows that the scientist observed 15–16 icebergs per day more often than she observed 11–14 icebergs per day.

Frequency Tables

A **frequency table** shows the total for each category or group. You can add a cumulative frequency column to the table. **Cumulative frequency** is a running total of the frequencies.

EXAMPLE 2

The local mall surveyed 75 teens regarding the overall quality of food at the mall's food court. Use the data from the survey to make a frequency table. Do you think most teens like the quality of food at the food court?

Quality of Food at Food Court

Quality	Tally
Very Good	ЖЖ ЖЖ ЖЖ ЖЖ
Good	ЖЖ ЖЖ ЖЖ ЖЖ ЖЖ ЖЖ
Fair	ЖЖ ЖЖ ЖЖ
Poor	ЖЖ ЖЖ

List the categories of quality in one column. Record the total for each category in the frequency column. Record the running total in the cumulative frequency column.

QUALITY OF FOOD AT FOOD COURT		
Quality	**Frequency**	**Cumulative Frequency**
Very Good	20	20
Good	30	50
Fair	15	65
Poor	10	75

← 20 + 30 = 50
← 50 + 15 = 65
← 65 + 10 = 75

So, most teens liked the quality of food at the food court: 50 out of 75 teens surveyed rated the food as good or very good.

A local mall took a survey to find the number of times people go to the mall in one year. The survey results are shown below.

NUMBER OF TRIPS TO THE MALL									
12	20	5	7	13	11	3	9	19	16
8	17	17	16	1	6	8	5	14	9

EXAMPLE 3

Remember that the range of a set of data is the difference between the greatest number and the least number in the set of data.

Use the data about the number of trips to the mall to make a cumulative frequency table with intervals.

20 − 1 = 19   *Find the range.*

Since 4 × 5 = 20, which is close to 19, make 4 intervals that include 5 consecutive numbers.   *Use the range to determine intervals.*

TRIPS TO THE MALL		
Number of Trips	**Frequency**	**Cumulative Frequency**
1-5	4	4
6-10	6	10
11-15	4	14
16-20	6	20

Complete the cumulative frequency table.

Cumulative frequency tables with intervals can also be used to organize data about different age groups. The data in the table below show the ages of joggers at the Get Fit exercise trail.

AGES OF JOGGERS									
27	31	11	22	25	42	44	36	33	24
19	34	48	17	28	33	32	30	40	31
38	18	22	23	24	27	36	19	26	32

EXAMPLE 4

Use the data about the ages of joggers at the Get Fit exercise trail to make a cumulative frequency table with intervals.

$48 - 11 = 37$ *Find the range.*

Since $4 \times 10 = 40$, which is close to 37, make 4 intervals that include 10 consecutive numbers.   *Use the range to determine intervals.*

Complete the cumulative frequency table.

AGES OF JOGGERS		
Age	Frequency	Cumulative Frequency
10–19	5	5
20–29	10	15
30–39	11	26
40–49	4	30

CHECK FOR UNDERSTANDING

Think and Discuss ▶ Look back at the lesson to answer each question.

1. **REASONING** Name the intervals you could have in Example 3 if you made 2 intervals. Explain.

2. **Explain** how to find the size of the survey sample by using the frequency table in Example 4.

Guided Practice ▶ For 3–5, use the data in the table.

3. Make a line plot.

4. Make a cumulative frequency table.

5. Find the range.

AGES OF POP MUSIC LISTENERS						
12	13	17	16	15	13	25
28	27	18	15	12	14	23
13	15	17	13	22	18	15

PRACTICE AND PROBLEM SOLVING

Independent Practice ▶ For 6–8, use the data in the table.

6. Make a line plot.

7. Make a cumulative frequency table.

8. Find the range.

NUMBER OF KILOMETERS BIKED						
6	11	7	8	8	5	15
14	12	10	11	6	7	9
15	15	8	9	12	17	10

Kim asked people in her neighborhood about their favorite type of movie. Use the data for 9–10.

FAVORITE MOVIES	
Movies	Tally
Drama	JHT JHT
Action	JHT III
Comedy	JHT JHT III
Mystery	JHT IIII

9. Organize the data into a frequency table. Include a cumulative frequency column.

10. What is the size of the sample?

For 11–14, use the data in the box at the right.

11. Make a line plot.

12. Find the range.

13. How many numbers would be in each interval if you used the range to make 4 intervals?

WEIGHTS OF STUDENTS' PET DOGS (LB)					
40	10	22	33	44	40
41	42	12	35	20	10
30	25	20	18	46	35
34	35	40	40	35	38

14. Make a cumulative frequency table using 4 intervals for the weights.

Problem Solving ▶ Applications

15. *REASONING* Survey your classmates about their favorite hobbies. Organize the data in a tally table and a frequency table. Can you make a line plot of the data? Explain.

16. Decide on appropriate intervals and make a cumulative frequency table using intervals for the data in the table below.

POUNDS OF VEGETABLES CONSUMED YEARLY												
12	16	18	12	9	21	14	17	18	5	23	25	32
17	16	8	9	22	21	14	18	7	23	24	19	13

17. ❓ **What's the Question?** In a set of data, the greatest value is 54 and the least value is 30. The answer is 24.

MIXED REVIEW AND TEST PREP

Tell whether the question is *biased* or *unbiased*. (p. 98)

18. Do you feel that sixth graders should do two hours of homework each night instead of watching TV?

19. Whom do you intend to vote for in the upcoming Student Government election?

20. Tell if the expression is numerical or algebraic. $a + 3.9$ (p. 28)

21. **TEST PREP** Which is the value of 4^5? (p. 40)

   A 9   **B** 20   **C** 625   **D** 1,024

22. **TEST PREP** Twenty-seven sixth graders read a total of 135 books. If each student reads the same number of books, how many more books must each student read to reach a total of 243 books read? (p. 22)

   F 2 books   **G** 4 books   **H** 50 books   **J** 108 books

EXTRA PRACTICE page H36, Set C

Measures of Central Tendency

Learn how to find the mean, median, and mode of a set of data and which best describes the set of data.

Three measures of central tendency are mean, median, and mode. Measures of central tendency can help you to describe a set of data.

In April 1998, amateur rocket builder Gib Reynolds set an American record for 7- to 13-year-olds. His model rocket soared to a height of 47 meters. Data for heights some model rockets traveled, including Gib's, are shown at the right. Find the mean, median, and mode for the data.

ROCKET HEIGHTS (M)				
47	28	33	35	28

Mean: (47 + 28 + 33 + 35 + 28) ÷ 5 = 171 ÷ 5 = 34.2 m

Median: 28 28 33 35 47; 33 m　　　　**Mode:** 28 m

Sometimes a line plot can help you to find the mode and the median.

EXAMPLE 1

Remember that the mean is the sum of a group of numbers divided by the number of addends.

The median is the middle number in a group of numbers arranged in numerical order.

The mode is the number that occurs most often.

Use the data to make a line plot. Find the mode and the median.

DAILY TEMPERATURES (°F)										
72	82	83	78	81	78	73	74	75	73	76
71	75	80	83	72	72	78	81	79	82	76

```
                X                       X
                X   X       X   X       X           X   X   X
        X       X   X   X   X   X       X   X   X   X   X   X
        +---+---+---+---+---+---+---+---+---+---+---+---+---+---+
       70  71  72  73  74  75  76  77  78  79  80  81  82  83  84
```

Mode: The modes are 72 and 78 since each occurs three times.

Median: Since there are 22 temperatures, the median is the mean of the 11th and 12th temperatures.　　(76 + 78) ÷ 2 = 77

Math Idea ▶ When you want to summarize a set of data as one value, you can use one of the three measures of central tendency.

EXAMPLE 2

Pedro jogged 6 mi, 5 mi, 2 mi, 2 mi, and 4 mi over 5 days. Which measure of central tendency is most useful to describe the data?

Mean: (6 + 5 + 2 + 2 + 4) ÷ 5 = 19 ÷ 5 = 3.8

Median: 2 2 4 5 6; 4　　　　**Mode:** 2

The mean, or 3.8, and the median, 4, are close to most of the data while the mode, or 2, is closer to the lower end of the data. So, the mean or median is most useful to describe the data.

Sometimes the mean is not the best measure of central tendency to describe a set of data.

EXAMPLE 3

The bald eagle, America's national bird, was once a greatly endangered species. In recent decades, the species has made a dramatic recovery. The map at the right shows the number of nests in five zones of Montana. Which measure of central tendency is most useful to describe the data?

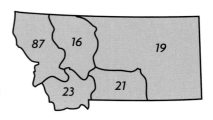

Numbers of nests: 87, 16, 19, 21, 23

Mean: (87 + 16 + 19 + 21 + 23) ÷ 5 = 166 ÷ 5 = 33.2

Median: 16 19 21 23 87 → 21

Mode: No value occurs more than any other, so there is no mode.

The median, 21, is close to most of the data. The high value of 87 makes the mean greater than any of the four other values. So, the median is most useful to describe the data.

CHECK FOR UNDERSTANDING

Think and Discuss ▶ **Look back at the lesson to answer the question.**

1. **What if,** in Example 3, the region with 87 nests had only 27 nests? Which of the measures would change? Which measure would be most useful to describe the data?

Guided Practice ▶ **For 2–4, use the table below.**

Day	Sun	Mon	Tue	Wed	Thu	Fri	Sat
Hours of Sleep	8	6	7	7	4	10	7

2. Find the mean. **3.** Find the median. **4.** Find the mode.

Find the mean, median, and mode.

5. 22, 24, 22, 29, 33, 14 **6.** 124, 120, 132, 133, 119, 90, 87

PRACTICE AND PROBLEM SOLVING

Independent Practice ▶ **For 7–9, use the table below.**

Game	1	2	3	4	5	6	7	8
Points Scored	10	12	10	17	18	12	20	24

7. Find the mean. **8.** Find the median. **9.** Find the mode.

Find the mean, median, and mode.

10. 76, 63, 40, 52, 52, 40, 6, 15 **11.** 365, 180, 360, 720, 59

107

Problem Solving ▶ Applications

12. Use the data below to make a line plot. Use your line plot to find the median and the mode. Then use the data to find the mean. Which measure of central tendency would be most useful to describe the data?

YEARLY RAINFALL (IN.)									
47	11	8	14	15	16	13	13	17	22

13. Over the past 8 days, Fernando spent $5, $6, $9, $8, $10, $15, $15, and $2. Which measure of central tendency would be most useful to describe the data?

14. **ALGEBRA** Reggie's average score on five math tests is 92. On his first four tests, Reggie's scores were 88, 97, 93, and 82. What was Reggie's score on his fifth test?

15. ⦿ **What's the Error?** Tim wrote $(7 + 3 + 10 + 4) \div 3 = 8$ to find the mean of 7, 3, 10, and 4. What is his error? What is the correct mean?

MIXED REVIEW AND TEST PREP

Use the table for 16–17. (p. 102)

16. How many named blue or red as their favorite color?

17. What is the size of the sample?

FAVORITE COLOR	FREQUENCY	CUMULATIVE FREQUENCY
Blue	32	32
Red	17	49
Green	22	71

Write the decimal as a percent. (p. 60)

18. 0.65

19. 0.9

⭐ **20. TEST PREP** Which is the product of 0.56 and 2.5? (p. 70)

    A 0.14   **B** 1.4   **C** 14   **D** 140

PROBLEM SOLVING  to Social Studies

World Population Population is growing at different rates around the world. In North America, population is growing fairly slowly. In Latin America and Africa, population is growing much more rapidly. The table shows how median ages can differ.

Median Ages				
Africa	Asia	Europe	Latin America and Caribbean	North America
18.3 years	26.0 years	37.4 years	24.2 years	35.6 years

1. Explain what it means to say the median age in Asia is 26.0 years.

2. Use < to order the median ages from least to greatest.

EXTRA PRACTICE page H36, Set D

Outliers and Additional Data

Learn how outliers and additional data affect the mean, median, and mode.

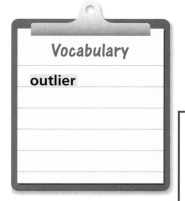

Vocabulary

outlier

Find the mean.

1. 6, 8, 10   **2.** 20, 32, 41   **3.** 15, 17, 19, 41

4. 25, 25, 25, 25   **5.** 100, 200, 300, 400

When you add data to a data set, some measures of central tendency for the set change. How they change depends on how the new data are related to the original data.

EXAMPLE 1

Gina saved dimes for 8 weeks. She recorded the number of dimes in a table. In the ninth week, her parents gave her 25 dimes. In the tenth week, her grandparents gave her 52 dimes. Find the mean, median, and mode for Gina's data, using the values for the 8 weeks of her savings. Then find the measures for all 10 weeks.

DIMES GINA SAVED								
Week	1	2	3	4	5	6	7	8
Dimes	17	18	30	15	1	6	19	6

Data for 8 Weeks
Mean: $(17 + 18 + \ldots + 19 + 6) \div 8 = 14$ **Mode:** 6
Median: 1 6 6 15 17 18 19 30; $(15 + 17) \div 2 = 16$

Data for 10 Weeks
Mean: $(17 + 18 + \ldots + 25 + 52) \div 10 = 18.9$ **Mode:** 6
Median: 1 6 6 15 17 18 19 25 30 52; $(17 + 18) \div 2 = 17.5$

EXAMPLE 2

On May 12, 1999, Armenian Lev Sarkisov became the oldest person to climb Mount Everest.

At age 61, Lev Sarkisov climbed Mount Everest. Suppose he climbed with a group whose ages were 30, 31, 33, 30, 29, 35, 33, 30, and 28. Which measure of central tendency is most affected by adding Sarkisov's age to the group?

Data without Sarkisov's age

Mean: $(30 + 31 + \ldots + 30 + 28) \div 9 = 31$ **Mode:** 30

Median: 28 29 30 30 30 31 33 33 35; 30

Data with Sarkisov's age

Mean: $(30 + 31 + \ldots + 61) \div 10 = 34$ **Mode:** 30

Median: 28 29 30 30 30 31 33 33 35 61; $(30 + 31) \div 2 = 30.5$

The mean increases by 3, the median increases by 0.5, and the mode remains the same. So, the mean is the measure most affected by adding Sarkisov's age.

Bethany used a line plot to record the number of concert tickets she sold on 10 different days. Most of the data are between 11 and 16. The data values at 32 and 34 are outliers, or extreme values. An **outlier** is a data value that stands out from other data values in a set. Outliers can significantly affect measures of central tendency.

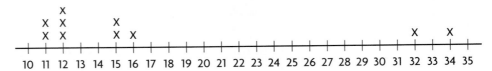

EXAMPLE 3

A. Find the measures of central tendency for the 8 data values in the line plot if the outliers are not included.

Mean: $(11 + 11 + 12 + 12 + 12 + 15 + 15 + 16) \div 8 = 104 \div 8 = 13$
Median: 11 11 12 12 12 15 15 16; 12   **Mode:** 12

B. Find the measures of central tendency for all the data values with the outliers included.

Mean: $(104 + 32 + 34) \div 10 = 170 \div 10 = 17$
Median: 11 11 12 12 12 15 15 16 32 34; $(12 + 15) \div 2 = 13.5$
Mode: 12

C. Suppose Bethany sold 1 ticket the next day and 0 tickets the final day. Include these data values with the data in the line plot and find the measures of central tendency for all 12 data values.

Mean: $(170 + 1 + 0) \div 12 = 171 \div 12 = 14.25$
Median: 0 1 11 11 12 12 12 15 15 16 32 34; 12   **Mode:** 12

• How do the outliers in part B affect the mean, median, and mode?

CHECK FOR UNDERSTANDING

Think and Discuss ▶ Look back at the lesson to answer the question.

1. **What if** Gina's grandparents had given her 32 dimes in Example 1? How would the measures of central tendency be affected?

Guided Practice ▶ 2. Find the mean, median, and mode for the ages in the table.

AGES OF MEMBERS OF MODEL AIRPLANE CLUB			
15	12	9	13
22	19	12	14

3. Find the mean, median, and mode if a 28-year-old joins the club.

Jeff received these scores on his science quizzes: 75, 70, 70, 45, 100, 70, 70, 80, 75, 80.

4. **a.** Use the scores to make a line plot. Circle the outliers.

   b. Find the mean, median, and mode with and without the outliers.

   c. Why did these outliers have so little effect on the mean, median, and mode?

Independent ▶ Practice

The table shows the points scored by two basketball players in the first six games of the season.

POINTS SCORED		
Game	Abe	Bart
1	7	15
2	5	0
3	4	0
4	20	12
5	7	13
6	2	5

5. **a.** Find the mean, median, and mode of each player's points for six games.

   b. In the seventh game, Abe scored 20 points and Bart scored 0 points. Find the mean, median, and mode of each player's points for seven games.

   c. How are the measures of central tendency for Abe's and Bart's scores affected by the points scored in the seventh game?

The table shows the number of pets owned by children in a pet club.

6. **a.** Draw a line plot for the data about the number of pets children own. Circle the outliers.

   b. Find the mean, the median, and the mode for the data with and without the outliers.

NUMBER OF PETS OWNED			
1	1	2	1
3	12	4	4
3	3	10	1

   c. How does including the outliers affect the mean? the median? the mode?

Problem Solving ▶ Applications

7. **a.** Emily scored 75, 85, 35, 85, 70, 10 on her first 6 math quizzes. What score does Emily need on her seventh quiz so that after 7 quizzes she will have a median score of 75?

   b. *REASONING* The greatest score Emily can make on a math quiz is 100. Explain whether it is possible for her to have a mean score of 70 after taking the seventh quiz.

8. *REASONING* The high temperatures in °F over 7 days in town were 72°, 73°, 70°, 68°, 70°, 71°, and 39°. Explain why the mean would not be a good measure to describe the temperatures.

MIXED REVIEW AND TEST PREP

Find the mean, median, and mode of the data. (p. 106)

9. 5, 6, 8, 6, 5, 7, 5, 4
10. 42, 56, 44, 38, 10
11. 10, 12, 12, 10, 8, 5

12. A company claims that its 8-year warranty lasts 4 times as long as any other warranty. What is the longest of the other warranties? (p. 22)

⭐ **13. TEST PREP** There are 180 players in a baseball league. Each team has 12 players. How many teams are in the league? (p. 22)

   A 12 teams   **B** 15 teams   **C** 16 teams   **D** 18 teams

Data and Conclusions

Learn how to decide whether conclusions based on data are valid.

QUICK REVIEW

Write each decimal as a percent.

1. 0.75 **2.** 0.02 **3.** 0.2 **4.** 0.185 **5.** 0.085

Would you take $1,000 to give up watching television for three months? In a recent survey of adults, 87% answered "Yes!"

Kendra used the same question to survey students at her middle school. She randomly selected 60 of these students to participate in the survey. She found that 90% of the students she surveyed answered "Yes." So, Kendra concluded that about 90% of all students at her middle school would take $1,000 to give up watching television for three months.

EXAMPLE 1

Is Kendra's conclusion valid?

To determine if Kendra's conclusion is valid, use the checklist below.

• Who are the people you are interested in studying (the population)?	✓ The population Kendra is interested in is the students at her middle school.
• Was the sample selected from the correct population?	✓ Yes. Kendra surveyed students from her middle school.
• Was the sample randomly selected?	✓ Yes. The sample was randomly selected.
• Was the question unbiased?	✓ Yes. The question was unbiased.

Remember that when you want to collect data by asking a question, how you state the question is very important. The question must be unbiased.

So, Kendra's conclusion is valid.

• **What if** Kendra had surveyed 60 students who participate in school athletics? Would her conclusion be valid? Explain.

Math Idea ▶ In order to draw a valid conclusion from a set of data, you must select a random sample from the correct population and ask an unbiased question.

Barry read the results of a survey conducted ten years ago at his school. A random sampling of sixth graders produced the results displayed in the circle graph at the right.

Barry wanted to see if a survey of sixth graders at his school today would produce the same results. He randomly selected 200 sixth graders from his school and asked them the same question.

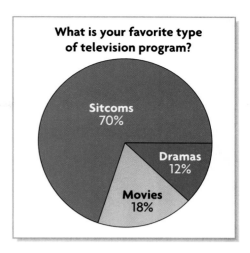

What is your favorite type of television program?

Sitcoms 70%
Dramas 12%
Movies 18%

EXAMPLE 2

What conclusions can Barry draw from the results of his survey shown in the circle graph below?

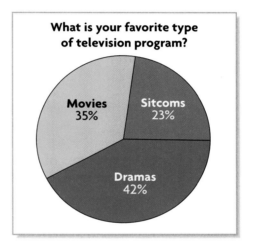

What is your favorite type of television program?

Movies 35%
Sitcoms 23%
Dramas 42%

Barry can conclude the following:

Type of Show	Conclusion
Sitcoms	Interest in sitcoms decreased.
Movies	Interest in movies increased.
Dramas	Interest in dramas increased.

• How do you know Barry's conclusions are valid?

CHECK FOR UNDERSTANDING

Think and Discuss ▶ **Look back at the lesson to answer each question.**

1. **Tell** if Kendra's conclusions in Example 1 would still be valid if she had surveyed students from her first two classes of the day. Explain.

2. **What if,** in Example 2, Barry had asked, "Do you agree with me that a drama is the best kind of television show to watch?" Would he be able to make the same conclusions? Explain.

For 3, write *yes* or *no* to tell whether the conclusion is valid. Explain your answer.

3. Five hundred randomly selected teenagers from Fresno are asked to name their favorite subject. Eighty percent of these teenagers respond that their favorite subject is math. You conclude that math is the most popular subject among teenagers in Fresno.

Use Data For 4–5, use the data in the graph.

4. The graph at the right shows the results of a survey of 600 randomly selected middle school students. Jen concludes that more middle school students are interested in football than in any other sport. Is her conclusion valid? Explain.

5. Robert concludes that more high school students are interested in football than in any other sport. Is his conclusion valid? Explain.

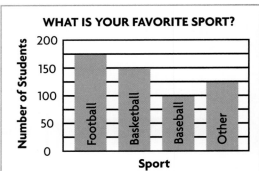

WHAT IS YOUR FAVORITE SPORT?

PRACTICE AND PROBLEM SOLVING

For 6–7, write *yes* or *no* to tell whether the conclusion is valid. Explain your answer.

6. All the members of the science club were asked if they wanted to take more science classes. Each of them answered yes. You decide that science is becoming more popular among all students.

7. A sample of your friends shows that their favorite color is green. You decide that the same is probably true for the entire school.

Use Data For 8–9, use the data in the graph.

8. The graph at the right shows the results of a survey of 1,000 randomly selected teenagers from Kentucky. José concludes that corn is the favorite vegetable of teenagers from New York. Is his conclusion valid? Explain.

9. Jerome concludes that carrots are the second favorite vegetable among Kentucky teenagers. Is his conclusion valid? Explain.

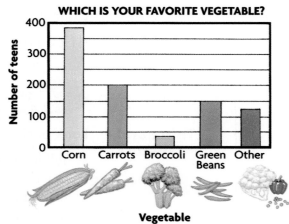

WHICH IS YOUR FAVORITE VEGETABLE?

Problem Solving ▶
Applications

For a survey completed five years ago, middle school students from Minneapolis were asked, "What is your favorite hobby?"

FAVORITE HOBBIES	
Reading	36%
Crafts	7%
Coin Collecting	2%
Stamp Collecting	2%
Building Models	12%
Trading Cards	39%
Other	2%

A research company randomly selected 750 current middle school students from Minneapolis and asked them the same question. The results are in the table.

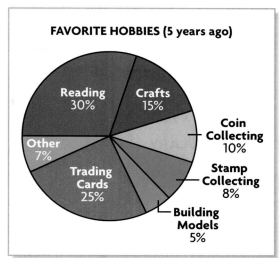

FAVORITE HOBBIES (5 years ago)

Reading 30%
Crafts 15%
Coin Collecting 10%
Other 7%
Stamp Collecting 8%
Trading Cards 25%
Building Models 5%

10. What conclusion can you draw about student interest in reading?

11. What conclusion can you draw about student interest in crafts?

12. What conclusion can you draw about student interest in coin collecting?

13. What conclusion can you draw about student interest in trading cards?

14. What conclusion can you draw about student interest in models?

15. How can you tell that your conclusions are valid?

16. For which of the hobbies did interest increase the most? decrease the most?

17. **? What's the Error?** A survey of middle school athletes showed that their favorite exercise is jogging. Cathy looked at the survey results and concluded that all students prefer jogging. Explain Cathy's error.

18. Robert spent $60 at the sports equipment store. He bought a glove for $35 and 5 baseballs. What was the price for each baseball?

MIXED REVIEW AND TEST PREP

Find the mean, median, and mode of the data. (p. 106)

19. 2, 4, 4, 3, 6, 8, 1, 6   20. 22, 44, 31, 46, 22   21. 6, 8, 11, 10, 2, 5

⭐ 22. **TEST PREP** Which is the solution of $x + 3 = 10$? (p. 30)

   A $x = 3$ **B** $x = 7$ **C** $x = 10$ **D** $x = 13$

⭐ 23. **TEST PREP** Which is the solution of $x \div 3 = 6$? (p. 30)

   F $x = 2$ **G** $x = 3$ **H** $x = 9$ **J** $x = 18$

EXTRA PRACTICE page H36, Set F

For 1–2, determine the type of sample. Write *convenience, random,* or *systematic.* (pp. 94–97)

1. A cereal company surveys a person in every third apartment in an apartment building.

2. Alicia wants to find out what snack foods students in her school prefer. She randomly surveys 100 students.

3. **VOCABULARY** In a survey, if individuals or groups in the population are not represented by the sample, the sample is __?__ .
(p. 98)

Tell whether the sample is biased or unbiased. (pp. 98–99)

4. A store randomly surveys 10 out of every 100 customers about the quality of its service.

5. A teacher surveys girls about the best day to give a test.

For 6–9, use the table at the right. (pp. 102–105)

6. What is the sample size?

7. Find the range.

8. Make a line plot.

9. Make a frequency table.

STUDENTS' HEIGHTS (in inches)									
62	72	63	62	69	70	60	64	66	63
71	62	65	68	63	70	64	62	67	70

Find the mean, median, and mode. (pp. 106–109)

10. 17, 12, 23, 19, 23

11. 6.2, 5.5, 8.4, 5.5

12. 265, 235, 171, 253

13. Ten people in Ana's neighborhood formed a bike club. The table shows the ages of the members. (pp.106–109)

   a. Find the mean, median, and mode for the ages of the ten members.

   b. Find the mean, median, and mode if the adult advisor for the club is 27 years old and is a club member. Compare with part *a*.

AGES OF MEMBERS OF BIKE CLUB				
13	14	10	11	13
12	13	9	11	10

14. Janyce wanted to find out where visitors to the local mall come from. She surveyed visitors, tallied her results, and came up with the following percents: California—34%; Idaho—14%; Nevada— 11%; Oregon—16%; Montana—8%; Washington—11%; other states—6%. Make a table to organize the information from greatest to least percent. (pp.100–101)

15. If 50% of the people who visit the mall in Exercise 14 come from within 20 miles of the mall, which is a reasonable conclusion? Why?
(pp.100–101)

   A The mall where Janyce took her survey is in Colorado.

   B The mall where Janyce took her survey is in eastern Nevada.

   C The mall where Janyce took her survey is in southern Washington.

   D The mall where Janyce took her survey is in northern California.

Eliminate choices.

See item **4.**

Keep the question in mind as you read each answer choice. Look for the most important characteristic of a survey sample to decide what was wrong with the survey.

Also see problem **5**, p. H64.

Choose the best answer.

1. Which symbol makes this true?

    8.34 ⬤ 8.342

    A <   **C** =

    B >   **D** ÷

2. Which of the following is equivalent to 27 + 43?

    F 25 + 50

    G 30 + 40

    H 27 + 40 + 12

    J 20 + 60

3. Which of the following numbers is between 8.07 and 8.3?

    A 8.06

    B 8.017

    C 8.29

    D 8.312

4. Before Warren ran for class president, he surveyed 20 of his friends, asking "Would you vote for me for class president?" Of those surveyed, 100% answered "yes." Warren was very surprised when he lost the election. What was wrong with his survey?

    F Nothing; his friends changed their minds and didn't vote for him.

    G His survey sample did not represent the student body.

    H He should have included teachers in his survey.

    J His survey did not ask the right question.

5. The ages of students in the Hiking Club are 13, 11, 15, 14, 12, 13, 13. What is the mode of this set of data?

    A 11   **B** 12   **C** 13   **D** 14

6. What is $2 \times 2 \times 2 \times 2 \times 2 \times 2 \times 2$ expressed in exponential notation?

    F 2^2 **G** 7^2 **H** 2^7 **J** 2^6

7. $883.03 \div 22.7$

    A 389   **C** 3.89

    B 38.9   **D** Not here

8. Jon had the following scores for 5 rounds of golf: 78, 80, 69, 75, 73. What is the mean of the scores?

    F 50   **H** 74

    G 71   **J** 75

Write What You Know

9. Look at the data in Problem 8. Jon thought the range of his golf scores was 5. Explain the mistake Jon made. Then explain the correct way to compute range.

10. The student council wants to conduct a survey to get an idea of what students think about changing the school colors. One council member suggested asking members of the football team. Explain why this might not be a representative sample.

CHAPTER 6

Graph Data

In July 1994, fragments of Comet Shoemaker-Levy bombarded Jupiter. One impact left a crater about the same diameter as Earth, which is approximately 12,756 km wide. Evidence of past impacts of meteors, asteroids, or comets also appears on Earth and the Moon. One newly discovered crater on the floor of Chesapeake Bay has a diameter of about 80 km.

PROBLEM SOLVING On a bar graph, about how many times as long would the bar for the Jupiter crater be than the bar for the Chesapeake Bay crater?

DATA LINK

EARTH'S AND MOON'S LARGEST CRATERS

Location	Approximate Diameter (km)
Popigri, Siberia	97
Manicouagan, Canada	100
Chicxulub, Mexico	180
Orientale basin, Moon	1,300
Ibrium basin, Moon	1,800
Aitken basin, Moon	2,500

Check What You Know

Use this page to help you review and remember important skills needed for Chapter 6.

 Read Bar Graphs (For Intervention, see p. H6.)

For 1–4, use the bar graph at the right.

1. List the four launch vehicles in order of size from tallest to shortest.

2. About how tall is the Saturn V?

3. About how much taller is the Saturn V than the Ariane IV?

4. About how much taller is the tallest launch vehicle than the shortest?

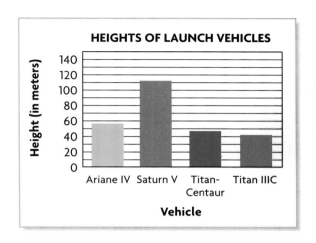

For 5–8, use the bar graph at the left.

5. Name two Great Lakes whose combined area is less than the area of Lake Superior.

6. What is the area of the third-largest Great Lake?

7. How much larger is the area of Lake Michigan than the area of Lake Ontario?

8. Which Great Lake has an area of about 22,000 sq mi more than the area of Lake Erie?

✓ **Read Stem-and-Leaf Plots** (For Intervention, see p. H7.)

For 9–15, use the stem-and-leaf plot at the right.

9. What is the score shown by the third stem and fourth leaf?

10. What is the median of Michael's golf scores?

11. What is the range of Michael's golf scores?

12. What is the mode of Michael's golf scores?

13. How many scores of 90 will Michael need in order to have a mode of 90 for all of his rounds?

14. How many rounds of golf did Michael play?

15. If Michael plays two more rounds and has scores of 86 and 81, what would be the median of the scores?

MICHAEL'S GOLF SCORES

Stems	Leaves
7	6 7 8 9 9
8	0 2 3 5 5 6 8 9
9	0 0 1 2 3 8

LOOK AHEAD

In Chapter 6 you will

- make and analyze stem-and-leaf plots, box-and-whisker graphs, histograms, and line plots
- find unknown values
- analyze graphs

Make and Analyze Graphs

Learn how to display and analyze data in bar graphs, line graphs, and circle graphs.

Vocabulary

multiple-bar graph

multiple-line graph

QUICK REVIEW

QUICK REVIEW

Compare. Use < or > .

1. 5 ● 9   **2.** 25 ● 20   **3.** 87 ● 88

4. 19 ● 23   **5.** 309 ● 304

When data are grouped in categories, a bar graph is a good way to display the data. Look at the bar graph below.

HEAVIEST MARINE MAMMALS

Weight (in tons) — y-axis: 0, 25, 50, 75, 100, 125, 150

Mammal — x-axis: Blue Whale, Fin Whale, Right Whale, Sperm Whale, Gray Whale

The data in the table below show the weights of other mammals.

WEIGHTS OF SOME MAMMALS				
Mammal	**Lion**	**Gorilla**	**Tiger**	**Grizzly Bear**
Male (lb)	400	450	420	500
Female (lb)	300	200	300	400

A **multiple-bar graph** shows two or more sets of data.

EXAMPLE 1

Use the data in the table above to make a double-bar graph. How do the weights of males compare to the weights of females?

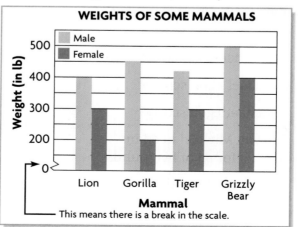

WEIGHTS OF SOME MAMMALS

Male / Female

Weight (in lb): 0, 200, 300, 400, 500

Mammal: Lion, Gorilla, Tiger, Grizzly Bear

This means there is a break in the scale.

Determine an appropriate scale.

Use bars of equal width. Use the data to determine the heights of the bars.

Title the graph and both axes. Include a key.

For each of the mammals in the graph above, the males weigh more than the females.

Line Graphs

A bar graph and a line graph are often used to display the same data. However, a line graph is the better choice when the data show change over time. The line graph below shows the change in the price of a stock every five years from 1970 to 2000.

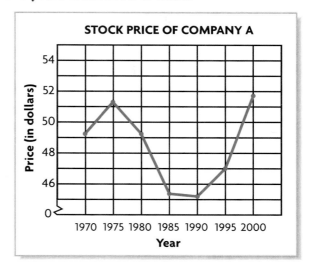

Like a multiple-bar graph, a **multiple-line graph** can show two or more different sets of data on one graph.

MONEY IN SAVINGS ACCOUNTS					
	March	**April**	**May**	**June**	**July**
Bob	$163	$172	$151	$138	$102
Jan	$43	$55	$76	$79	$96

EXAMPLE 2

Use the data in the table above to make a double-line graph. If the trends continue, how would you describe the amount of money that Bob is saving? that Jan is saving?

Determine an appropriate scale.

Mark a point for each amount saved for Bob and connect the points.

Mark a point for each amount saved for Jan and connect the points.

Title the graph and both axes. Include a key.

The money in Bob's savings account is decreasing. The money in Jan's savings account is increasing.

- Use the graph to predict what will happen to each savings account in August.

121

Graphs show pictures of data. Bar graphs and line graphs use axes to help you analyze data. A circle graph helps you compare parts to the whole or to other parts.

EXAMPLE 3

In 1987, compact discs (CDs) were new. The circle graph shows the way recorded music was sold in the United States in 1987. About how many cassettes were sold for every CD sold that year?

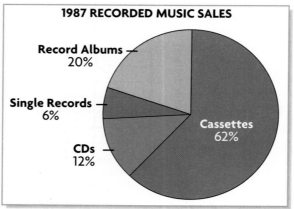

Find the parts that represent cassettes and CDs.

Compare the percent of recorded music sold on cassettes to the percent sold on CDs.

Cassettes made up 62% of recorded music sales, while CDs made up 12%. $62 \div 12 \approx 60 \div 12$, or 5.

So, about 5 cassettes were sold for every CD sold.

• By 1997, the percent of recorded music sold in the form of CDs was about 6 times as great as it had been in 1987. About what percent of recorded music was sold on CDs in 1997?

CHECK FOR UNDERSTANDING

Think and Discuss ▶ **Look back at the lesson to answer each question.**

1. **Explain** why the graphs in Example 1 and Example 2 have a key.

2. **Describe** the kind of data you would display in a bar graph, a line graph, and a circle graph.

Guided Practice ▶ **Tell if you would use a bar, line, or circle graph to display the data.**

3. The average monthly rainfall in your city over two years

4. A family budget divided into types of expenses

PRACTICE AND PROBLEM SOLVING

Independent Practice ▶ **Tell if you would use a bar, line, or circle graph to display the data.**

5. The heights of five different students

6. The price of a stock over a period of several months

7. The way you spend your weekly allowance

8. The population of six different cities in your state

9. Make a multiple-bar graph using the data in the table at the right.

NUMBER OF RAINY DAYS				
	April	**May**	**June**	**July**
1999	12	3	13	5
2000	6	7	9	11

10. Make a multiple-line graph using the data in the table at the right.

AVERAGE STOCK PRICES				
	Sep	**Oct**	**Nov**	**Dec**
Stock A	$26	$29	$32	$27
Stock B	$14	$10	$8	$19

Problem Solving ▶ Applications

Use Data For 11–13, use the double-line graph below.

11. How do the comedy video rentals compare to the action video rentals?

12. For the months of March and April combined, about how many more comedy rentals were there than action rentals?

13. *REASONING* If the trends continue, what will happen to the number of comedy and action video rentals in July?

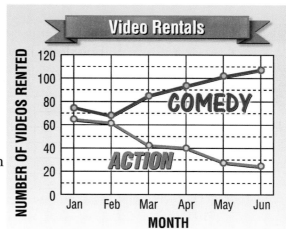

Use Data For 14–16, use the circle graph at left.

U.S. POPULATION BY AGE (1997)

19-64 61%

26%

13%

18 or younger   65 or older

14. The number of people aged 19–64 is about how many times the number who are 65 or older?

15. *REASONING* Which of the three age groups do you think spends the most money? Explain.

16. ❓ **What's the Error?** Jay looked at the circle graph and said that 61% of the U.S. population is aged 19 or older. What is his error?

MIXED REVIEW AND TEST PREP

17. Fifteen students in the gym are asked what their favorite sport is. All of them say "basketball." Ron concludes that the favorite sport of all students is basketball. Is Ron's conclusion valid? Explain. (p. 112)

Find the product. (p. 70)

18. 6×0.4

19. 0.8×0.7

⭐**20.** **TEST PREP** Which is 0.27×0.4? (p. 70)

A 0.108   **B** 0.675   **C** 1.08   **D** 10.8

⭐**21.** **TEST PREP** Find the value of 21^2. (p. 40)

F 21   **G** 2^3 **H** 42   **J** 441

Find Unknown Values

Learn how to estimate unknown values from a graph and solve for the values by using logic, arithmetic, and algebra.

For fitness training this week, Karen wants to walk a total of 24 mi. The table above shows how long it will take her to walk certain distances.

Time (hr)	1	2	3	4	5
Distance (mi)	4	8	12	16	20

EXAMPLE 1

Make a line graph with the data in the table and use it to estimate how long it will take Karen to walk 24 mi.

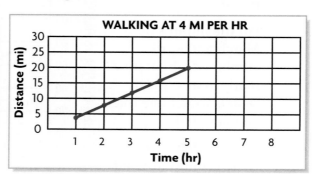

Look at the graph. If the pattern continues, it looks as if it will take Karen 6 hr to walk 24 mi.

EXAMPLE 2

Use logical reasoning and arithmetic to find how long it will take Karen to walk 24 mi.

Look at the data in the table. Notice that whenever the time increases by 1 hr, the distance increases by 4 mi.

$5 + 1 = 6$, or 6 hr   *Add 1 to the last time in the table.*

$20 + 4 = 24$, or 24 mi   *Add 4 to the last distance in the table.*

So, it will take Karen 6 hr to walk 24 mi.

You can also use the formula $d = rt$ to solve problems about distance (*d*), rate (*r*), and time (*t*). The distance traveled is a product of the rate of speed and the amount of time.

EXAMPLE 3

Use the formula $d = rt$ to find how long it will take Karen to walk 32 mi if she walks at a rate of 4 mi per hr.

$d = rt$ *Write the formula.*

$32 = 4 \times t$ *Replace d with 32 and r with 4. What number multiplied by 4 gives 32?*

$8 = t$ *The solution is t = 8.*

So, it will take Karen 8 hr to walk 32 mi.

Think and Discuss ▶ Look back at the lesson to answer the question.

1. **Explain** how you used the graph in Example 1 to predict how long it would take Karen to walk 24 mi.

Guided Practice ▶ **Ty averages 50 mi per hr on long car trips. Use the table for 2–4.**

2. Make a line graph. Use the line graph to find how long it will take Ty to drive 250 mi.

Time (hr)	1	2	3	4
Distance (mi)	50	100	150	200

3. Use logical reasoning and arithmetic to find how long it will take Ty to drive 250 mi.

4. Use the formula $d = rt$ to find how long it will take Ty to drive 350 mi.

PRACTICE AND PROBLEM SOLVING

Independent Practice ▶ **Juan averages 8 mi per hr while biking. Use the table for 5–7.**

5. Use the data in the table to make a line graph. Use the line graph to estimate how long it will take Juan to bike 48 mi.

Time (hr)	1	2	3	4
Distance (mi)	8	16	24	32

6. Use logical reasoning and arithmetic to find how long it will take Juan to bike 48 mi.

7. Use the formula $d = rt$ to find how long it will take Juan to bike 64 mi.

Problem Solving Applications ▶ **Calvin is training for a marathon. The total distance he has run each week is shown in the table.**

8. Make a line graph using the data in the table.

9. If the trend in Calvin's training continues, about how far will he run in the sixth week of training?

Distance Run

Week	1	2	3	4	5
Miles	14	20	26	33	39

10. ✎ **Write a problem** using Calvin's training data. Explain your solution.

MIXED REVIEW AND TEST PREP

Find the mean, median, and mode. (p. 106)

11. 9, 6, 4, 4, 5   **12.** 12, 14, 32, 28   **13.** 2, 9, 9, 9, 6, 5, 8, 4

14. Order 0.76, 0.765, and 0.076 from least to greatest. (p. 52)

⭐ **15. TEST PREP** Replace ● with the missing operation. 56 ● 32 ÷ (4 × 2) = 52 (p. 44)

    A ÷   **B** ×   **C** −   **D** +

Stem-and-Leaf Plots and Histograms

Learn how to display and analyze data in stem-and-leaf plots and histograms.

Vocabulary

stem-and-leaf plot

histogram

You can use a **stem-and-leaf plot** to organize data when you want to see each item in the data. For a stem-and-leaf plot, choose the stems first and then write the leaves.

EXAMPLE 1

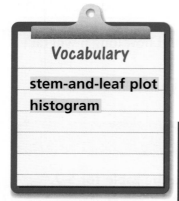

Bryan Berg built a 19 ft 16$\frac{1}{2}$ in. high card tower with 102 levels.

The table shows the number of levels reached, without cards falling, at a card-tower building competition. Use the data to make a stem-and-leaf plot.

CARD-TOWER COMPETITION					
21	18	32	47	50	33
19	21	11	54	31	18
33	42	21	29	16	12

11 12 16 18 18 19
21 21 21 29
31 32 33 33
42 47
50 54

First, group the data by tens digits. Then, order the data from least to greatest.

Card Tower Competition

Stems	Leaves
1	1 2 6 8 8 9
2	1 1 1 9
3	1 2 3 3
4	2 7
5	0 4

Use the tens digits as stems. Use the ones digits as leaves. Write the leaves in increasing order.

The line 4 | 2 7 means 42 and 47.

EXAMPLE 2

Use the data from a domino stacking competition to make a stem-and-leaf plot. Then use the stem-and-leaf plot to help find the mode and the median.

NUMBER OF DOMINOES STACKED									
97	88	74	96	98	58	68	90	80	90
	72	86	69	78	93	84	99	92	85

Stems	Leaves
5	8
6	8 9
7	2 4 8
8	0 4 5 6 8
9	0 0 2 3 6 7 8 9

90 occurs more than any other number.

There are 19 scores. The median is the 10th score.

Mode: 90   Median: 86

A **histogram** is a bar graph that shows the frequency, or the number of times, data occur within intervals. The bars in a histogram are connected, rather than separated.

BAR GRAPH

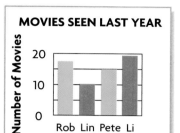

HISTOGRAM

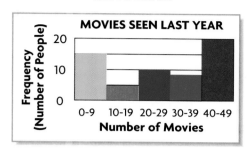

*R*emember that you can use the range of a set of data to help determine intervals.

This bar graph is used to show information about individual movie customers. The histogram is used to show information about groups of movie customers.

EXAMPLE 3

The table below shows the number of sit-ups students in gym class did in one minute. Make a histogram for the data.

NUMBER OF SIT-UPS									
28	19	32	45	44	12	24	32	35	47
55	59	24	25	37	36	38	36	42	41

First, make a frequency table with intervals of 10. Start with 10.

Interval	10-19	20-29	30-39	40-49	50-59
Frequency	2	4	7	5	2

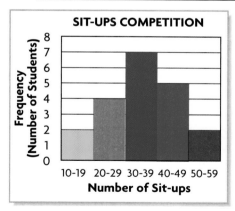

Title the graph and label the scales and axes.

Graph the number of students who did sit-ups within each interval.

CHECK FOR UNDERSTANDING

Think and ▶ Discuss

Look back at the lesson to answer each question.

1. **Tell** how the stem-and-leaf plot in Example 1 is useful in showing how well people did in the competition.

2. **Explain** how data displayed in a histogram are different from data displayed in a bar graph.

3. Make a stem-and-leaf plot of the data 32, 24, 44, 57, 31, 25, 41, 26.

Tell whether a bar graph or a histogram is more appropriate.

4. number of customers at different intervals of time

5. populations of five different states

PRACTICE AND PROBLEM SOLVING

6. Make a stem-and-leaf plot of the data 89, 74, 63, 65, 68, 74, 71, 80.

Tell whether a bar graph or a histogram is more appropriate.

7. heights of the tallest mountains in the United States

8. ages of 75 ice-skating competitors at a competition

Use the data in the table for 9.

9. a. Make a stem-and-leaf plot of the data.

PLANT HEIGHTS IN INCHES					
28	36	25	24	20	32
15	18	28	12	19	26

b. Use the stem-and-leaf plot to help find the median and mode.

c. *REASONING* Can you tell what the mean is for this set of data by looking at the stem-and-leaf plot? Explain.

AGES OF MARIA'S MUSIC CUSTOMERS									
10	25	33	14	54	62	29	44	41	40
11	31	41	24	65	16	39	50	51	55
19	22	17	26	31	42	17	18	42	37

10. Use the data in the table above to make a histogram.

11. 📝 **Write About It** Write a question that can be answered using the data from Exercise 10. Explain your answer.

MIXED REVIEW AND TEST PREP

12. Thirty students in the gym are asked their favorite sport. They all respond by saying "basketball." Is this valid for all students? (p. 112)

13. What type of sample does Ron get if he randomly surveys 175 people? (p. 94)

14. Evaluate $3k$ for $k = 6.5$. (p. 82)   **15.** Write 0.73 as a percent. (p. 60)

⭐**16. TEST PREP** Mt. McKinley has a height of 20,320 ft. Mt. Whitney has a height of 14,494 ft. If Brian climbs both mountains and wants to climb a total of 40,000 ft, how many more feet does he need to climb? (p. 20)

A 34,814 ft   **B** 25,506 ft   **C** 19,680 ft   **D** 5,186 ft

EXTRA PRACTICE page H37, Set C

Explore Box-and-Whisker Graphs

Explore how to make a box-and-whisker graph and understand its parts.

You need at least eleven 3 in. × 5 in. cards, marker.

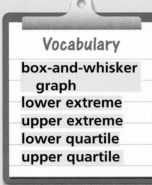

Vocabulary
box-and-whisker graph
lower extreme
upper extreme
lower quartile
upper quartile

A **box-and-whisker graph** shows how far apart and how evenly data are distributed.

Activity

• Write each of the data in the table on a separate card.

NUMBER OF TROPHIES WON									
35	21	24	32	36	20	24	29	27	30

• Order the data from least to greatest. Draw a star on the card with the least value, or **lower extreme**, and the card with the greatest value, or **upper extreme**.

• Find the median of the data. If the median is not one of the numbers already written, write it on a card, put it in the middle of the data, and circle it.

• Find the median of the lower half of the data. This median is called the **lower quartile.** Circle it. Separate the data to the left of the lower quartile from the rest of the data.

• Find the median of the upper half of the data. This median is the **upper quartile** . Circle it. Separate the data to the right of the upper quartile from the rest of the data.

Think and Discuss

• Look at your cards. Into how many parts do the lower quartile, the median, and the upper quartile separate the data?

• What fraction of the data are to the left of the lower quartile? to the right of the upper quartile? What fraction of the data are between the lower quartile and the upper quartile?

You have found all you need to make this box-and-whisker graph.

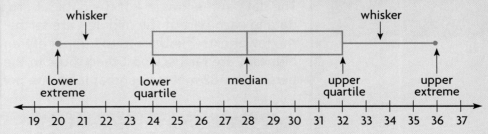

Practice
• Make a box-and-whisker graph of the data.

POINTS SCORED									
10	12	9	22	17	7	14	8	11	19

Box-and-Whisker Graphs

Learn how to analyze a box-and-whisker graph.

The only actual data you can identify from the data set in a box-and-whisker graph are the extremes.

EXAMPLE 1

The data in the box-and-whisker graph represent the diameters in kilometers of some asteroids observed by a scientist. What was the diameter of the largest asteroid? of the smallest asteroid?

The upper extreme is 12 and the lower extreme is 4. So, the diameter of the largest asteroid was 12 km, and the diameter of the smallest asteroid was 4 km.

- Look again at the graph. Which of the following can you determine—mean, median, mode, range?

EXAMPLE 2

Data about the number of meteors seen during a particular hour at different locations during the 1999 Leonids meteor shower are shown in the box-and-whisker graph. What does the graph show about how the data are distributed?

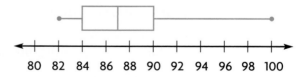

The data in the lowest $\frac{1}{4}$ of the data set are very close together. The data in each part of the middle $\frac{1}{2}$ are farther apart. They are closer to the lower extreme than to the upper extreme. The data in the highest $\frac{1}{4}$ are farther apart than those in the middle. At least one person saw 82 meteors, and at least one person saw 100 meteors.

CHECK FOR UNDERSTANDING

Think and ▶ Discuss

Look back at the lesson to answer the question.

1. Explain why you can't find the mean and the mode when looking at a box-and-whisker graph.

For 2–3, use the box-and-whisker graph.

2. What are the median, lower quartile, and upper quartile?

3. What are the lower and upper extremes and the range?

PRACTICE AND PROBLEM SOLVING

Independent ▶
Practice

For 4–5, use the box-and-whisker graph.

4. What are the median, lower quartile, and upper quartile?

5. What are the lower and upper extremes and the range?

For 6–8, use the data in the table below.

6. What are the median, lower quartile, and upper quartile?

7. What are the lower and upper extremes and the range?

WEIGHTS OF PANTHERS FOOTBALL PLAYERS					
85	102	89	86	104	92
103	97	91	100	100	93

8. Make a box-and-whisker graph and a histogram. Compare them.

Problem Solving ▶
Applications

The box-and-whisker graph shows data about the numbers of points scored by the basketball team each game.

9. What are the least and greatest numbers of points?

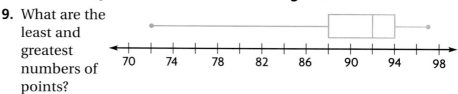

10. What does the graph show about how the data are distributed?

11. ❓ **What's the Question?** Use the box-and-whisker graph for 9–10. The answers are 88 and 94.

MIXED REVIEW AND TEST PREP

12. If you are displaying data about the number of cars that cross an intersection during different intervals of time, is a bar graph or a histogram more appropriate? (p. 126)

13. Sue purchases three items that cost $3.73, $9.21, and $2.99. If she pays with $20.00, how much change will she receive? (p. 66)

Write the value of the blue digit. (p. 52)

14. 9.64

15. 124.024

⭐ 16. **TEST PREP** Which is $3 \times 3 \times 3 \times 3$ in exponent form? (p. 40)

A 3×4 **B** 3^2 **C** 3^3 **D** 3^4

Analyze Graphs

Learn how to analyze data displays and determine how results and conclusions may have been influenced.

QUICK REVIEW

Write the corresponding decimal or percent.
1. 75%   2. 3%   3. 4.6%   4. 0.36   5. 0.07

Data can be displayed in many different ways. Sometimes, the way a question is asked can influence the results that are displayed.

Rita took a survey asking the following question: Do you agree with me that George Washington was the greatest U.S. President, or would you choose Thomas Jefferson or Abraham Lincoln?

EXAMPLE 1

The results of Rita's survey are displayed in the bar graph shown at the right. Could the way the question was asked have influenced the results? Explain.

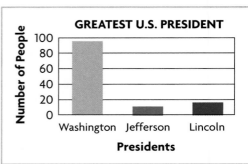

Yes. Rita's question is biased, since it leads people to agree with her that George Washington was the greatest U.S. President. As a result, the graph is misleading.

EXAMPLE 2

The results of two other surveys are shown below. Which graph is more likely to come from which question? Explain your reasoning.

Question 1:   Would you rather visit the Grand Canyon, Mount Rushmore, or the Statue of Liberty?

Question 2:   Would you rather visit the spectacular Grand Canyon, or would you rather visit Mount Rushmore or the Statue of Liberty?

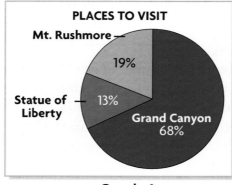

Graph A

Graph B

Graph A probably goes with Question 2, since the question is biased and leads people to choose the Grand Canyon. Graph B probably goes with Question 1, since that question is not biased.

Graphs can communicate information quickly. That's why they are used by advertisers on television and in magazines and newspapers. Some graphs can be misleading and influence conclusions that are drawn.

EXAMPLE 3

Jon looked at the bar graph below and concluded that the Mississippi River is twice as long as the Missouri River. Explain Jon's mistake and tell why his conclusion is wrong.

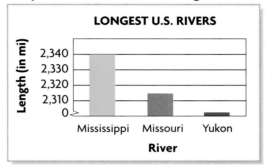

LONGEST U.S. RIVERS

The bar for the Mississippi is twice as long as the bar for the Missouri. However, if you look at the scale, you see that the rivers are about the same length. Because the lower part of the scale is missing, the differences are exaggerated.

When two graphs with different scales show two similar sets of data, comparing the graphs can sometimes be misleading.

EXAMPLE 4

The weekly ticket sales for the Mississippi River tour boat cruise are shown in the graphs below. Deb looked at the graphs and concluded that more tickets were sold in April than in March. Explain Deb's mistake.

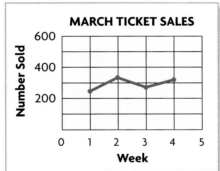

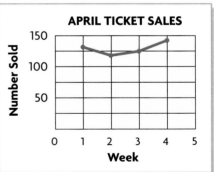

It appears that the April ticket sales were greater than in March since the line for April is higher than the line for March. However, if you look at the scale for each graph, you can see that ticket sales were much greater in March than in April.

CHECK FOR UNDERSTANDING

Think and ▶ Discuss

Look back at the lesson to answer each question.

1. **Tell** how you could rewrite the question in Example 1 so it would not influence the results of the survey.

133

2. **Explain** how you could change the graph in Example 3 so it is not misleading.

Guided ▶
Practice

Rosa took a survey, asking the following question: Don't you think that the Mustangs are the best baseball team, or would you choose the Wildcats or Cougars? The results of her survey are displayed in the bar graph at the right.

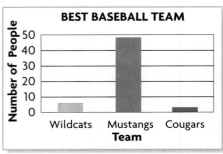

BEST BASEBALL TEAM

3. Could the way the question was asked have influenced the results? Explain.

For 4–6, use the graph at the right. The graph is misleading.

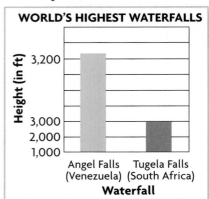

WORLD'S HIGHEST WATERFALLS

4. About how many times as high is the bar for Angel Falls than for Tugela Falls?

5. Is Angel Falls 3 times as high as Tugela Falls? Explain.

6. How could you change the graph so it is not misleading?

PRACTICE AND PROBLEM SOLVING

Independent ▶
Practice

Miguel took a survey, asking the following question: What is your favorite fruit—apples, bananas, or delicious, juicy Florida navel oranges? The results of Miguel's survey are displayed in the circle graph.

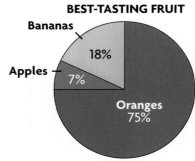

BEST-TASTING FRUIT

7. Could the way the question was asked have influenced the results? Explain.

For 8–10, use the graph at the right. The graph is misleading.

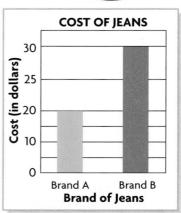

COST OF JEANS

8. About how many times as high is the bar for Brand B than for Brand A?

9. Does Brand B cost twice as much as Brand A? Explain.

10. How can you change the graph so it is not misleading?

Problem Solving ▶ Applications

11. Jeff looked at the bar graph at the right and concluded that Los Angeles has three times the population of Chicago. Explain Jeff's mistake and tell why his conclusion is wrong.

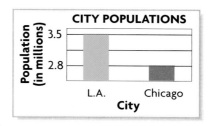

12. How could you fix the graph so Jeff would not make a mistake?

13. Lin looked at the graphs below and concluded that from June to September, the temperatures in Seattle are about the same as the temperatures in Boston. Explain Lin's mistake.

Boston, Massachusetts

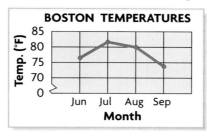

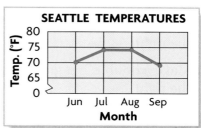

MIXED REVIEW AND TEST PREP

For 14–15, use the box-and-whisker graph. (p. 129)

14. What is the median?

15. What are the least and greatest values?

16. Find the mean of the data set 18, 12, 10, 8, 9, 14, and 6. (p. 106)

⭐**17. TEST PREP** Which is the value of $2n + 3.7$ for $n = 2.9$? (p. 82)

    A 8.6   **B** 9.5   **C** 10.4   **D** 21.46

⭐**18. TEST PREP** Replace ● with the missing operation. $(3 + 24) ● 3 \times 2 - 1 = 17$ (p. 44)

    F ×   **G** ÷   **H** +   **J** −

PROBLEM SOLVING   LiNKUP to Reading

Strategy • Classify and Categorize

To classify information means to group together similar information. To categorize the information, label the groups. Paige wants to buy a piece of Indian pottery. She can choose Navajo, Hopi, or Zuni style and select small, medium, or large size. By classifying and categorizing the data, you can see that she has 9 different choices.

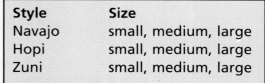

Style	Size
Navajo	small, medium, large
Hopi	small, medium, large
Zuni	small, medium, large

• Jerome is buying an Indian drum. He can choose Navajo, Hopi, or Zuni style and select miniature, small, medium, large, or extra-large size. How many choices does he have?

1. **VOCABULARY** A bar graph that shows frequencies within intervals is a(n) __?__ . (p. 127)

2. **VOCABULARY** A graph that shows how far apart and how evenly data are distributed is a(n) __?__ . (p. 129)

3. What type of graph would best show the highest and lowest temperatures for each of the last four years? (pp. 120–123)

4. What type of graph would best show high and low temperatures for a week? (pp. 120–123)

5. Make a double-line graph with the data at the right. (pp. 120–123)

6. Make a double-bar graph with the data at the right. (pp. 120–123)

END-OF-MONTH STOCK PRICES					
	Sep	**Oct**	**Nov**	**Dec**	**Jan**
Stock A	$80	$74	$45	$50	$52
Stock B	$50	$52	$52	$50	$45

For 7–9, use the graph at the right. (pp. 124–125)

7. About how much did the stock price decrease from Monday to Tuesday?

8. Describe the pattern in the graph.

9. If the trend continues, what do you think the stock price will be on Friday?

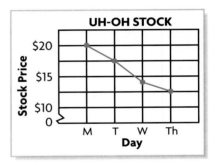

10. Make a stem-and-leaf plot for the data. (pp. 126–128)

11. Make a histogram for the data. (pp. 126–128)

POINTS SCORED					
33	52	45	47	34	52
34	58	48	52	46	59

HEIGHTS OF BUILDINGS (IN FT)						
20	50	80	20	40	45	85
25	30	80	60	70	75	55

For 12–17, use the following data: 14, 16, 9, 21, 35, 2, 26, 8, 17.
(pp.129–131)

12. Find the upper extreme.

13. Find the lower extreme.

14. Find the upper quartile.

15. Find the lower quartile.

16. Find the median.

17. Make a box-and-whisker graph.

For 18–20, use the graph at the right. (pp. 132–135)

18. About how many times as high is the bar for green peppers as the bar for carrots?

19. Were three times as many pounds of carrots sold as pounds of green peppers? Explain.

20. How could you change the graph so that it is not misleading?

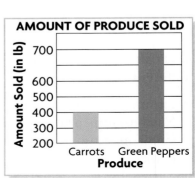

Check your work.

See item **5**.

Think about the different kinds of graphs and the most appropriate data for them. Check that your answer choice could be made into a bar graph.

Also see problem **7**, p. H65.

Choose the best answer.

1. The owner of a music shop made a line graph to show the number of CDs sold during a 4-week period. Which trend does the graph show?

SALES OF CDs

A Sales are increasing.

B Sales are even.

C Sales are decreasing.

D No trend is shown.

2. 22.8 − 3.11

F 18.79   H 19.79

G 19.69   J Not here

3. Which of the following is best suited for display in a multiple-bar graph?

A Average temperatures recorded in a town during a 1-year period

B Number of CDs owned by sixth-grade students

C Number of hours sixth- and seventh-grade students spend reading each week

D Changes in water temperature over a 24-hour period

4. Which kind of graph is shown?

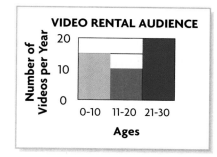

VIDEO RENTAL AUDIENCE

F Circle graph

G Histogram

H Line graph

J Stem-and-leaf plot

5. Which of the following is best suited for display in a bar graph?

A Heights of the tallest buildings in a city

B Frequency of cars stopping at a tollbooth

C Changes in a person's weight over a year

D Part of a day a person spends reading

Write What You Know

6. Suppose you wanted to show the favorite sports of a sample of teenagers. Would you use a line graph, bar graph, histogram, or stem-and-leaf plot? Justify your choice.

7. Explain how you would find the value of $\frac{12 - 2^3}{4}$. Then find the value.

137

MATH DETECTIVE

Measure by Measure

REASONING Use the clues and your knowledge of the measures of central tendency to find the set of numbers described. Be prepared to explain how you solved the mystery.

Mystery Number 1

Clues:
1. There are three numbers in the set.
2. The mean is 10.
3. The mode is 8.
What is the set of numbers?

Mystery Number 2

Clues:
1. There are three numbers in the set.
2. One number in the set is 5.
3. The median is 9.
4. The range is 16.
What is the set of numbers?

Mystery Number 3

Clues:
1. There are four numbers in the set.
2. The median is 12.
3. Two numbers in the set are 3 and 10.
4. The range is 15.
What is the set of numbers?

Think It Over!

- 📝 **Write About It** If you know that the sum of a set of six numbers is 150, which of the following measures could you find: mean, median, mode, range? Explain.

- **Stretch Your Thinking** A set of numbers forms the pattern 1, 3, 5, 7, ... If the median of the set is 15, what is the range?

Explore Scatterplots

Learn how to read and interpret a scatterplot.

Do you think there is any relationship between the number of people who paint a large building and the number of hours it takes them to finish the job?

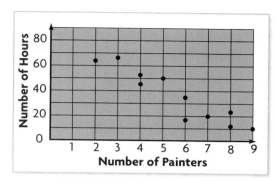

The *scatterplot* displays data for 11 large buildings that were painted. It shows that as the number of painters increased, the time it took them to finish the job tended to decrease.

A scatterplot shows the relationship between two variables.

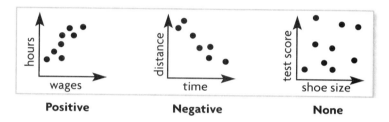

| Positive | Negative | None |

When the values of the two variables increase or decrease together, there is a **positive correlation**.

When the values of one variable increase while the others decrease, there is a **negative correlation**.

When the data points show no pattern of increase or decrease, there is **no correlation**.

TALK ABOUT IT

• Tell if the relationship between the speed of a car and the number of hours needed to drive 500 miles has a positive correlation, a negative correlation, or no correlation.

TRY IT

Sketch a scatterplot that could represent the situation. Then identify the type of correlation between the variables.

1. amount of time walking *and* total distance that you walk

2. number of rooms in house *and* street address of house

Write *positive correlation*, *negative correlation*, or *no correlation* to describe the relationship shown in the scatterplot.

3.

4.
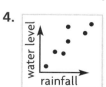

VOCABULARY

1. Everyone in the population has the same chance of being selected in a(n) __?__ . (p. 95)

2. A bar graph that shows the frequency at which data occur within intervals is a(n) __?__ . (p. 127)

3. Individuals in the population are not represented in the sample if the sample is __?__ . (p. 98)

EXAMPLES

EXERCISES

Chapter 5

- **Identify the type of sample.** (pp. 94–97)

  Determine the sampling method used if workers on an assembly line check every tenth tire.

  This is a systematic sample.

Determine the type of sample. Write *convenience*, *random*, or *systematic*.

4. From a computer list of students, each with an equal chance of being selected, 100 students are chosen.
5. Every tenth person walking down the street is surveyed about the President.

- **Record and organize data.** (pp. 102–105)

  What is the size of the sample?

SCORES ON MATH TEST		
Score Interval	Frequency	Cumulative Frequency
91–100	5	5
81–90	8	13
71–80	6	19
Below 71	5	24

  There are 24 people in the sample.

For 6–7, use the following data.

NUMBER OF MOVIES SEEN IN ONE YEAR			
8	12	16	12
9	10	9	8
9	15	20	14
12	7	9	15

6. Make a line plot.
7. Make a cumulative frequency table with intervals.

- **Find the mean, median, and mode.**
  (pp. 106–111)

  24, 20, 24, 21, 26
  Mean: 24 + 20 + 24 + 21 + 26 = 115;
  $\qquad$ 115 ÷ 5 = 23

  Median: 24

  Mode: 24

Find the mean, median, and mode.

8. 2, 7, 9, 4, 6, 8, 6
9. 5.3, 8.8, 4.7, 6.5, 4.7
10. 79, 87, 90, 100, 96, 89, 92, 87
11. Suppose you added the number 10.6 to the data in Exercise 9. Which measure(s) of central tendency would change?

Chapter 6

- **Display data in graphs.** (pp. 120–123)

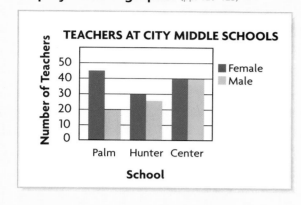

TEACHERS AT CITY MIDDLE SCHOOLS

For 12, use the graph at the left.

12. Which schools have more female teachers than male teachers?

13. Would you use a bar graph, line graph, or circle graph to display a city's temperature readings for 1 month? Explain.

14. Make a multiple-line graph with the data below.

HIGH AND LOW TEMPERATURES					
	Mon	**Tues**	**Wed**	**Thurs**	**Fri**
Highs	45°	53°	41°	48°	50°
Lows	34°	39°	35°	40°	41°

- **Use stem-and-leaf plots and histograms.**
(pp. 126–128)

In a stem-and-leaf plot, the tens digits of the data are stems, and the ones digits are leaves.

Ages

Stem	Leaves
1	1 3 3
2	0 3 5 5
3	7 7 8

Some students are participating in a jump-rope benefit. Their ages are 10, 15, 18, 20, 15, 11, 13, 10, 14, 14, 18, 19, 22, and 21.

15. Use the data to make a stem-and-leaf plot.

16. Use the data to make a histogram.

- **Make a box-and-whisker graph.** (pp. 129–131)

Heights of Students (in cm)

135 168 148 160 159 148 163 165 167

Draw a box-and-whisker graph.

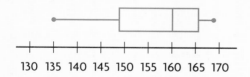

130 135 140 145 150 155 160 165 170

For 17–20, use the table below.

HEIGHTS OF TREES (in ft)								
8	12	6	9	15	9	16	20	24

17. What is the median?

18. What are the lower and upper quartiles?

19. What are the lower and upper extremes?

20. Make a box-and-whisker graph.

PROBLEM SOLVING APPLICATIONS

21. Roberto has to pay $27 for his CDs at the record store. In how many ways can he pay, using only bills of $10, $5, and $1? (pp. 100–101)

23. The machines in the exact change lane of a tollway accept any combination of coins that total exactly 75¢, but they do not accept pennies or half dollars. In how many different ways can a driver pay the toll in an exact change lane?
(pp. 100–101)

22. The sixth-grade students are having a car wash. They charge $4 for cars and $6 for SUVs. In how many ways can they earn exactly $100? (pp. 100–101)

24. The debate club has 10 members. Each member will debate each of the other members only once. How many debates will they have? (pp. 100–101)

Performance Assessment

TASK A • Too Much Homework?

For homework, Scott has 3 social studies questions, 3 English questions, 2 science problems, 5 Spanish questions, and 20 math exercises. Scott's friend Beth asks him about how much homework he has in each subject. Would he give the mean, median, or mode of the number of items to make his homework assignment seem as long as possible?

a. Explain your choice. Write down your thinking as you make your decision.

b. Suppose your friend asks you the same question about the homework you really have. How would you respond?

TASK B • Graph Analysis

Elements in the Earth's Crust: Aluminum 8.1%, Calcium 3.6%, Iron 5.0%, Oxygen 46.6%, Silicon 27.7%, All Others 9.0%

a. Organize the data in a table. Use the data to make a graph. Then write a short description of the graph. Include the following:

- What the title and axes represent

- What the parts represent, if it is a circle graph

- What stands out as important or obvious

b. Describe how the data can be used to make a different kind of graph.

c. Write a question that can be answered from either graph.

E-Lab • Exploring Box-and-Whisker Graphs

Janie kept track of the lengths of her workouts. She made a table like the one below. Then she made a box-and-whisker graph.

LENGTHS OF WORKOUTS (min)								
45	60	35	46	70	42	63	38	41

You can use E-Lab to make box-and-whisker graphs.

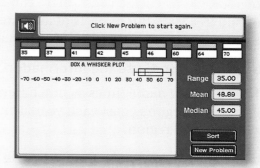

- Click on *Exploring Box-and-Whisker Graphs*.
- Click on *New Problem*.
- Type the length of each of Janie's workouts.
- Press *Enter* after each one.
- Click *Sort* to order the data from least to greatest.
- Copy the box-and-whisker graph.

What are the median and lower and upper extremes?

What are the lower and upper quartiles?

Practice and Problem Solving

Use E-Lab to make a box-and-whisker graph for each table. Copy each graph and answer the questions.

1.

NUMBER OF PEOPLE AT DIFFERENT PICNIC PAVILIONS							
15	30	48	24	40	26	55	52

2. What are the median, lower quartile, and upper quartile?

3. What are the lower and upper extremes and the range?

4.

NUMBER OF STUDENTS IN DIFFERENT AFTER-SCHOOL CLUBS							
10	24	16	30	21	14	26	19

5. What are the median, lower quartile, and upper quartile?

6. What are the lower and upper extremes and the range?

Multimedia Math Glossary www.harcourtschool.com/mathglossary

7. Vocabulary Locate *frequency table* and *line plot* in the Multimedia Math Glossary. Use the frequency table shown to make a line plot. Use the line plot shown to make a cumulative frequency table.

PROBLEM SOLVING ON LOCATION
In Oregon

Water Sources and Uses

Water is a critical resource everywhere in the world. In the United States, we have some very dry states, as well as states that receive an abundance of rainfall. Oregon has both wet and arid regions. On its Pacific coast, there can be 120 inches of rain in a year. But in parts of its eastern region, there might be only 8 inches of rain in a year.

Use Data **For 1, use the information above.**

1. About how many times as much rainfall can fall on Oregon's coast each year as in parts of its eastern region?

Use Data **For 2–4, use the first circle graph.**

2. Which category uses the most freshwater?

3. Which category uses the least freshwater?

4. Would you say that thermoelectric and home/commercial uses account for a little less than half or more than half of the freshwater?

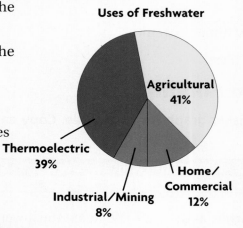

Uses of Freshwater

Agricultural 41%

Thermoelectric 39%

Home/ Commercial 12%

Industrial/Mining 8%

Use Data **For 5–6, use both circle graphs.**

5. About how many times as much freshwater is supplied by surface water as is supplied by ground water?

6. Could ground water meet the freshwater needs of agriculture? Explain your thinking.

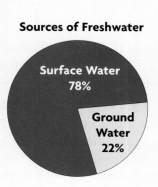

Sources of Freshwater

Surface Water 78%

Ground Water 22%

Dams

Suppose you live in the country. To get water, you would probably dig a well or use a nearby stream. But if you live in a city, you would use other means. Dams, reservoirs, and aqueducts help to collect, store, and transport water to city homes and businesses. In Oregon, more than 20 dams help meet the people's water needs.

Oregon's Bonneville Dam, located 40 miles east of Portland on the Columbia River, allows boats to travel upriver 188 miles.

Use Data **For 1–7, use the table.**

Oregon Dams		
Name	**Storage Capacity**	**Crest Elevation**
Gerber Dam	94,270 acre-feet	4,842 feet
Owyhee Dam	1,183,300 acre-feet	2,675 feet
Thief Valley Dam	26,000 acre-feet	3,143 feet
Warm Springs Dam	192,400 acre-feet	3,409 feet

1. Find the range of storage capacities.

2. Find the mean storage capacity.

3. Find the median storage capacity.

4. **REASONING** Why is the mean so much greater than the median?

5. Is the mean or median a better measure of central tendency for the storage capacities? Explain your thinking.

6. **Mental Math** One acre-foot is equivalent to 43,560 cubic feet. Use mental math to estimate the storage capacity of Gerber Dam in cubic feet.

7. Make and label a bar graph showing the crest elevations of the four dams. Explain how you used rounding or estimation in making your graph.

CHAPTER 7

Number Theory

Tulips are grown commercially in Woodburn, Oregon. As far as the eye can see, bright colors cover the surrounding fields each spring. Bulbs from these plants are collected, packaged, and sold to home gardeners who start tulip gardens of their own on a smaller scale.

PROBLEM SOLVING Suppose 1,200 tulip bulbs are put into packages of 20. How many packages could be made? Would there be any bulbs left over?

DATA LINK

PACKAGES OF TULIP BULBS	
Tulip Bulbs per Package	Complete Package— Yes or No?
25	Yes
30	
35	
40	Yes
45	
50	Yes

Check What You Know

Use this page to help you review and remember important skills needed for Chapter 7.

 Vocabulary

Choose the best term from the box.

1. A number whose only factors are 1 and itself is a __?__ .

2. A number that has more than two factors is a __?__ .

> composite
> number
> multiple
> prime number

 Prime Numbers (For Intervention, see p. H7.)

Decide whether the number is a prime number. Write *yes* or *no*.

3. 2	**4.** 5	**5.** 4	**6.** 9
7. 11	**8.** 21	**9.** 37	**10.** 26
11. 13	**12.** 7	**13.** 45	**14.** 70

 Composite Numbers (For Intervention, see p. H7.)

Decide whether the number is a composite number. Write *yes* or *no*.

15. 6	**16.** 15	**17.** 19	**18.** 81
19. 24	**20.** 53	**21.** 3	**22.** 25

 Multiples (For Intervention, see p. H8.)

Write the next three multiples.

23. 4   4, 8, 12, ■, ■, ■   **24.** 10   10, 20, 30, ■, ■, ■   **25.** 12   12, 24, 36, ■, ■, ■

26. 8   8, 16, 24, ■, ■, ■   **27.** 5   5, 10, 15, ■, ■, ■   **28.** 11   11, 22, 33, ■, ■, ■

Write the first five multiples of each number.

29. 6	**30.** 22	**31.** 30
32. 7	**33.** 9	**34.** 13

Factors (For Intervention, see p. H8.)

Write all of the factors of each number.

35. 8	**36.** 9	**37.** 11
38. 18	**39.** 54	**40.** 32

> **LOOK AHEAD**
>
> **In Chapter 7 you will**
> - use divisibility rules
> - find prime factors
> - determine the LCM and GCF of whole numbers

Divisibility

Learn how to tell if a number is divisible by 2, 3, 4, 5, 6, 8, 9, or 10.

QUICK REVIEW

Write *even* or *odd* for each.

1. 34,526   2. 5,437
3. 6,230   4. 27,343
5. 1,468

Remember that a number is divisible by another number if the remainder is zero.

Chris knows that a number is divisible by 2 if the last digit is 0, 2, 4, 6, or 8. He also knows that a number is divisible by 3 if the sum of the digits is divisible by 3. He wonders if there is a rule for numbers that are divisible by 6.

MATH LAB

Activity

You need: hundred chart

• Shade all the numbers divisible by 2.

• Circle all the numbers that are divisible by 3.

• Look at the numbers that are both shaded and circled. Divide these numbers by 6. What do you notice?

• What rule can Chris write about numbers divisible by 6?

Math Idea ▶ You can use divisibility rules to help you decide if a number is divisible by another number.

A number is divisible by	Divisible	Not Divisible
2 if the last digit is even (0, 2, 4, 6, or 8).	11,994	2,175
3 if the sum of the digits is divisible by 3.	216	79
4 if the last two digits form a number divisible by 4.	1,024	621
5 if the last digit is 0 or 5.	15,195	10,007
6 if the number is divisible by 2 and 3.	1,332	44
8 if the last three digits form a number divisible by 8.	5,336	3,180
9 if the sum of the digits is divisible by 9.	144	33
10 if the last digit is 0.	2,790	9,325

EXAMPLES

Tell whether each number is divisible by 2, 3, 4, 5, 6, 8, 9, or 10.

A. 610 is divisible by
   2; the last digit is even.
   5; the last digit is 0 or 5.
   10; the last digit is 0.

B. 459 is divisible by
   3; the sum of the three digits is divisible by 3.
   9; the sum of the digits is divisible by 9.

Think and ▶ Discuss

Look back at the lesson to answer each question.

1. **Compare** the divisibility rules for 3 and 9.

2. **Tell** the advantages of knowing the divisibility rules.

Guided ▶ Practice

Tell whether each number is divisible by 2, 3, 4, 5, 6, 8, 9, or 10.

3. 56　　　**4.** 200　　　**5.** 784　　　**6.** 2,345　　　**7.** 3,009

PRACTICE AND PROBLEM SOLVING

Independent ▶ Practice

Tell whether each number is divisible by 2, 3, 4, 5, 6, 8, 9, or 10.

8. 75　　**9.** 324　　**10.** 45　　**11.** 812　　**12.** 501

13. 615　　**14.** 936　　**15.** 744　　**16.** 5,188　　**17.** 4,335

18. 1,407　　**19.** 48,006　　**20.** 7,064　　**21.** 12,111　　**22.** 1,044

For 23–25 write *T* or *F* to tell whether each statement is true or false. If it is false, give an example that shows it is false.

23. All even numbers are divisible by 2.

24. All odd numbers are divisible by 3.

25. Some even numbers are divisible by 5.

Problem Solving ▶ Applications

26. A number is between 80 and 100 and is divisible by both 5 and 6. What is the number?

27. *REASONING*　Use the divisibility rules you know to write a divisibility rule for 15.

28. Kirk has 20 trading cards. He has 10 more hockey cards than baseball cards. How many does Kirk have of each type of card?

29. Amber charges $8.25 per hour to baby-sit. One Saturday she baby-sat for 3.5 hr in the morning and 2.25 hr in the evening. How much did she earn baby-sitting on Saturday? Round your answer to the nearest cent.

MIXED REVIEW AND TEST PREP

30. What is the lower quartile of the data 24, 26, 28, 29, 30, 32, 34, 36, 37? (p. 129)

31. Write the percent for 0.05. (p. 60)

32. $36.4 \div 0.28$ (p. 76)

33. 858×19 (p. 22)

34. **TEST PREP**　Which is the value of $3^2 + 2 \times 5 - 6$? (p. 44)

A 13　　**B** 24　　**C** 34　　**D** 49

Prime Factorization

Learn how to write a composite number as the product of prime numbers.

Vocabulary

prime factorization

QUICK REVIEW

Write the equal factors for each.

1. 3^2 **2.** 2^4 **3.** 4^3 **4.** 9^2 **5.** 5^4

A prime number is a whole number greater than 1 whose only factors are itself and 1. Here are the prime numbers less than 50.

2, 3, 5, 7, 11, 13, 17, 19, 23, 29, 31, 37, 41, 43, 47

A composite number, like 104, has more than two factors. You can write a composite number as the product of prime factors. This is called the **prime factorization** of the number.

You can divide to find the prime factors of a composite number.

EXAMPLE 1

Find the prime factorization of 104.

```
 2|104
 2|52
 2|26
13|13
   1
```
Repeatedly divide by the smallest possible prime factor until the quotient is 1.

$2 \times 2 \times 2 \times 13$ *List the prime numbers you divided by. These are the prime factors.*

So, the prime factorization of 104 is $2 \times 2 \times 2 \times 13$, or $2^3 \times 13$.

Use a factor tree to find the prime factors of a composite number.

EXAMPLE 2

Find the prime factorization of 156.

Choose any two factors of 156. Continue until only prime factors are left.

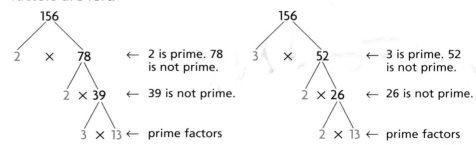

So, the prime factorization of 156 is $2 \times 2 \times 3 \times 13$, or $2^2 \times 3 \times 13$.

Math Idea ▶ Every composite number can be written as a product of two or more prime factors. No matter how you find the prime factors, you will get the same factors, but maybe in a different order.

Think and ▶
Discuss

Look back at the lesson to answer each question.

1. **Tell** what the prime factorization would be in Example 2 if you started with 4 and 39.

2. **Tell** how you know when you have finished the prime factorization of a number.

Guided ▶
Practice

Use division or a factor tree to find the prime factorization.

3. 12   **4.** 65   **5.** 16   **6.** 42

Write the prime factorization in exponent form.

7. 21   **8.** 28   **9.** 254   **10.** 908

PRACTICE AND PROBLEM SOLVING

Independent ▶
Practice

Use division or a factor tree to find the prime factorization.

11. 128   **12.** 50   **13.** 76   **14.** 108

Write the prime factorization in exponent form.

15. 18   **16.** 302   **17.** 49   **18.** 217

19. 532   **20.** 45   **21.** 746   **22.** 99

Solve for n to complete the prime factorization.

23. $2 \times n \times 5 = 20$ **24.** $44 = 2^2 \times n$ **25.** $75 = 3 \times 5 \times n$

Problem Solving ▶
Applications

26. *REASONING* The prime factorization of 50 is 2×5^2. Without dividing or using a factor tree, tell the prime factorization of 100.

27. *REASONING* A number, c, is a prime factor of both 12 and 60. What is c?

28. Chris bathed his pet every eighth day in June, beginning on June 8. On what other dates did he bathe his pet in June?

29. ✎ **Write About It** Do the prime factors of a number differ depending on which factors you choose first? Explain.

MIXED REVIEW AND TEST PREP

30. Is 3,543 divisible by 2, 3, or 9? (p. 146)

31. Write the percent for 0.6. (p. 60)

32. $18.3 + 22.6 + 17.03 + 21.99$ (p. 66)

33. Find the median for the data 13, 8, 9, 16, 18. (p. 106)

★ **34. TEST PREP** A sign is placed in a toy store, asking shoppers to complete a survey about their favorite video game. Which type of sampling is this? (p. 94)

A convenience   **C** systematic

B random   **D** compatible

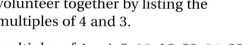

LESSON 7.3

Least Common Multiple and Greatest Common Factor

Learn how to find the LCM and GCF of numbers and use them to solve problems.

Vocabulary

least common multiple (LCM)

greatest common factor (GCF)

QUICK REVIEW

List the first five multiples of each number.

1. 4　　　　　**2.** 2　　　　　**3.** 3

4. 9　　　　　**5.** 6

Kirk and Amber volunteer during December. Kirk volunteers every fourth day beginning December 4. Amber volunteers every third day beginning December 3. Find the first day they will volunteer together by listing the multiples of 4 and 3.

multiples of 4:　4, 8, 12, 16, 20, 24, 28

multiples of 3:　3, 6, 9, 12, 15, 18, 21, 24, 27, 30

The multiples that appear in blue are called common multiples. The smallest of the common multiples is called the **least common multiple**, or **LCM**. The LCM of 4 and 3 is 12, the product of the two numbers. So, the first day they volunteer together is December 12.

Examples 1 and 2 show two ways to find the LCM.

EXAMPLE 1

One Way　Find the LCM of 12 and 8.

12:　12, 24, 36, 48, 60, 72, 84, 96　　*List the first eight multiples.*

8:　8, 16, 24, 32, 40, 48, 56, 64　　*Find the common multiples.*

So, the LCM of 12 and 8 is 24.　　*Find the LCM.*

EXAMPLE 2

Another Way　Find the LCM of 6, 9, and 18.

Write the prime factorizations.

$6 = 2 \times 3$　　　　$9 = 3 \times 3 = 3^2$　　　　$18 = 2 \times 3 \times 3 = 2 \times 3^2$

Write a product using each prime factor only once.

$$2 \times 3$$

For each factor, write the greatest exponent used with that factor in any of the prime factorizations. Multiply.

$$2 \times 3^2 = 18$$

So, the LCM of 6, 9, and 18 is 18.

Greatest Common Factor

Factors shared by two or more numbers are called common factors. The largest of the common factors is called the **greatest common factor**, or **GCF**.

To find the GCF of two or more numbers, list all the factors of each number, find the common factors, and then find the greatest common factor.

> 45: 1, 3, 5, 9, 15, 45   *The common factors are 1, 3, and 9.*
>
> 27: 1, 3, 9, 27   *The GCF of 45 and 27 is 9.*

The GCF can be used to solve problems.

EXAMPLE 3

Carlyn has 12 pens and 36 pencils. She is making packages with the same number of each item. What is the greatest number of packages she can make without any items left over? How many of each item will be in each package?

You can find the greatest number of packages by finding the GCF of 12 and 36.

12: 1, 2, 3, 4, 6, 12   *List the factors.*
36: 1, 2, 3, 4, 6, 9, 12, 18, 36   *Find the common factors.*

The GCF of 12 and 36 is 12.   *Find the GCF.*

So, Carlyn can make 12 packages without any items left over.

To find the number of each item, divide the number of pens and the number of pencils by the number of packages.

> Pens: $12 \div 12 = 1$ Pencils: $36 \div 12 = 3$

So, there will be 1 pen and 3 pencils in each package.

To find the GCF of two numbers, you can also use their prime factors. List the prime factors, find the common prime factors, and then find their product.

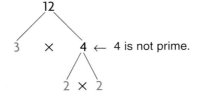
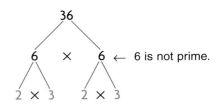

The prime factors of 12 are $2 \times 2 \times 3$.

The prime factors of 36 are $2 \times 2 \times 3 \times 3$.

The common prime factors are 2, 2, and 3.

Find the product of the common factors: $2 \times 2 \times 3$, or $2^2 \times 3 = 12$.

The GCF of 12 and 36 is 12.

151

EXAMPLE 4

Use prime factors to find the GCF of 48 and 72.

48: $2 \times 2 \times 2 \times 2 \times 3$ *Find the prime factors.*
72: $2 \times 2 \times 2 \times 3 \times 3$

2, 2, 2, and 3   *Find the common prime factors.*

$2 \times 2 \times 2 \times 3 = 24$ *Multiply the common factors.*

So, the GCF of 48 and 72 is 24.

CHECK FOR UNDERSTANDING

Think and Discuss

Look back at the lesson to answer each question.

1. **Explain** why you are able to find a greatest common factor but not a greatest common multiple of 2 or more numbers.

2. **Tell** the number of pens and pencils there would be in each package in Example 3 if Carlyn made 6 packages.

Guided Practice

List the first five multiples of each number.

3. 3   **4.** 7   **5.** 11   **6.** 15

Find the LCM of each set of numbers.

7. 3, 7   **8.** 2, 3   **9.** 6, 9   **10.** 5, 8, 20

Find the GCF of each set of numbers.

11. 6, 9   **12.** 4, 20   **13.** 9, 24   **14.** 12, 16, 20

PRACTICE AND PROBLEM SOLVING

Independent Practice

List the first five multiples of each number.

15. 4   **16.** 8   **17.** 16   **18.** 55

19. 10   **20.** 27   **21.** 14   **22.** 39

Find the LCM of each set of numbers.

23. 3, 8, 24   **24.** 32, 128   **25.** 12, 20   **26.** 40, 105

27. 24, 30   **28.** 18, 21, 36   **29.** 12, 27   **30.** 48, 116

Find the GCF of each set of numbers.

31. 16, 18   **32.** 15, 18   **33.** 21, 306   **34.** 16, 24, 40

35. 25, 33   **36.** 200, 215   **37.** 24, 32, 40   **38.** 630, 712

Find a pair of numbers for each set of conditions.

39. The LCM is 36. The GCF is 3.

40. The LCM is 24. The GCF is 2.

Problem Solving ▶ Applications

41. Peter will distribute cereal samples with pamphlets about good nutrition. The samples come in packages of 15. The pamphlets come in packages of 20.
   a. What is the least number of cereal samples and pamphlets needed to have equal amounts?
   b. How many packages of each does he need?

42. Ruth has 36 markers and 48 erasers. She will put them in bags with the same number of each item. What is the greatest number of bags she can make?

43. **?** **What's the Error?** Jan says the LCM of 10 and 15 is 5. Find her error and the correct answer.

MIXED REVIEW AND TEST PREP

44. Write the prime factorization for 45 in exponent form. (p. 148)

45. Evaluate $a \div c$, for $a = 4{,}602$ and $c = 37$. (p. 28)

46. Find the value of 5^4. (p. 40)

47. **TEST PREP** Which is the exponential notation for $3 \times 3 \times 3 \times 4 \times 4 \times 5 \times 5$? (p. 40)

   A $3 \times 4 \times 5^7$ **B** $3^3 \times 2^4 \times 2^5$ **C** $3^3 \times 4^2 \times 5^2$ **D** $3^3 \times 4^4 \times 5^5$

48. **TEST PREP** Which shows the decimal for 46%? (p. 60)

   F 0.046   **G** 0.46   **H** 4.6   **J** 46

PROBLEM SOLVING  **LiNKUP** to Careers

NASA Scientist When astronauts service and repair the Hubble Space Telescope, NASA scientists must set the orbit of the space shuttle so it will meet the Hubble Space Telescope in its orbit. The Hubble Space Telescope takes about 95 minutes to make one orbit around Earth. The space shuttle takes about 90 minutes to go around Earth.

One way to find when the two objects will meet is to find the LCM of the orbit times.

90: 90, 180, 270, 360, 450, 540, … 1,350; 1,440; 1,530; 1,620; 1,710

95: 95, 190, 285, 380, 475, 570, … 1,330; 1,425; 1,520; 1,615; 1,710

So, the space shuttle will meet the Hubble Space Telescope about every 1,710 minutes, or $28\frac{1}{2}$ hours.

• The Russian Mir space station orbits Earth about every 93 minutes. How often would Mir and the space shuttle meet?

EXTRA PRACTICE page H38, Set C

153

PROBLEM SOLVING STRATEGY
Make an Organized List

Learn how to solve problems by making an organized list.

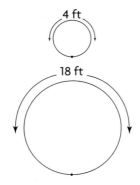

4 ft

18 ft

QUICK REVIEW

List the first four multiples of each number.

1. 3   **2.** 5   **3.** 9

4. 11   **5.** 20

Like many early bicycles, the Boneshaker had different sized wheels. Suppose the circumferences of the wheels were 4 ft and 18 ft. How many revolutions would each wheel make before marks on their rims would both be in the same positions again?

The Boneshaker was invented in 1865. The wooden wheels made for an uncomfortable ride.

Analyze   What are you asked to find?

What facts are given?

Choose   What strategy will you use?

You can use the strategy *make an organized list*. List the total distance traveled by each wheel for every complete turn.

Solve   How will you solve the problem?

Make a list of multiples of 4 and 18.
  multiples of 4: 4, 8, 12, 16, 20, 24, 28, 32, 36, 40
  multiples of 18: 18, 36, 54

The least common multiple is 36. When the wheels have traveled 36 ft, the marks will both be in the same positions again.

36 is the ninth multiple of 4, so the small wheel makes 9 revolutions.

36 is the second multiple of 18, so the large wheel makes 2 revolutions.

Check   How can you check your answer?

What if the smaller wheel had a circumference of 8 ft? How many revolutions would each wheel make before the marks were both in the same positions again?

PROBLEM SOLVING PRACTICE

PROBLEM SOLVING STRATEGIES

Draw a Diagram or Picture
Make a Model
Predict and Test
Work Backward
▶ **Make an Organized List**
Find a Pattern
Make a Table or Graph
Solve a Simpler Problem
Write an Equation
Use Logical Reasoning

Solve the problem by making an organized list.

1. Justin and Amy help with the shopping for their individual families. Justin goes to the store every 3 days and Amy goes every 5 days. They see each other at the store on September 30. On what date will Justin and Amy see each other at the store again?

For 2–3, use the information below.

Susanna buys one bag with 40 snacks and another with 32. She wants to make up small snack packs for a party. All snack packs must have the same number of each snack.

2. If you make lists to find the greatest number of snack packs Susanna can make, what will you include in the lists?

   A addends   **B** factors   **C** multiples   **D** fractions

3. What is the greatest number of snack packs Susanna can make?

   F 2 packs   **G** 4 packs   **H** 8 packs   **J** 15 packs

MIXED STRATEGY PRACTICE

4. Ray begins an eight-week exercise program. He exercises 30 min during the first week, 45 min the second week, and 60 min the third week. If this pattern continues, how many hours and minutes will Ray exercise the sixth week?

5. DeAnn has a total of 74 in. of yarn to make gifts. She uses 13 in. of yarn for each of the 5 gifts she is making. Find how many inches of yarn DeAnn will have left after making the 5 gifts.

6. A total of 316 middle school teachers are going to a meeting. There are 30 more male teachers than female. How many teachers are male?

7. Christi rides her bicycle 7 blocks south, 3 blocks east, 5 blocks north, and 8 blocks west. How many blocks has she ridden when she crosses her own path?

8. Use the graph at the right. Look at the area in square miles for each lake. How much larger is Lake Michigan than Lake Erie and Lake Ontario together?

9. **? What's the Question?** Laura wants to buy the same number of plums and kiwis. Plums are sold in bags of 4 and kiwis are sold in bags of 7. The answer is 28 plums.

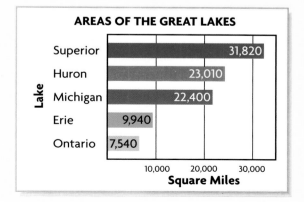

AREAS OF THE GREAT LAKES

Lake	Square Miles
Superior	31,820
Huron	23,010
Michigan	22,400
Erie	9,940
Ontario	7,540

1. **VOCABULARY** Writing a composite number as the product of prime factors is called __?__. (p. 148)

2. **VOCABULARY** The largest of the common factors of two or more numbers is the __?__. (p. 150)

3. **VOCABULARY** The smallest of the multiples of two or more numbers is the __?__. (p. 150)

Tell whether each number is divisible by 2, 3, 4, 5, 6, 8, 9, or 10. (pp. 146–147)

4. 42	**5.** 64	**6.** 96	**7.** 225
8. 330	**9.** 963	**10.** 450	**11.** 2,385

Use division or a factor tree to find the prime factorization. Write the prime factorization in exponent form. (pp. 148–149)

12. 9	**13.** 8	**14.** 14	**15.** 18
16. 80	**17.** 12	**18.** 33	**19.** 50
20. 49	**21.** 98	**22.** 504	**23.** 891

24. List the first twenty multiples of 4 and the first twenty multiples of 10. What is the least common multiple? What multiples greater than 80 do they have in common? (pp. 150–153)

Find the LCM and the GCF of each set of numbers. (pp.150–153)

25. 3, 9	**26.** 2, 6	**27.** 6, 4	**28.** 10, 15
29. 8, 12	**30.** 9, 27	**31.** 15, 25	**32.** 25, 115
33. 27, 189	**34.** 6, 8, 12	**35.** 6, 9, 12	**36.** 8, 16, 20

Solve.

37. A number is between 60 and 70. It is divisible by 3 and 9. What is the number? (pp. 146–147)

38. Meat patties are sold in packages of 12. Buns are sold in packages of 8. What is the least number of meat patties and buns needed to have an equal number of each? (pp. 150–153)

39. Cashews are sold in 8-oz jars, almonds in 12-oz jars, and peanuts in 16-oz jars. What is the least number of ounces of each type of nuts you can buy to make mixed nuts with equal amounts of each? How many jars of each would you need? (pp. 150–153)

40. Alissa jogs in the park every 3 days. Erin jogs in the park once a week on Saturday. If they met in the park on Saturday, April 30, when will they meet in the park again? (pp. 154–155)

Get the information you need.

See item **1.**

You know that juice boxes are sold in packages of 8 and rice snacks are sold in packages of 10. Use the least common multiple to find an equal number of treats before you determine how many packages to buy.

Also see problem **3**, p. H63.

1. Hu is buying treats for a party. Juice boxes are sold in packs of 8. Rice snacks are sold in packs of 10. What is the least number of packs of juice and rice snacks he should buy to have an equal number of juice packs and rice snacks?

   A 4 juice packs, 5 rice snack packs

   B 5 juice packs, 4 rice snack packs

   C 9 juice packs, 9 rice snack packs

   D 10 juice packs, 8 rice snack packs

2. What is the least common multiple of 18 and 27?

   F 3   **H** 54

   G 9   **J** 486

3. Which number is divisible by both 6 and 9?

   A 45   **C** 243

   B 216   **D** 768

4. Carni earned the following amounts baby-sitting: $17, $15, $12, $17, $14. What was the mean amount that Carni earned?

   F $14   **H** $16

   G $15   **J** Not here

5. What is the greatest common factor of 16 and 24?

   A 4   **B** 8   **C** 48   **D** 384

6. Marcus has a baseball game every fifth day in April. His first game is on April 5. How many games will there be in April?

   F 3   **G** 4   **H** 5   **J** 6

7. A theater holds 381 people. All the available tickets for 18 shows were sold. Which is a good estimate for the number of people who attended the shows?

   A Less than 4,000 people

   B Between 4,000 and 6,000 people

   C Between 6,000 and 8,000 people

   D More than 8,000 people

8. Eleni bakes brownies for a bake sale at school. She wants to make 4 batches of brownies and package the brownies in boxes of 8. How many boxes will she need if there are 32 brownies in each batch?

   F 12   **H** 20

   G 16   **J** 32

Write What You Know

9. Explain how to use a factor tree to find the prime factorization of a number. Then use your method to find the prime factorization of 120.

10. If a number is divisible by both 3 and 4, then it is divisible by 12. Use this fact to write a divisibility rule for 12. Explain your reasoning.

CHAPTER 8

Fraction Concepts

Around the world, there are more than 30 different species of cats living in forests, grasslands, deserts, and mountains. The body lengths of cats range from 2 feet to 10 feet. The average tiger is $9\frac{1}{4}$ feet long, and the average lion is $6\frac{1}{2}$ feet long. The average bobcat is $3\frac{1}{4}$ feet long.

PROBLEM SOLVING Compare the average length of a tiger to that of a lion.

AVERAGE BODY LENGTHS OF CATS

Average Body Length (in feet)

Cheetah $4\frac{1}{2}$ ft · Snow Leopard 5 ft · Tiger $9\frac{1}{4}$ ft · Jaguar 6 ft · Lion $6\frac{1}{2}$ ft · Bobcat $3\frac{1}{4}$ ft

Type of Cat

Check What You Know

Use this page to help you review and remember important skills needed for Chapter 8.

✅ Vocabulary

Choose the best term or symbol from the box.

1. The symbol for "is greater than" is ___?___ .

2. The top number of a fraction is called the ___?___ .

3. The bottom number of a fraction is called the ___?___ .

> denominator
> <
> numerator
> >

✅ Compare and Order Whole Numbers

(For Intervention, see p. H3.)

Write <, >, or = for each ●.

4. 408 ● 480

5. 4,279 ● 4,277

6. 30 tens ● 3 hundreds

7. 9,315 ● 9,351

8. 18,808 ● 18,880

9. 356,782 ● 356,482

Order the numbers from greatest to least.

10. 3,400; 3,439; 3,399

11. 61,060; 62,000; 61,600

12. 98,450; 98,405; 98,540

✅ Model Fractions (For Intervention, see p. H8.)

Write the fraction for the shaded part.

13.

14.

15.

16.

✅ Model Percents (For Intervention, see p. H9.)

Write the percent for the shaded part.

17.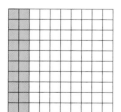

18.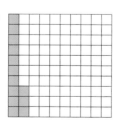

19.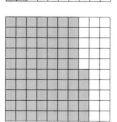

20.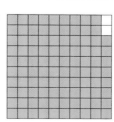

LOOK AHEAD

In Chapter 8 you will

- find equivalent fractions
- write fractions in simplest form
- write fractions as mixed numbers and mixed numbers as fractions
- compare and order fractions
- find relationships among fractions, decimals, and percents

Equivalent Fractions and Simplest Form

Learn how to identify and write equivalent fractions, and how to write fractions in simplest form.

Vocabulary

equivalent fractions

simplest form

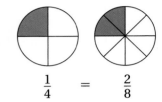

Fractions that name the same amount or the same part of a whole are called **equivalent fractions**. The figures show that the fractions $\frac{1}{4}$ and $\frac{2}{8}$ are equivalent, because they name the same part of a whole circle.

$$\frac{1}{4} = \frac{2}{8}$$

There are several ways to find equivalent fractions.

Activity

You need: fraction bars

Find how many eighths are equivalent to $\frac{1}{2}$.

- Place $\frac{1}{8}$ fraction bars along the $\frac{1}{2}$ bar until the lengths are equal. How many $\frac{1}{8}$ bars are there?

- Complete: $\frac{1}{2} = \frac{\blacksquare}{8}$

- Use fraction bars. How many fourths are equivalent to $\frac{1}{2}$?

- Complete: $\frac{1}{2} = \frac{\blacksquare}{4}$

Another way to find an equivalent fraction is to multiply or divide the numerator and denominator of a fraction by the same number, except 0. Doing this does not change the fraction's value, because this is the same as multiplying or dividing by 1.

EXAMPLE 1

Complete: $\frac{2}{4} = \frac{\blacksquare}{12}$

THINK: To get the denominator 12, I need to multiply the denominator 4 by 3. So, to get the missing numerator, I should multiply the numerator 2 by 3.

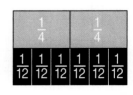

$$\frac{2}{4} = \frac{6}{12}$$

$$\frac{2 \times 3}{4 \times 3} = \frac{6}{12}$$

$\frac{3}{3} = 1$, so the product is still equal to $\frac{2}{4}$.

EXAMPLE 2

Complete: $\frac{6}{10} = \frac{\blacksquare}{5}$

THINK: I can get the denominator 5 by dividing the denominator 10 by 2. So, to get the missing numerator, I should divide the numerator 6 by 2.

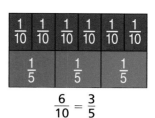

$$\frac{6}{10} = \frac{3}{5}$$

$$\frac{6 \div \boxed{2}}{10 \div \boxed{2}} = \frac{3}{5}$$ *$\frac{2}{2} = 1$, so the quotient is still equal to $\frac{6}{10}$.*

Math Idea ▶ When the numerator and denominator of a fraction have no common factors other than 1, the fraction is in **simplest form**.

$\frac{9}{16}$ **is** in simplest form because 9 and 16 have no common factors other than 1.

$\frac{9}{15}$ **is not** in simplest form because 9 and 15 have the common factor 3.

EXAMPLE 3

The sun is much farther from the Earth than the moon is. So, even though the sun is much larger than the moon, the sun's tide-raising force is only $\frac{12}{30}$ of the moon's force. Write $\frac{12}{30}$ in simplest form.

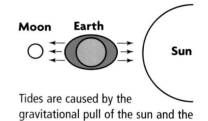

Tides are caused by the gravitational pull of the sun and the moon on Earth's oceans.

12: 1, 2, 3, 4, 6, 12   *Find the common factors of 12 and 30.*
30: 1, 2, 3, 5, 6, 10, 15, 30

$$\frac{12}{30} = \frac{12 \div 3}{30 \div 3} = \frac{4}{10}$$ *Divide the numerator and denominator by a common factor.*

$$\frac{4}{10} = \frac{4 \div 2}{10 \div 2} = \frac{2}{5}$$ *Repeat until the fraction is in simplest form.*

So, $\frac{2}{5}$ is the simplest form of $\frac{12}{30}$.

Bay of Fundy during low tide

In Example 3, the numerator and denominator were divided by common factors twice to find the simplest form. You can find it in just one step if you divide by the greatest common factor.

EXAMPLE 4

The world's highest tides occur in Canada's Bay of Fundy. Tides in Seattle, Washington, average only about $\frac{12}{42}$ of tidal heights in the Bay of Fundy. Write $\frac{12}{42}$ in simplest form.

12: 1, 2, 3, 4, 6, 12
42: 1, 2, 3, 6, 7, 14, 21, 42   *Find the GCF of 12 and 42.*
GCF = 6

$$\frac{12}{42} = \frac{12 \div 6}{42 \div 6} = \frac{2}{7}$$ *Divide the numerator and denominator by the GCF.*

So, $\frac{2}{7}$ is the simplest form of $\frac{12}{42}$.

Bay of Fundy during high tide

CHECK FOR UNDERSTANDING

Think and Discuss ▶ Look back at the lesson to answer each question.

1. **Describe** what you must do to find a fraction equivalent to $\frac{2}{3}$ if you first multiply the numerator by 5. What is the equivalent fraction?

2. **Explain** how you know that $\frac{5}{12}$ is in simplest form.

Guided Practice ▶ Complete.

3. $\frac{3}{5} = \frac{\blacksquare}{20}$

4. $\frac{3}{24} = \frac{\blacksquare}{8}$

5. $\frac{8}{12} = \frac{\blacksquare}{6}$

6. $\frac{3}{4} = \frac{\blacksquare}{24}$

7. $\frac{4}{8} = \frac{\blacksquare}{2}$

8. $\frac{9}{24} = \frac{\blacksquare}{8}$

9. $\frac{8}{18} = \frac{\blacksquare}{36}$

10. $\frac{12}{54} = \frac{\blacksquare}{9}$

Write the factors common to the numerator and denominator.

11. $\frac{4}{8}$

12. $\frac{9}{24}$

13. $\frac{8}{18}$

14. $\frac{12}{54}$

Write the fraction in simplest form.

15. $\frac{4}{32}$

16. $\frac{14}{21}$

17. $\frac{9}{54}$

18. $\frac{48}{54}$

19. $\frac{22}{8}$

20. $\frac{18}{5}$

21. $\frac{9}{30}$

22. $\frac{48}{32}$

PRACTICE AND PROBLEM SOLVING

Independent Practice ▶ Complete.

23. $\frac{1}{2} = \frac{\blacksquare}{10}$

24. $\frac{10}{15} = \frac{\blacksquare}{3}$

25. $\frac{16}{20} = \frac{\blacksquare}{5}$

26. $\frac{9}{27} = \frac{\blacksquare}{3}$

27. $\frac{2}{12} = \frac{1}{\blacksquare}$

28. $\frac{\blacksquare}{36} = \frac{2}{9}$

29. $\frac{21}{24} = \frac{7}{\blacksquare}$

30. $\frac{40}{\blacksquare} = \frac{5}{8}$

31. $\frac{9}{\blacksquare} = \frac{3}{4}$

32. $\frac{3}{\blacksquare} = \frac{12}{16}$

33. $\frac{16}{28} = \frac{4}{\blacksquare}$

34. $\frac{25}{40} = \frac{\blacksquare}{8}$

Write the factors common to the numerator and denominator.

35. $\frac{1}{7}$

36. $\frac{9}{30}$

37. $\frac{6}{27}$

38. $\frac{9}{63}$

39. $\frac{10}{35}$

40. $\frac{16}{40}$

41. $\frac{3}{5}$

42. $\frac{8}{10}$

Write the fraction in simplest form.

43. $\frac{4}{24}$

44. $\frac{9}{12}$

45. $\frac{6}{48}$

46. $\frac{10}{15}$

47. $\frac{10}{18}$

48. $\frac{20}{15}$

49. $\frac{18}{90}$

50. $\frac{28}{42}$

51. $\frac{22}{33}$

52. $\frac{24}{30}$

53. $\frac{60}{42}$

54. $\frac{24}{28}$

55. $\frac{16}{64}$

56. $\frac{8}{12}$

57. $\frac{18}{90}$

58. $\frac{48}{18}$

59. $\frac{4^2}{32}$

60. $\frac{3^2}{12}$

61. $\frac{2^3}{12}$

62. $\frac{3^2}{6^2}$

TECHNOLOGY LINK

More Practice: Use E-Lab, *Equivalent Fractions.* www.harcourtschool.com/ elab2002

Problem Solving ▶
Applications

AMERICAN METEOR SOCIETY FIREBALL REPORT, 1998	
Month	Number
January	13
February	8
March	20
April	18
May	8
June	13

63. **Use Data**   Look at the table at the left. A "fireball" is an extremely bright meteor occasionally seen streaking across the night sky. What fraction of the fireballs reported to the American Meteor Society during the first 6 months of 1998 occurred in March?

64. **Technology**   Some calculators have a **Simp** key that can be used to simplify fractions. What fraction would this key sequence give? 10 **n** 15 **d** **Simp** **Enter**

65. **(?) What's the Question?**   Esteban has 6 apple muffins, 2 corn muffins, and 4 bran muffins. The answer is $\frac{1}{3}$ of the muffins.

66. **Write a problem** about everyday life that involves finding the simplest form of a fraction.

MIXED REVIEW AND TEST PREP

Evaluate each expression.

67. $(6 \div 3)^3 + 2^4$ (p. 44)   **68.** $(6^2 \div 3^2) + 1$ (p. 44)   **69.** 3.5×0.01 (p. 70)

⭐ **70. TEST PREP** 0.5×1.2 (p. 70)

   A 0.06   **B** 0.6   **C** 6   **D** 6.2

⭐ **71. TEST PREP**   An adult's movie ticket costs $7.50, and a child's ticket costs $5.50. Find the total cost for 2 adults and 3 children. (p. 70)

   F $13.00   **G** $20.50   **H** $24.00   **J** $31.50

PROBLEM SOLVING | 💡 Thinker's Corner

a÷b ALGEBRA You can use what you know about equivalent fractions to solve some types of algebraic equations.

A. $\frac{2}{3} = \frac{x-5}{27}$

THINK: I need to find a fraction equivalent to $\frac{2}{3}$ that has 27 as its denominator. I'll multiply by 9.

$\frac{2}{3} = \frac{\blacksquare}{27}$ $\frac{2 \times 9}{3 \times 9} = \frac{18}{27}$

So, $x - 5 = 18$. If 5 less than some number is 18, that number must be 23. So, $x = 23$.

B. $\frac{3}{s+2} = \frac{18}{24}$

THINK: I need to find a fraction equivalent to $\frac{18}{24}$ that has 3 as its numerator. I'll divide by 6.

$\frac{3}{\blacksquare} = \frac{18}{24}$ $\frac{3}{4} = \frac{18 \div 6}{24 \div 6}$

So, $s + 2 = 4$. If some number plus 2 equals 4, that number must be 2. So, $s = 2$.

Solve the equation.

1. $\frac{x+3}{10} = \frac{1}{2}$

2. $\frac{5}{8} = \frac{c-5}{40}$

3. $\frac{3}{b-4} = \frac{9}{15}$

4. $\frac{1}{(2+y)} = \frac{6}{24}$

5. $\frac{4}{9} = \frac{(w+4)}{18}$

6. $\frac{(k-5)}{8} = \frac{12}{32}$

EXTRA PRACTICE page H39, Set A

163

LESSON 8.2

Mixed Numbers and Fractions

Learn how to write fractions as mixed numbers and mixed numbers as fractions.

Vocabulary

mixed number

QUICK REVIEW

1. $4 \times 7 + 2$ 2. $19 + 11$

3. $2 \times 7 + 5$ 4. $35 + 9$

5. $8 \times 4 + 6$

Total solar eclipses are rare, with only three visible in most of the U.S. since 1963. Eclipses last different times. In the graph, each quarter-section of a circle represents $\frac{1}{4}$ minute. How long will the 2017 eclipse last? It is represented by 11 sections, or $\frac{11}{4}$ minutes.

A total solar eclipse occurs when the moon passes between the sun and the Earth, "eclipsing" the sun's light.

The graph shows that 11 sections equal 2 whole minutes plus $\frac{3}{4}$ of another minute. So, $\frac{11}{4} = 2\frac{3}{4}$. The 2017 eclipse will last $2\frac{3}{4}$ minutes.

The fraction $\frac{11}{4}$ has a value greater than 1 because the numerator is greater than the denominator. Sometimes a fraction such as $\frac{11}{4}$ is called an "improper fraction." Any such fraction can be written as a **mixed number**, like $2\frac{3}{4}$.

APPROXIMATE LENGTH OF U.S. TOTAL SOLAR ECLIPSES

1963 ⊕ ◖

1970 ⊕ ⊕ ⊕ ◖

1979 ⊕ ⊕ ⊕

2017 ⊕ ⊕ ◖

◿ $= \frac{1}{4}$ minute

Math Idea ▶ A mixed number has a whole-number part that is not 0 and a fraction part.

EXAMPLE 1

The longest total solar eclipse in the next 200 years will take place in 2186. It will last about $\frac{15}{2}$ minutes. Write $\frac{15}{2}$ as a mixed number.

$$\frac{15}{2} \rightarrow \begin{array}{r} 7\frac{1}{2} \\ 2\overline{)15} \\ -14 \\ \hline 1 \end{array}$$

Divide the numerator by the denominator. For the fraction part of the quotient, use the remainder as the numerator and the divisor as the denominator. Write the fraction in simplest form.

So, $\frac{15}{2} = 7\frac{1}{2}$. The 2186 eclipse will last $7\frac{1}{2}$ minutes.

You can also write a mixed number as a fraction.

EXAMPLE 2

Write the mixed number $3\frac{2}{5}$ as a fraction.

$$3\frac{2}{5} = \frac{3 \times 5}{5} + \frac{2}{5} = \frac{(3 \times 5) + 2}{5} = \frac{17}{5}$$

Multiply the whole number by the denominator. Add the numerator. Use the same denominator.

So, $3\frac{2}{5} = \frac{17}{5}$.

Think and ▶ Discuss

Look back at the lesson to answer the question.

1. **Tell** how you know that a given number is a mixed number.

Write the fraction as a mixed number or a whole number.

2. $\frac{5}{3}$　　3. $\frac{7}{2}$　　4. $\frac{15}{5}$　　5. $\frac{11}{3}$　　6. $\frac{13}{4}$

Guided ▶ Practice

Write the mixed number as a fraction.

7. $1\frac{1}{4}$　　8. $1\frac{3}{5}$　　9. $2\frac{2}{3}$　　10. $3\frac{4}{5}$　　11. $5\frac{2}{7}$

PRACTICE AND PROBLEM SOLVING

Independent ▶ Practice

Write the fraction as a mixed number or a whole number.

12. $\frac{7}{4}$　　13. $\frac{9}{2}$　　14. $\frac{11}{2}$　　15. $\frac{23}{4}$　　16. $\frac{27}{3}$

17. $\frac{31}{6}$　　18. $\frac{18}{11}$　　19. $\frac{90}{7}$　　20. $\frac{104}{13}$　　21. $\frac{150}{9}$

22. $\frac{x}{y}$ for $x = 18$ and $y = 12$　　　　23. $\frac{a}{b}$ for $a = 55$ and $b = 15$

Write the mixed number as a fraction.

24. $3\frac{2}{3}$　　25. $6\frac{1}{2}$　　26. $5\frac{1}{3}$　　27. $1\frac{9}{10}$　　28. $4\frac{1}{9}$

29. $9\frac{1}{4}$　　30. $2\frac{3}{8}$　　31. $4\frac{9}{11}$　　32. $11\frac{4}{9}$　　33. $18\frac{3}{5}$

Problem Solving ▶

34. ❓ **What's the Error?**　Marti changed $3\frac{1}{4}$ to $12\frac{1}{4}$. What mistake did she make? What is the correct answer?

35. 📝 **Write About It**　Can any fraction be written as a mixed number? Explain.

36. **Astronomy**　On June 20, 1955, a total solar eclipse lasted 7 min 7 sec. On June 20, 1974, a total solar eclipse lasted 5 min 8 sec. Which lasted longer? How much longer?

MIXED REVIEW AND TEST PREP

Write the prime factorization of each number. (p.148)

37. 36　　　　　　38. 42　　　　　　39. 23

40. Is the following question biased? If so, rewrite it so that it is unbiased: The film *Time Warp* is great, isn't it? (p. 98)

⭐41. **TEST PREP**　Noella is hiking a 25-km trail. She has hiked 3.8 km to the first overlook and another 6.5 km to the second overlook. How many kilometers does she have left to hike? (p. 66)

A 9.3 km　　　**B** 10.3 km　　　**C** 14.7 km　　　**D** 15.7 km

EXTRA PRACTICE page H39, Set B

165

Compare and Order Fractions

Learn how to compare and order fractions.

Remember that values increase as you move to the right on a number line. Values decrease as you move left.

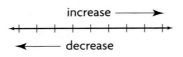

$\mathbf{I}$f two fractions have the same denominator, the fraction with the greater numerator is greater. So, $\frac{7}{12} > \frac{5}{12}$ because $7 > 5$.

If fractions do not have common denominators, you can use a number line to compare and order the fractions. The number line shows that $\frac{1}{4} < \frac{3}{8} < \frac{1}{2}$. From least to greatest the order is $\frac{1}{4}, \frac{3}{8}, \frac{1}{2}$.

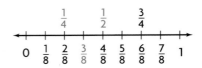

You can also use the least common multiple (LCM) to compare and order fractions.

EXAMPLE

George Washington Carver made over 500 useful agricultural products using peanuts, sweet potatoes, and pecans. About $\frac{7}{10}$ of the products used peanuts and about $\frac{1}{4}$ used sweet potatoes. Did Carver make more products with sweet potatoes or with peanuts?

To compare $\frac{7}{10}$ and $\frac{1}{4}$, find the LCM of the denominators, 10 and 4.

10: 10, 20, 30, 40   *Write multiples of 10.*

 4: 4, 8, 12, 16, 20, 24, 28   *Write multiples of 4.*

LCM = 20

$\frac{7}{10} = \frac{7 \times 2}{10 \times 2} = \frac{14}{20}$ *Rewrite the fractions, using the LCM*
                                        as a common denominator.
$\frac{1}{4} = \frac{1 \times 5}{4 \times 5} = \frac{5}{20}$

$\frac{14}{20} > \frac{5}{20}$, so $\frac{7}{10} > \frac{1}{4}$.   *Compare $\frac{14}{20}$ and $\frac{5}{20}$.*

So, Carver made more products with peanuts.

George Washington Carver, one of America's most honored scientists, was born a slave in 1864.

CHECK FOR UNDERSTANDING

Think and ▶
Discuss

Look back at the lesson to answer each question.

1. **Explain** how to compare $\frac{4}{9}$ and $\frac{5}{9}$.

Guided ▸ Practice

Compare the fractions. Write <, >, or = for each ⬤.

2. $\frac{13}{20}$ ⬤ $\frac{9}{20}$ 3. $\frac{1}{4}$ ⬤ $\frac{9}{20}$ 4. $\frac{5}{6}$ ⬤ $\frac{2}{3}$ 5. $\frac{3}{8}$ ⬤ $\frac{6}{16}$

Use the number line to order the fractions from least to greatest.

$$0 \quad \frac{1}{12} \quad \frac{1}{6} \quad \frac{1}{4} \quad \frac{1}{3} \quad \frac{5}{12} \quad \frac{1}{2} \quad \frac{7}{12} \quad \frac{2}{3} \quad \frac{3}{4} \quad \frac{5}{6} \quad \frac{11}{12} \quad 1$$

6. $\frac{3}{4}, \frac{1}{3}, \frac{11}{12}$ 7. $\frac{2}{3}, \frac{1}{4}, \frac{6}{12}$ 8. $\frac{1}{3}, \frac{5}{12}, \frac{2}{4}$

PRACTICE AND PROBLEM SOLVING

Independent ▸ Practice

Compare the fractions. Write <, >, or = for each ⬤.

9. $\frac{6}{7}$ ⬤ $\frac{4}{7}$ 10. $\frac{3}{11}$ ⬤ $\frac{4}{11}$ 11. $\frac{4}{12}$ ⬤ $\frac{1}{3}$ 12. $\frac{17}{20}$ ⬤ $\frac{3}{5}$

13. $\frac{5}{6}$ ⬤ $\frac{15}{18}$ 14. $\frac{7}{9}$ ⬤ $\frac{11}{12}$ 15. $\frac{3}{4}$ ⬤ $\frac{5}{8}$ 16. $\frac{11}{15}$ ⬤ $\frac{2}{3}$

Use the number line to order the fractions from least to greatest.

$$0 \quad \frac{1}{12} \quad \frac{1}{6} \quad \frac{1}{4} \quad \frac{1}{3} \quad \frac{5}{12} \quad \frac{1}{2} \quad \frac{7}{12} \quad \frac{2}{3} \quad \frac{3}{4} \quad \frac{5}{6} \quad \frac{11}{12} \quad 1$$

17. $\frac{9}{12}, \frac{1}{2}, \frac{2}{6}$ 18. $\frac{4}{12}, \frac{4}{6}, \frac{7}{12}$ 19. $\frac{7}{12}, \frac{5}{6}, \frac{1}{2}$

Order the fractions from least to greatest.

20. $\frac{5}{8}, \frac{1}{2}, \frac{3}{4}$ 21. $\frac{2}{5}, \frac{3}{10}, \frac{1}{2}$ 22. $\frac{11}{16}, \frac{3}{4}, \frac{5}{8}$

23. $\frac{1}{3}, \frac{1}{6}, \frac{1}{2}$ 24. $\frac{2}{3}, \frac{2}{9}, \frac{2}{6}$ 25. $\frac{3}{4}, \frac{1}{12}, \frac{5}{8}$

Problem Solving ▸ Applications

26. During a physical education class, $\frac{1}{3}$ of the students chose to play basketball, $\frac{4}{15}$ chose flag football, and the rest chose tetherball. Which activity was chosen by the most students?

27. ❓ **What's the Error?** For dinner, a mushroom pizza is cut into eighths and a cheese pizza into twelfths. After the meal there are 3 pieces of each left. Pablo tells his mother that the same amount of each pizza is left. What mistake did he make?

28. **REASONING** Find a fraction that has a denominator of 15 and is between $\frac{2}{3}$ and $\frac{4}{5}$.

MIXED REVIEW AND TEST PREP

Evaluate each expression for $a = 2.3$, $b = 0.7$, and $c = 5.4$. (p. 82)

29. $a - b + c$

30. $(b \times 5) + a$

31. Find the LCM of 8 and 12. (p. 150)

32. Find the GCF of 16 and 40. (p. 150)

⭐ 33. **TEST PREP** There are 12 cans of soup in 1 case. How many cases should you order if you need 132 cans of soup? (p. 22)

A 10   B 11   C 1,200   D 1,584

Explore Fractions and Decimals

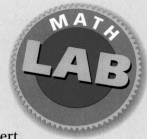

Explore how to convert fractions to decimals.

You need graph paper, scissors, colored pencils.

You can use decimal squares to help you convert fractions to decimals.

Activity

• Cut out a 10 × 10 grid from graph paper. Fold it into 2 equal parts. Then use a colored pencil to shade one of the equal parts. What fraction of the grid is shaded?

• How many small squares are in the whole grid? How many of these squares are shaded? What fraction compares these shaded squares to those in the whole grid? What decimal can you write for this fraction?

• How many columns are in the whole grid? How many columns are shaded? What fraction compares the shaded columns to the whole grid? What decimal can you write for this fraction?

• Are the fractions $\frac{1}{2}$, $\frac{5}{10}$, and $\frac{50}{100}$ equivalent? How do you know? What are two ways to write the fraction $\frac{1}{2}$ as a decimal?

• Cut out another 10 × 10 grid. Fold the grid into 5 equal parts and shade one part. How many rows or columns are shaded? How can you write $\frac{1}{5}$ as a decimal?

Think and Discuss

• How can you show tenths in a 10 × 10 grid?

• How can you show hundredths in a 10 × 10 grid?

• What fractions are easiest to write as decimals?

• How is the 10 × 10 grid helpful for writing fractions as decimals?

TECHNOLOGY LINK

More Practice: Use E-Lab, *Equivalent Fractions, Decimals, and Mixed Numbers.* www.harcourtschool.com/ elab2002

Practice

Write the fraction as a decimal. Use decimal squares.

1. $\frac{3}{10}$ 2. $\frac{60}{100}$ 3. $\frac{7}{10}$ 4. $\frac{3}{5}$ 5. $\frac{90}{100}$

6. $\frac{2}{10}$ 7. $\frac{2}{5}$ 8. $\frac{6}{10}$ 9. $\frac{85}{100}$ 10. $\frac{4}{5}$

Fractions, Decimals, and Percents

Learn how to convert fractions to decimals, decimals to fractions, and fractions to percents.

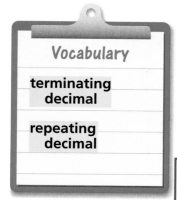

QUICK REVIEW

1. 24 ÷ 4   **2.** 144 ÷ 6   **3.** 155 ÷ 5

4. Write 0.73 in words.   **5.** Write 0.026 in words.

There are many ways to write numbers. Some ways are as fractions, decimals, and percents.

Sometimes you may have to rewrite a given number in a different form. The easiest conversion is from a decimal to a fraction.

EXAMPLE 1

Remember that *percent* means "out of one hundred." For example, 25% means "25 out of 100."

Write each decimal as a fraction.

A. 0.7   **B.** 0.29

Use the decimal's place value to write each fraction.

$0.7 = \frac{7}{10}$ **THINK:** *"seven tenths"* $0.29 = \frac{29}{100}$ **THINK:** *"twenty-nine hundredths"*

To rewrite a fraction as a decimal, use long division or a calculator.

EXAMPLE 2

A newborn koala is about 19 mm long. This is about $\frac{3}{4}$ in. Change $\frac{3}{4}$ to a decimal.

Use long division.

$$\begin{array}{r} 0.75 \\ 4\overline{)3.00} \end{array}$$ *Divide the numerator by the denominator.*

So, $\frac{3}{4} = 0.75$.

Use a calculator.

3 ÷ 4 Enter =

$$\boxed{\uparrow 3 \div 4 = \qquad 0.75}$$

The decimal 0.75 is an example of a **terminating decimal**. The decimal comes to an end at 5. You know that a decimal terminates if you reach a remainder of zero when you are using long division.

The decimal for the fraction $\frac{4}{11}$ does not terminate. When you divide 4 by 11, you never reach a remainder of zero. This decimal is a **repeating decimal** because it shows a pattern of repeating digits.

To write a repeating decimal, show three dots or draw a bar over the repeating part.

$$\frac{4}{11} = 0.363636\ldots \qquad \frac{4}{11} = 0.\overline{36}$$

$$\begin{array}{r} 0.3636 \\ 11\overline{)4.0000} \\ -3\,3 \\ \hline 70 \\ -66 \\ \hline 40 \\ -33 \\ \hline 70 \\ -66 \\ \hline 4 \end{array}$$

To compare a fraction and a decimal, you can first rewrite the fraction as a decimal. Then compare the decimals.

EXAMPLE 3

A newborn panda weighs about $\frac{1}{4}$ lb. A newborn cocker spaniel weighs about 0.4 lb. Which animal weighs less at birth?

Solve by using long division.

$$\begin{array}{r} 0.25 \\ 4\overline{)1.00} \\ -\underline{8} \\ 20 \\ -\underline{20} \\ 0 \end{array}$$

Divide the numerator by the denominator.

$0.25 < 0.4$, so $\frac{1}{4} < 0.4$.

So, a newborn panda weighs less than a newborn cocker spaniel.

Solve by using a calculator.

1 ÷ 4 = [0.25]

To write a fraction as a percent, first convert the fraction to a decimal. Then write the decimal as a percent.

EXAMPLE 4

The barrow ground squirrel of Point Barrow, Alaska, is the world's longest-hibernating animal. The squirrel hibernates $\frac{9}{12}$ of the year. What percent of the year does it hibernate?

$\frac{9}{12} = 0.75$ *Use long division or a calculator to rewrite the fraction as a decimal.*

$0.75 = \frac{75}{100}$ THINK: *75 hundredths. Write the decimal as a fraction.*

$\quad = 75\%$ THINK: *Percent means "out of one hundred." So, 75 hundredths is 75 percent.*

So, the barrow ground squirrel hibernates 75% of the year.

CHECK FOR UNDERSTANDING

Think and ▶ Discuss

Look back at the lesson to answer each question.

1. **Explain** how to use place value to change 0.026 to a fraction.

2. **Compare** a repeating decimal with a terminating decimal.

Guided ▶ Practice

Write the decimal as a fraction.

3. 0.7   **4.** 0.39   **5.** 0.105   **6.** 0.007

Write as a decimal. Tell whether the decimal terminates or repeats.

7. $\frac{1}{4}$ **8.** $\frac{7}{20}$ **9.** $\frac{2}{3}$ **10.** $\frac{8}{11}$

Compare. Write $<$, $>$, or $=$ for each ●.

11. 0.62 ● $\frac{1}{2}$ **12.** $\frac{12}{20}$ ● 0.9 **13.** $\frac{1}{8}$ ● 0.125

Write the fraction as a percent.

14. $\frac{7}{10}$ **15.** $\frac{1}{5}$ **16.** $\frac{1}{4}$ **17.** $\frac{40}{100}$

Independent ▶ Practice

Write the decimal as a fraction.

18. 0.4   **19.** 0.06   **20.** 0.35   **21.** 0.61

22. 0.115   **23.** 0.205   **24.** 0.079   **25.** 0.009

Write as a decimal. Tell whether the decimal terminates or repeats.

26. $\frac{1}{5}$ **27.** $\frac{1}{6}$ **28.** $\frac{1}{15}$ **29.** $\frac{5}{8}$

30. $\frac{11}{20}$ **31.** $\frac{3}{10}$ **32.** $\frac{5}{12}$ **33.** $\frac{1}{9}$

34. $\frac{11}{12}$ **35.** $\frac{9}{25}$ **36.** $\frac{17}{33}$ **37.** $\frac{15}{99}$

TECHNOLOGY LINK

More Practice: Use **Mighty Math Calculating Crew**, *Nautical Number Line*, Level Q

Compare. Write <, >, or = for each ●.

38. $\frac{1}{10}$ ● 0.04   **39.** 0.15 ● $\frac{3}{20}$ **40.** $\frac{1}{2}$ ● 0.52

41. 0.65 ● $\frac{3}{4}$ **42.** $\frac{1}{20}$ ● 0.1   **43.** 0.58 ● $\frac{7}{12}$

Write the fraction as a percent.

44. $\frac{9}{10}$ **45.** $\frac{3}{4}$ **46.** $\frac{1}{2}$ **47.** $\frac{6}{100}$

48. $\frac{3}{5}$ **49.** $\frac{25}{50}$ **50.** $\frac{3}{2}$ **51.** $\frac{1}{200}$

Problem Solving ▶ Applications

52. The goal of the East Side Animal Shelter is to have 0.8 of its animals adopted. One week, the shelter found homes for 20 of its 24 animals. Did the shelter reach its goal? Explain.

Use Data For 53–55, use the table.

53. Write Brian's math test score as a decimal.

54. Did Juan get a higher score on the math or science test?

55. Which student got a higher score on the math test than on the science test?

STUDENT	MATH SCORE	SCIENCE SCORE
Brian	$\frac{18}{25}$	0.95
Juan	$\frac{21}{25}$	0.85
Sabina	$\frac{17}{25}$	0.75
Megan	$\frac{23}{25}$	0.90

56. 📖 **Write About It** The decimal for $\frac{1}{9}$ is $0.\overline{1}$; for $\frac{2}{9}$, $0.\overline{2}$; and for $\frac{3}{9}$, $0.\overline{3}$. Explain how you could use this information to predict the decimal for $\frac{8}{9}$. Use division to check your method.

57. Write $4\frac{1}{2}$ as a fraction. (p. 164)

58. Write $\frac{36}{5}$ as a mixed number. (p. 164)

59. $79.02 - 2.13$ (p. 66)

60. $48.541 + 11$ (p. 66)

61. TEST PREP On his last six math quizzes, Rashard scored 86, 88, 92, 88, 96, and 84. His mean score increased by 1 after his next quiz. What was his seventh score? (p. 109)

A 96   **B** 95   **C** 90   **D** 89

EXTRA PRACTICE page H39, Set D

1. **VOCABULARY** When the numerator and denominator of a fraction have no common factors other than 1, the fraction is in __?__ . (p. 161)

2. **VOCABULARY** A number that is made up of a whole number and a fraction is called a __?__ . (p. 164)

Write the fraction in simplest form. (pp. 160–163)

3. $\frac{6}{12}$

4. $\frac{12}{16}$

5. $\frac{25}{30}$

Complete. (pp. 160–163)

6. $\frac{3}{5} = \frac{\blacksquare}{20}$

7. $\frac{2}{\blacksquare} = \frac{10}{35}$

8. $\frac{24}{32} = \frac{\blacksquare}{8}$

Write the fraction as a mixed number or a whole number. (pp. 164–165)

9. $\frac{7}{3}$

10. $\frac{30}{6}$

11. $\frac{19}{4}$

Write the mixed number as a fraction. (pp. 164–165)

12. $1\frac{5}{6}$

13. $3\frac{1}{3}$

14. $5\frac{7}{8}$

Compare the fractions. Write <, >, or = for each ●. (pp. 166–167)

15. $\frac{7}{8}$ ● $\frac{5}{8}$

16. $\frac{2}{3}$ ● $\frac{8}{12}$

17. $\frac{1}{3}$ ● $\frac{1}{2}$

18. $\frac{1}{2}$ ● $\frac{11}{20}$

19. $\frac{3}{4}$ ● $\frac{3}{8}$

20. $\frac{7}{25}$ ● $\frac{1}{5}$

Write the decimal as a fraction. (pp. 169–171)

21. 0.27

22. 0.1

23. 0.089

Write as a decimal. Tell whether the decimal terminates or repeats.

(pp. 169–171)

24. $\frac{1}{4}$

25. $\frac{5}{6}$

26. $\frac{7}{20}$

Write the fraction as a percent. (pp. 169–171)

27. $\frac{3}{4}$

28. $\frac{9}{100}$

29. $\frac{11}{25}$

30. In the election for class president, Marcus received $\frac{5}{12}$ of the votes, Denise received $\frac{1}{4}$ of the votes, and Alonzo received $\frac{1}{3}$ of the votes. Who won the election? (pp. 166–167)

31. **Use Data** Use the table to find the fraction of the new November films that are rated PG-13. Write your answer in simplest form. What percent is this? (pp. 160–163)

32. Of all U.S. car tunnels longer than 1 mile, $\frac{3}{8}$ are in Pennsylvania. Change $\frac{3}{8}$ to a decimal. (pp. 169–171)

33. On Library Day, $\frac{13}{20}$ of the students at Pine Street School checked books out of the library. What percent of the students checked out books? (pp. 169–171)

NOVEMBER FILMS	
Rating	Number
G	3
PG-13	8
PG	5
R	4

Understand the problem.

See item **7**.

To compare fractions and decimals, change them to the same form. Use the form that will be easier to work with. Then find the numbers that are in the proper order.

Also see problem **1**, p. H62.

Choose the best answer.

1. Which group contains fractions that are all equivalent to $\frac{1}{4}$?

   A $\frac{2}{8}, \frac{4}{20}, \frac{11}{44}$

   B $\frac{3}{6}, \frac{20}{80}, \frac{3}{12}$

   C $\frac{6}{24}, \frac{15}{60}, \frac{50}{200}$

   D $\frac{3}{12}, \frac{25}{100}, \frac{5}{9}$

2. Which is equivalent to 0.36?

   F $\frac{18}{100}$

   G $\frac{9}{25}$

   H $\frac{3}{6}$

   J Not here

3. Which pair contains numbers that are equivalent?

   A $3\frac{2}{3}, \frac{11}{3}$

   B $\frac{13}{4}, 2\frac{3}{4}$

   C $4\frac{2}{5}, \frac{21}{5}$

   D $\frac{16}{7}, 3\frac{1}{2}$

4. Eldora measured the length of a remote control sailboat course as $\frac{5}{8}$ of a mile. What is the decimal equivalent of $\frac{5}{8}$?

   F 0.5

   G 0.58

   H 0.625

   J 0.8

5. In which pair are both numbers equivalent to $\frac{3}{4}$?

   A 0.25; 25%

   B 30%; $\frac{3}{10}$

   C 80%; $\frac{12}{15}$

   D 75%; $\frac{12}{16}$

6. Evan has a bag of fruit. He has 9 apples, 4 oranges, and 3 bananas. What fraction represents the pieces of fruit that are oranges?

   F $\frac{4}{18}$

   G $\frac{1}{4}$

   H $\frac{5}{16}$

   J $\frac{3}{4}$

7. Which numbers are in order from least to greatest?

   A $0.3, \frac{3}{8}, \frac{2}{5}$

   B $\frac{2}{5}, 0.3, \frac{3}{8}$

   C $\frac{3}{8}, \frac{2}{5}, 0.3$

   D $\frac{2}{5}, \frac{3}{8}, 0.3$

8. If 5 packages of hot dogs cost $9.25, what is the cost of 1 package?

   F $0.92

   G $1.15

   H $1.85

   J $2.10

Write What You Know

9. Explain how the GCF can help you write a fraction in simplest form. Then write $\frac{135}{144}$ in simplest form.

10. Explain how the LCM can help you compare two fractions. Then compare $\frac{5}{6}$ and $\frac{7}{9}$.

CHAPTER 9

Add and Subtract Fractions and Mixed Numbers

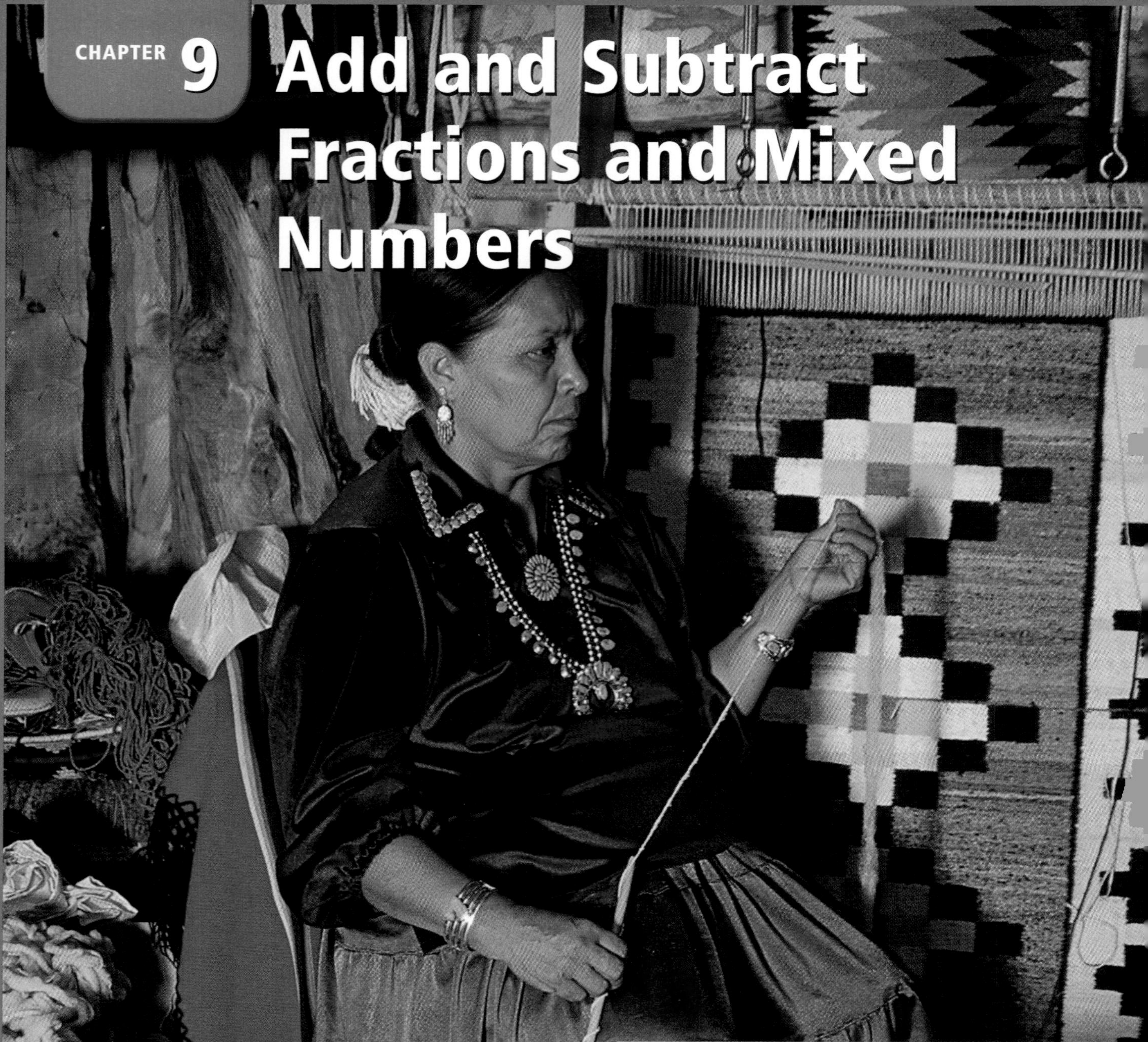

Traditional Navajo rugs often have geometric patterns woven into them. Completed by hand, a 3 ft x 5 ft rug can require hundreds of hours of work.

PROBLEM SOLVING One of the patterns from the rug in the picture is shown to the right. What fraction of the squares in the picture are dark brown border squares? What benchmark fraction is this closest to?

DATA LINK

38. Estimate the difference between 8 and $2\frac{7}{18}$.

39. About how much more is $8\frac{7}{10}$ feet than $2\frac{3}{8}$ feet?

40. Estimate the sum of $2\frac{1}{8}$, $7\frac{7}{9}$, $1\frac{7}{8}$, and $8\frac{1}{4}$.

41. Estimate the sum of 8.5, $5\frac{1}{4}$, $4\frac{2}{3}$, and 6.75.

42. Estimate the difference between $15\frac{3}{5}$ and $4\frac{2}{9}$.

43. Estimate the sum of $18\frac{3}{10}$, $5\frac{7}{8}$, $2\frac{1}{4}$, and $3\frac{4}{5}$.

44. About how much more in feet is $24\frac{1}{3}$ yd than $11\frac{5}{6}$ yd?

Problem Solving ▶ Applications

Use Data The table shows data on the longest ski trails at five different mountains in the United States. For 45–46, use the table.

Mountain	Longest Ski Trail (mi)
Beaver Creek, CO	$2\frac{3}{4}$
Killington, VT	$10\frac{1}{5}$
Mammoth, CA	$2\frac{1}{2}$
Taos, NM	$5\frac{1}{4}$
Whiteface, NY	3

Ski Trails

45. About how much longer is the trail at Killington than the trails at Beaver Creek and Taos?

46. Find the median of the trail lengths.

47. Leo needs $\frac{3}{4}$ yd of blue fabric, $\frac{7}{10}$ yd of purple fabric, and $\frac{1}{5}$ yd of white fabric for a sewing project. About how much fabric does he need for the sewing project?

48. **Write a problem** about everyday life at home that can be solved by estimating fractions. Exchange with a classmate and solve.

49. Merika mowed her yard every ninth day, beginning on March 6 and continuing through April. On what other days did she mow her yard in March? in April?

MIXED REVIEW AND TEST PREP

Write the fraction or decimal as a percent. (p. 169)

50. $\frac{3}{4}$

51. 0.45

52. A survey question reads "Is blue your favorite color?" Is the question biased or unbiased? (p. 98)

53. TEST PREP The band members are setting up chairs for a concert in the school auditorium. They can put 25 chairs in each row. How many rows will they need to seat 350 people? (p. 22)

A 10   **B** 14   **C** 18   **D** 25

54. TEST PREP Find the GCF for 80, 96, and 112. (p.150)

F 6   **G** 8   **H** 14   **J** 16

EXTRA PRACTICE page H40, Set A

Model Addition and Subtraction

Explore how to use fraction bars to add and subtract fractions with unlike denominators.

You need fraction bars.

Vocabulary

unlike fractions

Remember that the LCM is the least of the common multiples of two or more numbers.
2: 2, 4, 6, 8, 10, 12, . . .
3: 3, 6, 9, 12, 15, . . .
The LCM of 2 and 3 is 6. The LCM can be used to write common denominators of two or more fractions.

QUICK REVIEW

Find the LCM for each set of numbers.

1. 2, 8   **2.** 6, 9

3. 4, 15   **4.** 4, 10

5. 2, 3, 10

Fractions with the same denominator, such as $\frac{5}{9}$ and $\frac{4}{9}$, are called like fractions. Fractions with different denominators are called **unlike fractions**. You can use fraction bars to rename the denominators before adding.

Activity 1

Find $\frac{1}{6} + \frac{1}{4}$.

• Use fraction bars to show both fractions.

• Which fraction bars fit exactly across $\frac{1}{6}$ and $\frac{1}{4}$? Think about the LCM of 6 and 4.

• What is $\frac{1}{6} + \frac{1}{4}$?

$\frac{1}{6} = \frac{2}{12}$ $\frac{1}{4} = \frac{3}{12}$

Think and Discuss

• Look at the model for $\frac{1}{6} + \frac{1}{4}$. What do you know about $\frac{1}{6}$ and $\frac{2}{12}$? about $\frac{1}{4}$ and $\frac{3}{12}$?

• How are the denominators of $\frac{1}{6}$, $\frac{1}{4}$, and $\frac{1}{12}$ related? (HINT: Think about common multiples.)

Practice

Use fraction bars to find the sum. Draw a diagram of your model.

1. $\frac{1}{4} + \frac{1}{2}$ **2.** $\frac{1}{2} + \frac{1}{3}$ **3.** $\frac{1}{2} + \frac{2}{5}$ **4.** $\frac{2}{3} + \frac{1}{6}$

5. $\frac{1}{3} + \frac{1}{4}$ **6.** $\frac{3}{8} + \frac{1}{4}$ **7.** $\frac{1}{6} + \frac{1}{2}$ **8.** $\frac{3}{8} + \frac{1}{2}$

9. $\frac{1}{5} + \frac{1}{2}$ **10.** $\frac{3}{4} + \frac{1}{6}$ **11.** $\frac{1}{3} + \frac{1}{6}$ **12.** $\frac{5}{8} + \frac{1}{4}$

Fraction bars also can be used to subtract unlike fractions.

TECHNOLOGY LINK

More Practice: Use E-Lab, *Addition and Subtraction of Unlike Fractions.* **www.harcourtschool.com/ elab2002**

Activity 2

Find $\frac{1}{2} - \frac{1}{5}$.

- Use fraction bars to show $\frac{1}{2}$ and $\frac{1}{5}$.

- Which fraction bars fit exactly across $\frac{1}{2}$ and $\frac{1}{5}$? Think about the LCM.

- Compare $\frac{5}{10}$ and $\frac{2}{10}$. How much more is $\frac{5}{10}$ than $\frac{2}{10}$?

- What is $\frac{5}{10} - \frac{2}{10}$? What is $\frac{1}{2} - \frac{1}{5}$?

Think and Discuss

- How are the denominators of $\frac{1}{2}$, $\frac{1}{5}$, and $\frac{1}{10}$ related?

- Look at the model of $\frac{3}{4} - \frac{1}{3}$. Which fraction bars will fit exactly across $\frac{3}{4}$ and $\frac{1}{3}$? Explain.

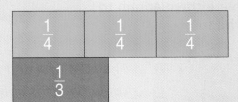

- Which fraction bars will fit exactly across $\frac{1}{2} - \frac{1}{4}$?

Practice
Use fraction bars to subtract. Draw a diagram of your model.

1. $\frac{3}{4} - \frac{1}{3}$ **2.** $\frac{2}{5} - \frac{1}{10}$ **3.** $\frac{1}{3} - \frac{1}{4}$ **4.** $\frac{1}{2} - \frac{1}{3}$

5. $\frac{1}{2} - \frac{2}{5}$ **6.** $\frac{1}{2} - \frac{5}{12}$ **7.** $\frac{1}{4} - \frac{1}{6}$ **8.** $\frac{1}{3} - \frac{1}{6}$

9. $\frac{1}{4} - \frac{1}{8}$ **10.** $\frac{1}{2} - \frac{1}{4}$ **11.** $\frac{5}{6} - \frac{1}{3}$ **12.** $\frac{7}{8} - \frac{3}{4}$

MIXED REVIEW AND TEST PREP

Compare. Write <, >, or = for each ●. (p. 169)

13. $\frac{3}{8}$ ● 0.375   **14.** 0.15 ● $\frac{1}{4}$ **15.** 0.6 ● $\frac{11}{20}$

16. Multiply. 9.25×3.2 (p. 70)

⭐ **17. TEST PREP** Of the 80 Moorers at their family reunion, there are 8 fewer females than males. How many males are at the reunion? (p. 22)

     A 32   **B** 36   **C** 40   **D** 44

Add and Subtract Fractions

Learn how to add and subtract fractions.

Vocabulary

least common denominator (LCD)

Solve. Write your answer in simplest form.

1. $\frac{1}{7} + \frac{5}{7}$ 2. $\frac{5}{6} - \frac{1}{6}$ 3. $\frac{7}{16} - \frac{5}{16}$ 4. $\frac{4}{12} + \frac{8}{12}$ 5. $\frac{8}{9} - \frac{3}{9}$

$\mathbf{Y}$ou can use a diagram to add and subtract fractions. To help you, think about the LCM of the denominators and about equivalent fractions.

EXAMPLE 1

Kayla is making two recipes. One recipe calls for $\frac{1}{4}$ c raisins, and the other recipe calls for $\frac{1}{3}$ c raisins. How many total cups of raisins does Kayla need?

Complete the diagram to find the sum of $\frac{1}{4}$ and $\frac{1}{3}$.

The LCM of 4 and 3 is 12.
Draw twelfths under $\frac{1}{4}$ and $\frac{1}{3}$.

THINK: $\frac{1}{4} = \frac{3}{12}$ and $\frac{1}{3} = \frac{4}{12}$.

So, Kayla needs $\frac{7}{12}$ c of raisins.

Math Idea ▶ To add fractions without using diagrams, you can write equivalent fractions by using the **least common denominator**, or **LCD**. The LCD is the LCM of the denominators.

EXAMPLE 2

Find $\frac{1}{2} + \frac{3}{5}$.

Estimate. Each fraction is close to $\frac{1}{2}$, so the sum is about 1.

$\frac{1}{2} = \frac{1 \times 5}{2 \times 5} = \frac{5}{10}$ *The LCM of 2 and 5 is 10, so the LCD*
$+\frac{3}{5} = +\frac{3 \times 2}{5 \times 2} = +\frac{6}{10}$ *of $\frac{1}{2}$ and $\frac{3}{5}$ is 10. Multiply to write equivalent fractions using the LCD.*

$\frac{1}{2} = \frac{5}{10}$ *Add the numerators.*
$+\frac{3}{5} = +\frac{6}{10}$ *Write the sum over the denominator.*

$\frac{11}{10}$, or $1\frac{1}{10}$ *Write the answer as a fraction or as a mixed number.*

Compare the answer to your estimate. Since $1\frac{1}{10}$ is close to the estimate of 1, the answer is reasonable. So, $\frac{1}{2} + \frac{3}{5} = 1\frac{1}{10}$.

You can use a similar method to subtract unlike fractions.

TECHNOLOGY LINK

More Practice: Use **Mighty Math Number Heroes**, *Fraction Fireworks*, Level W.

EXAMPLE 3

Kayla is preparing a pasta dish for a small dinner party. Kayla has $\frac{1}{2}$ c of grated mozzarella cheese. The recipe calls for $\frac{2}{3}$ c of grated mozzarella cheese. How much more mozzarella cheese does Kayla need to grate?

Find $\frac{2}{3} - \frac{1}{2}$.

Estimate. $\frac{2}{3}$ is a little more than $\frac{1}{2}$, so the difference is close to 0.

$$\frac{2}{3} = \frac{2 \times 2}{3 \times 2} = \frac{4}{6}$$ *The LCD of $\frac{2}{3}$ and $\frac{1}{2}$ is 6.*

$$-\frac{1}{2} = -\frac{1 \times 3}{2 \times 3} = -\frac{3}{6}$$ *Multiply to find the equivalent fractions using the LCD.*

$$\begin{array}{r} \frac{2}{3} = \frac{4}{6} \\ -\frac{1}{2} = -\frac{3}{6} \\ \hline \frac{1}{6} \end{array}$$ *Subtract the numerators. Write the difference over the denominator.*

Compare the answer to your estimate. Since $\frac{1}{6}$ is close to the estimate of 0, the answer is reasonable.

So, Kayla needs to grate $\frac{1}{6}$ c more cheese.

EXAMPLE 4

A. $$\begin{array}{r} \frac{5}{6} = \frac{5 \times 3}{6 \times 3} = \frac{15}{18} \\ -\frac{7}{9} = -\frac{7 \times 2}{9 \times 2} = -\frac{14}{18} \\ \hline \frac{1}{18} \end{array}$$

B. $$\begin{array}{r} \frac{5}{12} = \frac{5}{12} \\ -\frac{1}{4} = -\frac{1 \times 3}{4 \times 3} = -\frac{3}{12} \\ \hline \frac{2}{12} = \frac{1}{6} \end{array}$$

CHECK FOR UNDERSTANDING

Think and Discuss ▶ Look back at the lesson to answer each question.

1. **Tell** how much more cheese Kayla would need to grate if she had $\frac{1}{4}$ c of grated cheese.

2. **REASONING** **Tell** when the LCD of two fractions is equal to the product of the denominators.

Guided Practice ▶ Use the LCD to rewrite the problem by using equivalent fractions.

3. $\frac{7}{10} + \frac{1}{5}$ 4. $\frac{1}{3} + \frac{1}{8}$ 5. $\frac{4}{5} - \frac{1}{3}$

Write the sum or difference in simplest form. Estimate to check.

6. $\frac{1}{5} + \frac{3}{5}$　　7. $\frac{7}{9} - \frac{4}{9}$　　8. $\frac{7}{9} - \frac{1}{6}$　　9. $\frac{2}{3} + \frac{3}{4}$

10. $\frac{3}{4} - \frac{3}{8}$　　11. $\frac{2}{5} - \frac{1}{3}$　　12. $\frac{2}{5} + \frac{2}{4}$　　13. $\frac{4}{9} + \frac{1}{3}$

PRACTICE AND PROBLEM SOLVING

Independent ▶ Practice

Use the LCD to rewrite the problem by using equivalent fractions.

14. $\frac{9}{10} - \frac{1}{5}$　　15. $\frac{6}{7} - \frac{3}{4}$　　16. $\frac{1}{4} + \frac{5}{8}$

Write the sum or difference in simplest form. Estimate to check.

17. $\frac{1}{6} + \frac{2}{3}$　　18. $\frac{4}{7} - \frac{1}{7}$　　19. $\frac{1}{2} + \frac{3}{10}$　　20. $\frac{1}{3} - \frac{1}{4}$

21. $\frac{5}{7} - \frac{1}{2}$　　22. $\frac{1}{3} + \frac{2}{3}$　　23. $\frac{6}{10} - \frac{4}{10}$　　24. $\frac{3}{8} + \frac{1}{3}$

25. $\frac{3}{4} + \frac{5}{8}$　　26. $\frac{7}{12} + \frac{2}{3}$　　27. $1 - \frac{3}{8}$　　28. $\frac{1}{2} - \frac{2}{5}$

29. $\frac{4}{5} - \frac{1}{3}$　　30. $\frac{1}{4} + \frac{2}{3}$　　31. $\frac{5}{9} - \frac{1}{3}$　　32. $\frac{3}{8} + \frac{3}{20}$

33. Find the sum of $\frac{1}{8}$, $\frac{3}{8}$, and $\frac{2}{8}$.　　34. Find $\frac{1}{2} + \frac{2}{3} + \frac{1}{6}$.

35. Find $\frac{3}{4} + 0.5 + 0.75$.　　36. Find $0.6 - \frac{3}{8}$.

37. How much longer than $\frac{1}{4}$ mile is $\frac{2}{3}$ mile?

Solve each equation mentally. Write the answer in simplest form.

38. $p + \frac{1}{4} = \frac{3}{4}$　　39. $r = \frac{5}{12} + \frac{7}{12}$　　40. $\frac{4}{5} - q = \frac{2}{5}$

41. $c = \frac{7}{10} - \frac{1}{10}$　　42. $\frac{3}{7} + s = \frac{5}{7}$　　43. $m - \frac{1}{6} = \frac{5}{6}$

Problem Solving ▶ Applications

Use Data **For 44–46, use the recipe at the right.**

44. Yuji has $\frac{7}{8}$ c of orange juice. How much orange juice does he have left to drink after he doubles the recipe for fruit cups?

45. How many total teaspoons of vanilla and orange extract does Yuji need to make the fruit cups?

46. Yuji used $\frac{1}{4}$ tsp of orange extract. By how much did he exceed the amount of orange extract in the recipe?

Fruit Cups
2 cups of orange sections
1 cup blueberries
$\frac{1}{4}$ cup orange juice
1 tbsp sugar
1 tsp lemon juice
$\frac{1}{2}$ tsp vanilla extract
$\frac{1}{8}$ tsp orange extract

47. Each week, Reina spends $\frac{2}{3}$ of her allowance on school lunches and saves $\frac{1}{5}$ of it. What fraction of her allowance is left? Which operation(s) did you use? Why?

48. **? What's the Question?** In Mrs. Lucero's class, $\frac{1}{10}$ of the students are wearing blue shirts and $\frac{3}{5}$ of the students are wearing white shirts. The answer is $\frac{3}{10}$ of the class.

49. *REASONING* Joaquin is thinking of two numbers. Each number is between 21 and 30. The GCF of the numbers is 4. What are the numbers?

50. One cup of whole milk contains 166 calories and one cup of skim milk contains 88 calories. How many more calories are there in 4 cups of whole milk than in 1 quart of skim milk?

MIXED REVIEW AND TEST PREP

For 51-52, write in simplest form. (p.160)

51. $\frac{36}{81}$

52. $\frac{95}{200}$

53. Subtract. $9{,}285 - 3{,}153$ (p. 20)

54. **TEST PREP** A teacher selects 50 students by picking names out of a box without looking. What type of sample is this? (p. 94)

   A biased   **B** convenience   **C** random   **D** systematic

55. **TEST PREP** Eric skated 500 meters in 37.14 seconds. Andre skated the same distance in 37.139 seconds, and Al in 37.12 seconds. What is the correct order of their times in seconds, from fastest to slowest times? (p. 52)

   F 37.14, 37.139, 37.12   **H** 37.12, 37.139, 37.14

   G 37.12, 37.14, 37.139   **J** 37.139, 37.12, 37.14

PROBLEM SOLVING Thinker's Corner

DOMINO FRACTIONS
Materials: set of dominoes

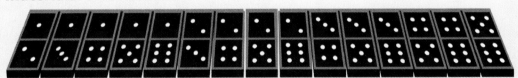

Dominoes have been played for centuries throughout the world. The Chinese played with them in the 12th century, and a set was even found in a tomb from Ancient Egypt. To play this domino game, think of each of the 15 dominoes shown above as a fraction. If a domino shows 2 circles and 4 circles, the fraction is $\frac{2}{4}$. Play with a partner to combine 5 dominoes at a time to form a fraction sum of $2\frac{1}{2}$. Then break the 15 dominoes into 3 sets of 5, each of which has a sum of $2\frac{1}{2}$.

EXTRA PRACTICE page H40, Set B

Add and Subtract Mixed Numbers

Learn how to add and subtract mixed numbers.

Write in simplest form.

1. $\frac{4}{16}$ 2. $\frac{14}{21}$ 3. $\frac{10}{16}$

4. $\frac{10}{30}$ 5. $\frac{25}{30}$

$\mathbf{A}$ flock of birds is migrating from Canada to Florida for the winter. One day they fly $2\frac{3}{4}$ hours before stopping to rest. Then they fly $1\frac{1}{6}$ more hours. For how long do the birds fly?

One Way   You can draw a diagram.

EXAMPLE 1

Remember that to write a fraction as a mixed number, you divide the numerator by the denominator. The quotient is the mixed number.

$$\frac{13}{3} \rightarrow 3\overline{)13} \quad \begin{array}{r} 4\frac{1}{3} \\ \underline{-12} \\ 1 \end{array}$$

Show $2\frac{3}{4} + 1\frac{1}{6}$.

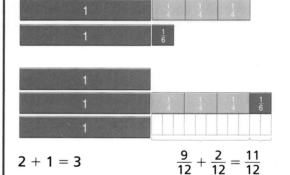

Combine whole numbers. Combine fractions. Draw equivalent fractions with the LCD, 12.

$2 + 1 = 3$ $\frac{9}{12} + \frac{2}{12} = \frac{11}{12}$ *Add fractions. Add whole numbers.*

So, the birds fly for $3\frac{11}{12}$ hours.

Another Way   You can use the LCD to write equivalent fractions.

EXAMPLE 2

Find $4\frac{2}{3} + 5\frac{4}{5}$.

$$\begin{array}{r} 4\frac{2}{3} = 4\frac{10}{15} \\ +5\frac{4}{5} = +5\frac{12}{15} \\ \hline 9\frac{22}{15} = 9 + 1\frac{7}{15} = 10\frac{7}{15} \end{array}$$

Write equivalent fractions, using the LCD, 15.
Add fractions. Add whole numbers.
Rename the fraction as a mixed number. Rewrite the sum.

So, $4\frac{2}{3} + 5\frac{4}{5} = 10\frac{7}{15}$.

Subtract Mixed Numbers

One Way You can use diagrams to subtract mixed numbers.

Find $1\frac{1}{2} - 1\frac{1}{10}$.

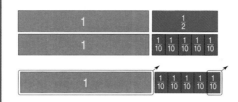

Draw $1\frac{1}{2}$.
Find the LCD for $\frac{1}{2}$ and $\frac{1}{10}$.
Change the half to tenths.

Subtract $1\frac{1}{10}$ from $1\frac{5}{10}$.

So, $1\frac{1}{2} - 1\frac{1}{10} = \frac{4}{10}$, or $\frac{2}{5}$.

- Draw a diagram to find $2\frac{1}{2} - 1\frac{1}{3}$.

Another Way When you subtract mixed numbers with unlike fractions, you can use the LCD to write equivalent fractions.

The rusty-spotted cat, the smallest meat-eating feline, lives in southern India and Sri Lanka. An adult has a head-and-body length of $13\frac{2}{5}$ in. to $18\frac{9}{10}$ in. Find the difference between the greatest and least lengths.

Find $18\frac{9}{10} - 13\frac{2}{5}$.

Estimate. $18\frac{9}{10}$ is close to 19 and $13\frac{2}{5}$ is close to 13. So, the difference is about $19 - 13$, or 6.

$$18\frac{9}{10} = \ \ 18\frac{9}{10}$$

Write equivalent fractions, using the LCD, 10.

$$-13\frac{2}{5} = -13\frac{4}{10}$$

Subtract the fractions.

$$5\frac{5}{10} = 5\frac{1}{2}$$

Subtract the whole numbers.

The exact answer is reasonable because it is close to the estimate of 6. So, the difference between the greatest and least lengths is $5\frac{1}{2}$ in.

CHECK FOR UNDERSTANDING

Think and Discuss ▶ Look back at the lesson to answer each question.

1. **Explain** why you must find equivalent fractions to add $1\frac{1}{5} + 1\frac{1}{2}$.

2. **Tell** how you know that $6\frac{1}{2} - 4\frac{1}{4}$ is more than 2.

Guided Practice ▶ Draw a diagram to find each sum or difference. Write the answer in simplest form.

3. $2\frac{3}{5} + 1\frac{1}{5}$ 4. $1\frac{1}{4} + 2\frac{2}{3}$ 5. $2\frac{4}{5} - 1\frac{1}{2}$

Write the sum or difference in simplest form.
Estimate to check.

6. $1\frac{1}{8} + 1\frac{5}{8}$ 7. $2\frac{1}{4} + 4\frac{1}{3}$ 8. $5\frac{3}{8} - 1\frac{1}{4}$

9. $4\frac{1}{3} - 3\frac{1}{6}$ 10. $3\frac{3}{4} + 4\frac{5}{12}$ 11. $6\frac{5}{6} - 5\frac{7}{9}$

PRACTICE AND PROBLEM SOLVING

Independent Practice ▶ Draw a diagram to find each sum or difference. Write the answer in simplest form.

12. $1\frac{5}{12} + 1\frac{1}{4}$ 13. $1\frac{1}{3} + 1\frac{1}{6}$ 14. $4\frac{1}{2} - 2\frac{2}{5}$

Write the sum or difference in simplest form.
Estimate to check.

15. $4\frac{1}{2} + 3\frac{4}{5}$ 16. $4\frac{1}{3} - 2\frac{1}{4}$ 17. $5\frac{5}{6} + 4\frac{2}{9}$

18. $3\frac{1}{4} - 1\frac{1}{6}$ 19. $7\frac{1}{2} - 3\frac{2}{5}$ 20. $3\frac{2}{7} + 8\frac{1}{3}$

21. $7\frac{3}{4} + 3\frac{2}{5}$ 22. $5\frac{5}{6} - 2\frac{7}{9}$ 23. $4\frac{5}{7} + 3\frac{1}{2}$

24. How much greater is $5\frac{3}{4}$ than 3?

25. What is the sum of $25\frac{3}{8}$ and $2\frac{3}{4}$?

26. What is the sum of $4\frac{5}{8}$ and 7.8?

Find the missing number and identify which property of addition you used.

27. $3\frac{7}{8} + \blacksquare = 2\frac{1}{4} + 3\frac{7}{8}$ 28. $3\frac{3}{4} + 0 = \blacksquare$ 29. $\left(\frac{2}{3} + 1\frac{5}{6}\right) + \frac{1}{6} = \frac{2}{3} + \left(\blacksquare + \frac{1}{6}\right)$

Problem Solving Applications ▶ **Use Data** The graph shows the head-and-body lengths of five small mammals. For 30–31, use the graph.

30. How much longer is the harvest mouse than the Kitti's hognosed bat?

31. The masked shrew is $1\frac{2}{3}$ in. long. Is it longer or shorter than the little brown bat? How much? Which operation did you use? Why?

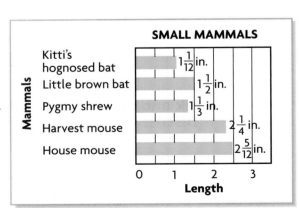

SMALL MAMMALS

Mammals:
- Kitti's hognosed bat — $1\frac{1}{12}$ in.
- Little brown bat — $1\frac{1}{2}$ in.
- Pygmy shrew — $1\frac{1}{3}$ in.
- Harvest mouse — $2\frac{1}{4}$ in.
- House mouse — $2\frac{5}{12}$ in.

Length: 0, 1, 2, 3

32. On its way to the shore, a sea turtle traveled $4\frac{1}{4}$ hr the first day. The second day, the turtle traveled $3\frac{1}{2}$ hr. How many hours did the sea turtle travel in the two days?

33. Mrs. Myers used $1\frac{1}{2}$ c of flour to make muffins, $4\frac{1}{4}$ c to make bread, and $\frac{3}{4}$ c to make gravy. If she had $9\frac{3}{4}$ c before she started the meal, how much flour does Mrs. Myers have left?

34. **?** **What's the Error?** Izumi added $3\frac{1}{4}$ and $2\frac{2}{3}$ and got $5\frac{3}{12}$. Explain the error. What is the correct sum?

35. Betty, Jim, Manuel, and Rosa won the first four prizes in a design contest. Jim won second prize. Manuel did not win third prize. Rosa won fourth prize. What prize did Betty win?

36. Alexis needs a new blender. She found the same blender on sale at five different stores. The prices are $22.95, $21.85, $22.05, $20.95, and $21.99. Order the prices from least to greatest.

MIXED REVIEW AND TEST PREP

37. Find the sum of $\frac{2}{3}$ and $\frac{2}{5}$. (p. 182)

38. Subtract. $425.2 - 51.05$ (p. 66)

39. Order $\frac{1}{2}$, $\frac{4}{5}$, and $\frac{2}{3}$ from greatest to least. (p. 166)

40. TEST PREP Find the value of $(6 + 4)^2 \div 5$. (p. 44)

  A 4   **B** 4.4   **C** 9.2   **D** 20

41. TEST PREP Which is the solution of $x + 3 = 10$? (p. 30)

  F $x = 3$ **G** $x = 7$ **H** $x = 10$ **J** $x = 13$

PROBLEM SOLVING LINKUP to Reading

Strategy • Choose Relevant Information

Sometimes a word problem contains more information than you need. You must decide which information is relevant, or needed to solve the problem.

The koala of eastern Australia feeds mostly on eucalyptus leaves. It nibbles on about 6 of the 500 species of eucalyptus per day and selects certain trees and leaves over others to find the $1\frac{1}{4}$ pounds of food that it needs. Suppose the koala finds only $1\frac{1}{8}$ pounds of food by the end of the day. How many more pounds of eucalyptus leaves does the koala need?

1. What does the problem ask you to find?

2. Identify relevant information.

3. What information is not relevant?

4. Solve the problem.

EXTRA PRACTICE page H40, Set C

Rename to Subtract

MATH LAB

Explore how to use fraction bars to subtract mixed numbers.

You need fraction bars.

QUICK REVIEW

Write the difference in simplest form.

1. $\frac{4}{7} - \frac{1}{7}$ **2.** $\frac{3}{4} - \frac{1}{4}$ **3.** $\frac{7}{10} - \frac{5}{10}$

4. $\frac{8}{12} - \frac{4}{12}$ **5.** $\frac{5}{9} - \frac{2}{9}$

Sometimes you need to rename mixed numbers before you can subtract.

Activity

A. Find $2\frac{1}{5} - 1\frac{4}{5}$.

- Use fraction bars to model $2\frac{1}{5}$.

 ← $2\frac{1}{5}$

- Here is another way to model $2\frac{1}{5}$.

 ← $1\frac{6}{5}$

- From which model can you subtract $1\frac{4}{5}$?

- Subtract $1\frac{4}{5}$ from $1\frac{6}{5}$. What is $2\frac{1}{5} - 1\frac{4}{5}$?

B. Find $2\frac{1}{6} - 1\frac{5}{12}$.

- Use fraction bars to model $2\frac{1}{6}$.

 ← $2\frac{1}{6}$

- Since you are subtracting twelfths, think of the LCD for $\frac{1}{6}$ and $\frac{5}{12}$. Change the sixths to twelfths.

 ← $2\frac{2}{12}$

- Can you subtract $1\frac{5}{12}$ from either of these models?

- Here is another way to model $2\frac{2}{12}$.

 ← $1\frac{14}{12}$

- Subtract $1\frac{5}{12}$ from $1\frac{14}{12}$. What is $2\frac{1}{6} - 1\frac{5}{12}$?

C. Find $2\frac{1}{4} - 1\frac{3}{8}$.

- Use fraction bars to model $2\frac{1}{4}$.

| 1 | 1 | $\frac{1}{4}$ | $\leftarrow 2\frac{1}{4}$ |

TECHNOLOGY LINK

More Practice: Use E-Lab, *Subtraction of Mixed Numbers.*
www.harcourtschool.com/elab2002

- Since you are subtracting eighths, think of the LCD for $\frac{1}{4}$ and $\frac{3}{8}$. Change the fourth to eighths.

| 1 | 1 | $\frac{1}{8}$ $\frac{1}{8}$ | $\leftarrow 2\frac{2}{8}$ |

- Can you subtract $1\frac{3}{8}$ from either of these models?

- Here is another way to model $2\frac{2}{8}$.

| 1 | $\frac{1}{8}$ $\frac{1}{8}$ $\frac{1}{8}$ $\frac{1}{8}$ $\frac{1}{8}$ $\frac{1}{8}$ $\frac{1}{8}$ $\frac{1}{8}$ $\frac{1}{8}$ $\frac{1}{8}$ | $\leftarrow 1\frac{10}{8}$ |

- Subtract $1\frac{3}{8}$ from $1\frac{10}{8}$. What is $2\frac{1}{4} - 1\frac{3}{8}$?

Think and Discuss

- Think about $2\frac{5}{6} - 1\frac{1}{6}$. Do you need to rename before you subtract? Explain.

- Think about $5\frac{2}{5} - 3\frac{4}{5}$. Do you need to rename before you subtract? Explain.

Practice

Use fraction bars to subtract. Draw a diagram of your model.

1. $3\frac{1}{3} - 1\frac{2}{3}$ **2.** $3\frac{3}{8} - \frac{3}{4}$ **3.** $3\frac{1}{9} - 1\frac{2}{3}$ **4.** $2\frac{3}{8} - 1\frac{1}{2}$

5. $1\frac{1}{2} - \frac{4}{5}$ **6.** $3\frac{1}{5} - 1\frac{3}{10}$ **7.** $2\frac{2}{3} - 1\frac{3}{4}$ **8.** $3\frac{1}{12} - 2\frac{5}{6}$

MIXED REVIEW AND TEST PREP

9. Draw a diagram to find the sum of $2\frac{1}{2}$ and $1\frac{1}{4}$. Write the answer in simplest form. (p. 186)

10. Write $\frac{2}{5}$ as a decimal and as a percent. (p. 169)

11. Evaluate $a \times b$ for $a = 4.2$ and $b = 5.1$. (p. 82)

12. Write the prime factorization of 245 in exponent form. (p. 148)

⭐**13. TEST PREP** Tyra earns $5.50 per hour working part-time for a surf shop. Last week she worked 3 hr on Tuesday and twice as long on Saturday. How much did she earn last week? (p. 22)

   A $33.00   **B** $44.00   **C** $49.50   **D** $55.50

LESSON 9.6

Subtract Mixed Numbers

Learn how to subtract mixed numbers involving renaming.

QUICK REVIEW

Write the number that makes the fraction equivalent to $\frac{1}{2}$.

1. $\frac{\blacksquare}{8}$ 2. $\frac{\blacksquare}{6}$ 3. $\frac{\blacksquare}{20}$ 4. $\frac{\blacksquare}{14}$ 5. $\frac{\blacksquare}{100}$

A 1993 blizzard called The Storm of the Century brought record cold, wind, and snow to many cities in the southern and eastern United States.

How much more snow did Mount LeConte receive than Birmingham received?

Find $4\frac{1}{3} - 1\frac{5}{12}$.

STORM OF THE CENTURY	
City	**Snow (ft)**
Birmingham, AL	$1\frac{5}{12}$
Asheville, NC	$1\frac{7}{12}$
Pittsburgh, PA	$2\frac{1}{2}$
Syracuse, NY	$3\frac{7}{12}$
Mount LeConte, TN	$4\frac{1}{3}$

Estimate. $4\frac{1}{3}$ is close to $4\frac{1}{2}$ and $1\frac{5}{12}$ is close to $1\frac{1}{2}$. So, the difference is about $4\frac{1}{2} - 1\frac{1}{2}$, or 3.

Subtract.

$$
\begin{array}{rcl}
4\frac{1}{3} & = & 4\frac{4}{12} \\
-1\frac{5}{12} & = & -1\frac{5}{12}
\end{array}
$$

The LCD of $\frac{1}{3}$ and $\frac{5}{12}$ is 12.
Write equivalent fractions, using the LCD, 12.

$$
\begin{array}{rcccl}
4\frac{1}{3} & = & 4\frac{4}{12} & = & 3\frac{16}{12} \\
-1\frac{5}{12} & = & -1\frac{5}{12} & = & -1\frac{5}{12} \\
\hline
& & & & 2\frac{11}{12}
\end{array}
$$

Since $\frac{5}{12}$ is greater than $\frac{4}{12}$, rename $4\frac{4}{12}$.
$4\frac{4}{12} = 3 + \frac{12}{12} + \frac{4}{12} = 3\frac{16}{12}$.
Subtract the fractions.
Subtract the whole numbers.

The answer is reasonable because it is close to the estimate of 3 ft. So, Mount LeConte received $2\frac{11}{12}$ ft more snow than Birmingham.

EXAMPLE

Jake is running in a 5-km race. So far he has run $1\frac{4}{5}$ km. How far does he have to go?

Find the difference. $5 - 1\frac{4}{5}$

$$
\begin{array}{rcl}
5 & = & 4\frac{5}{5} \\
-1\frac{4}{5} & = & -1\frac{4}{5} \\
\hline
& & 3\frac{1}{5}
\end{array}
$$

Since you are subtracting fifths, rename 5 as $4\frac{5}{5}$.

Subtract the fractions.
Subtract the whole numbers.

So, Jake has $3\frac{1}{5}$ km to go.

Think and Discuss ▶ Look back at the lesson to answer each question.

1. **Explain** how to rename $3\frac{1}{3}$ so you could subtract $1\frac{2}{3}$.

2. **What if** you wanted to find $4\frac{5}{12} - 2\frac{3}{8}$? What equivalent fractions would you write using the LCD?

Guided Practice ▶ Write the difference in simplest form. Estimate to check.

3. $4\frac{1}{3} - 2\frac{1}{4}$

4. $6 - 2\frac{2}{3}$

5. $7\frac{3}{10} - 3\frac{2}{5}$

6. $6\frac{1}{5} - 3\frac{7}{10}$

7. $8\frac{5}{6} - 4\frac{8}{9}$

8. $16\frac{3}{8} - 7\frac{1}{2}$

PRACTICE AND PROBLEM SOLVING

Independent Practice ▶ Write the difference in simplest form. Estimate to check.

9. $3\frac{1}{6} - 1\frac{1}{4}$

10. $5\frac{1}{2} - 3\frac{7}{10}$

11. $12\frac{1}{9} - 7\frac{1}{3}$

12. $4\frac{1}{4} - 2\frac{2}{5}$

13. $8\frac{1}{4} - 5\frac{2}{3}$

14. $7\frac{5}{9} - 2\frac{5}{6}$

15. $11\frac{1}{4} - 9\frac{7}{8}$

16. $5.25 - 2\frac{3}{8}$

17. $6.2 - 3\frac{1}{2}$

Evaluate each expression for $j = 5\frac{1}{2}$, $k = 4\frac{3}{5}$, and $m = 2\frac{7}{10}$.

18. $j - k$

19. $j - m$

20. $k - m$

Problem Solving Applications ▶

21. Whit usually drives $4\frac{7}{8}$ mi on the expressway to work. Sometimes traffic is bad due to weather conditions and he takes another route which is $5\frac{3}{4}$ mi long. How much shorter is his usual route?

22. ✎ **Write About It**   Why do you write equivalent fractions before you rename? Can you rename before you write equivalent fractions?

23. **Number Sense** Prime numbers that differ by 2, such as 3 and 5, or 59 and 61, are called twin primes. Write two other pairs of twin primes between 1 and 50.

MIXED REVIEW AND TEST PREP

24. Find the sum of $1\frac{1}{3}$ and $2\frac{1}{6}$. (p. 186)

25. Tell which are equivalent numbers.
$\frac{11}{3}$, $7\frac{1}{3}$, $6\frac{1}{3}$, $\frac{22}{3}$ (p. 160)

26. Find the median for the data. 30, 36, 39, 38, 36, 33 (p. 106)

27. Evaluate $b \div d$, for $b = 2,260$ and $d = 41$.
(p. 28)

⭐ 28. **TEST PREP**   Alexander buys 3 boxes of computer paper for $10.79 per box, including tax. How much change will he get from $50? (p. 70)

A $39.21

B $32.37

C $17.63

D $11.00

EXTRA PRACTICE page H40, Set D

PROBLEM SOLVING STRATEGY
Draw a Diagram

Learn how to use the strategy *draw a diagram* to solve problems.

A taxicab travels $1\frac{1}{3}$ mi west, 4 mi north, $8\frac{1}{4}$ mi east, $1\frac{1}{3}$ mi south, and then 10 mi west. How far has the taxicab traveled when it crosses its own path?

Analyze

What are you asked to find?

What information is given?

Is there numerical information you will not use? If so, what?

Choose

What strategy will you use?

You can draw a diagram that shows the taxicab's route.

Draw a diagram and label the distances and locations.

Solve

How will you solve the problem?

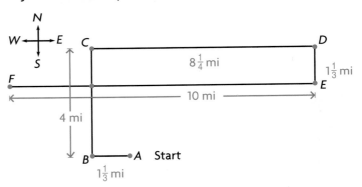

Add all the distances from A to E. Then add the distance from E to the point where the taxicab crosses its own path. This will be the same as the distance from C to D.

$$1\frac{1}{3} + 4 + 8\frac{1}{4} + 1\frac{1}{3} + 8\frac{1}{4}$$

Write equivalent fractions by using the LCD, 12.

$$1\frac{4}{12} + 4 + 8\frac{3}{12} + 1\frac{4}{12} + 8\frac{3}{12} = 22 + \frac{14}{12} = 22 + 1\frac{2}{12} = 23\frac{1}{6}$$

So, the taxicab has traveled $23\frac{1}{6}$ mi when it crosses its own path.

Check

How can you check to see if your answer is reasonable?

PROBLEM SOLVING PRACTICE

PROBLEM SOLVING STRATEGIES

▶ **Draw a Diagram or Picture**
Make a Model
Predict and Test
Work Backward
Make an Organized List
Find a Pattern
Make a Table or Graph
Solve a Simpler Problem
Write an Equation
Use Logical Reasoning

Solve by drawing a diagram.

1. A tour bus travels $7\frac{1}{2}$ mi south, $3\frac{1}{3}$ mi east, $4\frac{1}{3}$ mi north, and $11\frac{1}{2}$ mi west. How far has the tour bus traveled when it crosses its own path?

2. Tanisha needs a fence that is 33 ft long to separate her two gardens. If she puts one post in the ground every $5\frac{1}{2}$ ft, how many posts will Tanisha need for the fence?

For 3–4, use the information below.

Carla drives south $2\frac{1}{2}$ mi from her home. Next, she drives east $\frac{1}{3}$ mi. Then, she drives south 3 mi.

3. How many total miles does Carla drive from her home?

   A 5 mi   **C** $5\frac{5}{6}$ mi

   B $5\frac{1}{5}$ mi   **D** 6 mi

4. In which directions must Carla drive to return home?

   F east and north   **H** west and north

   G east and south   **J** west and south

MIXED STRATEGY PRACTICE

5. A display in a store has 24 cans in the bottom row, 21 in the second row, and 18 in the third row. If the pattern continues, how many cans are in the fifth row?

6. Tickets to see an Irish dance group cost $52 and $28. Mario bought 7 tickets for his family and paid a total of $316. How many tickets at each price did he buy?

7. Ms. Lopez travels north from home to pick up Maria at school. Then she travels $3\frac{1}{2}$ mi east to pick up Marcos and $4\frac{1}{4}$ mi south to pick up Carter. If Ms. Lopez has driven a total of $12\frac{1}{2}$ mi, what is the distance from home to Maria's school?

8. Tim has a job in the city. He has to commute $6\frac{1}{2}$ km to work. The bus he takes to work travels about $3\frac{1}{4}$ km in 10 minutes. About how much time does Tim spend on the bus going to and from work?

9. In a recent contest, Gary scored more points than Catherine, who scored more points than Clara. Christopher scored more points than Clara but fewer than Gary. Who had the most points?

10. Use the table below. What is Joshua's total bill if he rented 3 videos and kept them for 6 days?

11. ✍ **Write a problem** that can be solved by using the strategy *draw a diagram*. Explain the steps you would use to solve the problem, and draw the diagram.

Video Rental Prices	
1 movie for 1 day	$2.99
Each additional day	$0.99
5 additional days	$3.79

Estimate the sum or difference. (pp. 176–179)

1. $\frac{7}{12} + \frac{1}{4}$

2. $\frac{4}{5} - \frac{2}{7}$

3. $\frac{4}{5} + \frac{3}{8}$

4. $\frac{3}{4} - \frac{1}{3}$

5. $4\frac{2}{9} - \frac{1}{7}$

6. $\frac{9}{20} + 1\frac{4}{5}$

7. $6\frac{1}{4} + 3\frac{2}{9}$

8. $9\frac{4}{5} - 2\frac{7}{9}$

Write the sum or difference in simplest form.
Estimate to check. (pp. 182–185)

9. $\frac{1}{2} + \frac{1}{3}$

10. $\frac{3}{4} + \frac{1}{6}$

11. $\frac{2}{5} + \frac{2}{4}$

12. $\frac{2}{3} + \frac{3}{4}$

13. $\frac{5}{6} + \frac{2}{3}$

14. $\frac{1}{3} + \frac{5}{6}$

15. $\frac{3}{8} + \frac{3}{4}$

16. $\frac{5}{8} + \frac{1}{6}$

17. $\frac{3}{4} - \frac{1}{3}$

18. $\frac{7}{8} - \frac{1}{4}$

19. $\frac{5}{6} - \frac{2}{9}$

20. $\frac{8}{9} - \frac{1}{6}$

21. $\frac{7}{8} - \frac{5}{6}$

22. $\frac{7}{12} - \frac{5}{12}$

23. $\frac{3}{5} - \frac{1}{4}$

24. $\frac{4}{5} - \frac{1}{10}$

Draw a diagram to find the sum or difference. Write the answer in simplest form. (pp. 186–189)

25. $2\frac{3}{8} + 1\frac{1}{4}$

26. $6\frac{2}{3} - 3\frac{1}{4}$

27. $5\frac{2}{5} - 3\frac{3}{10}$

28. $2\frac{1}{6} + 1\frac{1}{3}$

Write the sum or difference in simplest form.
Estimate to check. (pp. 186–189, 192–193)

29. $1\frac{1}{6} + 3\frac{2}{3}$

30. $2\frac{3}{4} + 3\frac{1}{8}$

31. $1\frac{1}{2} + 2\frac{1}{4}$

32. $1\frac{1}{5} + 1\frac{3}{10}$

33. $7\frac{3}{4} - 5\frac{1}{3}$

34. $8\frac{1}{3} - 3\frac{1}{8}$

35. $3\frac{1}{4} - 2\frac{1}{2}$

36. $4\frac{1}{2} - 1\frac{2}{3}$

37. $7\frac{1}{5} - 5\frac{4}{9}$

Solve each problem by drawing a diagram. (pp. 194–195)

38. A minibus leaves the garage and travels $9\frac{5}{6}$ mi north to pick up Tanya. Then it travels $3\frac{1}{6}$ mi west to pick up Luis, $4\frac{1}{4}$ mi south to pick up Alissa, and $4\frac{5}{12}$ mi east to the school. How far does the minibus travel before crossing its own path?

39. Satoko has a board that is 9 ft long. She needs to cut the board into $2\frac{1}{4}$-ft sections. How many cuts will she have to make?

40. Del drives $3\frac{1}{4}$ mi north from his home. Next, he drives west $\frac{3}{4}$ mi. Then, he drives north $\frac{1}{2}$ mi. In which directions must Del drive to return home?

Decide on a plan.

See item **2**.

Show the hours practiced as an addition sentence. Find the *least common denominator* before you add.

Also see problem **4**, p. H63.

Choose the best answer.

1. Bryce had $\frac{13}{15}$ gallon of paint. He used $\frac{1}{3}$ gallon for a project. How much paint did he have left?

A $\frac{8}{15}$ gal   **C** $\frac{1}{3}$ gal

B $\frac{1}{2}$ gal   **D** $\frac{1}{5}$ gal

2. Tirzah practiced the piano for $1\frac{2}{3}$ hours on Monday, $\frac{3}{4}$ hour on Wednesday, and $1\frac{1}{2}$ hours on Friday. Which is the total amount of time she spent practicing the piano?

F $3\frac{2}{5}$ hr   **H** $3\frac{11}{12}$ hr

G $3\frac{3}{4}$ hr   **J** 4 hr

3. Last week Jessie's swimming practice lasted for $\frac{3}{5}$ hour on Monday, $\frac{5}{6}$ hour on Wednesday, and $\frac{9}{10}$ hour on Friday. How many hours did she have swimming practice during the week?

A $1\frac{1}{3}$ hr   **C** 2 hr

B $1\frac{2}{3}$ hr   **D** $2\frac{1}{3}$ hr

4. What is the value of $n + \frac{1}{2}$ for $n = \frac{5}{6}$?

F $\frac{1}{3}$ **H** $1\frac{1}{6}$

G $\frac{11}{12}$ **J** $1\frac{1}{3}$

5. In which pair are both numbers equivalent to $\frac{2}{5}$?

A 0.2; 20%   **C** 40%; 0.4

B 25%; $\frac{4}{10}$ **D** 0.6; $\frac{6}{15}$

6. $14\frac{2}{3} - 9\frac{5}{12}$

F $4\frac{1}{3}$ **H** $5\frac{1}{4}$

G $4\frac{1}{2}$ **J** Not here

7. The mean of 5 numbers is 25.6. What is the sum of the numbers?

A 128   **C** 26.1

B 30.6   **D** Not here

8. Mel hiked $\frac{9}{16}$ mile on Saturday and $\frac{7}{8}$ mile on Sunday. Which is a good estimate for how far Mel hiked in all?

F about 2 mi

G about $1\frac{1}{2}$ mi

H about 1 mi

J about $\frac{1}{2}$ mi

9. Which is a reasonable estimate of the sum of the fractions $\frac{1}{12}$, $\frac{4}{9}$, $\frac{5}{8}$, and $\frac{11}{12}$?

A 1   **C** 4

B 2   **D** $4\frac{1}{2}$

10. $115,371.9 + 22,671.25$

F 138,043.015   **H** 138,043.259

G 138,043.15   **J** Not here

Write What You Know

11. Bart passes a sign that read "City Limit $5\frac{1}{2}$ miles." If he drives $2\frac{1}{4}$ miles farther, how far will he be from the city limit? Draw a diagram for the problem and explain how you found your answer.

12. Explain why you must rename to find the difference $1\frac{7}{8} - \frac{9}{10}$. Then find the difference.

10 Multiply and Divide Fractions and Mixed Numbers

The 86th Tour de France began on July 3, 1999, and ended on July 25. The cyclists rode for 21 days and rested for 2 days. The first day of the race is called the Prologue. On that day, riders traveled 6.8 km. Each of the days after the Prologue is called a stage. Each stage covers a different distance, from about 50 km to nearly 230 km. The total distance covered in the race was about 3,690 km.

PROBLEM SOLVING Estimate the number of kilometers the cyclists had ridden after they had completed $\frac{1}{3}$ of the race.

DATA LINK

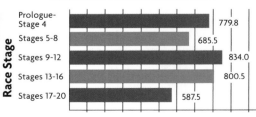

TOUR DE FRANCE DISTANCES

Race Stage	Kilometers Covered
Prologue–Stage 4	779.8
Stages 5-8	685.5
Stages 9-12	834.0
Stages 13-16	800.5
Stages 17-20	587.5

0 100 200 300 400 500 600 700 800 900 1000

Kilometers Covered

Check What You Know

Use this page to help you review and remember important skills needed for Chapter 10.

✅ Vocabulary

Choose the best term from the box.

> fraction
> mixed number
> equation

1. An algebraic or numerical sentence that shows two quantities are equal is a(n) __?__ .

2. A number that is made up of a whole number and a fraction is a(n) __?__ .

✅ Round Fractions (For Intervention, see p. H10.)

Round each fraction to 0, $\frac{1}{2}$, or 1.

3. $\frac{2}{9}$ 4. $\frac{7}{8}$ 5. $\frac{7}{15}$ 6. $\frac{5}{8}$ 7. $\frac{1}{4}$

8. $\frac{2}{15}$ 9. $\frac{3}{8}$ 10. $\frac{5}{6}$ 11. $\frac{1}{3}$ 12. $\frac{10}{11}$

13. $\frac{11}{20}$ 14. $\frac{7}{12}$ 15. $\frac{1}{6}$ 16. $\frac{4}{7}$ 17. $\frac{2}{6}$

✅ Mental Math and Equations (For Intervention, see p. H11.)

Use mental math to solve.

18. $9.3 + x = 12.5$ 19. $4c = 128$ 20. $x - 160 = 520$ 21. $5.11 = 5.28 - x$

22. $0.12 = \frac{x}{0.6}$ 23. $35 - t = 21$ 24. $6y = 0.24$ 25. $r + 3.7 = 6.3$

✅ Fractions and Mixed Numbers (For Intervention, see p. H11.)

Write each fraction as a mixed number.

26. $\frac{18}{5}$ 27. $\frac{7}{6}$

28. $\frac{16}{15}$ 29. $\frac{4}{3}$

Write each mixed number as a fraction.

30. $1\frac{5}{8}$ 31. $7\frac{1}{2}$

32. $3\frac{2}{3}$ 33. $2\frac{4}{5}$

> **LOOK AHEAD**
>
> **In Chapter 10 you will**
> - estimate products and quotients
> - multiply and divide fractions and mixed numbers
> - use fractions in expressions and equations

Estimate Products and Quotients

Learn how to estimate products and quotients of fractions and mixed numbers.

In a landfill, bulldozers spread and compact the garbage into 10-foot layers. Every layer is covered with clean soil.

Remember that when rounding fractions, round to 0, $\frac{1}{2}$, or 1. When rounding mixed numbers, round to the nearest whole number.

The landfills in more than half of the states in the United States will soon be full. It is estimated that each person in the United States produces $4\frac{2}{5}$ pounds of garbage a day. If the population of the United States was $249\frac{7}{10}$ million, about how many pounds of garbage would be produced every day?

One way to estimate the answer is to round the mixed numbers to the nearest whole number.

Estimate. $4\frac{2}{5} \times 249\frac{7}{10}$

$4\frac{2}{5} \times 249\frac{7}{10}$ *Round to the nearest whole number.*

$\downarrow$ $\downarrow$ THINK: *$\frac{2}{5}$ rounds to 0, and $\frac{7}{10}$ rounds to 1.*

$4 \times 250 = 1,000$ *Multiply.*

So, about 1,000 million pounds would be produced each day.

You can also estimate by averaging two estimates.

EXAMPLE

Estimate. $8 \div \frac{3}{4}$

Since $\frac{3}{4}$ is halfway between $\frac{1}{2}$ and 1, find the two estimates and then find their average.

Round up.   *Round down.*

$8 \div \frac{3}{4} \rightarrow 8 \div 1 = 8$ $8 \div \frac{3}{4} \rightarrow 8 \div \frac{1}{2} = 16$

$8 + 16 = 24; \ 24 \div 2 = 12$ So, $8 \div \frac{3}{4}$ is about 12.

You can use compatible numbers to estimate a product or quotient.

$23\frac{3}{4} \div 4\frac{1}{2} \rightarrow 25 \div 5 = 5$

So, $23\frac{3}{4} \div 4\frac{1}{2}$ is about 5.

Think and Discuss ▶ Look back at the lesson to answer each question.

1. **What if** each person in the United States produced $2\frac{1}{5}$ pounds of garbage? About how many pounds of garbage would be produced?

2. **Tell** what compatible numbers you could use to find $81\frac{3}{5} \div 12\frac{7}{8}$.

Guided Practice ▶ Estimate each product or quotient.

3. $\frac{7}{8} \times \frac{7}{16}$

4. $10\frac{8}{11} \div 2\frac{1}{5}$

5. $78\frac{3}{7} \div 4\frac{1}{6}$

6. $\frac{3}{5} \times 38$

7. $1\frac{3}{4} \times 35$

8. $21\frac{3}{8} \div 17\frac{1}{3}$

9. $58\frac{3}{4} \times 1\frac{5}{6}$

10. $98\frac{7}{8} \div 23\frac{1}{5}$

Independent Practice ▶ Estimate each product or quotient.

11. $\frac{7}{9} \times \frac{1}{3}$

12. $10\frac{8}{9} \times \frac{5}{6}$

13. $\frac{5}{6} \div \frac{11}{12}$

14. $67\frac{9}{12} \div 2\frac{7}{10}$

15. $24\frac{9}{10} \div 6\frac{2}{3}$

16. $36\frac{5}{8} \div 13\frac{3}{5}$

17. $67\frac{2}{3} \div 23\frac{1}{8}$

18. $97\frac{2}{9} \div 52\frac{5}{8}$

19. $3\frac{11}{12} \times 4\frac{6}{7}$

20. $12\frac{5}{24} \div \frac{8}{12}$

21. $\frac{5}{9} \times \frac{7}{12}$

22. $\frac{2}{5} \div \frac{10}{21}$

Problem Solving Applications ▶ Estimate to compare. Write < or > for each ●.

23. $4\frac{1}{6} \times 3\frac{2}{3}$ ● $7\frac{5}{8} \div 2\frac{1}{3}$

24. $7\frac{2}{3} \div 5$ ● $2\frac{4}{8} \times 3\frac{1}{8}$

25. Cal runs $5\frac{3}{10}$ miles in $33\frac{4}{5}$ minutes. About how many minutes does it take for Cal to run one mile?

26. ✍ **Write About It** Explain how you would estimate a quotient of mixed numbers.

27. Doris picked up used newspapers around her neighborhood for six days. She picked up 0.5 kg the first day, 2 kg the second day, and 3.5 kg the third day. If this pattern continued, how much newspaper did she pick up on the sixth day?

MIXED REVIEW AND TEST PREP

28. Write $\frac{7}{20}$ as a percent. (p. 169)

29. $5\frac{7}{9} - 2\frac{1}{3}$ (p. 186)

30. Find the mean, median, and mode for the data.
27, 48, 83, 76, 48, 27 (p. 110)

31. Evaluate the expression $m - 4^2$ for $m = 3^3$. (p. 28)

⭐ 32. **TEST PREP** How much greater is the LCM of 18 and 24 than the LCM of 12 and 8? (p.150)

A 2   **B** 8   **C** 48   **D** 128

Multiply Fractions

Learn how to multiply fractions.

Carolyn asked $\frac{3}{4}$ of her classmates what time they leave for school in the morning. Of those she asked, $\frac{1}{2}$ leave at 7:00 A.M. What fractional part of the class told her that they leave for school at 7:00 A.M.?

One way to find the fractional part of a fraction is to make a model.

Find $\frac{1}{2}$ of $\frac{3}{4}$, or $\frac{1}{2} \times \frac{3}{4}$.

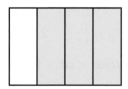

Fold a piece of paper into 4 equal parts.

Shade 3 parts to show $\frac{3}{4}$.

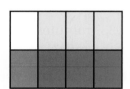

Fold the paper in half. Shade $\frac{1}{2}$ of the paper.

Of the $2 \times 4 = 8$ parts, $1 \times 3 = 3$ are shaded, so $\frac{3}{8}$ of the paper is shaded twice. These parts represent $\frac{1}{2} \times \frac{3}{4}$.

$$\frac{1}{2} \times \frac{3}{4} = \frac{3}{8}$$

So, $\frac{3}{8}$ of the students Carolyn asked leave at 7:00 A.M.

• Compare the numerator and denominator of the product with the numerators and denominators of the factors. What relationship do you see?

You can see this relationship in the solution to the problem.

$$\frac{\text{numerator} \times \text{numerator}}{\text{denominator} \times \text{denominator}} = \frac{\text{numerator}}{\text{denominator}}$$

$\qquad\qquad \uparrow \qquad\qquad\quad \uparrow \qquad\qquad\quad \uparrow$

$\qquad\quad$ factor $\qquad\qquad$ factor $\qquad\qquad$ product

• Make a model to find $\frac{1}{3} \times \frac{3}{4}$.

You can use this relationship to multiply fractions without making a model.

EXAMPLE 1

Remember that to write a fraction in simplest form, divide the numerator and the denominator by the greatest common factor (GCF).

Find $\frac{1}{3} \times \frac{3}{7}$. Write the product in simplest form.

$$\frac{1}{3} \times \frac{3}{7} = \frac{1 \times 3}{3 \times 7} = \frac{3}{21}$$

Multiply the numerators.
Multiply the denominators.

$$= \frac{3 \div 3}{21 \div 3}$$

Divide the numerator and the denominator by the GCF, 3.

$$= \frac{1}{7}$$

Write the product in simplest form.

So, $\frac{1}{3} \times \frac{3}{7} = \frac{1}{7}$.

• Explain why the product, $\frac{1}{7}$, is less than the factor $\frac{3}{7}$.

You can also multiply a whole number and a fraction without making a model.

EXAMPLE 2

Ms. Jones's car is being repaired after being in an accident. Her daughter Cindy will have to walk a total of $\frac{9}{10}$ mi to and from school every day for 11 days. How far will Cindy walk in all?

Find $11 \times \frac{9}{10}$.

Estimate. $11 \times 1 = 11$

$$11 \times \frac{9}{10} = \frac{11}{1} \times \frac{9}{10}$$

Write the whole number as a fraction.

$$= \frac{11 \times 9}{1 \times 10}$$

Multiply the numerators.
Multiply the denominators.

$$= \frac{99}{10}, \text{ or } 9\frac{9}{10}$$

Write the answer as a fraction or as a mixed number in simplest form.

Compare the product to your estimate. $9\frac{9}{10}$ is close to the estimate of 11. The product is reasonable.

So, Cindy will walk $9\frac{9}{10}$ mi.

• **What if** Cindy walked to school for 21 days? How far would Cindy walk in all?

When a numerator and a denominator have a common factor, you can simplify before you multiply.

EXAMPLE 3

Cheryl has $\frac{3}{4}$ of a box of snacks left from a school party. She gives $\frac{2}{5}$ of the snacks to the people in the school office. What part of the box of snacks does she give the people in the office?

Find $\frac{2}{5} \times \frac{3}{4}$.

Estimate. $\frac{1}{2} \times 1 = \frac{1}{2}$

$\frac{2}{5} \times \frac{3}{4}$ ← The GCF of 2 and 4 is 2.

Look for a numerator and denominator with common factors. Find the GCF.

$\frac{\overset{1}{2}}{5} \times \frac{3}{\underset{2}{4}}$ ← $2 \div 2 = 1$
         ← $4 \div 2 = 2$

Divide the numerator and denominator by the GCF, 2.

$\frac{\overset{1}{2}}{5} \times \frac{3}{\underset{2}{4}} = \frac{1 \times 3}{5 \times 2} = \frac{3}{10}$ *Multiply.*

So, Cheryl gives away $\frac{3}{10}$ of the box of snacks.

• What is $\frac{1}{8} \times \frac{6}{7}$? Simplify the fractions before you multiply.

EXAMPLE 4

Find $\frac{8}{9} \times \frac{3}{4}$. Use the GCF to simplify the fractions before you multiply.

$\frac{8}{9} \times \frac{3}{4}$ ← The GCF of 8 and 4 is 4.
        ← The GCF of 3 and 9 is 3.

$\frac{\overset{2}{8}}{\underset{3}{9}} \times \frac{\overset{1}{3}}{\underset{1}{4}} = \frac{2 \times 1}{3 \times 1} = \frac{2}{3}$ *Divide the numerators and denominators by the GCFs, 3 and 4. Multiply.*

So, $\frac{8}{9} \times \frac{3}{4} = \frac{2}{3}$.

CHECK FOR UNDERSTANDING

Think and Discuss   **Look back at the lesson to answer each question.**

1. **Explain** how to make a model to show $\frac{1}{3} \times \frac{3}{8}$.

2. **Tell** how you can rewrite a whole number before you multiply it by a fraction.

Guided Practice   **Make a model to find the product.**

3. $\frac{3}{4} \times \frac{1}{2}$ 4. $\frac{1}{3} \times \frac{5}{8}$ 5. $\frac{2}{5} \times \frac{1}{2}$ 6. $\frac{1}{3} \times \frac{1}{2}$

Multiply. Write the answer in simplest form.

7. $\frac{3}{4} \times \frac{2}{5}$ **8.** $\frac{2}{5} \times \frac{7}{8}$ **9.** $2 \times \frac{6}{7}$ **10.** $\frac{2}{3} \times 16$

11. $\frac{2}{3} \times 4$ **12.** $9 \times \frac{2}{3}$ **13.** $\frac{5}{6} \times \frac{2}{3}$ **14.** $\frac{3}{5} \times \frac{5}{6}$

PRACTICE AND PROBLEM SOLVING

Independent ▶
Practice

Make a model to find the product.

15. $\frac{3}{4} \times \frac{1}{4}$ **16.** $\frac{3}{4} \times \frac{2}{3}$ **17.** $\frac{1}{8} \times \frac{1}{2}$ **18.** $3 \times \frac{1}{4}$

Multiply. Write the answer in simplest form.

19. $\frac{1}{3} \times \frac{2}{3}$ **20.** $\frac{3}{4} \times \frac{1}{3}$ **21.** $\frac{1}{5} \times \frac{2}{3}$ **22.** $\frac{1}{4} \times \frac{2}{7}$

23. $\frac{4}{5} \times \frac{7}{8}$ **24.** $\frac{2}{9} \times \frac{3}{4}$ **25.** $\frac{1}{8} \times \frac{4}{5}$ **26.** $\frac{5}{9} \times \frac{3}{10}$

27. $\frac{4}{9} \times \frac{3}{5}$ **28.** $\frac{2}{3} \times 21$ **29.** $24 \times \frac{1}{12}$ **30.** $\frac{1}{8} \times 16$

Compare. Write < , >, or = for ●.

31. $\frac{2}{9} \times \frac{3}{10}$ ● $\frac{2}{9}$ **32.** $\frac{5}{6} \times 5$ ● $\frac{5}{6}$ **33.** $8 \times \frac{1}{9}$ ● 8

Problem Solving ▶
Applications

34. Sandra takes $\frac{1}{2}$ hr to walk to school. She spends $\frac{1}{2}$ of that time walking down her street. What part of an hour does Sandra spend walking down her street? How many minutes is this?

35. There are 144 registered voters in Booker County. In the last election, $\frac{1}{4}$ of them did not vote. How many voters did vote?

36. **❓ What's the Question?** Natalie runs $\frac{2}{3}$ the distance her brother runs in one week. Her brother runs 15 miles one week. The answer is 10 miles.

37. *REASONING* Scott chose a number, added 2, multiplied the sum by 4, and divided the product by 8. The final number was 4. What number had Scott chosen?

MIXED REVIEW AND TEST PREP

38. $3\frac{2}{3} \times 2\frac{1}{7}$ (p. 206)   **39.** Solve. $5b = 60.45$ (p. 82)   **40.** 267.45×2.8 (p. 70)

⋆ 41. TEST PREP The manager of a grocery store asks customers in the store on a given day to fill out a card naming their favorite cookie. Which type of sample is this? (p. 98)

     A convenience   **B** biased   **C** random   **D** systematic

⋆ 42. TEST PREP Which shows the GCF for 130 and 75? (p. 150)

     F 3   **G** 4   **H** 5   **J** 6

EXTRA PRACTICE page H41, Set B

Multiply Mixed Numbers

Learn how to multiply mixed numbers.

Write the missing numerator.

1. $1\frac{2}{3} = \frac{\blacksquare}{3}$ 2. $6\frac{3}{5} = \frac{\blacksquare}{5}$

3. $7\frac{3}{7} = \frac{\blacksquare}{7}$ 4. $9\frac{3}{8} = \frac{\blacksquare}{8}$

5. $3\frac{2}{5} = \frac{\blacksquare}{5}$

Remember that you can write a mixed number as a fraction.

$$2\frac{3}{4} = \frac{(2 \times 4) + 3}{4}$$

$$= \frac{11}{4}$$

Ann and Sheri are training for a bicycle race. On one day, Ann rides $3\frac{1}{5}$ mi. Sheri rides $2\frac{1}{2}$ times as far as Ann. How many miles does Sheri ride?

Find $2\frac{1}{2} \times 3\frac{1}{5}$.   Estimate. $3 \times 3 = 9$

$2\frac{1}{2} \times 3\frac{1}{5} = \frac{5}{2} \times \frac{16}{5}$ *Write the mixed numbers as fractions.*

$= \frac{\overset{1}{\cancel{5}}}{\underset{1}{\cancel{2}}} \times \frac{\overset{8}{\cancel{16}}}{\underset{1}{\cancel{5}}}$ *Simplify the fractions. Multiply.*

$= \frac{8}{1}$, or 8   *Write the answer in simplest form or as a whole or mixed number.*

So, Sheri rides 8 mi. The answer is reasonable since it is close to the estimate of 9 mi.

You can use the Distributive Property to multiply a whole number by a mixed number.

EXAMPLE

Multiply. $5 \times 2\frac{3}{8}$

$5 \times 2\frac{3}{8} = 5 \times \left(2 + \frac{3}{8}\right)$

$= (5 \times 2) + \left(5 \times \frac{3}{8}\right)$ *Use the Distributive Property.*

$= (5 \times 2) + \left(\frac{5}{1} \times \frac{3}{8}\right)$ *Write the whole number as a fraction. Find 5×2 and $\frac{5}{1} \times \frac{3}{8}$.*

$= 10 + \frac{15}{8}$

$= 10 + 1\frac{7}{8} = 11\frac{7}{8}$ *Write the fraction as a mixed number. Find the sum.*

So, $5 \times 2\frac{3}{8} = 11\frac{7}{8}$.

Think and ▶
Discuss

Look back at the lesson to answer each question.

1. Show how to use the Distributive Property to find $3 \times 4\frac{2}{3}$.

2. Discuss whether the product of two mixed numbers is greater than or less than the factors. Give two examples.

Guided ▶
Practice

Multiply. Write your answer in simplest form.

3. $\frac{3}{4} \times 1\frac{1}{2}$ **4.** $\frac{1}{2} \times 2\frac{1}{3}$ **5.** $1\frac{1}{2} \times 1\frac{1}{2}$ **6.** $1\frac{2}{5} \times 2\frac{1}{4}$

Use the Distributive Property to multiply.

7. $6\frac{1}{8} \times 3$ **8.** $3 \times 9\frac{4}{5}$ **9.** $1\frac{1}{8} \times 2$ **10.** $6 \times 4\frac{1}{4}$

Independent ▶
Practice

Multiply. Write your answer in simplest form.

11. $4\frac{2}{3} \times 1\frac{3}{4}$ **12.** $1\frac{3}{8} \times 4\frac{2}{3}$ **13.** $5\frac{1}{2} \times 6$ **14.** $2 \times 3\frac{1}{7}$

15. $4\frac{1}{6} \times 3\frac{3}{5}$ **16.** $1\frac{3}{4} \times 3$ **17.** $10\frac{1}{5} \times 8\frac{1}{3}$ **18.** $5 \times 1\frac{5}{6}$

Use the Distributive Property to multiply.

19. $3 \times 2\frac{2}{5}$ **20.** $4 \times 8\frac{5}{6}$ **21.** $3\frac{3}{4} \times 6$ **22.** $1\frac{1}{2} \times 12$

Compare. Write $<$, $>$, or = for each .

23. $3\frac{1}{3} \times 2\frac{1}{7}$ ● $3\frac{1}{4} \times 5$ **24.** $7 \times 7\frac{3}{7}$ ● $6\frac{3}{4} \times 4\frac{4}{5}$

Problem Solving ▶
Applications

25. Mr. Jackson rides his bicycle $1\frac{2}{3}$ mi every day. His wife rides $1\frac{1}{4}$ times as far as he does. How many miles does Mrs. Jackson ride her bicycle?

26. John works part time for $6.50 an hour. He works $3\frac{1}{2}$ hr each on Monday, Tuesday, and Thursday afternoons. How much does he earn those three days?

27. ✏ **Write About It** Without multiplying, tell whether the product $\frac{2}{3} \times \frac{3}{4}$ is a fraction, a whole number, or a mixed number.

28. $\frac{4}{9} \times \frac{2}{3}$ (p. 202) **29.** $8\frac{3}{4} - 2\frac{2}{5}$ (p. 186) **30.** Simplify. $\frac{27}{45}$ (p. 160) **31.** $414,089 - 62,036$ (p. 20)

⭐ **32. TEST PREP** Which is the mean of the data? (p. 110)

90, 94, 65, 90, 84, 94, 85

A 29 **B** 86 **C** 90 **D** 94

EXTRA PRACTICE page H41, Set C

Division of Fractions

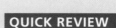

Explore how to model division of fractions.

You need fraction circles.

Vocabulary

reciprocal

TECHNOLOGY LINK

More Practice: Use E-Lab, *Exploring Division of Fractions.*
www.harcourtschool.com/elab2002

QUICK REVIEW

1. $3 \times \frac{1}{3}$ **2.** $\frac{1}{4} \times 4$ **3.** $7 \times \frac{3}{2}$

4. $2 \times \frac{8}{12}$ **5.** $\frac{2}{5} \times \frac{5}{2}$

Using models will help you understand division of fractions.

Activity 1

A. Use fraction circles to find $4 \div \frac{1}{3}$, or the number of thirds in 4 wholes.

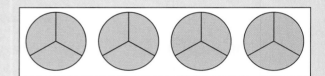

- Trace 4 whole circles on your paper.

- Model $4 \div \frac{1}{3}$ by tracing $\frac{1}{3}$-circle pieces on the 4 circles.

One whole equals three thirds.

- How many thirds are in 4 wholes? What is $4 \div \frac{1}{3}$?

B. Use fraction circles to find $\frac{1}{3} \div \frac{1}{6}$, or the number of sixths in $\frac{1}{3}$.

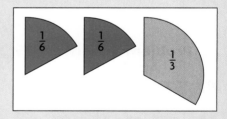

- Place as many $\frac{1}{6}$ pieces as you can on the $\frac{1}{3}$ piece.

- How many sixths are in $\frac{1}{3}$? What is $\frac{1}{3} \div \frac{1}{6}$?

Think and Discuss

- If $4 \div \frac{1}{3} = 12$, explain what $8 \div \frac{1}{3}$ must be.
- If $2 \div \frac{1}{6} = 12$, explain what $8 \div \frac{1}{6}$ must be.

Practice

Use fraction circles to model each problem. Draw a diagram of your model.

1. $3 \div \frac{1}{3}$ **2.** $4 \div \frac{1}{4}$ **3.** $\frac{1}{2} \div \frac{1}{8}$ **4.** $\frac{3}{4} \div \frac{1}{8}$

Two numbers are **reciprocals** if their product is 1.

$$\frac{1}{2} \times 2 = 1 \qquad\qquad \frac{3}{4} \times \frac{4}{3} = 1 \qquad\qquad 6 \times \frac{1}{6} = 1$$

$$\underset{\text{reciprocals}}{\uparrow\ \uparrow} \qquad\qquad \underset{\text{reciprocals}}{\uparrow\ \uparrow} \qquad\qquad \underset{\text{reciprocals}}{\uparrow\ \uparrow}$$

By using inverse operations, you can write related number sentences.

$$1 \div \frac{1}{2} = 2 \qquad\qquad 1 \div \frac{3}{4} = \frac{4}{3} \qquad\qquad 1 \div 6 = \frac{1}{6}$$

$$1 \div 2 = \frac{1}{2} \qquad\qquad 1 \div \frac{4}{3} = \frac{3}{4} \qquad\qquad 1 \div \frac{1}{6} = 6$$

You can use reciprocals and inverse operations when you divide.

Activity 2

• Study these problems.

Find $6 \div \frac{1}{2}$.

$1 \div \frac{1}{2} = 2$ *Think of the reciprocal of $\frac{1}{2}$.*

Since $1 \div \frac{1}{2} = 1 \times 2$, $6 \div \frac{1}{2} = 6 \times 2$.

So, $6 \div \frac{1}{2} = 6 \times 2 = 12$.

Find $\frac{3}{4} \div \frac{1}{3}$.

$1 \div \frac{1}{3} = 3$ *Think of the reciprocal of $\frac{1}{3}$.*

Since $1 \div \frac{1}{3} = 1 \times 3$, $\frac{3}{4} \div \frac{1}{3} = \frac{3}{4} \times 3$.

So, $\frac{3}{4} \div \frac{1}{3} = \frac{3}{4} \times 3 = \frac{9}{4}$, or $2\frac{1}{4}$.

Think and Discuss

• When you divide 1 by a number, what is the quotient?

• If $1 \div \frac{1}{3} = 3$, then what is $2 \div \frac{1}{3}$? What is $5 \div \frac{1}{3}$?

• If $1 \div \frac{2}{5} = \frac{5}{2}$, then what is $2 \div \frac{2}{5}$? What is $\frac{1}{2} \div \frac{2}{5}$?

Practice

Find the value of *n*.

1. If $1 \div \frac{1}{2} = 2$, then $2 \div \frac{1}{2} = n$.

2. If $1 \div \frac{4}{5} = \frac{5}{4}$, then $4 \div \frac{4}{5} = n$.

3. If $1 \div \frac{2}{3} = \frac{3}{2}$, then $\frac{1}{2} \div \frac{2}{3} = n$.

4. If $1 \div \frac{3}{5} = \frac{5}{3}$, then $\frac{3}{4} \div \frac{3}{5} = n$.

Find the quotient.

5. $6 \div \frac{3}{4}$ 　　　　 **6.** $4 \div \frac{1}{2}$ 　　　　 **7.** $\frac{1}{2} \div \frac{1}{3}$ 　　　　 **8.** $\frac{2}{3} \div \frac{1}{8}$

MIXED REVIEW AND TEST PREP

9. $2\frac{4}{5} \times 3\frac{3}{8}$ (p. 206)

10. Complete. $\frac{3}{8} = \frac{\blacksquare}{24}$ (p. 160)

11. Compare 606.64 and 606.074. Use <, >, or =. (p. 52)

12. $46.08 - 19.204$ (p. 66)

⭐**13.** **TEST PREP** How many times greater is the GCF of 15 and 18 than the GCF of 25 and 33? (p. 150)

A 2 　　　　 **B** 3 　　　　 **C** 5 　　　　 **D** 6

Divide Fractions and Mixed Numbers

Learn how to divide with fractions and mixed numbers.

Write each mixed number as a fraction.

1. $2\frac{1}{3}$ 2. $1\frac{3}{8}$ 3. $4\frac{1}{2}$

4. $3\frac{9}{10}$ 5. $6\frac{2}{3}$

Stacey is bringing juice to be served after the school council meeting. Stacey said she would bring enough juice for everyone at the meeting to have a glass. Each glass holds $\frac{1}{5}$ liter. How many glasses can be served if Stacey brings 8 liters of juice?

Find $8 \div \frac{1}{5}$.

Math Idea ▶ When you divide by a fraction, you can use multiplication to find the quotient.

Rewrite the division problem as a multiplication problem by using the reciprocal.

$8 \div \frac{1}{5} = \frac{8}{1} \div \frac{1}{5}$ *Write the whole number as a fraction.*

$= \frac{8}{1} \times \frac{5}{1}$ *Use the reciprocal of the divisor to write a multiplication problem.*

$= \frac{8}{1} \times \frac{5}{1} = \frac{40}{1}$, or 40   *Multiply.*

So, 8 liters can fill 40 glasses.

You can also use the reciprocal of the divisor to divide fractions and mixed numbers.

EXAMPLE 1

Find $\frac{2}{3} \div \frac{4}{7}$.

$\frac{2}{3} \div \frac{4}{7} = \frac{2}{3} \times \frac{7}{4}$ *Use the reciprocal of the divisor to write a multiplication problem.*

$= \frac{\overset{1}{\cancel{2}}}{3} \times \frac{7}{\underset{2}{\cancel{4}}}$ *Divide the numerator and denominator by the GCF, 2.*

$= \frac{7}{6}$, or $1\frac{1}{6}$ *Multiply.*

So, $\frac{2}{3} \div \frac{4}{7} = 1\frac{1}{6}$.

EXAMPLE 2

Each member of the school council will write his or her name on a strip of paper at the meeting. Jeremy has pieces of paper $5\frac{1}{4}$ in. long. Each strip of paper should be $1\frac{3}{4}$ in. How many strips can Jeremy cut from the length of one piece of paper?

Find $5\frac{1}{4} \div 1\frac{3}{4}$.

Estimate. $5 \div 2 = 2\frac{1}{2}$

$5\frac{1}{4} \div 1\frac{3}{4} = \frac{21}{4} \div \frac{7}{4}$ *Write the mixed numbers as fractions.*

$= \frac{21}{4} \times \frac{4}{7}$ *Use the reciprocal of the divisor to write a multiplication problem.*

$= \frac{\overset{3}{\cancel{21}}}{\underset{1}{\cancel{4}}} \times \frac{\overset{1}{\cancel{4}}}{\underset{1}{\cancel{7}}}$ *Simplify and multiply.*

$= \frac{3}{1}$, or 3

Compare the product to your estimate. 3 is close to the estimate of $2\frac{1}{2}$. The product is reasonable.

So, Jeremy can cut 3 strips from each piece of paper.

• What is $2\frac{3}{4} \div 1\frac{2}{5}$?

Sometimes you can use mental math to divide whole numbers and fractions.

EXAMPLE 3

Use mental math to solve.

A $9 \div \frac{1}{2}$ **THINK:** $9 \times 2 = 18$.   *Dividing by $\frac{1}{2}$ is the same as multiplying by 2.*

So, $9 \div \frac{1}{2} = 18$.   *There are 18 halves in 9.*

B $13 \div \frac{1}{3}$ **THINK:** $13 \times 3 = 39$.   *Dividing by $\frac{1}{3}$ is the same as multiplying by 3.*

So, $13 \div \frac{1}{3} = 39$.   *There are 39 thirds in 13.*

C $20 \div \frac{2}{5}$ **THINK:** $20 \times 5 = 100$.   *Dividing by $\frac{2}{5}$ is the same as multiplying by $\frac{5}{2}$.*

$100 \div 2 = 50$.

So, $20 \div \frac{2}{5} = 50$.

• Use mental math to find $15 \div \frac{1}{6}$.

Think and ▶ Discuss

Look back at the lesson to answer each question.

1. **Tell** what the reciprocal of a number is. Give an example.

2. **Give** an example of a fraction or mixed-number division problem where the quotient is greater than the dividend.

Guided ▶ Practice

Write the reciprocal of the number.

3. $\frac{2}{3}$ 4. $\frac{3}{4}$ 5. 7   6. $2\frac{3}{8}$ 7. $4\frac{1}{3}$

Find the quotient. Write the answer in simplest form.

8. $\frac{1}{3} \div \frac{1}{2}$ 9. $\frac{1}{5} \div \frac{1}{4}$ 10. $\frac{1}{4} \div \frac{1}{2}$ 11. $\frac{1}{8} \div 3$

12. $4 \div \frac{2}{3}$ 13. $3\frac{3}{5} \div 1\frac{1}{5}$ 14. $2\frac{3}{5} \div 4$ 15. $3\frac{1}{4} \div 2\frac{2}{3}$

PRACTICE AND PROBLEM SOLVING

Independent ▶ Practice

Write the reciprocal of the number.

16. $\frac{5}{8}$ 17. 10   18. $\frac{1}{6}$ 19. $\frac{2}{9}$ 20. $3\frac{1}{2}$

21. $\frac{15}{7}$ 22. $\frac{1}{5}$ 23. 9   24. $1\frac{5}{6}$ 25. $2\frac{3}{5}$

Find the quotient. Write the answer in simplest form.

26. $\frac{3}{8} \div \frac{1}{2}$ 27. $\frac{2}{3} \div \frac{4}{7}$ 28. $\frac{7}{8} \div \frac{1}{3}$ 29. $8 \div \frac{6}{7}$

30. $12 \div \frac{3}{5}$ 31. $\frac{3}{4} \div \frac{1}{3}$ 32. $\frac{4}{9} \div \frac{3}{5}$ 33. $4 \div \frac{4}{5}$

34. $3\frac{2}{5} \div 1\frac{1}{5}$ 35. $3\frac{4}{5} \div \frac{3}{4}$ 36. $4\frac{1}{2} \div \frac{1}{4}$ 37. $4\frac{1}{5} \div 2\frac{3}{5}$

Use mental math to find each quotient.

38. $10 \div \frac{1}{2}$ 39. $12 \div \frac{2}{3}$ 40. $6 \div \frac{1}{4}$ 41. $8 \div \frac{1}{4}$

42. $3 \div \frac{1}{6}$ 43. $9 \div \frac{1}{9}$ 44. $7 \div \frac{1}{2}$ 45. $11 \div \frac{1}{4}$

ALGEBRA Evaluate the expression.

46. $4 \div a$ for $a = \frac{2}{3}$ 47. $b \div 2\frac{1}{3}$ for $b = 5\frac{1}{2}$

48. $1\frac{4}{5} \div a$ for $a = 2\frac{3}{5}$ 49. $c \div 7\frac{4}{5}$ for $c = 2\frac{1}{6}$

50. $b \div 7$ for $b = \frac{1}{5}$ 51. $3\frac{1}{5} \div b$ for $b = 4$

Problem Solving ▶ Applications

52. The members of the school council were told they could divide the 15 acres of land next to the school into $1\frac{1}{2}$-acre sections to be used for new building projects for the school. How many $1\frac{1}{2}$-acre sections will there be?

53. Katie bought 2-lb, 7-lb, and 3-lb packages of ground turkey. How many $\frac{1}{4}$-lb turkey burgers can she make?

54. In a $\frac{1}{4}$-mi relay, each runner on a team runs $\frac{1}{16}$ mi. How many runners are in the relay?

55. Gerald has $8\frac{3}{4}$ yd of fabric. This is 7 times the amount he needs to make one costume for the school play. How much fabric does he need for each costume?

56. **? What's the Error?** Jamie worked the problem below. What mistake did Jamie make? What is the correct answer in simplest form?

$$3\frac{1}{3} \div 2\frac{2}{9} = \frac{10}{3} \div \frac{20}{9} = \frac{3}{10} \times \frac{20}{9} = \frac{2}{3}$$

57. Jacki wants to buy a blouse for $12.95, a T-shirt for $15.95, a book for $10.50, and two pens for $3.75 each. What is the total cost?

MIXED REVIEW AND TEST PREP

58. $3\frac{1}{3} \times 2\frac{2}{5}$ (p. 206)

59. $10\frac{1}{2} + 2\frac{7}{8}$ (p. 186)

60. Evaluate $5.2x$ for $x = 3.41$. (p. 82)

61. **TEST PREP** Which is the sum for $234,607 + 84,395$? (p. 20)

    A 218,002   **B** 218,992   **C** 318,902   **D** 319,002

62. **TEST PREP** Which shows $4\frac{3}{8}$ as a fraction? (p. 172)

    F $\frac{11}{8}$ **G** $\frac{15}{8}$ **H** $\frac{35}{8}$ **J** $\frac{39}{8}$

PROBLEM SOLVING   LiNKUP to Careers

Architecture Horace King was born in South Carolina in 1807. As an adult, he built more than 100 covered bridges in Georgia and nearby states. A bridge builder must first plan on paper a structure that will support heavy loads and withstand forces of nature. The mathematics that a bridge builder uses must be exact and correct at all times.

- The panels used on a covered bridge are $\frac{3}{4}$ ft wide. How many panels are needed to cover a side of a bridge that is 90 ft long?

EXTRA PRACTICE page H41, Set D

PROBLEM SOLVING SKILL
Choose the Operation

Analyze
Choose
Solve
Check

Learn how to choose the operation needed to solve a problem.

Use the chart to help you decide how the numbers in a problem are related.

Then choose the operation needed to solve each problem.

ADD	• Joining groups of different sizes
SUBTRACT	• Taking away or comparing groups
MULTIPLY	• Joining equal-sized groups
DIVIDE	• Separating into equal-sized groups • Finding out how many in each group

Read each problem and decide how you would solve it.

A In one week, José drove $3\frac{1}{4}$ mi to play practice, $3\frac{1}{4}$ mi to the hardware store to buy supplies, and $7\frac{4}{5}$ mi to Jim's house to work on the costumes for the play. How many miles did he drive in all?	**B** Carla bought 6 yd of terry cloth to make collars for some of the costumes for the play. Each collar takes $\frac{2}{3}$ yd of fabric. How many collars can Carla make?
C Leanne makes props for the school play. She uses $4\frac{1}{8}$ lb of clay to make each vase and $2\frac{3}{4}$ lb to make each bowl. How much more clay does Leanne use to make one vase than one bowl?	**D** Sean bought $12\frac{2}{3}$ yd of pine lumber for $6 per yd to build some scenery for the school play. How much money did he spend?

• What operation would you use to solve each problem?

• Which problems could you solve by using a combination of operations?

• How did the chart above help you decide which operation to use for each problem?

Solve. Name the operations used.

1. It takes Karen $5\frac{1}{2}$ minutes to walk to the community center. It takes Jackie $7\frac{1}{3}$ minutes. How much longer does it take Jackie?

2. Rob makes candles that weigh $1\frac{3}{16}$ lb each. How much do 24 of them weigh?

3. Katie decorates travel bags. Each bag requires $1\frac{1}{3}$ yd of trim around the top and 1 yd of trim on the handle.

   a. Which operations could you use to find how much trim is needed for a number of bags?

      A subtraction and division

      B addition and multiplication

      C division and multiplication

      D addition and division

   b. How much trim would it take to make 4 bags?

      F $8\frac{1}{3}$ yd

      G 8 yd

      H $9\frac{1}{3}$ yd

      J 9 yd

MIXED APPLICATIONS

4. Marcie has 18 oz of dough left in a container. She needs $3\frac{3}{5}$ oz of dough to make one ornament.

   a. Write the expression you would use to find the number of ornaments she can make. Solve the problem.

   b. How many ounces of dough are left in the container if she makes 4 ornaments?

Use Data For 5–8, use the map.

5. Darrin hiked the shortest route from the trailhead to Hart Mountain in $7\frac{3}{4}$ hours. What is the average number of miles he hiked in 1 hour?

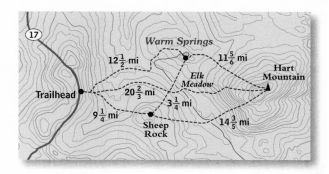

6. How much shorter was the trail that Darrin took than the trail through Warm Springs?

7. Elk Meadow is halfway along the trail from the trailhead to Hart Mountain. How far is it from the trailhead to Elk Meadow?

8. Sharlene left the trailhead at 8:30 A.M. for a daylong hike. She returned to the trailhead $7\frac{1}{2}$ hours after she left. What time did she return?

9. ✏️ **Write About It** Explain how you decide what operation to use when solving a problem.

ALGEBRA
Fraction Expressions and Equations

Learn how to evaluate expressions and solve equations with fractions.

Venus's-flytrap is found in the coastal regions of North and South Carolina.

Ann will be mailing some brochures about the different types of plants that eat insects. The price to mail them depends on the total weight. Each brochure weighs $\frac{3}{4}$ ounce. How much do the brochures weigh in all?

You can write and evaluate an expression. Choose a variable to represent the number of brochures.

Let b = the number of brochures.

$b \times \dfrac{3}{4}$ or $\dfrac{3}{4}b$ *Write the expression.*

The number of brochures Ann mails changes every week. What was the total weight of the 23 brochures she mailed last week?

$b \times \dfrac{3}{4}$ or $\dfrac{3}{4}b$ *Write the expression.*

$23 \times \dfrac{3}{4}$ *Replace b with 23.*

$\dfrac{69}{4}$, or $17\dfrac{1}{4}$ *Multiply.*

So, the total weight is $17\frac{1}{4}$ oz.

You have solved equations with whole numbers and decimals by using mental math. You can use mental math to solve some equations with fractions.

EXAMPLE

Solve the equation $n \div \frac{1}{4} = 8$ by using mental math.

$n \div \dfrac{1}{4} = 8$ *Remember, when dividing by $\frac{1}{4}$, you multiply by the reciprocal, 4.*

$n = 2$ *What number times 4 equals 8?*

$2 \div \dfrac{1}{4} = 8$ *Check your answer. Replace n with 2.*

$\dfrac{2}{1} \times \dfrac{4}{1} = \dfrac{8}{1}$, or 8   So, $n = 2$.

• Use mental math to solve $x + \frac{2}{3} = \frac{5}{6}$.

Think and Discuss ▶ Look back at the lesson to answer each question.

1. **What if** Ann mails 32 brochures? Show how you would find the total weight of the brochures.

2. **Explain** how you would solve the equation $d \div \frac{1}{3} = 9$.

Guided Practice ▶ Evaluate the expression.

3. $x - 1\frac{1}{5}$ for $1x = 3\frac{2}{5}$ 4. $\frac{1}{4}y$ for $y = \frac{1}{2}$ 5. $t + 2\frac{1}{4}$ for $t = 3\frac{3}{8}$

Use mental math to solve the equation.

6. $x - 4\frac{1}{4} = 2\frac{1}{2}$ 7. $\frac{2}{3}y = \frac{1}{4}$ 8. $2\frac{5}{8} = t + 1\frac{1}{4}$

Independent Practice ▶ Evaluate the expression.

9. $c + 2\frac{2}{3}$ for $c = 5\frac{1}{6}$ 10. $4\frac{5}{8}m$ for $m = 3\frac{1}{3}$ 11. $b \div 2\frac{1}{2}$ for $b = 3\frac{1}{8}$

12. $4\frac{5}{8} + a$ for $a = 4\frac{4}{8}$ 13. $b \times \frac{1}{5}$ for $b = \frac{1}{4}$ 14. $\frac{5}{8} \div c$ for $c = 5$

15. $\frac{21}{24} - a$ for $a = \frac{1}{2}$ 16. $c \div 1\frac{1}{4}$ for $c = 1\frac{2}{3}$ 17. $\frac{1}{3} + \frac{1}{2} + d$ for $d = \frac{1}{6}$

Use mental math to solve the equation.

18. $x - 3\frac{1}{2} = 2\frac{1}{4}$ 19. $\frac{1}{5}x = \frac{2}{5}$ 20. $14\frac{2}{3} + x = 28$

21. $8x = 2$ 22. $\frac{2}{3}t = 2$ 23. $3\frac{1}{3} + x = 4\frac{2}{3}$

Problem Solving Applications ▶

24. Ann's brochures tell about the eating habits of pitcher plants, sundew plants, and Venus's-flytraps. Suppose one type of plant traps about $3\frac{1}{2}$ oz of insects each day. Choose an operation. Then write and solve an equation to find how many days the plant takes to trap a total of $17\frac{1}{2}$ oz of insects.

Sundew Plant

25. ❓ **What's the Error?** Robin evaluated $\frac{3}{4}h$ for $h = \frac{3}{8}$ in this way: $\frac{3}{4} \times \frac{3}{8} = \frac{6}{8} \times \frac{3}{8} = \frac{9}{8}$. Explain her mistake.

26. $4\frac{5}{8} \div 2\frac{1}{6}$ (p. 210)

27. Order from greatest to least. $\frac{3}{4}, \frac{3}{8}, \frac{3}{5}$ (p. 166)

28. 25.95×13.3 (p. 70)

29. Solve. $1.5 = \frac{a}{3}$ (p. 82)

⭐ **30. TEST PREP** Which decimal is 10 times as great as 34.62? (p. 70)

     A 0.34   **B** 3.4   **C** 3.46   **D** 346.2

1. VOCABULARY Two numbers are __?__ if their product is 1. (pp. 208–209)

Estimate each product or quotient. (pp. 200–201)

2. $\frac{2}{9} \times \frac{1}{6}$ **3.** $\frac{7}{8} \div \frac{11}{12}$ **4.** $\frac{7}{15} \div \frac{8}{9}$

5. $2\frac{4}{5} \times 3\frac{3}{10}$ **6.** $4\frac{2}{15} \times 5\frac{4}{7}$ **7.** $31\frac{3}{8} \div 4\frac{1}{2}$

Multiply. Write the answer in simplest form. (pp. 202–207)

8. $\frac{1}{6} \times \frac{3}{5}$ **9.** $\frac{2}{3} \times \frac{4}{7}$ **10.** $16 \times \frac{5}{12}$ **11.** $\frac{3}{8} \times 10$

12. $1\frac{1}{2} \times \frac{3}{4}$ **13.** $4\frac{1}{2} \times 2\frac{1}{3}$ **14.** $1\frac{1}{2} \times \frac{2}{3}$ **15.** $2\frac{1}{3} \times 3\frac{1}{7}$

Find the quotient. Write it in simplest form. (pp. 210–213)

16. $\frac{6}{7} \div \frac{3}{5}$ **17.** $\frac{3}{4} \div \frac{1}{3}$ **18.** $3\frac{1}{3} \div 2\frac{4}{5}$ **19.** $9\frac{1}{2} \div 1\frac{3}{8}$

20. $8 \div \frac{6}{7}$ **21.** $\frac{4}{5} \div 4$ **22.** $2\frac{3}{5} \div 4\frac{1}{5}$ **23.** $\frac{5}{8} \div 10$

Evaluate the expression. (pp. 216–217)

24. $x - 3\frac{1}{3}$ for $x = 6\frac{1}{5}$ **25.** $\frac{3}{4} \div r$ for $r = 4$ **26.** $25\frac{1}{8} + m$ for $m = 6\frac{2}{3}$

27. $x + 3\frac{1}{10}$ for $x = 1\frac{1}{2}$ **28.** $\frac{7}{8} a$ for $a = \frac{2}{5}$ **29.** $\frac{3}{4} b$ for $b = 1\frac{1}{6}$

Use mental math to solve the equation. (pp. 216–217)

30. $x - 6\frac{1}{9} = 3\frac{2}{3}$ **31.** $x \div \frac{1}{3} = 12$ **32.** $12\frac{3}{4} + x = 15\frac{7}{8}$

33. $1\frac{1}{8} b = 4\frac{1}{20}$ **34.** $c - \frac{1}{5} = \frac{1}{2}$ **35.** $\frac{3}{7} \div a = \frac{1}{28}$

Solve.

36. Mike rides his bicycle $6\frac{1}{2}$ min to school. Cami rides her bicycle $1\frac{1}{2}$ times as long. How long does it take Cami to ride to school? (pp. 206–207)

37. Eric practiced $\frac{3}{4}$ hr on Monday and $\frac{2}{5}$ hr on Saturday. How much longer did Eric practice on Monday? (pp. 214–215)

38. A race course is $\frac{3}{4}$ mi long. Racers want to run three equal sprints. How far apart should the markers be for the sprints? (pp. 210–213)

39. Sara has $\frac{5}{6}$ yd of ribbon and Mark has $\frac{3}{4}$ yd of ribbon. How much ribbon do Sara and Mark have altogether? (pp. 214–215)

40. The sum of Kiesha's and Brent's heights is $127\frac{1}{4}$ in. Kiesha's height is $64\frac{1}{2}$ in. Write an equation to find Brent's height. (pp. 216–217)

Understand the problem.

See item **2.**

You know that this week's playing time is $\frac{2}{3}$ of last week's playing time. Use a *variable* to write any relationship like this one as an equation.

Also see problem **1**, p. H62.

Choose the best answer.

1. Carla rode her bicycle $1\frac{2}{5}$ miles on Monday, $2\frac{1}{4}$ miles on Wednesday, and $\frac{4}{5}$ mile on Friday. Which is a reasonable estimate of the total distance Carla rode her bicycle?

   A 3 mi **B** $4\frac{1}{2}$ mi **C** $5\frac{1}{2}$ mi **D** $6\frac{1}{2}$ mi

2. Vince is trying to decrease the amount of time he spends playing video games. This week he spent $5\frac{2}{3}$ hours playing video games. That was $\frac{2}{3}$ of the amount of time he played video games last week. How long did Vince play video games last week?

   F $4\frac{1}{2}$ hr **G** $6\frac{1}{3}$ hr **H** 7 hr **J** $8\frac{1}{2}$ hr

3. Eva ordered $2\frac{1}{2}$ pounds of potato salad, $2\frac{3}{4}$ pounds of fruit salad, and $1\frac{5}{8}$ pounds of cole slaw. Which operation would be best to use to find out how much more fruit salad she ordered than cole slaw?

   A addition **C** subtraction

   B multiplication **D** division

4. Which expression can be used to find $\frac{3}{5} \div \frac{1}{3}$?

   F $\frac{3}{5} + \frac{1}{3}$ **H** $\frac{3}{5} \times \frac{3}{1}$

   G $\frac{3}{5} - \frac{1}{3}$ **J** $\frac{5}{3} \times \frac{1}{3}$

5. Which is a reasonable estimate for $21\frac{4}{5} \div 3$?

   A 3 **B** 4 **C** 5 **D** 7

6. Solve.
   $$b - 4\frac{1}{2} = 2\frac{2}{3}$$

   F $b = 1\frac{5}{6}$ **H** $b = 6\frac{5}{6}$

   G $b = 6\frac{3}{5}$ **J** $b = 7\frac{1}{6}$

7. Cal has 4 bags of apples with masses of 2.5 kilograms, 2.05 kilograms, 3.12 kilograms, and 2.18 kilograms. Which shows the masses in order from lightest to heaviest?

   A 2.05, 2.18, 2.5, 3.12

   B 2.5, 2.18, 3.12, 2.05

   C 3.12, 2.5, 2.18, 2.05

   D 2.18, 2.5, 2.05, 3.12

8. The sum of the prime factors of 12 is 7. What is the sum of the prime factors of 72?

   F 12 **H** 42

   G 27 **J** Not here

Write What You Know

9. Textbooks are $1\frac{1}{2}$ in. thick. They are packed in boxes with 2 stacks of books in each box. A box is 15 in. high. How many textbooks can one box hold? Explain the method you used to solve the problem.

10. Explain how the Distributive Property can be used to find the product of a whole number and a mixed number. Use what you wrote to find the product $8 \times 6\frac{1}{2}$.

MATH DETECTIVE

On a Roll

REASONING Roll a number cube ten times. Record the numbers that you roll. Use your knowledge of fractions to attempt to solve the problems below. You may use each of your ten numbers exactly once. Once you use a number, cross it out so that you don't use it again. Depending on which numbers you roll, you may or may not be able to solve all four mysteries.

Fraction Mystery 1

Write a pair of equivalent fractions.

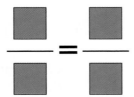

Fraction Mystery 2

Write a fraction with a value greater than 1.

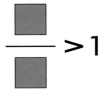

Fraction Mystery 3

Write a fraction with a value greater than $\frac{1}{2}$ and less than 1.

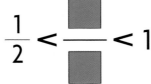

Fraction Mystery 4

Write a fraction that is not in simplest form.

Think It Over!

- 📓 **Write About It** Salvador rolled all 2's and 6's. Explain why he couldn't solve all four mysteries.

- **Stretch Your Thinking** Write a set of ten numbers with which you could solve only two of the mysteries.

Mixed Numbers and Time

Learn how to write times as mixed numbers, and how to add and subtract times.

The ticket agent at the airport told Mario that his flight from Portland, OR, to St. Paul, MN, would take "about $4\frac{1}{2}$ hours." The agent expressed the time as a mixed number.

To write a time as a mixed number, use the fact that there are 60 minutes in 1 hour.

EXAMPLE 1

Mario's flight from Portland to St. Paul took 4 hours 24 minutes. Write the time as a mixed number.

$24 \text{ min} = 24 \times \frac{1}{60} \text{ hr} = \frac{24}{60} \text{ hr}$ *Think: 60 min = 1 hr. So, 1 min = $\frac{1}{60}$ hr.*

$= \frac{2}{5} \text{ hr}$ *Write the fraction in simplest form.*

$4 \text{ hr } 24 \text{ min} = 4\frac{2}{5} \text{ hr}$ *Write the time.*

So, Mario's flight took $4\frac{2}{5}$ hr.

To add or subtract times, use the same methods you use to add or subtract mixed numbers.

EXAMPLE 2

Solve.

A.   2 hr 18 min
    + 3 hr 52 min
    ‾‾‾‾‾‾‾‾‾‾‾‾‾
     5 hr 70 min

$= 5 \text{ hr} + (60 + 10) \text{ min}$
$= 5 \text{ hr} + 1 \text{ hr} + \frac{10}{60} \text{ hr}$
$= 6 \text{ hr} + \frac{1}{6} \text{ hr}$
$= 6\frac{1}{6} \text{ hr}$

B.   8 hr 14 min →   7 hr (60 + 14) min
    −3 hr 20 min → −3 hr 20 min
    ‾‾‾‾‾‾‾‾‾‾‾‾‾‾‾‾‾‾‾‾‾‾‾‾‾‾‾‾‾‾‾‾‾

         7 hr 74 min
        −3 hr 20 min
        ‾‾‾‾‾‾‾‾‾‾‾‾
         4 hr 54 min

$= 4 \text{ hr} + \frac{54}{60} \text{ hr}$
$= 4\frac{9}{10} \text{ hr}$

TALK ABOUT IT

• Tell how you would write the time 7 hours 35 minutes as a mixed number.

TRY IT

Write as a mixed number.

1. 1 hr 30 min   **2.** 4 hr 55 min   **3.** 1 hr 19 min

4. 28 min   **5.** 45 min   **6.** 3 hr 42 min

Solve. Write as a mixed number.

7. 5 hr 12 min + 8 hr 18 min   **8.** 6 hr 10 min − 4 hr 20 min

VOCABULARY

1. The largest of the common factors of two numbers is the __?__ . (p. 151)

2. To add unlike fractions, you can write equivalent fractions by using the __?__ . (p. 182)

EXAMPLES	EXERCISES

Chapter 7

- **Write the prime factorization of a number.** (pp. 148–149)

  Find the prime factorization of 150.

  $2 \times 3 \times 5^2$

  150
  15 × 10
  3 × 5 × 2 × 5

Write the prime factorization of each number in exponent form.

3. 52   **4.** 125   **5.** 180

6. 27   **7.** 100   **8.** 72

- **Find the GCF and LCM of two or more numbers.** (pp. 150–153)

  Find the LCM of 9 and 15.

  9: 9, 18, 27, 36, 45   *Find multiples of 9*
  15: 15, 30, 45   *and 15.*
  LCM = 45   *Find the LCM.*

Find the GCF for each set of numbers.

9. 15, 30   **10.** 9, 12   **11.** 8, 16, 12

Find the LCM for each set of numbers.

12. 6, 10   **13.** 6, 8   **14.** 5, 9, 12

Chapter 8

- **Write fractions in simplest form.** (pp. 160–163)

  Write $\frac{18}{30}$ in simplest form.

  $\frac{18}{30} = \frac{18 \div 6}{30 \div 6} = \frac{3}{5}$ *Divide by the GCF.*

Write the fraction in simplest form.

15. $\frac{8}{12}$ **16.** $\frac{15}{20}$ **17.** $\frac{40}{50}$

18. $\frac{36}{48}$ **19.** $\frac{18}{21}$ **20.** $\frac{90}{40}$

- **Write mixed numbers as fractions and fractions as mixed numbers.** (pp. 164–165)

  Write $3\frac{5}{6}$ as a fraction.

  $3\frac{5}{6} = \frac{(3 \times 6)}{6} + \frac{5}{6} = \frac{23}{6}$

Write the mixed number as a fraction.

21. $4\frac{2}{3}$ **22.** $2\frac{8}{9}$ **23.** $8\frac{1}{2}$

Write the fraction as a mixed number.

24. $\frac{17}{2}$ **25.** $\frac{40}{7}$ **26.** $\frac{33}{4}$

- **Convert among fractions, decimals, and percents.** (pp. 169–171)

  Write $\frac{1}{8}$ as a percent.

  $\frac{1}{8} = 1 \div 8 = 0.125$ *Change $\frac{1}{8}$ to a decimal.*

  $0.125 = 12.5\%$
  So, $\frac{1}{8} = 12.5\%$.

Write the decimal as a fraction.

27. 0.75   **28.** 0.6   **29.** 0.43

Write the fraction as a percent.

30. $\frac{5}{8}$ **31.** $\frac{2}{5}$ **32.** $\frac{5}{2}$

Chapter 9

- **Add and subtract fractions and mixed numbers.** (pp. 186–193)

  Subtract. $4\frac{1}{4} - 2\frac{2}{3}$

  $$\begin{array}{rcl} 4\frac{1}{4} & = & 4\frac{3}{12} = 3\frac{15}{12} \\ -2\frac{2}{3} & = & -2\frac{8}{12} = -2\frac{8}{12} \\ \hline & & 1\frac{7}{12} \end{array}$$

  Write equivalent fractions. Rename as needed. Subtract fractions. Subtract whole numbers.

Add or subtract. Write the answer in simplest form.

33. $\frac{3}{4} + \frac{1}{3}$ 34. $\frac{3}{5} + \frac{1}{2}$

35. $\frac{9}{10} - \frac{3}{5}$ 36. $\frac{5}{6} - \frac{1}{4}$

37. $2\frac{3}{8} + 3\frac{3}{4}$ 38. $4\frac{3}{5} + 2\frac{1}{3}$

39. $4\frac{1}{3} - 1\frac{5}{6}$ 40. $5\frac{1}{8} - 3\frac{2}{3}$

Chapter 10

- **Multiply and divide fractions and mixed numbers.** (pp. 202–213)

  Divide. $3\frac{3}{4} \div 4\frac{1}{2}$

  $$3\frac{3}{4} \div 4\frac{1}{2} = \frac{15}{4} \div \frac{9}{2}$$

  Write mixed numbers as fractions.

  $$= \frac{15}{4} \times \frac{2}{9}$$

  Multiply by the reciprocal.

  $$= \frac{5}{6}$$

Multiply or divide. Write the answer in simplest form.

41. $\frac{1}{3} \times \frac{2}{5}$ 42. $\frac{5}{8} \times \frac{7}{10}$

43. $\frac{5}{9} \div \frac{1}{3}$ 44. $9 \div \frac{4}{5}$

45. $3\frac{3}{4} \times 3\frac{1}{3}$ 46. $3\frac{1}{8} \times 1\frac{1}{5}$

47. $1\frac{2}{3} \div 3\frac{5}{6}$ 48. $1\frac{7}{9} \div 2\frac{2}{5}$

- **Evaluate an algebraic expression using fractions.** (pp. 216–217)

  Evaluate $f + 5\frac{1}{8}$ for $f = 2\frac{1}{3}$.

  $f + 5\frac{1}{8}$

  $2\frac{1}{3} + 5\frac{1}{8}$ *Replace f with $2\frac{1}{3}$.*

  $2\frac{8}{24} + 5\frac{3}{24} = 7\frac{11}{24}$ *Add.*

Evaluate each expression.

49. $d + 2\frac{2}{3}$ for $d = 2\frac{5}{6}$

50. $k \times \frac{3}{4}$ for $k = 1\frac{3}{5}$

51. $m - 3\frac{4}{9}$ for $m = 6\frac{5}{18}$

52. $w \div 1\frac{1}{3}$ for $w = 2\frac{1}{2}$

PROBLEM SOLVING APPLICATIONS

53. Katelynn has two sheets of paper $8\frac{1}{2}$ in. by 11 in. She cuts out a 3-in. square from the center top edge of one of the sheets. Which has the greater perimeter, the sheet with the square cut out or the uncut sheet? How much greater?

    (pp. 194–195)

54. Juan goes to the fruit market every 6 days. Anita goes to the market every 4 days. They meet at the market on August 31. When will they meet at the market again? (pp. 154–155)

Performance Assessment

TASK A • Fraction Fun

Margo finds the product of $3\frac{1}{2}$ and $8\frac{2}{3}$ by writing each as a fraction and multiplying numerators and denominators. Chen said he can find the product by using the Distributive Property. Here's how he starts:

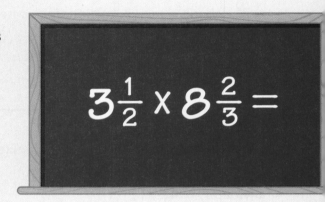

$3\frac{1}{2} \times 8\frac{2}{3} = (3 + \frac{1}{2}) \times (8 + \frac{2}{3})$

$= 3 \times (8 + \frac{2}{3}) + \frac{1}{2} \times (8 + \frac{2}{3})$

a. Finish Chen's solution to find the answer.

b. Use Margo's method to find $3\frac{1}{2} \times 8\frac{2}{3}$. Do you get the same answer as with Chen's method?

c. Tom said he thinks $(3 + \frac{1}{2}) \times (8 + \frac{2}{3}) = 3 \times 8 + \frac{1}{2} \times \frac{2}{3}$. Is he correct? Explain.

TASK B • Pen Pal

Assume that you have a pen pal and are going to start writing. This is your first letter. Write a short paragraph describing yourself and your activities. Include the following in the paragraph:

a. How you divide your time during a typical week

b. How your school day is divided by classes and other activities

c. What you like to do after school

As you write your letter, use fractions or decimals to name parts of classes, hours, days, or weeks. Each time you use a fraction, write the equivalent decimal in parentheses next to it. Do the reverse each time you use a decimal.

Technology Linkup

Mighty Math Astro Algebra

Fractions, Decimals, and Percents

Click [OFF DUTY], and then [ ].

- Choose Red Level D. Click [ON DUTY]. Then click [ ] and review the topics on translating among decimals, fractions, and percents.
- Close the AstroNet. Complete the mission by sorting the fractions, decimals, and percents.

Practice and Problem Solving

1. Click [ ]. Choose Red Level I. Click [OK] and then [ON DUTY]. Complete the mission by sorting the fractions, decimals and percents.

For 2–9, click [ ] and use the calculator for the problems. Write the fraction as a decimal. Tell whether the decimal terminates or repeats.

2. $\frac{5}{12}$
3. $\frac{7}{50}$
4. $\frac{15}{32}$
5. $\frac{5}{11}$

Write as a fraction.

6. 0.78
7. 0.045
8. 68%
9. 85%
10. In Ms. Dehmel's class, $\frac{4}{5}$ of the students ride the bus to school. In Mr. Eaton's class, 78% of the students ride the bus. Which class has a greater percent of students who ride the bus?
11. ***REASONING*** Erika needs to score 91% on her math test to get an "A" for the semester. She gets a score of $\frac{23}{25}$. Does Erika get an "A"? Explain.

Multimedia Math Glossary **www.harcourtschool.com/mathglossary**

12. **Vocabulary** Look up *prime factorization* in the Multimedia Math Glossary. use a factor tree to write the prime factorization of 256.

PROBLEM SOLVING ON LOCATION

In South Carolina

The palmetto tree is South Carolina's state tree.

Trees

South Carolina is a state known for its beautiful trees. In fact, the state takes its nickname, the Palmetto State, from a tree. The palmetto tree, a familiar sight in South Carolina, appears on the state flag and the state seal.

1. South Carolina's tallest sweetgum tree and water tupelo tree are found at Congaree Swamp National Monument. The water tupelo is 114 feet tall and the sweetgum is 159 feet tall. Express the water tupelo's height as a fraction of the sweetgum's height, in simplest form.

2. The diameter of the sweetgum tree is shown in the diagram. Write the diameter as a fraction.

$\leftarrow 5\frac{7}{25} \text{ ft} \rightarrow$

3. A butternut tree in Oconee County, South Carolina, is 71 feet tall. Classify the number 71 as prime or composite.

4. A live oak tree on John's Island, near Charleston, South Carolina, is believed to be about 1,400 years old. Write the prime factorization of 1,400.

Use Data For 5, use the table.

5. The table shows how the heights of three tall South Carolina trees compared with the height of the water tupelo tree at Congaree Swamp. Order the three trees from tallest to least tall.

Heights Compared to Water Tupelo Tree	
Black Locust	$\frac{27}{38}$
Sassafras	$\frac{25}{57}$
Southern Catalpa	$\frac{109}{114}$

Table Rock State Park has the most challenging hiking trails in South Carolina.

State Parks

Each year, thousands of visitors enjoy swimming, boating, fishing, and other outdoor activities in South Carolina's state parks. Many of the parks have hiking trails that allow close-up views of palmettos and others of the state's many varieties of trees.

Use Data   For 1–9, use the table.

1. Which trails are the same length?

2. Which park has the longest trail listed?

3. Suppose you hike both of the listed trails at Table Rock. How many miles will you hike in all?

4. Ari hiked the Sandhills Trail at an average rate of $\frac{3}{4}$ mile per hour. How long did the hike take?

5. How much longer is the Sulphur Springs Trail than the Brissey Ridge Trail?

Mr. Ristorcelli hiked the Pinnacle Mountain Trail every summer for the past 15 years.

6. How far has he hiked altogether?

7. How many times would he have to hike the Yemasee Trail in order to cover the same distance?

8. Kendra hiked four trails for a total of $8\frac{7}{10}$ miles. Three of the trails she hiked were Oconee Bell, Sandhills, and Pinnacle Mountain. What was the fourth trail?

9. One weekend Dean hiked the Sandhills Trail. The next weekend he hiked a trail that was a little more than twice as long. Which trail did he hike the second weekend?

South Carolina State Park Trails		
Park	**Trail**	**Approx. Length (mi)**
Devil's Fork	Oconee Bell	$1\frac{1}{2}$
Lake Warren	Yemasee	$\frac{1}{3}$
Paris Mountain	Sulphur Springs	4
Paris Mountain	Brissey Ridge	$2\frac{3}{10}$
Sesquicentennial	Sandhills	$\frac{3}{2}$
Table Rock	Carrick Creek	2
Table Rock	Pinnacle Mountain	$3\frac{2}{5}$

CHAPTER 11

Algebra: Number Relationships

The elevation of Mt. McKinley is 20,320 feet. It is the highest mountain in North America and has one of Earth's steepest vertical rises. With the high altitude, unpredictable weather, and steep, icy slopes, this mountain is a challenge to climb.

PROBLEM SOLVING Climbers are flown into base camp, where they start their climb at 7,200 ft. If a climb takes 20 days, what is the average altitude gained per day?

DATA LINK

Mt. McKINLEY		
Camp	**Altitude**	**Possible Time**
Base Camp	7,200 ft	0 day
Camp 2	9,500 ft	1st day
Camp 3	11,000 ft	3rd day
Camp 4	14,200 ft	10th day
Camp 5	16,200 ft	12th day
High Camp	17,200 ft	15th day
Summit	20,320 ft	20th day

Check What You Know

Use this page to help you review and remember important skills needed for Chapter 11.

✅ Vocabulary

Choose the best term from the box.

positive numbers
negative numbers
whole numbers

1. Numbers to the left of zero on the number line are ___?___.

2. Numbers greater than zero are ___?___.

✅ Locate Points on a Number Line (For Intervention, see p. H13.)

Copy each number line. Graph the numbers on the number line.

3. 4   **4.** 0   **5.** ⁻5   **6.** ⁻1

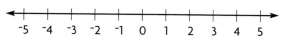

7. 0.5   **8.** $\frac{-1}{2}$ **9.** ⁻0.75   **10.** $1\frac{1}{4}$

✅ Sets of Numbers (For Intervention, see p. H12.)

Give four examples from each set of numbers.

11. whole numbers   **12.** counting numbers   **13.** odd numbers

✅ Compare Fractions (For Intervention, see p. H12.)

Compare. Write <, >, or = for each ●.

14. $9\frac{1}{2}$ ● $9\frac{3}{4}$ **15.** $\frac{2}{3}$ ● $\frac{3}{2}$ **16.** $\frac{3}{4}$ ● $\frac{3}{5}$

17. $\frac{16}{2}$ ● $\frac{24}{3}$ **18.** $2\frac{3}{5}$ ● $2\frac{1}{4}$ **19.** $\frac{12}{2}$ ● $2\frac{1}{12}$

✅ Temperature (For Intervention, see p. H12.)

Tell the temperature shown by each letter on the thermometer.

20. A   **21.** B   **22.** C

23. D   **24.** E   **25.** F

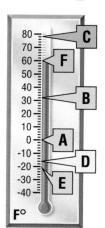

> **LOOK AHEAD**
>
> **In Chapter 11 you will**
> - identify integers and rational numbers
> - compare and order rational numbers
> - use number lines

Understand Integers

Learn how to identify integers and find absolute value.

Mount Sunflower in Kansas is 4,039 ft above sea level. New Orleans is 8 ft below sea level. Sea level equals 0 ft. You can use the integers ⁺4,039 and ⁻8 to represent these elevations.

New Orleans is located along the Mississippi River.

Vocabulary

integers
opposites
positive integers
negative integers
absolute value

Integers include all whole numbers and their **opposites**. Each integer has an opposite that is the same distance from 0 but on the opposite side of 0. The opposite of positive 8 (⁺8) is negative 8 (⁻8). The opposite of 0 is 0.

Integers greater than 0 are **positive integers**. Integers less than 0 are **negative integers**. The integer 0 is neither positive nor negative.

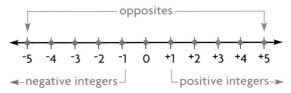

Negative integers are written with a negative sign, ⁻.

Positive integers can be written with or without a positive sign, ⁺.

EXAMPLE 1

Name an integer to represent the situation.

A. a gain of 12 yd   **B.** 30° below zero   **C.** a deposit of $100
⁺12   ⁻30°   ⁺100

The **absolute value** of an integer is its distance from 0. Look at ⁺3 and ⁻3. They are both 3 units from 0.

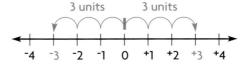

Write: $|{^-3}| = 3$ **Read:** The absolute value of negative three is three.
Write: $|{^+3}| = 3$ **Read:** The absolute value of positive three is three.

EXAMPLE 2

Use the number line to find each absolute value.

A. $|{^-4}|$ **B.** $|{^+2}|$ **C.** $|{^-1}|$ **D.** $|{^+3}|$
     4   2   1   3

CHECK FOR UNDERSTANDING

Think and Discuss ▶ Look back at the lesson to answer each question.

1. **Give an example** of a scale in real life in which zero is used together with other integers.

2. **Tell** what the absolute value of an integer is.

Guided Practice ▶ Write an integer to represent each situation.

3. 350 ft below sea level

4. an increase of 78 points

5. 14 degrees below zero

Write the opposite integer.

6. $^-289$ 7. $^-25$ 8. $^+315$ 9. $^+742$ 10. $^+993$

PRACTICE AND PROBLEM SOLVING

Independent Practice ▶ Write an integer to represent each situation.

11. a $5.00 decline in value

12. an increased attendance of 477

13. a gain of 12,000 ft in altitude

14. a decrease of 50 points

Write the opposite integer.

15. $^-2$ 16. $^-14$ 17. $^-31$ 18. $^+88$ 19. $^+207$

Find the absolute value.

20. $|^+390|$ 21. $|^-28|$ 22. $|^-727|$ 23. $|^+660|$ 24. $|^+795|$

Problem Solving Applications ▶

25. The elevation of the Dead Sea is about 1,310 ft below sea level. Write the elevation using an integer.

26. *REASONING* What values can n have if $|n| = 5$?

27. Chris has a stack of coins containing 5 quarters, 5 dimes, and 15 nickels. What fraction of the number of coins are quarters?

28.  **Write a problem** about two campers who camped at altitudes of 9,470 ft and 7,200 ft. Use positive and negative integers to describe the change in elevation.

The Dead Sea

MIXED REVIEW AND TEST PREP

29. Multiply. $\frac{4}{5} \times \frac{3}{4}$ (p. 202)

30. Lisa works for $6.50 an hour, 3 hours a day, 4 days a week. How many weeks will it take her to earn more than $300? (p. 76)

31. Find the value of the expression $9 \times (12 - 4) \div 2^3 + 4$ (p. 44)

32. Write the prime factorization of 72. (p.148)

33. **TEST PREP** Six soccer teams compete for the regional play-offs. Each team plays each of the other teams only once. How many games do they play? (p. 154)

     A 6   **B** 12   **C** 15   **D** 30

EXTRA PRACTICE page H42, Set A

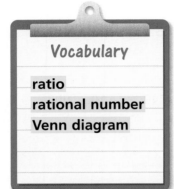

LESSON 11.2

Rational Numbers

Learn how to classify rational numbers and find another rational number between two rational numbers.

QUICK REVIEW

Write as a decimal and as a fraction.

1. eight tenths   2. fifty-four hundredths   3. three tenths

4. nineteen hundredths   5. forty thousandths

Vocabulary

ratio

rational number

Venn diagram

A **ratio** is a comparison of two numbers, a and b, written as a fraction $\frac{a}{b}$. A **rational number** is any number that can be written as a ratio $\frac{a}{b}$, where a and b are integers and $b \neq 0$. The numbers below are all rational numbers since they can be expressed as a ratio $\frac{a}{b}$.

$$3\frac{2}{5} \qquad 0.6 \qquad 42 \qquad {}^-2.5$$

Write each rational number as a ratio $\frac{a}{b}$.

EXAMPLE 1

A. $3\frac{2}{5}$ **B.** 0.6 **C.** 42 **D.** $^-2.5$

$$3\frac{2}{5} = \frac{17}{5} \qquad 0.6 = \frac{6}{10} \qquad 42 = \frac{42}{1} \qquad {}^-2.5 = \frac{^-5}{2}$$

The **Venn diagram** shows how the sets of rational numbers, integers, and whole numbers are related.

> **Rational Numbers**
> **Integers**
> **Whole Numbers**

EXAMPLE 2

Use the Venn diagram to determine in which set or sets each number belongs.

A. 80   The number 80 belongs in the sets of whole numbers, integers, and rational numbers.

B. $^-2$ The number $^-2$ belongs in the sets of integers and rational numbers but not in the set of whole numbers.

C. $6\frac{1}{2}$ The number $6\frac{1}{2}$ belongs in the set of rational numbers but not in the set of integers or the set of whole numbers.

D. 7.09   The number 7.09 belongs in the set of rational numbers but not in the set of integers or the set of whole numbers.

• Name two integers that are not also whole numbers.

Christopher is training to run in a 5-km road race. Yesterday he ran $4\frac{1}{4}$ km, and he plans to run $4\frac{1}{2}$ km tomorrow. What distance could he run today if he wants to run between $4\frac{1}{4}$ km and $4\frac{1}{2}$ km?

Think of the distances of Christopher's training runs as rational numbers.

One Way You can use a number line to find numbers between two rational numbers.

EXAMPLE 3

Find a distance between $4\frac{1}{4}$ km and $4\frac{1}{2}$ km, using the number line.

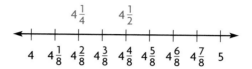

Notice that there is a mark between $4\frac{1}{4}$ and $4\frac{1}{2}$. That could be a distance Christopher could run.

So, Christopher could run $4\frac{3}{8}$ km today.

Another Way You can use a common denominator to find a number between two rational numbers.

EXAMPLE 4

Find a number between $4\frac{1}{4}$ and $4\frac{1}{2}$.

$4\frac{1}{4} = 4\frac{2}{8}$ $\qquad$ $4\frac{1}{2} = 4\frac{4}{8}$ $\qquad$ *Use a common denominator to write equivalent fractions.*

$4\frac{3}{8}$ is between $4\frac{2}{8}$ and $4\frac{4}{8}$. $\qquad$ *Find a rational number between the two numbers.*

So, $4\frac{3}{8}$ is between $4\frac{1}{4}$ and $4\frac{1}{2}$.

- What are some common denominators you could use to find other rational numbers between $4\frac{1}{4}$ and $4\frac{1}{2}$?

You can also find a number between two rational numbers in decimal form.

EXAMPLE 5

Find a rational number between $^{-}8.4$ and $^{-}8.5$.

$^{-}8.4 = ^{-}8.40$ $\qquad$ *Add a zero to each decimal.*

$^{-}8.5 = ^{-}8.50$

Use a number line to find a number between the two decimals.

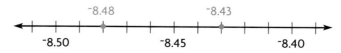

So, $^{-}8.43$, $^{-}8.45$, and $^{-}8.48$ are some of the numbers between $^{-}8.4$ and $^{-}8.5$.

Remember that you can add any number of zeros to the right of a decimal without changing its value.

231

CHECK FOR UNDERSTANDING

Think and ▶
Discuss

Look back at the lesson to answer each question.

1. **Tell** why every integer is a rational number. Give an example to support your answer.

2. **Tell** what numbers would be between $4\frac{1}{4}$ and $4\frac{1}{2}$ if you divided a number line into sixteenths.

Guided ▶
Practice

Write each rational number in the form $\frac{a}{b}$.

3. $^{-}0.37$　　**4.** $2\frac{4}{5}$　　　**5.** 0.889　　**6.** 7.31　　　**7.** $^{-}7\frac{1}{3}$

Use the number line to find a rational number between the two given numbers.

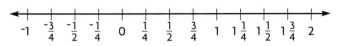

8. 1 and $1\frac{1}{2}$　　**9.** $^{-}\frac{3}{4}$ and $^{-}\frac{1}{4}$　　**10.** $\frac{1}{2}$ and 1　　**11.** $^{-}\frac{1}{2}$ and $\frac{1}{2}$

PRACTICE AND PROBLEM SOLVING

Independent ▶
Practice

Write each rational number in the form $\frac{a}{b}$.

12. $9\frac{2}{3}$　　**13.** $^{-}0.71$　　**14.** 80.4　　**15.** $^{-}2\frac{5}{8}$　　**16.** 3.18

Use the number line to find a rational number between the two given numbers.

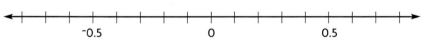

17. $^{-}0.2$ and $^{-}0.4$　　**18.** 0 and 0.2　　　**19.** $^{-}0.6$ and $^{-}0.8$

Find a rational number between the two given numbers.

20. $\frac{3}{4}$ and $\frac{1}{2}$　　　**21.** $^{-}7$ and $^{-}\frac{15}{2}$　　　**22.** 104.1 and $103\frac{7}{8}$

23. 16.1 and 16.01　　**24.** $3\frac{5}{8}$ and $\frac{27}{8}$　　**25.** $^{-}\frac{3}{4}$ and $^{-}\frac{3}{8}$

Tell if the first rational number is between the second and third rational numbers. Write *yes* or *no*.

26. $\frac{1}{3}$; $\frac{1}{2}$ and $\frac{3}{4}$　　　**27.** 0.97; 0.85 and 0.99　　**28.** 3.29; 3.20 and 3.25

29. 3.07; 3.1 and 3.01　　**30.** $^{-}8$; $^{-}\frac{29}{4}$ and $^{-}\frac{33}{4}$　　**31.** $^{-}\frac{1}{8}$; $^{-}\frac{1}{16}$ and $^{-}\frac{1}{4}$

Problem Solving ▶
Applications

32. Marie had already completed $\frac{1}{2}$ of the running race but had not yet reached the point $\frac{3}{4}$ of the way through the race. Could she have completed $\frac{5}{8}$ of the race? Explain.

33. Susan follows the directions of a treasure map and walks 45 steps north, then $\frac{24}{3}$ steps east, $\frac{75}{5}$ steps south, and 22 steps west. How many steps does Susan walk?

34. Is it easier to find a rational number between $\frac{1}{2}$ and $\frac{3}{4}$ or between 0.50 and 0.75? Explain your reasoning.

35. **?** **What's the Error?** Jeff says that every whole number is an integer and that every integer is a whole number. Explain his error.

MIXED REVIEW AND TEST PREP

36. Find the absolute value. $\left|{}^-88\right|$ (p. 228)

37. Write the decimal and fraction equivalents of 34%. (p. 169)

38. $\frac{2}{5} + \frac{4}{10} + \frac{4}{5}$ (p. 182)

39. **TEST PREP** What is the difference between the range and the mean of the number set 30, 8, 13, 20, and 24? (p. 106)

A 0 **B** 1 **C** 2 **D** 3

40. **TEST PREP** Which is the GCF of 54 and the number that is the LCM of 3, 5, 9, and 15? (p. 150)

F 6 **G** 9 **H** 18 **J** 45

PROBLEM SOLVING **LiNKÜP** to Reading

Strategy • Use Graphic Aids Graphic aids such as Venn diagrams, charts, and tables provide specific or important information in a visual form rather than in text. Sometimes, the information needed to solve a problem may be provided only in a graphic aid.

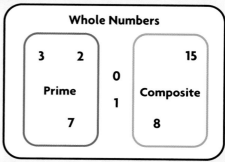

Look at the Venn diagram to the right. It shows the relationships among whole, prime, and composite numbers.

• Is a composite number always, sometimes, or never a whole number?

• Is a whole number always, sometimes, or never a prime number?

Use the Venn diagram to solve the following problems.

1. How would you describe the numbers 7 and 15? Are they whole numbers? Are they prime or composite?

2. Is 1 a whole number? Is it prime or composite? Explain.

EXTRA PRACTICE page H42, Set B

Compare and Order Rational Numbers

Learn how to compare and order rational numbers.

TECHNOLOGY LINK

To learn more about comparing rational numbers, watch the **Harcourt Math Newsroom Video** *Slow Down Light.*

QUICK REVIEW

Order the numbers from greatest to least.

1. 12, 14, 10 **2.** ⁻10, ⁻21, ⁻20 **3.** 3.10, 3.05, 3.15

4. $\frac{1}{4}, \frac{1}{2}, \frac{3}{4}$ **5.** $\frac{3}{5}, \frac{3}{6}, \frac{3}{4}$

Temperature commonly is measured on a scale in units of degrees. The scale contains both negative and positive numbers, like a number line. For example, the temperature in Death Valley has reached a high of 132°F, and the record low in Alaska is ⁻80°F.

You can use a number line to compare integers. On a number line, each number is greater than any number to its left and less than any number to its right. The number line below shows that ⁻80° < 132° and 132° > ⁻80°.

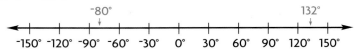

⁻150° ⁻120° ⁻90° ⁻60° ⁻30° 0° 30° 60° 90° 120° 150°

EXAMPLE 1

Compare the integers. Use < and >. Think about their positions on a number line.

A. 2 and ⁻3
⁻3 is to the left of 2 on the number line.
So, ⁻3 < 2, or 2 > ⁻3.

B. ⁻2 and ⁻4
⁻2 is to the right of ⁻4 on the number line.
So, ⁻2 > ⁻4, or ⁻4 < ⁻2.

It is easier to compare and order rational numbers when they are all expressed as decimals or as fractions with a common denominator.

EXAMPLE 2

Order $8\frac{1}{4}$, ⁻$8\frac{1}{2}$, and 8.3 from least to greatest.

Since ⁻$8\frac{1}{2}$ is the only negative number, it is the least number.

Compare $8\frac{1}{4}$ and 8.3.

$8\frac{1}{4} = 8.25$ and $8.3 = 8.30$ *Write the numbers as decimals with the same number of decimal places.*

$8.25 < 8.30$, or $8\frac{1}{4} < 8.3$ *Compare by looking at the place values.*

⁻$8\frac{1}{2} < 8\frac{1}{4} < 8.3$ *Order the three numbers.*

So, from least to greatest the numbers are ⁻$8\frac{1}{2}$, $8\frac{1}{4}$, and 8.3.

Think and ▶ Discuss

Look back at the lesson to answer each question.

1. **Describe** how you would compare 2.62 and $2\frac{3}{5}$.

2. **Give an example** of a rational number that is greater than $^-2.5$ and one that is less than $^-2.5$.

Guided ▶ Practice

Compare. Write $<$, $>$, or $=$ for each ⬤.

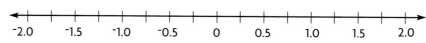

3. $^-1.5$ ⬤ $^-0.5$ 4. 0.5 ⬤ $^-1.0$ 5. $1\frac{1}{4}$ ⬤ 1.5 6. $^-2$ ⬤ $^-1\frac{1}{2}$

Independent ▶ Practice

Compare. Write $<$, $>$, or $=$ for each ⬤.

7. $\frac{1}{2}$ ⬤ $\frac{3}{4}$ 8. $\frac{1}{4}$ ⬤ $^-\frac{1}{2}$ 9. 0.5 ⬤ $\frac{3}{8}$ 10. $^-\frac{1}{4}$ ⬤ 0.25

11. 1.25 ⬤ 1.75 12. $^-\frac{1}{4}$ ⬤ $^-\frac{1}{3}$ 13. 2 ⬤ $^-3$ 14. $\frac{4}{5}$ ⬤ 0.9

15. $3.2 + 4.4$ ⬤ $4\frac{3}{4} + 2\frac{3}{4}$ 16. $2\frac{3}{4} \times 4$ ⬤ $3\frac{1}{4} + 8.5$

Order the rational numbers from least to greatest.

17. $0.6, 0.4, 0.46$ 18. $^-\frac{1}{8}, ^-\frac{1}{2}, ^-\frac{3}{8}, ^-\frac{3}{4}$ 19. $^-0.2, \frac{1}{4}, 0, ^-\frac{1}{4}$

Problem Solving ▶ Applications

20. Lynda's times for running a mile are $5\frac{1}{2}$ min, 5.48 min, 5.51 min, and $5\frac{2}{5}$ min. What is the longest she has taken to run a mile?

21. The mean temperatures for three days were $^-3°$, $^-5°$, and $1°$. Order the temperatures from highest to lowest.

22. ✎ **Write About It**   Explain how you would order three numbers that include a positive fraction and decimal and a negative integer.

23. Write an integer to represent 450 ft below sea level. (p. 228)

24. Round 2.0955 to the nearest thousandth. (p. 58)

25. Write $3\frac{3}{7}$ as a fraction. (p. 164)

26. $\frac{7}{8} - \frac{1}{2}$ (p. 182)

⭐ 27. **TEST PREP**   The swim team practiced for 8 hours the first week, $1\frac{1}{4}$ times as long the second week, 12 hours the third week, and $\frac{24}{4}$ hours the fourth week. How many hours total did the swim team practice? (p. 206)

A 26   **B** 28   **C** 36   **D** 38

PROBLEM SOLVING STRATEGY
Use Logical Reasoning

Learn how to solve problems by using logical reasoning.

Stephen, Leticia, Shelley, and Ed emptied their banks. They found $7.75, $4.35, $5.00, and $10.00, but not necessarily in that order. Leticia has twice as much money as Shelley. Stephen has an amount between Shelley's and Leticia's. Who has $4.35?

Analyze

What are you asked to find?

What information are you given?

Choose

What strategy will you use?

You can *use logical reasoning*.

Solve

How will you solve the problem?

Take the clues one at a time. Use a table to help.

Only one box in each row and column can have a "yes."

Leticia has twice as much money as Shelley. So, Leticia must have $10.00 and Shelley $5.00. Fill in "yes" in those boxes, and fill in "no" in the rest of the boxes in those rows and columns.

	$4.35	$5.00	$7.50	$10.00
Stephen	no	no	yes	no
Leticia	no	no	no	yes
Shelley	no	yes	no	no
Ed	yes	no	no	no

Stephen has an amount between Shelley's and Leticia's. $7.50 is between $5.00 and $10.00, so Stephen must have $7.50. Fill in the rest of the boxes with "yes" and "no."

So, Ed has $4.35.

Check

How can you check your answer?

What if the amounts were $7.50, $4.35, $5.00, and $12.50? How could you change the clues in the problem?

PROBLEM SOLVING STRATEGIES

Draw a Diagram or Picture

Make a Model

Predict and Test

Work Backward

Make an Organized List

Find a Pattern

Make a Table or Graph

Solve a Simpler Problem

Write an Equation

▶ **Use Logical Reasoning**

Solve the problems by using logical reasoning.

1. Arthur, Victoria, and Jeffrey are in sixth, seventh, and eighth grades, although not necessarily in that order. Victoria is not in eighth grade. The sixth grader is in chorus with Arthur and in band with Victoria. Which student is in each grade?

2. Use the following information to tell which numbers in the box at the right are A, B, C, D, and E.

$^-3.5$	$2\frac{1}{2}$
	0.43
4.3	$^-0.43$

    - A is greater than D and less than C.
    - A and D are opposites.
    - E is the greatest number.

For 3–4, use this information.

Amir, Katherine, Patrick, and Lee are comparing their stamp collections. Lee has twice as many stamps as Amir. Patrick has 5 fewer stamps than Katherine, who has 5 fewer stamps than Lee. Their collections consist of 15, 20, 25, and 30 stamps.

3. If Amir has 15 stamps, how many stamps does Patrick have?

    A 30   **C** 20
    B 25   **D** 15

4. The four have a total of 90 stamps. How many stamps does Katherine have?

    F 30   **H** 20
    G 25   **J** 15

MIXED STRATEGY PRACTICE

5. Kathy, Arlene, and Helene are in line for concert tickets, though not necessarily in that order. Helene is behind Kathy. Kathy is not first in line. Tell the order of the three in line.

6. Mel sold 25 beanbag toys at a fair. He sold some for $8 and some for $5. He made $170. How many toys did he sell for $8?

7. A guidebook costs $3.75 in Canadian dollars in a Vancouver, B.C., bookstore. It also has a price of $3.50 in United States dollars. If the exchange rate is $1.12 Canadian dollars for each U.S. dollar, would you rather pay in Canadian or U.S. dollars?

8. Shelters are located at regular intervals along a wilderness trail. The distance from the first shelter to the fourth shelter is 6 mi. What is the distance from the seventh shelter to the thirteenth shelter?

9. Sandy has to be at the airport for a 5:20 P.M. flight. She wants to arrive 1 hr 15 min early. If it takes 50 min to drive to the airport, when should she leave for the airport?

10. ❓ **What's the Question?** Anne bought 28 fish. She bought three times as many goldfish as angelfish. The answer is 7.

1. **VOCABULARY** Positive whole numbers, their opposites, and 0 make up the set of __?__ . (p. 228)

2. **VOCABULARY** A number that can be written as a ratio $\frac{a}{b}$, where a and b are integers and $b \neq 0$, is a(n) __?__ . (p. 230)

Write an integer to represent each situation. (pp. 228–229)

3. an increase of 15 points

4. 6 degrees below zero

5. a loss of 20 pounds

Write the opposite integer. (pp. 228–229)

6. $^-32$

7. 12

8. $^-15$

9. $^-289$

10. 0

Find the absolute value. (pp. 228–229)

11. $\left|^-12\right|$

12. $\left|^-4\right|$

13. $\left|^+17\right|$

14. $\left|^+8\right|$

15. $\left|^-347\right|$

Write each rational number in the form $\frac{a}{b}$. (pp. 230–233)

16. 2

17. $^-0.89$

18. $3\frac{2}{3}$

19. 5.4

20. 14

21. 0.334

Find a rational number between the two given numbers. (pp. 230–233)

22. $\frac{1}{4}$ and $\frac{2}{3}$

23. 1.3 and 1.32

24. $\frac{2}{5}$ and $\frac{3}{4}$

25. $^-4.3$ and $^-4.4$

26. 3.4 and 3.52

27. $2\frac{1}{5}$ and $2\frac{1}{2}$

28. $3\frac{5}{8}$ and $3\frac{9}{10}$

29. 0.9 and 0.94

Compare. Write $<$, $>$, or $=$ for each ●. (pp. 234–235)

30. $7\frac{5}{8}$ ● $7\frac{10}{16}$

31. 1.01 ● 1.10

32. $^-3.4$ ● $^-4.3$

33. $\frac{^-25}{3}$ ● $\frac{^-17}{2}$

34. $4\frac{3}{4}$ ● 4.77

Order the rational numbers from greatest to least. (pp. 234–235)

35. 3.7, 3.2, $3\frac{5}{8}$

36. $^-8$, $^-3$, $\frac{^-77}{11}$

37. $^-1.2$, 1.2, 0.12

38. $\frac{5}{7}$, $\frac{9}{14}$, $\frac{11}{8}$

39. Elizabeth, Doria, Emily, and Claudia each earned money doing odd jobs. They earned $4.50, $6.50, $8.00, and $9.00. Doria earned twice as much as Elizabeth. Emily earned $1.50 more than Claudia. Claudia earned $2.00 more than Elizabeth. How much did each person earn? (pp. 236–237)

40. Louisa, Chris, and Vicki each have one pet: a dog, a cat, and an iguana. Louisa does not have a cat and Chris does not have a dog. Louisa's pet is a reptile. What pet does each girl have? (pp. 236–237)

Look for important words.

See item **7.**

Important words are *least to greatest*. Write the numbers as decimals and order them from least to greatest.

Also see problem **2**, p. H62.

Choose the best answer.

1. Which of the following can be used to represent the depth of a cave that goes 37 feet below ground level?

   A $^-37$

   B $^-3.7$

   C $^+3.7$

   D $^+37$

2. Which points on the number line show opposite integers?

   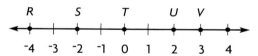

   F R and S

   G S and U

   H U and V

   J T and R

3. With respect to sea level, the average depth of the Atlantic Ocean is $^-11{,}730$ feet, the Pacific Ocean is $^-12{,}925$ feet, and the Gulf of Mexico is $^-5{,}297$ feet. Which shows these values in order from least to greatest?

   A $^-11{,}730, \ ^-12{,}925, \ ^-5{,}297$

   B $^-11{,}730, \ ^-5{,}297, \ ^-12{,}925$

   C $^-12{,}925, \ ^-11{,}730, \ ^-5{,}297$

   D $^-12{,}925, \ ^-5{,}297, \ ^-11{,}730$

4. Last week, the dance band practiced for $5\frac{1}{3}$ hours. The band usually practices for $8\frac{3}{4}$ hours per week. How many fewer hours did the band practice last week?

   F $2\frac{3}{4}$ hr

   G $3\frac{1}{3}$ hr

   H $3\frac{5}{12}$ hr

   J $4\frac{1}{5}$ hr

5. Earl read that $\frac{3}{8}$ of his town's budget is spent on road repairs. Which is that amount expressed as a decimal?

   A 0.375 **B** 0.4 **C** 0.67 **D** 0.875

6. Which symbol makes this number sentence true?

   $$^-3 \ \bullet \ 0$$

   F $<$ **G** $>$ **H** $=$ **J** $\geq$

7. Which shows the numbers in order from least to greatest?

   $$1.7, \ \frac{1}{2}, \ ^-0.3, \ \frac{^-3}{5}$$

   A $\frac{1}{2}, \ 1.7, \ \frac{^-3}{5}, \ ^-0.3$

   B $\frac{^-3}{5}, \ ^-0.3, \ 1.7, \ \frac{1}{2}$

   C $^-0.3, \ \frac{1}{2}, \ \frac{^-3}{5}, \ 1.7$

   D Not here

8. Vicki has 8 mysteries, 6 biographies, and 10 adventure books. What part of her book collection is biographies?

   F $\frac{1}{8}$ **G** $\frac{1}{6}$ **H** $\frac{1}{4}$ **J** $\frac{1}{3}$

Write What You Know

9. Four baseball players are standing in line. Owen is standing between Paul and Jim. Casey is in front of Paul. Jim is last. Find how they are arranged in line. Explain how you found your answer.

10. Explain what it means for a number to be rational. Then show that both 1.02 and $3\frac{1}{3}$ are rational.

CHAPTER 12

Algebra: Operations with Integers

NASA's Pathfinder and Sojourner space probes were built to withstand extreme temperatures as they traveled from Earth to Mars.

PROBLEM SOLVING The south pole on Mars is made not of frozen water but of frozen carbon dioxide, which has a temperature of ⁻193°F. The coldest day recorded on Earth was ⁻129°F, in Antarctica. Which temperature is lower? By about how much?

DATA LINK

Temperature	Earth	Mars
Lowest Recorded	⁻129°F	⁻220°F
Highest Recorded	136°F	68°F
Average	57°F	⁻81°F

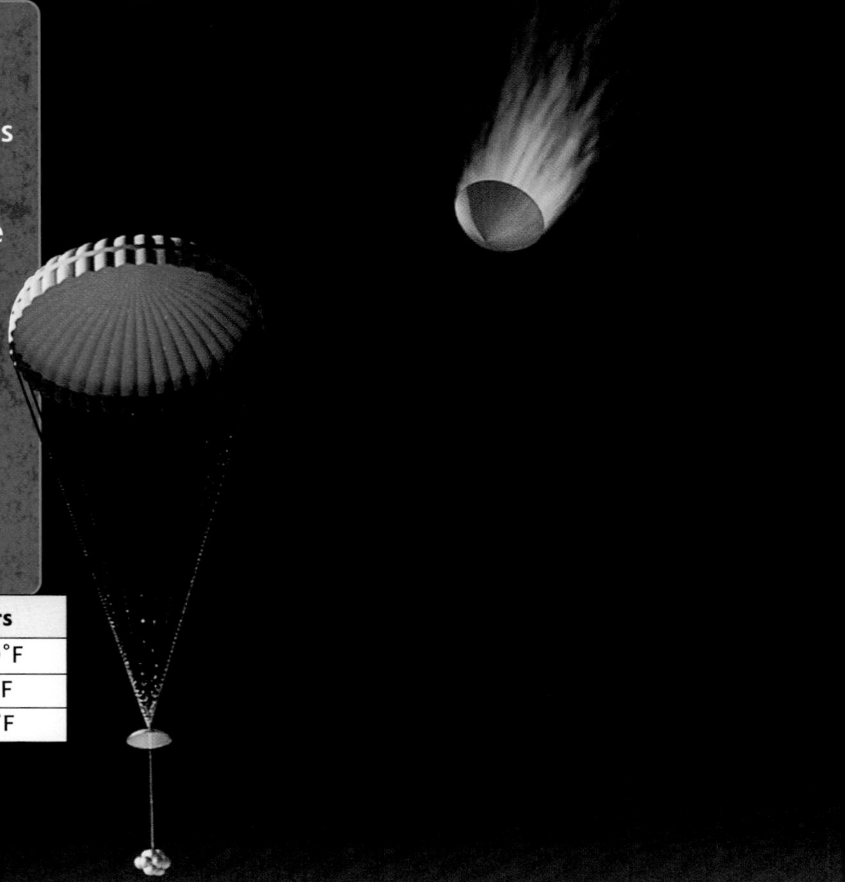

Check What You Know

Use this page to help you review and remember important skills needed for Chapter 12.

 Vocabulary

Choose the best term from the box.

1. The set of whole numbers and their opposites is the set of ___?___ .

> positive
> numbers
> integers

 Understand Integers (For Intervention, see p. H12.)

Write a positive or negative integer to represent each situation.

2. 14° below zero

3. 62 degrees above zero

4. 10 ft above sea level

5. 13 m below sea level

6. bottom of a well, 50 ft below the surface

7. a gain of 12 yards

8. a bank deposit of $280

9. 3 ft below ground level

 Number Lines (For Intervention, see p. H13.)

Name the integer that corresponds to the point.

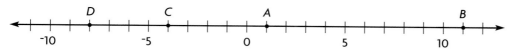

10. A

11. B

12. C

13. D

 Multiplication and Division Facts (For Intervention, see p. H13.)

Find the product or the quotient.

14. 4×6

15. 9×7

16. $81 \div 9$

17. $80 \div 8$

18. $24 \div 3$

19. 4×1

20. $120 \div 12$

21. 8×7

22. 12×6

23. $108 \div 9$

24. 3×11

25. $45 \div 9$

26. 7×12

27. $144 \div 12$

Solve for n.

28. $2 \times n = 14$

29. $n \times 5 = 15$

30. $n \div 6 = 9$

31. $88 \div n = 11$

32. $42 = n \times 6$

33. $11 = n \div 12$

> **LOOK AHEAD**
>
> **In Chapter 12 you will**
> - add and substract integers
> - multiply and divide integers
> - explore operations with rational numbers

Model Addition of Integers

M A T H
LAB

Explore how to use two-color counters to add integers.

You need two-color counters.

Vocabulary

additive inverse

Lin and Nina are playing a board game. To keep track of points, they are using yellow counters to represent positive points, or points gained, and red counters to represent negative points, or points lost.

Activity 1

• Lin earned 6 points during the first round. She earned 3 points during the second round. Use yellow counters to model Lin's total score.

First-round points   Second-round points

$6 + 3 = 9 \leftarrow$ total score

• Nina lost 2 points in the first round. Then she lost 5 points in the second round. Use red counters to model Nina's total score.

First-round points   Second-round points

          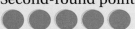

$^-2 + {}^-5 = {}^-7 \leftarrow$ total score

Think and Discuss

• How is adding the scores like adding whole numbers? How is it different?

• Would changing the order when adding Lin's or Nina's points change their scores? Why or why not?

• How would you model $2 + 7$? $^-2 + {}^-7$?

Practice

Use counters to find the sum.

1. $4 + 9$ **2.** $^-3 + {}^-7$ **3.** $^-6 + {}^-4$ **4.** $5 + 8$

The **additive inverse** of an integer is its opposite. 1 and ⁻1 are the additive inverses of each other. When you add an integer and its additive inverse, the sum is always 0. You can model this using counters.

- Model the sum of 1 and its additive inverse, ⁻1.

 = 0

- Model the sum of 5 and its additive inverse, ⁻5.

 = 0

- During a game, Carmen gained 8 points and then lost 5 points. To find her total score, Carmen paired points gained with points lost. Use yellow and red counters to model Carmen's total score. Remember that pairs of red and yellow counters equal 0.

 $8 + {}^{-}5 = 3$

- Robert gained 3 points and then lost 7 points. Use yellow and red counters to model Robert's total score.

$3 + {}^{-}7 = {}^{-}4$

Think and Discuss

- Why is Carmen's score positive?

- Why is Robert's score negative?

Practice

Use counters to find each sum.

1. $4 + {}^{-}6$ **2.** ${}^{-}2 + 6$ **3.** $7 + {}^{-}7$ **4.** ${}^{-}3 + 8$

5. $5 + 2$ **6.** $3 + 1$ **7.** ${}^{-}4 + {}^{-}5$ **8.** ${}^{-}3 + {}^{-}8$

MIXED REVIEW AND TEST PREP

Order the rational numbers from least to greatest. (p. 234)

9. ${}^{-}6.4, {}^{-}6.2, {}^{-}6.8$ **10.** $\frac{1}{3}, \frac{2}{5}, \frac{1}{4}, \frac{3}{5}$ **11.** ${}^{-}3\frac{4}{7}, {}^{-}3.6, {}^{-}3\frac{1}{2}$

12. Evaluate the expression $c + \frac{1}{2}$ for $c = \frac{1}{3}$. (p. 216)

⭐**13. TEST PREP** Which shows the difference $6\frac{5}{6} - 5\frac{3}{4}$? (p. 186)

A $\frac{1}{9}$ **B** $1\frac{1}{12}$ **C** $1\frac{1}{2}$ **D** $1\frac{3}{4}$

243

Add Integers

*L*earn how to use a number line to add integers.

*J*eb and Raul made up a game using a number line. Play starts at 0. A spinner is used to show positive moves and negative moves.

Jeb's first spin was ⁻3, and his second spin was ⁻2. What is Jeb's position on the number line?

⁻3 + ⁻2 = ⁻5 ← Jeb is at ⁻5.

*R*emember that you can write a positive number without the ⁺ sign.

⁺7 = 7

Raul's first spin was 4, and his second spin was ⁻9. Where is Raul on the number line?

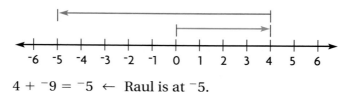

4 + ⁻9 = ⁻5 ← Raul is at ⁻5.

You can use a number line to find the sum of two integers.

EXAMPLE 1

Use a number line to find the sum 4 + ⁻6.

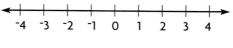

Draw a number line.

Start at 0. Move 4 units to the right to show 4.

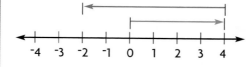

From 4, move 6 units to the left to show ⁻6. This takes you to ⁻2.

So, 4 + ⁻6 = ⁻2.

• When integers are added on a number line, when do the arrows point in the same direction and when do the arrows point in different directions?

Remember that the absolute value of an integer is its distance from 0 on the number line.

When adding integers, you can use their absolute values to find the sum.

> **Adding with the Same Sign**
> When adding integers with like signs, add the absolute values of the integers. Use the sign of the addends for the result.

EXAMPLE 2

Find the sum $^-7 + {}^-2$.

$^-7 + {}^-2$

$|^-7| + |^-2| = 7 + 2$ *Add the absolute values of the integers.*

$= 9$

So, $^-7 + {}^-2 = {}^-9$.   *Use the sign of the original addends.*

> **Adding with Different Signs**
> When adding integers with unlike signs, subtract the lesser absolute value from the greater absolute value. Use the sign of the addend with the greater absolute value for the result.

EXAMPLE 3

A. Find the sum $^-8 + 3$.

$^-8 + 3$

Subtract the lesser absolute value from the greater absolute value.

$|^-8| - |3| = 8 - 3$

$= 5$

Use the sign of the addend with the greater absolute value.

$|^-8| > |3|$ *The sum is negative.*

So, $^-8 + 3 = {}^-5$.

B. Find the sum $^-5 + 9$.

$^-5 + 9$

Subtract the lesser absolute value from the greater absolute value.

$|9| - |^-5| = 9 - 5$

$= 4$

Use the sign of the addend with the greater absolute value.

$|9| > |^-5|$ *The sum is positive.*

So, $^-5 + 9 = 4$.

• Find the sum $9 + {}^-12$.

EXAMPLE 4

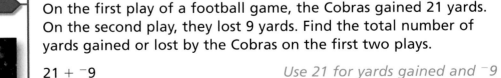

On the first play of a football game, the Cobras gained 21 yards. On the second play, they lost 9 yards. Find the total number of yards gained or lost by the Cobras on the first two plays.

$21 + {}^-9$ *Use 21 for yards gained and $^-9$ for yards lost.*

$|21| - |^-9| = 21 - 9 = 12$ *Subtract the lesser absolute value from the greater absolute value.*

$|21| > |^-9| \rightarrow 21 + {}^-9 = 12$ *Use the sign of the addend with the greater absolute value.*

So, the Cobras gained a total of 12 yards on the first two plays.

Think and ▶
Discuss
Look back at the lesson to answer each question.

1. **Explain** how you determine the sign of the sum of two integers with the same sign.

2. **Explain** how you determine if the sum of two integers with different signs is positive or negative.

3. **Tell** how you know the Cobras had a gain of 12 yards instead of a loss of 12 yards in Example 4.

Guided ▶
Practice
Write the addition problem modeled on the number line.

4.

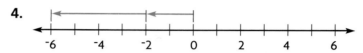

5.

Find the sum.

6. $^-9 + 6$ **7.** $^-3 + {}^-4$ **8.** $^-8 + 2$ **9.** $5 + {}^-7$

10. $^-3 + 7$ **11.** $^-8 + {}^-2$ **12.** $11 + {}^-5$ **13.** $6 + {}^-6$

Independent ▶
Practice
Write the addition problem modeled on the number line.

14.

15.

Find the sum.

16. $^-5 + 8$ **17.** $2 + {}^-3$ **18.** $7 + 2$ **19.** $^-1 + 4$

20. $8 + 7$ **21.** $^-12 + {}^-8$ **22.** $^-15 + {}^-10$ **23.** $^-17 + 25$

24. $^-2 + 5$ **25.** $^-12 + {}^-16$ **26.** $^-17 + 5$ **27.** $25 + {}^-37$

28. $24 + 12$ **29.** $30 + {}^-41$ **30.** $|16| + |{}^-9|$ **31.** $|{}^-64| + |36|$

ALGEBRA   **Use mental math to find the value of x.**

32. $^-5 + x = {}^-7$ **33.** $x + {}^-6 = {}^-13$

34. $x + {}^-10 = {}^-4$ **35.** $^-8 + x = {}^-3$

36. In the morning the temperature was ⁻6°F. By noon it had risen 11°F. What was the temperature at noon?

37. In the evening the temperature was ⁻9°F. By midnight it had dropped 3°F. What was the temperature at midnight?

38. Tina played a game in which she owned stock worth $33. During the game, the stock increased $11 in value and then decreased $15 in value. Write an addition sentence to find the new value of the stock.

39. On the first three plays of a football game, the Wildcats gained 15 yards, lost 9 yards, and lost 8 yards. Find the total number of yards gained or lost by the Wildcats on the first three plays.

40. **What's the Error?** Ken says that ⁻6 + 2 = 8. What is his error? What is the correct sum?

41. **Geometry** A rectangle with an area of 180 cm² has a length of 15 cm. Find the width.

MIXED REVIEW AND TEST PREP

42. Is ⁻1.5 less than, greater than, or equal to ⁻1$\frac{3}{6}$? (p. 234)

43. Is 8$\frac{3}{5}$ less than, greater than, or equal to 8.5? (p. 234)

44. Is ⁻3.4 less than, greater than, or equal to ⁻3$\frac{1}{5}$? (p. 234)

45. **TEST PREP** Find the quotient 4$\frac{2}{3}$ ÷ 3$\frac{1}{2}$. (p. 210)

A $\frac{3}{4}$ **B** 1$\frac{1}{4}$ **C** 1$\frac{1}{3}$ **D** 1$\frac{1}{2}$

46. **TEST PREP** Find the sum 5$\frac{3}{4}$ + 6$\frac{1}{3}$. (p. 186)

F 11$\frac{1}{12}$ **G** 11$\frac{4}{7}$ **H** 12$\frac{1}{12}$ **J** 12$\frac{11}{12}$

PROBLEM SOLVING | 🔆 Thinker's Corner

Math Fun • Opposites Distract Remember that every number has an opposite that is the same distance from zero but is on the opposite side on the number line. Use a number line to solve these riddles.

1. I am the opposite of a number between 2$\frac{2}{3}$ and 3$\frac{5}{6}$.

2. I am the first integer that is less than the opposite of a number between 2$\frac{1}{2}$ and 2$\frac{3}{4}$.

3. I am the opposite of a number between ⁻4.3 and ⁻4$\frac{3}{8}$.

4. We are between ⁻2.4 and ⁻2.6. If you add our opposites, you get 5.

5. I am the opposite of the integer between the sum ⁻5 + 2 and the sum ⁻15 + 10.

6. I am the second integer that is less than the sum ⁻22 + ⁻17.

7. I am the opposite of the even integer between the sum 35 + ⁻14 and the sum 12 + 6.

8. I am the integer that is 4 times the sum 42 + ⁻35.

Model Subtraction of Integers

Explore how to use two-color counters to subtract integers.

You need two-color counters.

QUICK REVIEW

1. 17 − 12   **2.** 42 + 20
3. 224 − 19   **4.** 132 − 18
5. 89 + 19

You can use red and yellow counters to subtract integers. Subtracting integers is similar to subtracting whole numbers.

Activity 1

• Find ⁻9 − ⁻4. First, make a row of 9 red counters.

• Then, take away 4 of them.

 → ⁻9 − ⁻4

• How many counters are left? What is ⁻9 − ⁻4?

Using red and yellow counters, model ⁻7 − 5.

• First, make a row of 7 red counters.

• Recall that a red counter paired with a yellow counter equals 0. Adding a red counter paired with a yellow counter does not change the value of ⁻7. Show another way to model ⁻7 that includes 5 yellow counters.

• Use your model to find ⁻7 − 5. Take away 5 yellow counters.

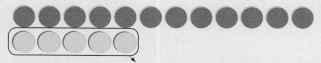

• What does your model show now? What is ⁻7 − 5?

Now model 7 − ⁻2.

• Model 7. Put down pairs of yellow and red counters until you can take away ⁻2. What is 7 − ⁻2?

Addition and subtraction of integers are related.

- Copy the model for ⁻7 below.

- Use the model to find ⁻7 − ⁻3. What is ⁻7 − ⁻3?

- Model ⁻7 again. Then add three yellow counters to find ⁻7 + 3. What is ⁻7 + 3?

TECHNOLOGY LINK

More Practice: Use E-Lab, *Modeling Subtraction of Integers.*
www.harcourtschool.com/elab2002

Think and Discuss

- The models above show that ⁻7 − ⁻3 = ⁻7 + 3. How are ⁻3 and 3 related? How are subtraction and addition related?

- How are the two models different?

You can write a subtraction problem as an addition problem by adding the opposite of the number you are subtracting.

$$6 - {}^-2 = 6 + 2 \leftarrow \text{Add the opposite of } {}^-2.$$

- How can you write ⁻6 − 2 as an addition problem?

Practice

Use counters to find the difference.

1. ⁻7 − 4   **2.** ⁻8 − ⁻5   **3.** ⁻13 − 9   **4.** ⁻9 − 4

Complete the addition problem.

5. ⁻6 − ⁻2 = ⁻6 + ▓   **6.** ⁻9 − ⁻3 = ⁻9 + ▓   **7.** ⁻7 − 5 = ⁻7 + ▓   **8.** ⁻19 − 12 = ⁻19 + ▓

MIXED REVIEW AND TEST PREP

Find the sum. (p. 244)

9. ⁻3 + 9   **10.** 2 + ⁻7   **11.** ⁻1 + ⁻7

12. Write an integer to represent 217 m below sea level. (p. 228)

13. TEST PREP Three friends shared a pizza that had 8 slices. There are $1\frac{1}{2}$ slices left. Alana had $2\frac{1}{2}$ slices, and Jill had $1\frac{1}{2}$ slices. How many slices did George have? (p. 182)

A $5\frac{1}{2}$ **B** 5   **C** 4   **D** $2\frac{1}{2}$

Subtract Integers

Learn how to use a number line to subtract integers.

1. $29 - 24$
2. $14 + 8$
3. $217 - 12$
4. $97 + 17$
5. $365 - 295$

During the summer of 1997, NASA landed the Mars Pathfinder on the planet Mars. On July 9, Pathfinder reported a temperature of $^-1°F$. On July 10, Pathfinder reported a temperature of $8°F$. Find the range of temperatures reported by Pathfinder from July 9 to July 10.

Pathfinder endured temperatures as low as $^-89°F$.

To find the range of temperatures, you need to find the difference of 8 and $^-1$, or $8 - {}^-1$. You can find the difference of two integers by adding the opposite of the integer you are subtracting. You can then use the rules for addition of integers.

The opposite of $^-1$ is 1. So, $8 - {}^-1$ becomes $8 + 1$.

$8 - {}^-1 = 8 + 1 = 9$

So, the range of temperatures was $9°F$.

EXAMPLE

During an experiment, a scientist recorded a high temperature of $^-9°C$ and a low temperature of $^-22°C$. What was the range of temperatures during the experiment?

$^-9 - {}^-22 = {}^-9 + 22$ *Write the subtraction problem as an addition problem. Use the rules for addition of integers.*

$^-9 + 22$

$|22| - |{}^-9| = 22 - 9$ *Subtract the lesser absolute value from the greater absolute value.*

$= 13$

$|22| > |{}^-9| \rightarrow {}^-9 - {}^-22 = 13$ *Use the sign of the addend with the greater absolute value.*

So, the range of temperatures during the experiment was $13°C$.

• During the afternoon, the temperature fell from $7°F$ to $^-5°F$. What was the range of temperatures?

Think and ▶
Discuss

Look back at the lesson to answer the question.

1. Tell how you would write the subtraction problem as an addition problem if the temperature at the end of the experiment in the example was $^-40°C$.

Guided ▶
Practice

Rewrite the subtraction problem as an addition problem.

2. $7 - 10$ **3.** $3 - {}^-6$ **4.** $^-1 - {}^-8$ **5.** $^-4 - 6$

Find the difference.

6. $4 - 8$ **7.** $^-7 - {}^-2$ **8.** $4 - {}^-5$ **9.** $1 - 8$

PRACTICE AND PROBLEM SOLVING

Independent ▶
Practice

Rewrite the subtraction problem as an addition problem.

10. $12 - 15$ **11.** $8 - {}^-11$ **12.** $^-6 - {}^-13$ **13.** $^-9 - 11$

Find the difference.

14. $6 - 11$ **15.** $^-9 - {}^-5$ **16.** $2 - {}^-1$ **17.** $3 - 5$

18. $7 - 11$ **19.** $^-5 - {}^-5$ **20.** $8 - {}^-3$ **21.** $4 - 9$

22. $31 - 37$ **23.** $35 - 39$ **24.** $|{}^-43| - |12|$ **25.** $|{}^-27| - |{}^-32|$

Evaluate.

26. $^-3 - {}^-5 + {}^-8$ **27.** $6 - {}^-4 + {}^-5$ **28.** $8 - {}^-6 - 10$

Problem Solving ▶
Applications

29. On Friday morning, the temperature in Anchorage, Alaska, was $^-12°F$. By that evening, the temperature had fallen 7°. The temperature on Saturday morning was 5° higher than the temperature on Friday evening. What was the temperature on Saturday morning?

30. The water level of a river was 3 ft above normal. After an unusually dry season, the water level is 6 ft below normal. Find the range of the water levels of the river.

31. **?** **What's the Question?** The temperature at noon was 15°F. By midnight the temperature was $^-3°F$. The answer is 18°F.

MIXED REVIEW AND TEST PREP

32. Find the sum $^-4 + 9$. (p. 244)

33. Write the opposite integer for 213. (p. 228)

34. Find the quotient $5\frac{3}{4} \div 2\frac{1}{4}$. (p. 210)

35. Find the prime factorization of 84. (p. 148)

⭐ **36. TEST PREP** Solve $x + 9 = 21$ using mental math. (p. 30)

    A $x = 9$ **B** $x = 11$ **C** $x = 12$ **D** $x = 30$

Multiply and Divide Integers

Learn how to multiply and divide integers.

QUICK REVIEW

1. 80 × 6   **2.** 4 × 25   **3.** 50 × 6   **4.** 3,600 ÷ 4   **5.** 540 ÷ 60

Use red and yellow counters to model multiplication of integers. A red counter represents ⁻1 and a yellow counter represents ⁺1.

Activity 1

You need: two-color counters

- Use yellow counters to model the product 2 × 3.

   ← 2 groups of ⁺3

  2 × 3 = 6

- Use red counters to model the product 2 × ⁻3.

  ← 2 groups of ⁻3

  2 × ⁻3 = ⁻6

- Use red counters to model the product ⁻2 × 3. Using the Commutative Property, you can write ⁻2 × 3 as 3 × ⁻2.

   ← 3 groups of ⁻2

  ⁻2 × 3 = 3 × ⁻2 = ⁻6

- How could you model the product 3 × 4?

- How could you model the product 3 × ⁻4?

- What do you notice about the product of two positive integers? of a positive integer and a negative integer?

The depth of a submarine is changing ⁻30 m every minute. If the submarine started at the surface of the ocean, how far below the surface is the submarine after 4 min?

Use a number line to find the product 4 × ⁻30.

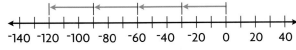

The number line shows that the depth of the submarine changed ⁻120 m. So, the submarine is 120 m below the surface.

You can use patterns to find rules to multiply integers.

EXAMPLE 1

Complete the pattern.

$4 \times 3 = 12$
$4 \times 2 = 8$
$4 \times 1 = 4$
$4 \times 0 = 0$
$4 \times {}^-1 = \blacksquare$
$4 \times {}^-2 = \blacksquare$
$4 \times {}^-3 = \blacksquare$

Study the pattern. As the second factor decreases by 1, the product decreases by 4. Use this to complete the pattern.

$4 \times 3 = 12$
$4 \times 2 = 8$
$4 \times 1 = 4$
$4 \times 0 = 0$
$4 \times {}^-1 = {}^-4$
$4 \times {}^-2 = {}^-8$
$4 \times {}^-3 = {}^-12$

So, the missing products are ⁻4, ⁻8, and ⁻12.

• What is the sign of the product of a positive integer and a negative integer?

EXAMPLE 2

Complete the pattern.

${}^-4 \times 3 = {}^-12$
${}^-4 \times 2 = {}^-8$
${}^-4 \times 1 = {}^-4$
${}^-4 \times 0 = 0$
${}^-4 \times {}^-1 = \blacksquare$
${}^-4 \times {}^-2 = \blacksquare$
${}^-4 \times {}^-3 = \blacksquare$

Study the pattern. As the second factor decreases by 1, the product increases by 4. Use this to complete the pattern.

${}^-4 \times 3 = {}^-12$
${}^-4 \times 2 = {}^-8$
${}^-4 \times 1 = {}^-4$
${}^-4 \times 0 = 0$
${}^-4 \times {}^-1 = 4$
${}^-4 \times {}^-2 = 8$
${}^-4 \times {}^-3 = 12$

So, the missing products are 4, 8, and 12.

• What is the sign of the product of two positive integers? two negative integers?

Examples 1 and 2 lead to the rules below.

> The product of two integers with like signs is positive.
> The product of two integers with unlike signs is negative.

Multiplication and division are inverse operations. To solve a division problem, think about the related multiplication problem.

$$42 \div 7 = \blacksquare \quad \rightarrow \quad 7 \times 6 = 42, \text{ so } 42 \div 7 = 6.$$

Use related multiplication problems to determine the sign of the quotient when dividing integers.

$8 \times 3 = 24$, so $24 \div 8 = 3$. ${}^-8 \times {}^-3 = 24$, so $24 \div {}^-8 = {}^-3$.

${}^-8 \times 3 = {}^-24$, so ${}^-24 \div {}^-8 = 3$. $8 \times {}^-3 = {}^-24$, so ${}^-24 \div 8 = {}^-3$.

In the problems above, look at the sign of the quotient of two positive integers and the sign of the quotient of two negative integers. Then look at the sign of the quotient of a positive integer and a negative integer. The rules below apply when dividing integers.

> The quotient of two integers with like signs is positive.
> The quotient of two integers with unlike signs is negative.

EXAMPLE 3

Find the quotient.

A. $^-84 \div ^-7$

   $^-84 \div ^-7 = 12$

Divide as with whole numbers. The quotient is positive since the integers have like signs.

B. $^-55 \div 11$

   $^-55 \div 11 = ^-5$

Divide as with whole numbers. The quotient is negative since the integers have unlike signs.

- Will the quotient $^-72 \div 8$ be positive or negative? Will the quotient $^-72 \div ^-8$ be positive or negative?

EXAMPLE 4

The low temperatures for five days in Fairbanks, Alaska, were $^-3°F$, $^-8°F$, $2°F$, $3°F$, and $^-4°F$. Find the average low temperature for the five days.

Find the sum. Divide the sum by 5.

$^-3 + ^-8 + 2 + 3 + ^-4 = ^-10$
$^-10 \div 5 = ^-2$

So, the average low temperature was $^-2°F$.

CHECK FOR UNDERSTANDING

Think and Discuss ▶ **Look back at the lesson to answer each question.**

1. **Find** the product $5 \times ^-7$. Find the product $^-5 \times ^-7$.

2. **Tell** how the rules for multiplying two integers compare with the rules for dividing two integers.

Guided Practice ▶ **Find the product or quotient.**

  3. $^-9 \times 6$ **4.** $^-3 \times ^-4$ **5.** $^-8 \times 2$ **6.** $8 \times ^-7$

  7. $^-21 \div 7$ **8.** $^-16 \div ^-2$ **9.** $120 \div ^-5$ **10.** $^-132 \div 6$

PRACTICE AND PROBLEM SOLVING

Independent Practice ▶ **Find the product or quotient.**

  11. $^-5 \times 8$ **12.** $2 \times ^-3$ **13.** 7×2 **14.** $^-9 \times 4$

  15. $84 \div 7$ **16.** $^-72 \div ^-8$ **17.** $^-63 \div ^-7$ **18.** $^-70 \div 7$

  19. 24×12 **20.** $30 \times ^-12$ **21.** $^-75 \times 4$ **22.** $^-62 \times ^-20$

  23. $432 \div 12$ **24.** $255 \div ^-15$ **25.** $^-960 \div ^-12$ **26.** $^-4,978 \div 19$

Use mental math to find the value of *y*.

27. $y \times {}^-6 = {}^-18$ **28.** $y \times {}^-10 = 40$ **29.** ${}^-8 \times y = 32$

30. $y \div {}^-6 = {}^-3$ **31.** $y \div {}^-7 = 5$ **32.** ${}^-18 \div y = 2$

Problem Solving ▶
Applications

33. Drought conditions caused the water level in a lake to change by ⁻3 in. per month in May and June and by ⁻4 in. per month in July and August. Write the change in water level over these four months as a negative number.

34. The low temperatures for the four days of a winter festival were ⁻8°F, ⁻6°F, ⁻9°F, and ⁻1°F. Find the average low temperature for the four days.

35. Use the Associative Property to help you find the product.
⁻3 × ⁻2 × ⁻5.

36. During the first 6 months of business, a sporting goods store showed its losses as ⁻$3,054. How would it show the average monthly loss?

37. The depth of a submarine changed ⁻630 ft over a period of 9 minutes. If the submarine descended at a constant rate, what was the change of depth per minute?

38. ❓ **What's the Error?** Beth says that ⁻5 × ⁻2 = ⁻10. What is her error? What is the correct product?

39. ❓ **What's the Question?** In a division problem, the divisor is ⁻250 and the dividend is 10. The answer is negative.

40. NUMBER SENSE Joe picked a number, added 5, multiplied the sum by 3, subtracted 10, and doubled the result. His final result was 28. What number had Joe picked?

MIXED REVIEW AND TEST PREP

41. $16 - {}^-6$ (p. 250)   **42.** ${}^-32 + {}^-42$ (p. 244)   **43.** $5\frac{2}{3} \times 6\frac{3}{4}$ (p. 206)

44. TEST PREP Evaluate $2k$ for $k = 9.6$. (p. 82)

    A 19.2   **B** 11.6   **C** 7.6   **D** 4.8

45. TEST PREP Write the fraction $\frac{5}{8}$ as a percent. (p. 169)

    F 0.625%   **G** 6.25%   **H** 62.5%   **J** 625%

EXTRA PRACTICE page H43, Set C

Explore Operations with Rational Numbers

Learn how to add, subtract, multiply, and divide with rational numbers.

QUICK REVIEW

1. $2 \times {}^-5$ 2. ${}^-7 \times {}^-3$

3. ${}^-4 \times 9$ 4. $18 \div {}^-6$

5. ${}^-24 \div {}^-8$

LOW TIDES				
DATE	**A.M.**	**Feet**	**P.M.**	**Feet**
Dec. 7	6:41	2.6	7:35	${}^-0.4$
Dec. 8	7:19	3.1	8:09	${}^-0.4$
Dec. 9	7:56	3.3	8:43	${}^-0.3$
Dec. 10	8:32	3.4	9:15	${}^-0.1$
Dec. 11	9:10	3.5	9:47	0.1
Dec. 12	9:51	3.6	10:22	0.3
Dec. 13	10:37	3.6	11:01	0.6

When you add, subtract, multiply, and divide rational numbers, use the same rules for determining signs as you use with integers.

EXAMPLE 1

Ocean tides are caused by the gravitational pull of the moon and the sun. On most days there are two high tides and two low tides. The table above shows the heights of the low tides, in feet, for one week in Astoria, Oregon. What is the range of the heights of the low tides?

The highest low tide measures 3.6 ft. The lowest measures ${}^-0.4$ ft.

$3.6 - {}^-0.4 = 3.6 + 0.4$ *Write the subtraction problem as an addition problem.*

$\phantom{3.6 - {}^-0.4} = |3.6| + |0.4|$ *Add the absolute values of the integers.*

$\phantom{3.6 - {}^-0.4} = 3.6 + 0.4$

$\phantom{3.6 - {}^-0.4} = 4.0$

So, the range of the heights of the low tides is 4.0 ft.

- Low tide on December 1 was 2.5 ft higher than on December 9. How high was the low tide on December 1?

EXAMPLE 2

Find the product. ${}^-3\frac{1}{2} \times {}^-4$

${}^-3\frac{1}{2} \times {}^-4 = \dfrac{{}^-7}{2} \times \dfrac{{}^-4}{1}$ *Write the mixed number as a fraction.*

$\phantom{{}^-3\frac{1}{2} \times {}^-4} = \dfrac{28}{2}$ *Multiply. The fractions have like signs, so the product is positive.*

$\phantom{{}^-3\frac{1}{2} \times {}^-4} = 14$ *Simplify.*

So, ${}^-3\frac{1}{2} \times {}^-4 = 14$.

EXAMPLE 3

As the tide went out, the water level dropped $15\frac{1}{2}$ in. in $2\frac{1}{2}$ hours. What was the average change in the water level per hour? Estimate to check that your answer is reasonable.

Find $^-15\frac{1}{2} \div 2\frac{1}{2}$.

Estimate. $^-16 \div 2 = {}^-8$

$^-15\frac{1}{2} \div 2\frac{1}{2} = \frac{^-31}{2} \times \frac{2}{5}$ *Multiply by the reciprocal of the divisor. The numbers have unlike signs, so the product is negative.*

$= \frac{^-31}{5}$ *Simplify.*

$= {}^-6\frac{1}{5}$ *Write the quotient as a mixed number.*

$^-6\frac{1}{5}$ is close to the estimate of $^-8$. The quotient is reasonable.

So, the average change in the water level was $^-6\frac{1}{5}$ in. per hour.

EXAMPLE 4

Remember the order of operations.

1. Operate inside parentheses.
2. Clear exponents.
3. Multiply and divide from left to right.
4. Add and subtract from left to right.

At 4 A.M. the air temperature was $^-5.2°$F. For the next 3 hours, the temperature fell 1.5° per hour. Then the sun came up and the temperature rose 4.6°. What was the final temperature?

Write an expression to find the change in temperature.

$^-5.2 + (3 \times {}^-1.5) + 4.6$

$^-5.2 + {}^-4.5 + 4.6$ *Operate inside parentheses.*

$^-9.7 + 4.6$ *Add from left to right.*

$^-5.1$

So, the final temperature was $^-5.1°$F.

CHECK FOR UNDERSTANDING

Think and Discuss ▶ **Look back at the lesson to answer each question.**

1. **Compare and contrast** multiplication of a positive decimal and a negative decimal with multiplication of a positive fraction and a negative fraction.

2. **Tell** if the order of operations process changes depending upon whether you are operating on positive or negative numbers. Give an example to illustrate your answer.

Find the sum or difference. Estimate to check.

3. $^-3\frac{4}{5} + 4\frac{1}{2}$ 4. $7\frac{3}{4} - 8\frac{1}{2}$ 5. $2.5 + ^-3.2$ 6. $2.4 - ^-1.5$

7. $9\frac{2}{5} - ^-3\frac{4}{5}$ 8. $\frac{1}{2} + ^-3\frac{1}{2}$ 9. $^-3.9 - 4.1$ 10. $^-10 + 8\frac{2}{3}$

Find the product or quotient. Estimate to check.

11. $^-4.5 \times 2.4$ 12. $^-6 \div ^-1\frac{1}{2}$ 13. $3.8 \div 0.2$ 14. $\frac{4}{5} \times ^-2\frac{1}{2}$

15. $4.9 \div 1.4$ 16. $^-\frac{1}{2} \times ^-6$ 17. $1\frac{1}{4} \times ^-1\frac{3}{5}$ 18. $^-4\frac{1}{2} \div \frac{1}{2}$

PRACTICE AND PROBLEM SOLVING

Find the sum or difference. Estimate to check.

19. $^-2.5 - ^-1.5$ 20. $^-1.5 + 5.8$ 21. $^-5\frac{1}{2} - 3\frac{3}{4}$

22. $6.5 + 0.9$ 23. $3\frac{1}{2} + ^-7$ 24. $10\frac{1}{2} - ^-12\frac{1}{2}$

25. $^-27.6 - ^-43.2$ 26. $^-15\frac{1}{2} + ^-22$ 27. $^-34.6 + ^-52.8$

Find the product or quotient. Estimate to check.

28. $\frac{3}{5} \times \frac{5}{6}$ 29. $6.8 \div ^-0.4$ 30. $^-1.4 \times ^-20$

31. $^-7\frac{1}{2} \div \frac{5}{8}$ 32. $0.6 \times ^-2.5$ 33. $^-\frac{2}{3} \div ^-3$

34. $8.8 \div 0.2$ 35. $^-8 \div \frac{1}{6}$ 36. $16\frac{2}{3} \times \frac{2}{5}$

Evaluate the expression.

37. $6.4 + 17.1 \div ^-3$ 38. $\left(\frac{^-1}{2}\right)^2 - \left(^-3\frac{1}{2} + 2\right)$

39. $(3.6 - 5.4) \div (^-2.5 + 3.7)$ 40. $^-7 - ^-5 - (^-3 \times ^-1)$

41. $8 \times ^-1.5 \div \left(\frac{^-2}{3} + \frac{1}{6}\right)$ 42. $\frac{2}{3} + \frac{8}{9} \div \frac{^-2}{3} + 1$

43. $4.4 \div (3.8 - 6) \div ^-0.5$ 44. $\left(\frac{^-1}{2} + \frac{1}{3}\right)^2 \times ^-72$

a+b⁄c ALGEBRA Evaluate the expression for $x = ^-2.4$.

45. $x - 8.1$ 46. $^-3.7 + x$ 47. $x - ^-6.8$

48. $10.7 - x$ 49. $x + ^-22.6$ 50. $12 + (x - 1.2)$

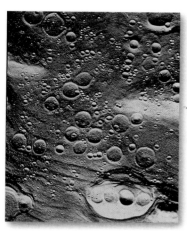

51. Science The daytime temperature on the moon can reach 265.9°F. At night the temperature can fall to ⁻291.5°F. Find the difference between the daytime and nighttime temperatures.

52. In a football game, Sean gained $4\frac{1}{2}$ yd on one play, lost 8 yd on the next play, and gained $6\frac{1}{2}$ yd on the third play. What was his total gain?

53. Ellen had $5\frac{1}{2}$ lb of flour. She used $3\frac{1}{4}$ lb and gave half of what was left to Ruben. How much flour did she give to Ruben?

54. During a storm, rain fell at a rate of 0.9 in. per hour from 9 P.M. until 1:30 A.M. From 1:30 A.M. until 4 A.M., the rate of rainfall was 0.6 in. per hour. What was the total amount of rainfall?

55. *REASONING* For what values of a will $^-2 \times (a - 3)$ be positive? Explain.

MIXED REVIEW AND TEST PREP

56. Find the sum. ⁻15 + ⁺9 (p. 244)

57. Write the percent for 0.085. (p. 60)

58. Evaluate the expression $n + \frac{5}{6}$ for $n = \frac{1}{2}$. (p. 216)

59. TEST PREP Mr. Tilson works 40 hours per week. One year he worked 50 weeks and earned $24,000. How much did he earn per hour? (p. 22)

   A $10.00   **B** $11.00   **C** $11.50   **D** $12.00

60. TEST PREP Ari's test scores were 88, 79, 74, 79, and 90. What is the median of his scores? (p. 106)

   F 16   **G** 74   **H** 79   **J** 82

PROBLEM SOLVING LiNKUP to Careers

Veterinarian A veterinarian is a doctor who treats animals. Even though a veterinarian spends most of his or her time tending to patients, there is also some mathematics to be done. For example, after a dog undergoes surgery, it may need medication. The amount to give depends on the weight of the dog. Too much medicine could harm the dog. Figuring out the right amount of medicine requires mathematics.

- A certain medication is given at a rate of 0.2 mg per pound daily. Georgia's dog Barkum is undergoing surgery. Barkum weighs $30\frac{1}{2}$ lb. If needed, how much medicine should Barkum receive daily?

1. **VOCABULARY** The opposite of an integer is its ___?___ . (p. 243)

Write the addition problem modeled on each number line. (pp. 244–247)

2.

3.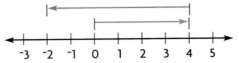

Find the sum. (pp. 244–247, 256–259)

4. $7 + {}^-6$

5. ${}^-5 + {}^-3$

6. ${}^-7 + 4$

7. $4 + {}^-1$

8. ${}^-37 + 24$

9. $17 + {}^-19$

10. ${}^-17 + {}^-41$

11. $21 + 17$

12. ${}^-7.2 + {}^-4$

13. $6\frac{1}{2} + {}^-2\frac{1}{4}$

14. ${}^-2\frac{4}{5} + 1\frac{1}{5}$

15. ${}^-9.7 + {}^-1.3$

Find the difference. (pp. 250–251, 256–259)

16. $7 - 11$

17. ${}^-2 - 8$

18. ${}^-3 - {}^-5$

19. $4 - {}^-4$

20. ${}^-1 - 4$

21. ${}^-6 - {}^-8$

22. $51 - {}^-23$

23. ${}^-41 - 18$

24. ${}^-8\frac{5}{6} - {}^-2\frac{1}{3}$

25. $5\frac{1}{4} + {}^-2\frac{1}{8}$

26. ${}^-16.2 + {}^-5.9$

27. $36.4 + {}^-25.1$

Find the product or quotient. (pp. 252–255, 256–259)

28. ${}^-9 \times {}^-5$

29. ${}^-12 \times 5$

30. ${}^-36 \div {}^-6$

31. $144 \div {}^-12$

32. $\frac{1}{3} \times \frac{{}^-1}{8}$

33. ${}^-4\frac{1}{2} \div \frac{1}{2}$

34. ${}^-9.6 \div 0.4$

35. ${}^-2.1 \times {}^-20$

Solve.

36. A submarine started one leg of its voyage at ${}^-300$ ft. At the end of that leg of the voyage, it was at ${}^-1,250$ ft. What was the difference between the two depths? (pp. 250–251)

37. Tasha needed to find the average temperature during three days in the winter. Her first step was to add the temperatures together. If the temperatures were ${}^-6°$, ${}^-8°$, and $11°$, what was the sum? (pp. 244–247)

38. At midnight the temperature was ${}^-15°$F. The temperature continued to fall until 7:00 A.M., when it was ${}^-22°$F. How many degrees had the temperature fallen during the seven hours? (pp. 250–251)

39. During the last glacial age, sea level in one body of water changed by an average of ${}^-3$ feet every 200 years. The glacial age lasted 10,000 years. What was the total change in sea level? (pp. 252–255)

40. During the first nine months of the year, a worldwide club showed its total loss of members as ${}^-2,718$. What integer shows the average loss of members per month? (pp. 252–255)

Get the information you need.

See item **3**.

Think about how you can write the subtraction of integers as addition. Then find the expression that has the same solution as the given subtraction problem.

Also see problem **3**, p. H63.

Choose the best answer.

1. $^-54 \div {}^-9$?

  A $^-486$ **C** 6

  B $^-6$ **D** Not here

2. What is the value of $29 + k$ for $k = {}^-26$?

  F $^-55$ **H** 3

  G $^-3$ **J** 55

3. Which problem has the same solution as $^-3 - {}^-5$?

  A $^-3 - 5$ **C** $^-3 + 5$

  B $^-3 + {}^-5$ **D** $3 + 5$

4. $^-1.8 \times 30$

  F $^-54$ **H** 54

  G 0.06   **J** Not here

5. Use mental math to find the value of x.

$$x + {}^-9 = 5$$

  A $x = {}^-14$ **C** $x = 4$

  B $x = {}^-4$ **D** $x = 14$

6. The graph shows the number of sixth-grade students at a middle school who are in each club listed. How many more students are in the Science Club than in the Chess Club?

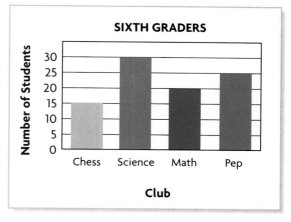

  F 2   **G** 15   **H** 20   **J** 30

7. How is $2 \times 2 \times 2 \times 2$ written in exponent form with a base of 2?

  A 2^4 **C** 4^2

  B 2^5 **D** 2^{10}

8. Ellen got a score of 50 on her history exam. The mean score on that exam was 45. Which statement must be true?

  F Ellen's score is above the average score.

  G Ellen got the top score.

  H Ellen's score is higher than most of the other students' scores.

  J Ellen's score is below the average score.

9. $0.053 - 0.019$

  A 3.4   **C** 0.034

  B 0.34   **D** Not here

Write What You Know

10. Explain how any integer subtraction problem can be rewritten as an integer addition problem. Use your method to find $^-4 - {}^+8$.

11. Estimate the quotient $9\frac{1}{6} \div 1\frac{7}{8}$ using mental math and explain the method you used. Then find the quotient and explain how you found your answer.

MATH DETECTIVE

Play by the Rules

REASONING Each function machine is missing something. For Machines 1, 2, and 3, an input or an output value is missing in each step. For Machine 4, the rule is missing. Use your knowledge of integer operations to find what is missing for each machine.

Function Machine 1

➡ Rule: Add ⁻6.

INPUT	OUTPUT
3	■
■	0
9	■
⁻3	■
■	-12

Function Machine 2

➡ Rule: Subtract 4.

INPUT	OUTPUT
5	■
■	7
1	■
⁻2	■
■	0

Function Machine 3

➡ Rule: Multiply by ⁻3. Then add 1.

INPUT	OUTPUT
3	■
1	■
0	■
⁻2	■
⁻10	■

Function Machine 4

➡ Rule: ?

INPUT	OUTPUT
⁻8	4
⁻4	2
6	⁻3
10	⁻5
0	0

Think It Over!

- 📓 **Write About It** Explain how you found a rule for Function Machine 4.

- **Stretch Your Thinking** Find a rule for this function machine.

INPUT	OUTPUT
0	3
1	5
2	7
3	9

Challenge

Negative Exponents

Learn how to work with negative exponents and how to write small numbers by using scientific notation.

Activity

Copy and complete the pattern. Use a calculator as needed.

$10^3 = 1,000$

$10^2 = 100$

$10^1 = 10$

$10^0 = \blacksquare$

$10^{-1} = \blacksquare$

$10^{-2} = \blacksquare$

$10^{-3} = \blacksquare$

Notice that as the value of the exponent decreases, each number is 0.1, or $\frac{1}{10}$, as great as the previous number.

• For powers of 10, how is the negative exponent related to the number of decimal places?

Negative exponents are used to write very small numbers in scientific notation.

$$0.004 = 4 \times 0.001 = 4 \times 10^{-3}$$

↑
Replace 0.001 with 10^{-3}.

Look at the relationship between 0.004 and 4×10^{-3}. To write 0.004 as 4, move the decimal point 3 places to the right, multiplying by 1,000. Use ⁻3 as the exponent of 10 to show the corresponding division by 1,000.

TECHNOLOGY LINK

You can use the y^x key on a calculator to help you compute exponents. To find the value of 3^4, press

3 y^x 4 = .

The display will show 81.

EXAMPLE

Write 0.0000245 in scientific notation.

0.0000245
5 places

Count the number of places the decimal point must be moved to the right to form a number that is at least 1 but less than 10.

2.45×10^{-5}

Write the number. Since the decimal point moved 5 places to the right, the exponent of 10 is 10^{-5}.

TALK ABOUT IT

• When writing a very small number in scientific notation, how can you tell what the exponent should be?

TRY IT

Write using scientific notation.

1. 0.002   **2.** 0.00034   **3.** 0.06   **4.** 0.005365

5. 0.0084   **6.** 0.0000794   **7.** 0.0000008   **8.** 0.000202

VOCABULARY

1. The distance an integer is from zero is its __?__ . (p. 228)

2. All whole numbers and their opposites are the set of __?__ . (p. 228)

3. A comparison of two numbers, a and b, written as a fraction $\frac{a}{b}$ is a __?__ . (p. 230)

EXAMPLES

EXERCISES

Chapter 11

- **Write the absolute value of an integer.**
  (pp. 228–229)

  $|{}^{-}5| = 5$ *Write the distance $^{-}5$ is from 0 on the number line.*

Write the absolute value.

4. $|{}^{-}8|$	5. $|{}^{+}6|$	6. $|{}^{-}28|$
7. $|{}^{-}73|$	8. $|{}^{+}49|$	9. $|0|$

- **Classify numbers as whole numbers, integers, and rational numbers.** (pp. 230–233)

  $^{-}5$ is an integer and a rational number.

  $\frac{3}{5}$ is a rational number.

Name the sets to which each number belongs.

10. $3\frac{5}{9}$	11. 0.35
12. $^{-}42$	13. $\frac{6}{3}$

- **Find a rational number between two rational numbers.** (pp. 230–233)

  Find a rational number between $^{-}8.5$ and $^{-}8.45$.

  $^{-}8.50$ $^{-}8.45$ *Write each decimal using hundredths.*

  $^{-}8.46$ is between $^{-}8.5$ and $^{-}8.45$. Some other numbers are $^{-}8.47$, $^{-}8.48$, and $^{-}8.482$.

Find a rational number between the two given numbers.

14. $\frac{1}{8}$ and $\frac{1}{2}$	15. $^{-}2.3$ and $^{-}2.4$
16. 18.01 and 18.02	17. $\frac{^{-}1}{8}$ and $\frac{1}{10}$
18. $\frac{1}{10}$ and 0.2	19. $^{-}1\frac{1}{2}$ and $^{-}1\frac{3}{4}$

- **Compare and order rational numbers.**
  (pp. 234–235)

  Compare. Write $<$, $>$, or $=$ for the ●.

  $^{-}5.8$ ● $^{-}5\frac{3}{4}$

  $^{-}5\frac{3}{4} = {}^{-}5.75$ *Write the fraction as a decimal.*

  $^{-}5.80 < {}^{-}5.75$ *Compare the decimals.*

  So, $^{-}5.8 < {}^{-}5\frac{3}{4}$.

Compare. Write $<$, $>$ or $=$ for each ●.

20. $\frac{2}{5}$ ● 0.38	21. $3\frac{3}{4}$ ● 3.89
22. $^{-}0.25$ ● $\frac{^{-}1}{4}$	23. $4\frac{3}{10}$ ● 4.03

Order from least to greatest.

24. 1.55, $\frac{2}{3}$, $^{-}3$, 4.2, $^{-}1.8$

25. $\frac{^{-}3}{4}$, 0, 6.5, $^{-}8$, $^{-}3.6$

Chapter 12

• **Add and subtract integers.** (pp. 242–251)

Subtract. $6 - {}^-4$

$6 - {}^-4 = 6 + 4$ *Write as an addition*
$\quad\quad = 10$ *sentence.*

Add. $15 + {}^-3$

$|15| - |{}^-3| = 15 - 3$ *Subtract the*
$\quad\quad\quad\quad\quad\quad\quad\quad$ *absolute values.*
$\quad\quad\quad\quad = 12$ *The sum is positive.*

Find the sum or difference.

26. $^-3 + 9$ 27. $^-8 + {}^-9$
28. $12 + {}^-5$ 29. $^-4 - 10$
30. $^-9 - {}^-13$ 31. $15 - {}^-6$

• **Multiply and divide integers.** (pp. 252–255)

$^-4 \times 7 = {}^-28$ *The product or quotient of*
a positive and a negative
integer is negative.

$^-72 \div {}^-9 = 8$ *The product or quotient*
of two positive or two
negative integers is
positive.

Find the product or quotient.

32. $^-16 \cdot 3$ 33. $^-12 \cdot {}^-12$
34. $15 \cdot {}^-6$ 35. $^-14 \cdot {}^-9$
36. $^-54 \div 9$ 37. $124 \div {}^-4$
38. $^-75 \div {}^-15$ 39. $^-119 \div 7$

• **Operate with rational numbers.**
(pp. 256–259)

Find the product.

$2\frac{2}{3} \times {}^-1\frac{1}{3} = \frac{8}{3} \times \frac{^-4}{3}$ *Write mixed*
numbers as
fractions.

$\quad\quad\quad = \frac{^-32}{9}$, or $^-3\frac{5}{9}$ *Multiply. The*
fractions have
unlike signs so
the product is
negative.

Add, subtract, multiply, or divide.

40. $36.7 + {}^-51.4$
41. $^-87.3 - 49.1$
42. $3\frac{3}{4} \times \frac{^-1}{3}$
43. $^-3\frac{1}{8} \div {}^-5$

PROBLEM SOLVING APPLICATIONS

44. Four students spent money on books. The amounts they spent were $2.50, $3.25, $5.00, and $5.75. Ari spent $3.25 more than Nina. Nina spent half as much as Jamal. Kenny spent $1.75 less than Jamal. How much did each person spend? (pp.236–237)

45. Natalie, Aaron, Toral, and Casey are each on a different sports team. Toral is not on the baseball team. Casey's team plays on the ice. Aaron's team practices in water. The four teams are soccer, hockey, swimming, and baseball. Who plays on each team? (pp. 236–237)

46. Rachel is the owner of a restaurant. Her profits and losses for the past four weeks are $^-$380, $420, $^-$145, and $620. How much was her profit or loss for the past four weeks? (pp.244–247)

47. Frances claimed that the average of the low temperatures last week was greater than $^-4°C$. Patty says that the average was less than $^-4°C$. The low temperatures last week were $^-6°C$, $^-8°C$, $^-3°C$, $^-4°C$, $0°C$, $^-9°C$, and $^-5°C$. Who is correct? Explain. (pp.252–255)

Performance Assessment

TASK A • Calculate This!

Materials: simple four-function calculator

Suppose the $+/-$ key on a simple calculator does not work. However, it does have memory $+$, memory $-$, and memory recall keys.

a. Explain how to use the calculator to find $7 - {}^-8$. What is the answer?

b. Explain how to use the calculator to find ${}^-9 + 7$. What is the answer?

c. Explain how to find the product or quotient of a positive and a negative number and of two negative numbers by using the calculator.

TASK B • Brrrr!

Materials: Local newspaper weather report

The graph shows the high and low temperatures in Barrow, Alaska, for one week.

a. Find the median high temperature.

b. Find the median low temperature.

c. Graph the low temperatures on a number line.

d. Write two statements comparing the high and low temperatures on Thursday.

Find a similar graph in your local newspaper for where you live. Repeat the activity using that graph.

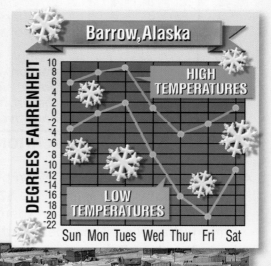

Technology Linkup

Mighty Math Calculating Crew
Operations with Integers

Click [OFF DUTY]. Then click [ ].

- Choose Red Level A. Click **OK**. Then click [ON DUTY]. Complete the mission by finding the value of each addition expression.

- Click [ ]. Choose Red Level G. Click **OK** and then [ON DUTY]. Find the value of each multiplication expression.

Practice and Problem Solving

1. Click [ ]. Choose Red Level F. Click **OK** and then [ON DUTY]. Find the value of each addition or subtraction expression.

2. Click [ ]. Choose Red Level M. Click **OK** and then [ON DUTY]. Find the value of each division expression.

Click [ ]. **Use the calculator to find the value of the expressions.**

3. $24 + {}^-18$	**4.** $34 + {}^-50$	**5.** ${}^-7 + {}^-21$
6. $14 - 51$	**7.** ${}^-17 - 5$	**8.** ${}^-22 - {}^-32$
9. 34×6	**10.** $18 \times {}^-12$	**11.** ${}^-21 \times 7$
12. $168 \div 12$	**13.** ${}^-128 \div 4$	**14.** ${}^-144 \div {}^-9$

Write and evaluate the expression.

15. The amount of dog food in the bag changes by ${}^-8$ oz a day. What is the total change after 2 weeks?

16. The high temperature was 12°F. The low temperature was ${}^-5$°F. What was the range of the temperatures?

Multimedia Math Glossary www.harcourtschool.com/mathglossary

17. **Vocabulary** Locate *additive inverse* in the Multimedia Math Glossary. Use a model to represent the sum of 4 and its additive inverse.

PROBLEM SOLVING ON LOCATION
In Kentucky

Temperatures

Kentucky has experienced some extreme weather in the past 100 years. In 1930, the temperature hit 114°F in Greensburg. More recently, Shelbyville residents braved temperatures of ⁻37°F on January 19, 1994, following a blizzard that closed state highways for two days.

1. Find the difference between Kentucky's record high temperature and record low temperature.

Use Data   For 2–3, use the table.

2. By how many degrees did Kentucky's record high February temperature exceed its record low February temperature?

3. Find the difference between Kentucky's record high March temperature and record high February temperature.

4. Here's a useful rule for estimating temperature change: temperature decreases by about 3.6°F for every 1,000 feet that you climb in elevation. Suppose you hiked 2,500 feet up Black Mountain. By about how many degrees does the temperature decrease as you hike?

Monthly Record Kentucky Temperatures		
	February	March
Low	⁻32°F	⁻14°F
High	86°F	94°F

Because of the "wind-chill factor," air temperatures feel colder when the wind is blowing. The stronger the wind, the colder it feels. The table below shows how much colder it feels.

Use Data   For 5–7, use the table.

5. Suppose the wind is blowing at 10 mph. How many degrees colder will the wind make it feel if the air temperature is 30°F?

6. Suppose the wind is blowing at 20 mph. How many degrees colder will the wind make it feel if the air temperature is 25°F?

7. Describe any pattern you see in the row for 20 mph.

Wind-Chill Factor				
	Air Temperature (°F)			
Wind speed	35	30	25	20
5 mph	32	27	22	16
10 mph	22	16	10	3
15 mph	16	9	2	⁻5
20 mph	12	4	⁻3	⁻10

Elevations

Being a hilly and mountainous state, all of Kentucky lies above sea level. Its highest point is Black Mountain in Harlan County, 4,139 feet above sea level. Its lowest point is along the Mississippi River in Fulton County, 257 feet above sea level. The lowest point in the United States is Death Valley, California, where the elevation is 282 feet below sea level ($^-$282). Closer to Kentucky, New Orleans, Louisiana, has an elevation of 8 feet below sea level ($^-$8).

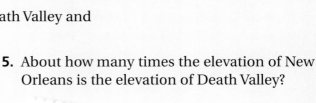

Use Data For 1–6, use the diagram.

1. What is the difference in elevation between the lowest point in Kentucky and the highest point in Kentucky?

2. What is the difference in elevation between Death Valley and the top of Black Mountain?

3. What is the difference in elevation between Death Valley and the lowest point in Kentucky?

4. About how many times the elevation of the Mississippi River in Fulton County is the elevation of Black Mountain?

5. About how many times the elevation of New Orleans is the elevation of Death Valley?

6. **REASONING** Evaluate this statement: "Death Valley is about as far below sea level as the Mississippi River is above sea level in Fulton County, Kentucky."

Black Mountain, Kentucky's highest point

13 Algebra: Expressions

Kruger National Park in South Africa was established in 1898. Many different types of vegetation and wildlife are contained within its 7,523 square miles. The wildlife includes more than 800 species of mammals, birds, reptiles, and amphibians.

PROBLEM SOLVING Suppose there is an average of n buffalo per square mile. Write an expression with n to show how many buffalo live in the park.

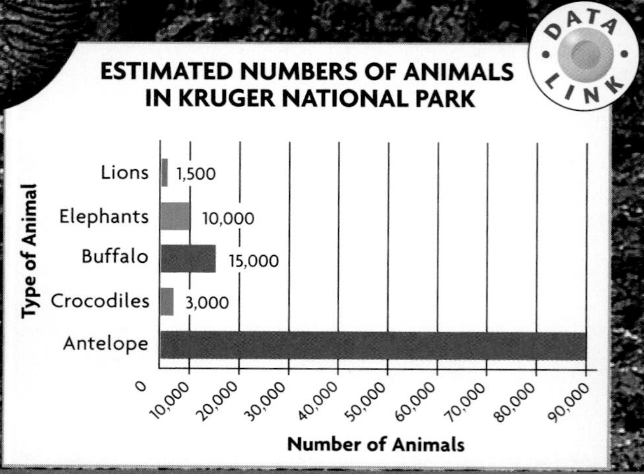

ESTIMATED NUMBERS OF ANIMALS IN KRUGER NATIONAL PARK

DATA LINK

Type of Animal	Number of Animals
Lions	1,500
Elephants	10,000
Buffalo	15,000
Crocodiles	3,000
Antelope	

Number of Animals: 0, 10,000, 20,000, 30,000, 40,000, 50,000, 60,000, 70,000, 80,000, 90,000

Check What You Know

Use this page to help you review and remember important skills needed for Chapter 13.

 Vocabulary

Choose the best term from the box.

numerical expression
algebraic expression
exponent
operation

1. The number that shows how many times the base is used as a factor is a(n) __?__.

2. A mathematical phrase that includes only numbers and operation symbols is a(n) __?__.

3. An expression that includes a variable is called a(n) __?__.

 Exponents (For Intervention, see p. H15.)

Find the value.

4. 4^3

5. 2^5

6. 3^4

7. 5^3

8. 3^3

9. 2^4

10. 8^2

11. 9^2

12. 2^8

13. 7^2

14. 12^2

15. 6^3

 Order of Operations (For Intervention, see p. H15.)

Evaluate each expression.

16. $6 + 3 \cdot 5$

17. $3^2 + 4^2$

18. $(5 + 2) \cdot 6$

19. $(8 + 4) - (3 + 2)$

20. $24 \div (8 - 5)$

21. $\frac{(15 - 7)}{8} \cdot 2$

22. $\frac{(5 + 9)}{7} + 4^2$

23. $5^2 - (4 - 2)$

24. $3^2 \div (5 - 2)$

25. $24 - 8 \div 2$

26. $2^3 \cdot (15 - 8)$

27. $8^2 \div (9 + 7)$

28. $3^2 + 4 - (2 + 1)$

29. $20 - 5 \cdot 3 + 16$

30. $3^3 \div (13 - 9) + 8$

 Factors (For Intervention, see p. H8.)

Find all the factors of the number.

31. 16

32. 24

33. 33

34. 28

35. 17

36. 64

37. 42

38. 81

39. 40

40. 39

> **LOOK AHEAD**
>
> **In Chapter 13 you will**
> - write algebraic expressions
> - evaluate algebraic expressions
> - evaluate expressions with squares and square roots

LESSON 13.1

Write Expressions

Learn how to write algebraic expressions.

QUICK REVIEW

1. $^-3 \cdot 6$
2. $^-8 \cdot {^-7}$
3. $5 \cdot {^-11}$
4. $9 \cdot 4$
5. $0 \cdot {^-10}$

Remember that $a \times b$ can be written as $a \cdot b$ or ab.

John mows lawns to earn spending money. He charges a flat rate of $15.75 for each lawn plus $2.00 for every hour it takes him to mow, trim, and edge the lawn. To find the total amount of money he earns, John uses the algebraic expression below.

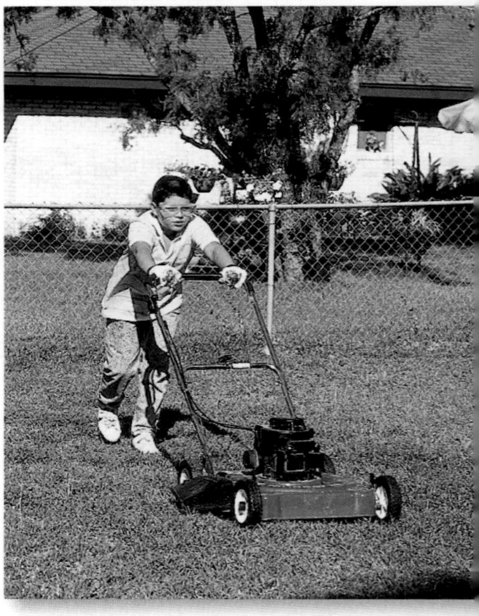

$15.75 for each lawn	plus	$2.00 for every hr
↓	↓	↓
15.75	+	$2 \cdot h$
		↑ number of hours

As shown above, word expressions can be written as algebraic expressions.

EXAMPLE 1

Write an algebraic expression for the word expression.

18 more than twice *s*	$2s + 18$
q divided by 8, added to the product of *t* and *r*	$tr + \frac{q}{8}$
5.6 less than $\frac{4}{5}m$	$\frac{4}{5}m - 5.6$
the sum of $\frac{1}{2}a$, $\frac{1}{6}b$, and $\frac{1}{4}c$	$\frac{1}{2}a + \frac{1}{6}b + \frac{1}{4}c$

You can also write word expressions for numerical and algebraic expressions.

EXAMPLE 2

Write a word expression for the numerical or algebraic expression.

a. $15 + 4.5$ — the sum of 15 and 4.5

b. abc — the product of *a*, *b*, and *c*

c. $\frac{2}{3}m + 8$ — 8 more than $\frac{2}{3}m$

d. $xy - \frac{z}{5}$ — the difference of the product of *x* and *y* and the quotient of *z* divided by 5

CHECK FOR UNDERSTANDING

Think and Discuss ▶ Look back at the lesson to answer the question.

1. Give another word expression for Example 2c.

Guided Practice ▶ Write an algebraic expression for the word expression.

2. 7 more than $\frac{3}{8}y$

3. y divided by 1.5

Write a word expression for each.

4. $19 + 8.7$

5. $19x - 54$

6. $p \cdot q \cdot r$

PRACTICE AND PROBLEM SOLVING

Independent Practice ▶ Write an algebraic expression for the word expression.

7. 2.9 more than 3 times c

8. d divided by 8, less $\frac{1}{3}$

9. the product of h, $4j$, and k

10. the quotient of a and 3.5, less 2

11. the difference between $6.3p$ and $5m$

12. $\frac{1}{2}$ multiplied by the sum of $2m$ and $4n$

Problem Solving Applications ▶ Write a word expression for each.

13. $42 + 2.5$

14. $8c - \frac{5}{6}$

15. $4.5 \cdot a \cdot b$

16. $yz + \frac{1}{2}x$

17. $a + b + c$

18. $98 \div 4.1y$

19. What if John charged a flat rate of $14.25 per lawn and $3.50 for each hour he mows, trims, and edges? What expression could he use to find the total amount he charges?

20. Let x represent José's height, in inches. His nephew is $\frac{1}{4}$ as tall as José. Write an algebraic expression for José's nephew's height.

21. Jill's niece is 2 inches taller than twice her nephew's height. What algebraic expression could you write for her niece's height?

22. A taxi charges $2.75 for the first mile and $2.25 for each additional mile. How much would a 15-mi trip cost?

23. ✏️ **Write a problem** in which you have to write an algebraic expression using the variable x.

MIXED REVIEW AND TEST PREP

Evaluate the expression. (p. 256)

24. $^-15 + 10 \div 2 \cdot {}^-6$

25. $(^-2 - {}^-5)^2 \cdot {}^-4$

26. $(14 - 47) \div (2 - 5)$

27. Find a pair of numbers that have a LCM of 24 and a GCF of 2. (p. 150)

⭐ **28. TEST PREP** Use mental math to tell which is the solution for $x + 2.5 = 10.5$. (p. 82)

A $x = 13$ **B** $x = 8.5$ **C** $x = 8$ **D** $x = 2.5$

Evaluate Expressions

Learn how to evaluate algebraic expressions.

Vocabulary

terms

like terms

Ricardo runs the concession stand at school. To find out how much money the stand makes or loses, he uses the algebraic expression $s - c - a$ where s is the amount of sales, c is the cost of food products, and a is advertising costs.

EXAMPLE 1

During the month of November, Ricardo spent $327 on food products and $50 on advertising for the concession stand. The amount of sales was $349. How much money did the concession stand make or lose during November?

Use the algebraic expression $s - c - a$.

$s - c - a$	*Replace s with 349, c with 327, and a with 50.*
$349 - 327 - 50$	*Subtract.*
$22 - 50$	*Subtract.*
$^-28$	*A negative result represents a loss.*

So, the concession stand lost $28 during November.

Algebraic expressions can be evaluated for different values.

EXAMPLE 2

Evaluate $^-4x + 7$ for $x = ^-2, ^-1, 0,$ and 1.

$x = ^-2$		$x = ^-1$	
$^-4x + 7$	*Replace x with $^-2$.*	$^-4x + 7$	*Replace x with $^-1$.*
$^-4 \cdot ^-2 + 7$	*Multiply.*	$^-4 \cdot ^-1 + 7$	*Multiply.*
$8 + 7$	*Add.*	$4 + 7$	*Add.*
15		11	
So, for $x = ^-2$, $^-4x + 7 = 15$.		So, for $x = ^-1$, $^-4x + 7 = 11$.	

$x = 0$		$x = 1$	
$^-4x + 7$	*Replace x with 0.*	$^-4x + 7$	*Replace x with 1.*
$^-4 \cdot 0 + 7$	*Multiply.*	$^-4 \cdot 1 + 7$	*Multiply.*
$0 + 7$	*Add.*	$^-4 + 7$	*Add.*
7		3	
So, for $x = 0$, $^-4x + 7 = 7$.		So, for $x = 1$, $^-4x + 7 = 3$.	

Remember that the order of operations is:

1. Operate inside parentheses.
2. Clear exponents.
3. Multiply and divide from left to right.
4. Add and subtract from left to right.

Some algebraic expressions are more complicated. To evaluate them, replace each variable with its given value. Then follow the order of operations.

EXAMPLE 3

Evaluate $2(a + b)^2 \div c$ for $a = {}^-5$, $b = 14$, and $c = {}^-3$.

$2(a + b)^2 \div c$	*Replace a, b, and c with their given values.*
$2({}^-5 + 14)^2 \div {}^-3$	*Operate inside parentheses.*
$2 \cdot 9^2 \div {}^-3$	*Clear the exponent.*
$2 \cdot 81 \div {}^-3$	*Multiply.*
$162 \div {}^-3$	*Divide.*
${}^-54$	

So, for $a = {}^-5$, $b = 14$, and $c = {}^-3$, the expression $2(a + b)^2 \div c = {}^-54$.

TECHNOLOGY LINK

More Practice: To learn more about *Evaluating Expressions,* watch the **Harcourt Math Newsroom Video** *Bangladesh Bank.*

The parts of an algebraic expression that are separated by an addition or subtraction sign are called **terms** . Before you evaluate some algebraic expressions, you can simplify them by combining like terms. **Like terms** have the same variable raised to the same power.

ALGEBRAIC EXPRESSION	LIKE TERMS
$6x + 5x + 17$	$6x$ and $5x$
$42 + 13y^2 - 10y^2$	$13y^2$ and $10y^2$

To make an algebraic expression simpler, combine like terms by adding or subtracting like terms.

ALGEBRAIC EXPRESSION	SIMPLIFIED
$6x + 5x + 17$	$11x + 17$
$42 + 13y^2 - 10y^2$	$42 + 3y^2$

EXAMPLE 4

Simplify $5x + 3x + 7$ by combining like terms. Then evaluate the expression for $x = {}^-2$.

$5x + 3x + 7 = 8x + 7$	*Simplify by combining 5x and 3x.*
$8 \cdot {}^-2 + 7$	*Replace x with ${}^-2$ and multiply.*
${}^-16 + 7$	*Add.*
${}^-9$	

So, $5x + 3x + 7 = {}^-9$ when $x = {}^-2$.

- Simplify $12a - 9a + 42$ by combining like terms. Then evaluate the expression for $a = {}^-4$.

You can use the Distributive Property to solve some problems mentally.

EXAMPLE 5

Evaluate pq for $p = {}^-8$ and $q = 57$.

pq
$^-8 \cdot 57$ *Replace p with $^-$8 and q with 57.*
$^-8(50 + 7)$ *Think of 57 as 50 + 7. Then*
$^-8 \cdot 50 + {}^-8 \cdot 7$ *use the Distributive Property*
$^-400 + {}^-56$ *and mental math.*
$^-456$

CHECK FOR UNDERSTANDING

Think and ▶ Discuss

Look back at the lesson to answer each question.

1. **Tell** why it can help to simplify an algebraic expression before evaluating it.

2. **Show** that evaluating the following expression before you simplify gives the same result as evaluating it after you simplify. $4x + 6x + 17$ for $x = {}^-2$

Guided ▶ Practice

Evaluate the expression for $x = {}^-2$, $^-1$, 0, and 1.

3. $^-6 + 5x$

4. $2x + 3$

5. $14 + 8x$

6. $^-6x - (4 - 9)$

7. $x^2 - 3$

8. $\dfrac{^-24}{3x - 2}$

Simplify the expression. Then evaluate the expression for the given value of the variable.

9. $4x + 3x - 8$ for $x = {}^-3$

10. $15y - 11y + 10$ for $y = {}^-3$

PRACTICE AND PROBLEM SOLVING

Independent ▶ Practice

Evaluate the expression for $x = 1$, 2, 3, and 4.

11. $^-12 - 4x$

12. $0.25x + 4$

13. $21 - \dfrac{1}{2}x$

14. $x^2 + 4$

15. $\dfrac{^-12}{x} + 7$

16. $\dfrac{^-48}{2x + 4}$

Simplify the expression. Then evaluate the expression for the given value of the variable.

17. $3x + 9x - 41$ for $x = {}^-4$

18. $22y - 19y + 17$ for $y = {}^-9$

19. $356 + 13a - 7a$ for $a = {}^-20$

20. $10x + 6x - 59 + 8y$ for $x = {}^-12$ and $y = 8$

21. $2a^2 + bc$ for $a = 7$, $b = {}^-25$, and $c = 4$

22. $\dfrac{1}{2}x - 2y + z$ for $x = 20$, $y = 8$, and $z = {}^-10$

Name a property you could use to help evaluate the expression. Then evaluate the expression.

23. $x + y + z$ for $x = 17$, $y = 58$, and $z = {}^-8$

24. ab for $a = {}^-3$ and $b = 87$

Find a value of x for which the expressions are equal.

25. $5x$; $x + 8$

26. $x - 3$; $17 - x$

27. $2x - 1$; $3x$

Problem Solving ▶
Applications

28. To determine if there is enough money in her checking account to write the last check, Julie uses the expression $a + d - c$, where a is the starting amount, d is the total of the deposits made, and c is the total of her checks. Does Julie have enough money in her checking account if she started with $334?

Deposits	Checks
$72	$177
$18	$53
$28	$70
$14	$48
	$122

29. **(?) What's the Error?** Rick evaluates $x + y \cdot z$ for $x = 3$, $y = 4$, and $z = {}^-2$. He says the answer is $^-14$. What is his error? What is the correct answer?

MIXED REVIEW AND TEST PREP

30. Write an algebraic expression for 3 more than twice x. (p. 270)

31. $^-15 \div 3$ (p. 252)

32. $9 \cdot {}^-4$ (p. 252)

33. TEST PREP Find the sum. $^-26 + 9$ (p. 244)

A $^-35$ **B** $^-17$ **C** 17 **D** 35

34. TEST PREP A satellite mission cost $4,500 per second. Find the cost from 19 seconds before launch until 11 seconds after launch. (p. 250)

F $150 **G** $36,000 **H** $81,000 **J** $135,000

PROBLEM SOLVING    to Reading

Strategy • Use Graphic Aids Sometimes a graphic aid, such as a table, can help you to see the solution to a problem. Chad is ordering CDs from a catalog. Each CD costs $11 and there is a $7 charge for shipping no matter how many CDs are ordered. If Chad has $57 to spend, what is the maximum number of CDs he can order?

Use the expression $11 \cdot x + 7$ and an input/output table.

INPUT (CDs)	RULE 11 · x + 7	OUTPUT (COST)
1	11 · 1 + 7	$18
2	11 · 2 + 7	$29
3	11 · 3 + 7	$40
4	11 · 4 + 7	$51
5	11 · 5 + 7	$62

So, Chad can order a maximum of 4 CDs.

• Suppose Chad had $90, each CD cost $13, and the shipping cost $5. What is the maximum number of CDs he could order?

EXTRA PRACTICE page H44, Set B

275

Squares and Square Roots

Explore how to use a model to find squares and square roots.

You need square tiles.

Vocabulary
square
square root

QUICK REVIEW

Find the value.

1. 10^2 2. 15^2 3. 3^3
4. 4^3 5. 5^3

A **square** is the product of a number and itself. The square of 4 is 16 since $4 \cdot 4 = 16$. A square can be expressed with the exponent 2. You read 4^2 as "4 squared." Use square arrays to model square numbers.

Activity 1

- Make a square array with 4 square tiles on each side. How many tiles did you use?

- Make a square array with 5 square tiles on each side.

- Make a square array with 6 square tiles on each side.

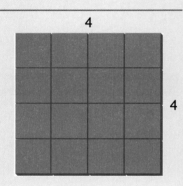

Think and Discuss

- How many tiles are in the array with 5 tiles on each side? Complete: $5^2 = \blacksquare$.

- How many tiles are in the array with 6 tiles on each side? Complete: $6^2 = \blacksquare$.

- Without making an array, tell how many tiles are in these: a 7×7 array, an 8×8 array, a 9×9 array, and a 10×10 array.

- Suppose you have a square with n tiles on each side. How many tiles would be in the array?

Practice

Use square tiles to find each square.

1. 2^2 2. 3^2 3. 7^2

You can also use a calculator to find squares.

16 $\boxed{x^2}$ $\boxed{ 256}$ *Enter the value of the base, 16. Then press* $\boxed{x^2}$.

So, $16^2 = 256$.

4. Use a calculator to find 17^2, 25^2, and 51^2.

When you find the two equal factors of a number, you are finding the **square root** of the number. The symbol for a positive square root is $\sqrt{}$.

Since $5^2 = 25$, $\sqrt{25} = 5$. It is also true that $(^-5)^2 = 25$, but we use $\sqrt{}$ for the positive square root. You read $\sqrt{25}$ as "the square root of 25."

Activity 2

You can think about the sides of a square array to help you find the square root of a number.

- Make a square array with 9 square tiles. How many tiles are on each side of your square array? What is $\sqrt{9}$?

- How many tiles are on each side of a square array with 16 tiles? What is $\sqrt{16}$?

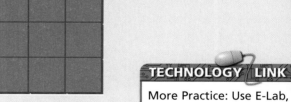

TECHNOLOGY LINK

More Practice: Use E-Lab, *Squares and Square Roots.*
www.harcourtschool.com/elab2002

Think and Discuss

- How many tiles do you think would be on each side of a square array with 225 square tiles? 400 squares tiles? Explain.

- Can you form a square with 10 square tiles? 12 square tiles? Explain.

You can use a calculator to find square roots.

289   **SHIFT** $\sqrt{}$ ⟦   *17* ⟧

So, $\sqrt{289}$ is 17.

Enter the number, 289. Then press **SHIFT** *and* $\sqrt{}$.

Practice

Find the square roots.

1. $\sqrt{64}$ **2.** $\sqrt{144}$ **3.** $\sqrt{196}$

4. $\sqrt{441}$ **5.** $\sqrt{484}$ **6.** $\sqrt{1,764}$

7. If $9 \cdot 9 = 81$, then $\sqrt{\blacksquare} = \blacksquare$.

MIXED REVIEW AND TEST PREP

8. Evaluate $3xy + z$ for $x = 2$, $y = ^-3$, and $z = ^-4$. (p. 272)

9. Simplify and evaluate $5a + 7a + 35$ for $a = ^-7$. (p. 272)

10. Write an algebraic expression for $\frac{1}{2}$ more than twice k. (p. 270)

11. Compare. Write $<$ or $>$ for ●. $4\frac{3}{4} \cdot 3$ ● $11\frac{1}{4} + 2.5$ (p. 234)

12. TEST PREP   Replace ● with the correct operation sign.

$2^4 \cdot (3 - 5)$ ● $(^-2 - 6) = 4$ (p. 256)

A $+$ **B** $-$ **C** $\times$ **D** $\div$

Expressions with Squares and Square Roots

Learn how to evaluate expressions with squares and square roots.

You can evaluate expressions containing squares and square roots by using the order of operations. Evaluate square roots at the same time you evaluate exponents.

EXAMPLE 1

Mayla Sue wants to put a fence around her square garden. Jay told her that the area of the garden is 64 ft². Find the perimeter of the garden.

Since the garden is square, the length of one side is $\sqrt{64}$.

$\sqrt{64} = 8$ *Evaluate $\sqrt{64}$. $\sqrt{64} = 8$ since $8 \cdot 8 = 64$.*

So, the length of one side is 8 ft. The formula for the perimeter of a square is $P = 4s$. Evaluate for $s = 8$.

$P = 4s = 4 \cdot 8$ *Replace s with 8. Then multiply.*

$P = 32$ So, the perimeter of the garden is 32 ft.

EXAMPLE 2

Evaluate $18 - (\sqrt{100} + 5^2)$.

$18 - (\sqrt{100} + 5^2)$ *Evaluate $\sqrt{100}$ and 5^2.*

$18 - (10 + 25)$ *Operate inside parentheses.*

$18 - 35$ *Subtract.*

$^-17$ So, $18 - (\sqrt{100} + 5^2) = {^-}17$.

Sometimes the value of a variable is a square root.

EXAMPLE 3

Evaluate $(x - 2)^2 + 8$ for $x = \sqrt{25}$.

$(x - 2)^2 + 8$ *Substitute $\sqrt{25}$ for x.*

$(\sqrt{25} - 2)^2 + 8$ *Evaluate $\sqrt{25}$ inside the parentheses.*

$(5 - 2)^2 + 8$ *Operate inside parentheses.*

$3^2 + 8$ *Clear the exponent.*

$9 + 8$ *Add.*

17 So, for $x = \sqrt{25}$, the expression $(x - 2)^2 + 8 = 17$.

CHECK FOR UNDERSTANDING

Think and ▶
Discuss

Look back at the lesson to answer the question.

1. **Explain** how to evaluate the expression $\sqrt{36} \cdot (7 - 5)$. Then evaluate.

Guided ▶
Practice

Evaluate the expression.

2. $\sqrt{25} - 8^2 + 42$

3. $10 + \sqrt{36} - 7^2$

4. $\sqrt{100} \cdot \sqrt{225}$

5. $45 \div \sqrt{81} \cdot {}^-7 + 59$

PRACTICE AND PROBLEM SOLVING

Independent ▶
Practice

Evaluate the expression.

6. $\sqrt{4} - 5^2 + 22$

7. $246 + \sqrt{16} - 10^2$

8. ${}^-25 (\sqrt{49} - 10)$

9. $88 \div \sqrt{121} \cdot {}^-3$

Evaluate the expression for the given value of the variable.

10. $(y + 2) - 42$ for $y = \sqrt{1}$

11. $(5^2 + \sqrt{p}) \div {}^-4$ for $p = 225$

12. $4a + b - 125$ for $a = \sqrt{9}$ and $b = 9$

13. $(x - 5)^2 + 7y$ for $x = \sqrt{144}$ and $y = \sqrt{16}$

Compare for $a = 5$, $b = 20$, and $c = 16$. Use $<$, $>$, or $=$.

14. $a^2 + 10 \, \bullet \, 10a^2$

15. $\sqrt{ab} \, \bullet \, {}^-c$

Problem Solving ▶
Applications

16. **REASONING** Insert one square root symbol so the value of the expression $25 + 64 - 16 \cdot 81$ is ${}^-55$.

17. A square flower garden and a sidewalk around the garden have an area of 400 ft^2. The sidewalk has a width of 3 ft. What is the length of one side of the flower garden?

18. **?** **What's the Question?** The expression is $\sqrt{x} + {}^-2$. The answer is 7.

MIXED REVIEW AND TEST PREP

Evaluate the expression for $x = 1, 2, 3,$ and $4.$ (p. 272)

19. ${}^-4x + 2$

20. $20 - 6x$

21. $\frac{12}{x} - 10$

22. Find $5\frac{2}{3} \div 2\frac{1}{3}$. (p. 210)

⭐ 23. **TEST PREP** A bag held $6\frac{3}{4}$ c of flour. Bill took out $2\frac{1}{2}$ c. He had too much, so he put back $\frac{3}{4}$ c. How much flour was left in the bag? (p. 186)

A $3\frac{1}{2}$ c
B 5 c
C $8\frac{1}{2}$ c
D 10 c

EXTRA PRACTICE page H44, Set C

279

1. **VOCABULARY**   When you find the two equal factors of a number, you are finding the ___?___ of the number. (p. 277)

2. **VOCABULARY**   The parts of an algebraic expression that are separated by an addition or subtraction sign are ___?___ . (p. 273)

Write an algebraic expression for the word expression.
(pp. 270–271)

3. the product of 2, $^-2$, and b

4. the quotient of s divided by 1.6

5. 15 more than twice p

6. $\frac{1}{2}$ less than the product of 6 and r

7. the sum of $4a$ and $5b$

8. the product of ^-2q and r

Evaluate the expression for $y = 5$. (pp. 272–275)

9. $^-4y + 12$

10. $y^3 + 12$

11. $4 + 2y$

12. $5 - 3y$

13. $\frac{10}{y} - 1$

14. $10 + 3y$

Evaluate the expression for $a = ^-4$, $b = 3$, and $c = ^-5$. (pp. 272–275)

15. $3b + (a - c)$

16. $12 - a + c$

17. $28 + (c + 2)^2$

18. $18c + \frac{8b}{a}$

19. $(3c - a)^2 - b$

20. $338 \div (a + ^-2bc)^2$

Simplify the expression. Then evaluate the expression for the given value of the variable. (pp. 272–275)

21. $6x - 2x$, $x = ^-2$

22. $3x + 2x + 6$, $x = 1$

23. $10x - 2x$, $x = ^-10$

Evaluate the expression. (pp. 278–279)

24. $\sqrt{49} - 2 + 5$

25. $3 \times (6^2 - 16)$

26. $4 \times \sqrt{81} - 3 \times 6$

27. $\sqrt{25} \cdot \sqrt{16}$

28. $^-124 \times (\sqrt{64} - 6)$

29. $10 + \sqrt{100} - 4^2$

Evaluate the expression for the given value of the variable. (pp. 278–279)

30. $4 - \sqrt{c}$ for $c = 49$

31. $^-(5^2) - d$ for $d = \sqrt{121}$

32. A cashier uses the expression $d + s - c$ to check the amount of money in her cash drawer, where d is the initial amount in the drawer, s is the amount of cash that customers pay, and c is the amount of change that customers receive. Suppose the initial amount is $125.00, customers pay $15.00, $18.25, and $24.00, and the amount of change returned is $1.50 and $1.95. How much money is in the cashier's drawer? (pp. 272–275)

33. Write an expression for the perimeter of the rectangle. Then evaluate the expression to find the perimeter when $a = 4$ cm and $b = 7$ cm. (pp. 272–275)

Choose the answer.

See item **4.**

Think about the symbols and what they tell you to do. Check your computation to make sure it is correct before you choose the answer.

Also see problem **6**, p. H64.

Choose the best answer.

1. Which of the following expressions is represented by the algebraic expression below?

$$m - 6$$

   A m decreased by 6

   B 6 more than m

   C the product of m and 6

   D m divided by 6

2. The attendance at a concert on Saturday night was 42 more than the attendance on Friday night. If a represents the attendance on Friday night, which expression represents the attendance for Saturday night?

   F $42a$

   G $42 - a$

   H $a + 42$

   J $a - 42$

3. The cost of a bag of apples was shared equally among 5 friends. If a represents the cost of the apples, which algebraic expression represents the amount each friend paid?

   A $\frac{5}{a}$ **C** $5 - a$

   B $a + 5$ **D** $\frac{a}{5}$

4. $6^2 + \sqrt{25}$

   F 11   **G** 37   **H** 41   **J** 61

5. What is the value of $c - 4$ for $c = {^-}5$?

   A $^-9$ **C** 2

   B $^-2$ **D** Not here

6. Martha kept track of her quiz scores. During one week she received these scores: 12, 17, 27, 23, 19. What is the difference between the mean and the median for the quiz scores?

   F 0.4   **G** 0.6   **H** 1.6   **J** 19

7. $8 \times {^-}4 + 17$

   A $^-544$ **C** 49

   B $^-15$ **D** Not here

8. Which algebraic expression represents "29.1 times a number, n"?

   F $29.1 - n$ **H** $n \div 29.1$

   G $29.1 + n$ **J** $29.1 \times n$

Write What You Know

9. Mac is buying items for a cookout. Hot dogs are sold 10 in a pack. Buns are sold 8 in a pack. What is the least number of packs of each he should buy to have an equal number of hot dogs and buns? Explain how you found your answer.

10. Juanita said that $3a + 4a = 7a^2$. Greg said $3a + 4a = 7a$. Pam said $3a + 4a = 12a$. Who was right? Explain.

CHAPTER 14

Algebra: Addition and Subtraction Equations

Beginning on January 27, 1984, Klaus Friedrich single-handedly set up and toppled the greatest number of dominoes ever. Not all of the dominoes he set up toppled on the first push.

PROBLEM SOLVING The expression $281,581 + x$ can be used to model the toppled and untoppled dominoes. How can you use the equation $320,236 = 281,581 + x$ to find the number of untoppled dominoes?

DATA LINK

KLAUS FRIEDRICH'S DOMINO TOPPLING		
	Set Up	**Toppled**
Amount of Time	18,600 min	13 min
Number of Dominoes	320,236	281,581

Check What You Know

Use this page to help you review and remember important skills needed for Chapter 14.

✓ Vocabulary

Choose the best term from the box.

1. A statement showing that two quantities are equal is a(n)   ?   .

2. The property which states that $5 + x = x + 5$ is the   ?   Property of Addition.

> Associative
> Commutative
> equation
> expression

✓ Add and Subtract (For Intervention, see p. H16.)

Find the sum or difference.

3. $25 + 12$

4. $256 + 125$

5. $49 - 26$

6. $321 - 265$

7. $6.25 + 5.12$

8. $9.14 + 4.27$

9. $8.4 - 5.7$

10. $14.72 - 11.82$

11. $\frac{5}{6} + \frac{3}{4}$

12. $8\frac{2}{3} + 5\frac{1}{8}$

13. $\frac{5}{7} - \frac{3}{14}$

14. $6\frac{1}{8} - 2\frac{3}{5}$

15. $4\frac{1}{4} + 3\frac{2}{3}$

16. $\frac{4}{5} - \frac{1}{3}$

17. $4\frac{4}{7} - 2\frac{1}{4}$

✓ Inverse Operations (For Intervention, see p. H16.)

Use the inverse operation to check the equation.

18. $14 + 5 = 19$

19. $24 + 19 = 43$

20. $125 + 219 = 344$

21. $19 - 3 = 16$

22. $37 - 14 = 23$

23. $242 - 196 = 46$

24. $8 \cdot 6 = 48$

25. $25 \cdot 5 = 125$

26. $14 \cdot 34 = 476$

27. $12 \div 6 = 2$

28. $72 \div 8 = 9$

29. $252 \div 12 = 21$

30. $52 \cdot 4 = 208$

31. $400 \div 16 = 25$

32. $19 \cdot 17 = 323$

✓ Words for Operations (For Intervention, see p. H17.)

Write the operation described by the phrase.

33. the sum of p and 12

34. the product of x and 3

35. the difference of 30 and k

36. the quotient of b and 4

37. 15 times y

38. m decreased by 4

39. n reduced by 72

40. 234 increased by z

> **LOOK AHEAD**
>
> **In Chapter 14 you will**
> • write equations
> • solve addition equations
> • solve subtraction equations

Connect Words and Equations

Learn how to translate words into numbers, variables, and operations.

Mt. McKinley in Alaska is the highest point in the United States.

Remember that an equation is a statement showing that two quantities are equal.

You can translate word expressions into equations by translating words into numbers, variables, and operations.

The lowest point in the United States is Death Valley, California, with an altitude of $^-86$ m. The altitude of Mt. McKinley, in Alaska, is 6,281 m greater than the altitude of Death Valley.

EXAMPLE 1

Write an equation to find the altitude of Mt. McKinley.

Choose a variable. Let a be the altitude of Mt. McKinley. Then translate the words into an equation.

altitude of Mt. McKinley	is	6,281 m	greater than	altitude of Death Valley
↓	↓	↓	↓	↓
a	$=$	6,281	$+$	$^-86$

So, the equation is $a = 6{,}281 + ^-86$.

• Write an equation to show that the altitude of Mt. Hood, in Oregon, is 2,768 m less than the altitude of Mt. McKinley, which is 6,195 m.

Equations can also include subtraction, multiplication, and division.

EXAMPLE 2

Write an equation for the following word sentence: The number of hours worked, less $3\frac{1}{2}$ hr, is $15\frac{1}{4}$ hr.

Choose a variable. Let h be the number of hours worked.

number of hours worked	less	$3\frac{1}{2}$ hr	is	$15\frac{1}{4}$ hr
↓	↓	↓	↓	↓
h	$-$	$3\frac{1}{2}$	$=$	$15\frac{1}{4}$

Translate the sentence into an equation.

So, the equation is $h - 3\frac{1}{2} = 15\frac{1}{4}$.

• Write an equation for the following word sentence: Four times the original price is $24.96.

CHECK FOR UNDERSTANDING

Think and ▶ Discuss

Look back at the lesson to answer each question.

1. **Tell** what the variable represents when you translate a word sentence into an algebraic equation.

2. **Tell** what kind of equation matches this sentence: The quotient of $265.75 and the number of payments is $53.15. Then write the equation.

Guided ▶ Practice

Write an equation for the word sentence.

3. 7 more than a number is 20.

4. 9 fewer than a number is $17\frac{5}{8}$.

5. 5 times the number of marbles is 35.

6. The quotient of 567 points and the number of test scores is 81.

PRACTICE AND PROBLEM SOLVING

Independent ▶ Practice

Write an equation for the word sentence.

7. 14 is 12 more than a number.

8. 8.09 less than a number equals 44.2.

9. A number divided by $2\frac{3}{4}$ is $\frac{5}{6}$.

10. $^-16$ times a number is $^-144$.

Problem Solving ▶ Applications

11. **Science** The diameter of Saturn is 72,000 miles. This is about 9 times as great as the diameter of Earth. Write an equation that describes this relationship.

12. Translate the following sentence into an equation: $^-2$ is 10 less than a number. Could 12 be the number? Explain.

13. **Write a problem** using facts found in your science book. Then translate the problem into an equation.

14. **Number Sense** Picnic forks come in packs of 20, and picnic knives come in packs of 32. What is the least number of packs of each that you could buy to have the same number of forks and knives?

MIXED REVIEW AND TEST PREP

Simplify the expression. Then evaluate the expression for the given value of the variable. (p. 272)

15. $5x + 3x - 15$ for $x = ^-2$

16. $10y - 8y + 22$ for $y = ^-6$

17. $25 + 22z - 13z$ for $z = ^-8$

18. Find the quotient $225 \div ^-25$. (p. 252)

⭐ 19. **TEST PREP** Find $^-25 - ^-16$. (p. 250)

   A $^-41$ **B** $^-9$ **C** 9   **D** 41

EXTRA PRACTICE page H45, Set A

Model and Solve One-Step Equations

Explore how to model the solutions to one-step equations.

You need algebra tiles, or green paper rectangles and yellow squares.

Math Idea ▶

Activity

- Model an equation by using algebra tiles. Use a green rectangle to represent the variable. Use a yellow square to represent 1. Model and solve the equation $x + 3 = 4$.

$$x \quad + \quad 3 \quad = \quad 4 \qquad \leftarrow x + 3 = 4$$

- To solve an equation, you must get the variable alone on one side. To do this, take away 3 ones from each side.

- Taking 3 ones from each side of the equation leaves the variable alone on one side. The equation is now solved.

$$x \quad = \quad 1 \qquad \leftarrow \textit{Green rectangle is alone.}$$

The value of the variable that makes the equation true is called the solution of the equation.

Think and Discuss

- What is the solution of the equation $x + 3 = 4$?
- How can you tell this from the model?
- What operation did you model in the second step?
- What would you do to solve the equation $x + 8 = 11$?

TECHNOLOGY LINK

More Practice: Use E-Lab, Modeling Addition and Subtraction Equations.
www.harcourtschool.com/elab2002

Practice

Solve each equation by using algebra tiles.

1. $x + 2 = 6$ **2.** $5 = x + 3$ **3.** $x + 1 = 8$ **4.** $x + 4 = 9$

5. $x + 1 = 4$ **6.** $x + 4 = 7$ **7.** $6 = x + 5$ **8.** $x + 2 = 7$

LESSON 14.3

Solve Addition Equations

Learn how to solve addition equations.

QUICK REVIEW

1. $9.7 - 4.3$ 2. $^-5 - 7$
3. $12.9 - 4.2$ 4. $9\frac{3}{4} - 4\frac{1}{4}$
5. $10 - 37$

Vocabulary

Subtraction Property of Equality

Ralf Laue of Leipzig, Germany, set a world record in February of 1999 for the most dominoes stacked. He stacked 555 dominoes on a single supporting domino.

How many more dominoes would you need to stack to match the world record if you have already stacked 123? One way to solve the problem is by representing it with an equation.

dominoes left to stack		dominoes stacked		total
↓		↓		↓
d	$+$	123	$=$	555

Addition and subtraction are inverse operations. To solve the addition equation, you will use the inverse operation, subtraction, to get the variable alone on one side of the equation. The Subtraction Property of Equality justifies that step.

Subtraction Property of Equality	$5 = 5$
If you subtract the same number from both sides of an equation, the two sides remain equal.	$5 - 2 = 5 - 2$ $3 = 3$

EXAMPLE 1

Solve the equation $d + 123 = 555$ to find the number of additional dominoes you need to stack. Check your solution.

$$d + 123 = 555 \qquad \textit{Write the equation.}$$
$$d + 123 - 123 = 555 - 123 \qquad \textit{Use the Subtraction Property of Equality.}$$
$$d + 0 = 432 \qquad \textit{Use the Identity Property of Zero.}$$
$$d = 432$$

$$d + 123 = 555 \qquad \textit{Check your solution.}$$
$$432 + 123 \stackrel{?}{=} 555 \qquad \textit{Replace d with 432.}$$
$$555 = 555 \checkmark \qquad \textit{The solution checks.}$$

So, you must stack 432 more dominoes to match the world record.

Remember that the Identity Property of Zero states: For all numbers a, $a + 0 = a$.

287

You can sometimes use the Commutative Property of Addition to help solve equations. Remember that equations can have decimals, fractions, or negative integers.

EXAMPLE 2

Solve and check. $8.4 + x = 12.7$

$$8.4 + x = 12.7$$ *Use the Commutative Property*
$$x + 8.4 = 12.7$$ *of Addition.*

$$x + 8.4 - 8.4 = 12.7 - 8.4$$ *Use the Subtraction Property of Equality.*

$$x + 0 = 4.3$$ *Use the Identity Property of Zero.*

$$x = 4.3$$

$$8.4 + x = 12.7$$ *Check your solution.*

$$8.4 + 4.3 \stackrel{?}{=} 12.7$$ *Replace x with 4.3.*

$$12.7 = 12.7 \checkmark$$ *The solution checks.*

So, $x = 4.3$.

• Solve. $1.4 + x = {}^-8$.

Sometimes the variable will be on the right side of the equation.

EXAMPLE 3

Solve and check. $14 = 8\frac{2}{3} + k$

$$14 = 8\frac{2}{3} + k$$ *Use the Commutative Property of*
$$14 = k + 8\frac{2}{3}$$ *Addition.*

$$14 - 8\frac{2}{3} = k + 8\frac{2}{3} - 8\frac{2}{3}$$ *Use the Subtraction Property of Equality.*

$$5\frac{1}{3} = k + 0$$ *Use the Identity Property of Zero.*

$$5\frac{1}{3} = k$$

$$14 = 8\frac{2}{3} + k$$ *Check your solution.*

$$14 \stackrel{?}{=} 8\frac{2}{3} + 5\frac{1}{3}$$ *Replace k with $5\frac{1}{3}$.*

$$14 = 14 \checkmark$$ *The solution checks.*

So, $k = 5\frac{1}{3}$.

• Solve. $6\frac{5}{6} = 3\frac{1}{4} + y$.

CHECK FOR UNDERSTANDING

Think and Discuss ▶ **Look back at the lesson to answer each question.**

1. **Explain** why the choice of a variable does not affect the solution of the equation.

2. **Describe** how to solve an addition equation.

Guided ▶
Practice

Solve and check.

3. $x + 10 = 17$ **4.** $3 + b = {}^-9$ **5.** $5.2 = c + 2.5$

6. $y + 1.2 = 18$ **7.** $21 = m + 5\frac{3}{4}$ **8.** $7\frac{1}{5} + d = 22$

PRACTICE AND PROBLEM SOLVING

Independent ▶
Practice

Solve and check.

9. $x + 4 = 12$ **10.** $29 = a + 17$ **11.** $9 + k = {}^-32$

12. $z + 6.3 = 11.7$ **13.** $4.2 = b + 1.3$ **14.** $14.9 = 12.5 + c$

15. $y + 9\frac{1}{4} = 18$ **16.** $16 = 3\frac{2}{5} + s$ **17.** ${}^-12 = t + 7$

Problem Solving ▶
Applications

For 18–19, choose a variable, write an equation, and solve it.

18. What if you stack 221 dominoes? How many more would you have to stack to match the world record?

19. **Geometry**   The perimeter of the triangle at the right is 29 cm. Find the unknown length.   ?

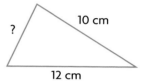

10 cm

12 cm

20. **What's the Question?**   Jay opened a savings account with a deposit of $124.50. Later he made two equal deposits and his account balance was $350.30. The answer is $112.90.

MIXED REVIEW AND TEST PREP

For 21–22, write an equation for the word sentence. (p. 284)

21. 11 is 8 more than a number.   **22.** 3.65 less than a number is 12.2.

23. Evaluate. $\sqrt{64} - 4 + 9$ (p. 278)   **24.** Evaluate. $25 + \sqrt{81} - 6^2$ (p. 278)

⭐**25.** **TEST PREP**   Evaluate $3(a - b)^2 \cdot c$ for $a = {}^-3$, $b = 1$, and $c = {}^-2$. (p. 272)

   A 96   **B** 46   **C** ${}^-24$ **D** ${}^-96$

PROBLEM SOLVING   ⭐ Thinker's Corner

Equation Relation   Solve the equations below. What is the relationship among the solutions of the blue equations? Which one of the green equations should be blue?

$x + 5 = 12$	$59 = a + 17$	$b + 19 = 22$
$10 = s + 2$	$b + 15 = 43$	$95 = a + 25$
$r + 6 = 29$	$78 = m + 1$	$c + 4 = 48$
$79 = p + 23$	$x + 12 = 103$	$105 = t + 42$

EXTRA PRACTICE page H45, Set B

Solve Subtraction Equations

Learn how to solve subtraction equations.

Vocabulary

Addition Property of Equality

Tamika was practicing the trick of catching dimes stacked on her elbow. On her first try, she caught 13 dimes. How many dimes were originally stacked on her elbow if 11 fell to the floor?

One way to solve this problem is by writing and solving an equation. Let d be the number of dimes originally stacked on Tamika's elbow.

dimes on elbow		dimes caught		dimes on floor
↓		↓		↓
d	$-$	13	$=$	11

Addition and subtraction are inverse operations. To solve the subtraction equation, use the inverse operation, addition, to get the variable alone on one side of the equation. The Addition Property of Equality justifies that step.

Addition Property of Equality	$5 = 5$
If you add the same number to both sides	$5 + 2 = 5 + 2$
of an equation, the two sides remain equal.	$7 = 7$

EXAMPLE

Solve the equation $d - 13 = 11$ to find the number of dimes originally stacked on Tamika's elbow. Check your solution.

$$d - 13 = 11 \qquad \textit{Write the equation.}$$
$$d - 13 + 13 = 11 + 13 \qquad \textit{Use the Addition Property of Equality.}$$
$$d - 0 = 24 \qquad \textit{Use the Identity Property of Zero.}$$
$$d = 24$$

$$d - 13 = 11 \qquad \textit{Check your solution.}$$
$$24 - 13 \stackrel{?}{=} 11 \qquad \textit{Replace d with 24.}$$
$$11 = 11 \checkmark \qquad \textit{The solution checks.}$$

So, Tamika had 24 dimes stacked on her elbow.

• Solve. $k - 4.7 = 5.8$

Think and ▸
Discuss
Look back at the lesson to answer the question.

1. **Explain** how you know what number to add to both sides of a subtraction equation.

Guided ▸
Practice
Solve and check.

2. $x - 8 = 15$

3. $b - 23 = {}^-29$

4. $7.9 = c - 3.4$

5. $y - 1.2 = 22$

6. $7\frac{1}{8} = m - 4\frac{3}{4}$

7. $d - 8\frac{4}{5} = 25\frac{2}{3}$

PRACTICE AND PROBLEM SOLVING

Independent ▸
Practice
Solve and check.

8. $x - 6 = 10$

9. $29 = a - 4$

10. $k - 5 = {}^-2$

11. $z - 5.8 = 11.2$

12. $7.2 = b - 1.9$

13. $14.5 = c - 8.8$

14. $y - 9\frac{1}{4} = 18$

15. $17\frac{2}{3} = s - 5\frac{1}{4}$

16. ${}^-8 = t - 5$

17. $2.8 = m - 8.3$

18. $h - 11 = {}^-3$

19. $f - \frac{3}{8} = \frac{5}{12}$

Problem Solving ▸
Applications
For 20–21, choose a variable, write an equation, and solve it.

JUANITA'S BACK-TO-
SCHOOL SHOPPING

$40 School Supplies
$90 New Bicycle
$110 New Clothes

20. **What if** Tamika had caught 17 dimes but 2 dimes fell on the floor? How many dimes would she have had stacked on her elbow?

21. **Use Data** Look at the graph. Juanita withdrew money from her savings account for back-to-school shopping. She has $527 left in her savings account. How much did she have in her account before she withdrew the money?

22. **❓ What's the Error?** Frank said the solution of the equation $x - 7 = 10$ is $x = 3$. Find his error and then solve the equation.

23. Change one operation sign so the value of the expression $61 - 12 - 13 - 14 - 5 - 7$ is doubled.

MIXED REVIEW AND TEST PREP

For 24–26, solve and check. (p. 287)

24. $x + 3 = 9$

25. $5.6 = b + 1.2$

26. $5\frac{1}{4} = 3 + p$

27. Evaluate ${}^-2x + 5$ for $x = {}^-2, {}^-1, 0,$ and 1. (p. 272)

⭐ **28. TEST PREP** The low temperatures for five days in Yorktown were ${}^-2°F, {}^-3°F, 4°F, {}^-1°F,$ and $2°F.$ Find the average low temperature for the five days. (p. 252)

A ${}^-2°F$ **B** $0°F$ **C** $2°F$ **D** $3°F$

1. **VOCABULARY**   If you subtract the same number from both sides of an equation, you are using the ___?___ Property of Equality. (p. 287)

2. **VOCABULARY**   Subtraction is the ___?___ of addition. (p. 287)

3. **VOCABULARY**   If you add the same number to both sides of an equation, you are using the ___?___ Property of Equality. (p. 290)

Write an equation for the word sentence. (pp. 284–285)

4. 72 is 15 more than a number.

5. 8.6 decreased by a number is 9.

6. ⁻18 times a number is ⁻45.

7. A number divided by $\frac{3}{8}$ is 16.

8. The product of $1\frac{3}{4}$ and a number is $21\frac{3}{5}$.

9. A number increased by 17 is 24.

10. 72 less than a number is 125.

11. The quotient of a number and 8 is 17.

Solve and check. (pp. 287–291)

12. $x + 7 = 10$

13. $9 + y = {}^{-}19$

14. $x - 6 = 3$

15. $3.78 = c - 2.33$

16. $21 + n = 52$

17. $x + 3\frac{1}{3} = 4\frac{2}{3}$

18. $3.4 + x = 19.5$

19. $1.9 = k - 1.2$

20. $y - 3\frac{1}{9} = 11\frac{5}{9}$

21. $d - 23 = {}^{-}18$

22. $42\frac{1}{5} + v = 51\frac{2}{3}$

23. $x - 33.7 = 10.9$

24. $w + 9\frac{3}{8} = 11\frac{2}{3}$

25. $2.3 + x = 5.4$

26. $q - 9.1 = 1.37$

27. $3\frac{1}{4} = x - 2\frac{1}{3}$

28. $17.5 = 2.7 + x$

29. $\frac{3}{5} + y = \frac{9}{10}$

For 30–33, choose a variable, write an equation, and solve it.

30. After soccer practice Li saw that there were 14 bottles of water left. During the practice the team consumed 18 bottles. How many bottles had there been at the start of practice? (p. 290)

31. Mirabelle has collected $37.50 so far for her favorite charity. The most she had collected in the past was $55.75. How much more does she need to collect to equal her record? (p. 287)

32. A triangle has a perimeter of 42 in. The sum of the lengths of two sides of the triangle is 29 in. Find the length of the third side of the triangle. (p. 287)

33. Thelma removed 12 cans of tomato paste from the pantry in her restaurant. There were 17 cans left. How many cans were originally in the pantry? (p. 290)

Check your work.

See item **5.**

Use number sense to decide if c will be greater or less than the difference. Write the inverse of the equation to find the value of c.
Also see problem **7**, p. H65.

Choose the best answer.

1. Solve for y.

$$y - 8 = 4$$

A $y = 2$ **C** $y = 12$

B $y = 4$ **D** $y = 32$

2. Solve for p.

$$6 + p = 18$$

F $p = 3$ **H** $p = 24$

G $p = 12$ **J** Not here

3. Solve for j.

$$j + 13.4 = {}^-26.8$$

A $j = {}^-40.2$ **C** $j = {}^-18.2$

B $j = {}^-22.2$ **D** $j = {}^-13.4$

4. Solve for k.

$$6.2 + k = 20$$

F $k = 13.8$ **H** $k = 18.8$

G $k = 14.2$ **J** $k = 26.2$

5. Solve for c.

$$c - 18.95 = 72.5$$

A $c = 53.55$ **C** $c = 90.145$

B $c = 54.45$ **D** $c = 91.45$

6. What is the value of $a \div b$ for $a = 25.5$ and $b = 1.5$?

F 17   **H** 170

G 24   **J** Not here

7. Terri blends $\frac{3}{4}$ cup of water, $\frac{5}{8}$ cup of milk, and $\frac{1}{3}$ cup of ice. Which is a reasonable total for the amount she blended?

A Less than 1 c

B Between $1\frac{1}{2}$ and 2 c

C More than 2 c

D About $2\frac{1}{2}$ c

8. What is the prime factorization of 210?

F $2 \times 3 \times 9$ **H** $2^2 \times 3 \times 5$

G $2 \times 3 \times 5 \times 7$ **J** $2 \times 3^2 \times 5$

Write What You Know

9. A backpack and a spiral notebook cost $12.78 before tax. The notebook costs $1.49. Let b represent the cost of the backpack. Write an addition equation you could use to find the cost of the backpack. Then find the cost and describe the steps you use.

10. Describe how you would find the GCF of two numbers. Include an example with your description.

15 Algebra: Multiplication and Division Equations

Travelers often take taxis in the cities they visit. Within the city of Washington, DC., taxi charges are based on a zone system. In other cities, taxi charges are based on a starting fare and a metered mileage charge.

PROBLEM SOLVING Suppose it costs $2.50 when you get into a cab. Added to that is a mileage charge of $1.50 per mile. You can use the equation $m = \$1.50x$ to find the mileage charge, m, for any number of miles, x. How would you find the mileage charge for a 2-mile trip? Find the mileage charge. How would you find the number of miles traveled if the mileage charge is $5.25? Find the number of miles.

TAXI CHARGES		
City	Starting Fare	Per-Mile Rate
Chicago, IL	$1.60	$1.40
Dallas, TX	$2.70	$1.20
New York City, NY	$2.00	$1.50
San Francisco, CA	$2.50	$1.80
Seattle, WA	$1.80	$1.80

Check What You Know

Use this page to help you review and remember important skills needed for Chapter 15.

✔ Vocabulary

Choose the best term from the box.

1. The temperature scale on which water freezes at 0° and boils at 100° is the __?__ scale.

2. A statement showing that two quantities are equal is a(n) __?__.

> Fahrenheit
> Celsius
> equation
> expression

✔ Words and Equations (For Intervention, see p. H17.)

Write an equation for the word expression.

3. 18 is 6 more than a number.

4. 12.5 less than a number equals 23.7.

5. A number divided by 2 is $\frac{2}{3}$.

6. 6 times a number is $^-90$.

7. Twice a number is 47.

8. The quotient of a number and 3 is $^-24$.

✔ Evaluate Expressions (For Intervention, see p. H18.)

Evaluate the expressions for the given values of the variables.

9. $2x$ for $x = 7$

10. $3a + 7$ for $a = {}^-5$

11. $5t - 13$ for $t = {}^-12$

12. pq for $p = 5$ and $q = 14$

13. $4mn$ for $m = {}^-5$ and $n = 7$

14. ^-8ab for $a = {}^-5$ and $b = 3$

15. $\frac{1}{2}x$ for $x = 18$

16. $\frac{2}{3}k$ for $k = 24$

17. $\frac{z}{4}$ for $z = 100$

18. $\frac{5}{9}f$ for $f = 27$

19. $\frac{9}{5}c$ for $c = 55$

20. $\frac{9}{5}c + 32$ for $c = 18$

21. $3.9b$ for $b = 7$

22. $^-7c + 3$ for $c = {}^-5$

23. $\frac{4}{5} + \frac{1}{2}c$ for $c = 6$

✔ Mental Math and Equations (For Intervention, see p. H11.)

Solve the equation by using mental math.

24. $5x = 20$

25. $3y = 15$

26. $42 = 6z$

27. $72 = 8a$

28. $\frac{b}{3} = 4$

29. $\frac{c}{5} = 6$

30. $4 = \frac{p}{4}$

31. $7 = \frac{q}{8}$

32. $\frac{4.5}{x} = 0.5$

33. $0.36 = 9y$

> **LOOK AHEAD**
>
> **In Chapter 15 you will**
> - solve multiplication and division equations
> - use formulas
> - explore two-step equations

Model Multiplication Equations

You can use algebra tiles to model solving multiplication equations.

Explore how to model the solutions of multiplication equations.

You need algebra tiles or paper rectangles and squares.

Activity

• A green rectangle represents a variable, and a yellow square represents the value 1. Here is a model of the multiplication equation $3y = 12$. To represent $3y$, use 3 green rectangles.

$$3y \qquad = \qquad 12$$

TECHNOLOGY LINK
More Practice: Use E-Lab, *Exploring Equations.*
www.harcourtschool.com/elab2002

• Divide each side of the model into 3 equal groups.

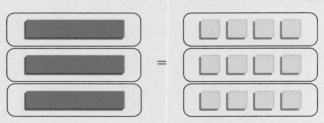

• Look at one group on each side. What is in each pair? What equation does this model?

Think and Discuss

• What is the solution of the equation $3y = 12$?

• What operation did you model in the second step? How does this operation relate to multiplication?

• *REASONING*   How do you think you would solve an algebraic equation involving division?

Practice

Use algebra tiles to model and solve each equation. Record each step.

1. $2c = 8$ **2.** $3a = 9$ **3.** $4b = 8$ **4.** $15 = 5y$

Solve Multiplication and Division Equations

Learn how to solve multiplication and division equations.

Vocabulary

Division Property of Equality

Multiplication Property of Equality

Over a period of 7 days, a koala can sleep as many as 154 hours. Excluding animals that hibernate, the koala averages the most hours of sleep per day of any animal. Write and solve an equation to find the average number of hours per day a koala sleeps.

Choose a variable to represent the average number of hours per day that a koala sleeps. Then write an equation.

number of days	×	average daily hr of sleep	=	total hr of sleep
↓		↓		↓
7	×	h	=	154

Remember that division can be written as a fraction.

$2 \div 3 = \frac{2}{3}$

$x \div 3 = \frac{x}{3}$

Multiplication and division are inverse operations. To solve the multiplication equation, use the inverse operation, division, so that the variable is alone on one side of the equation. The Division Property of Equality explains that step.

Division Property of Equality	$10 = 10$
If you divide both sides of an equation by the same nonzero number, the two sides remain equal.	$\frac{10}{2} = \frac{10}{2}$ $5 = 5$

EXAMPLE 1

Solve and check the equation $7h = 154$.

$\quad 7h = 154$ *Write the equation.*

$\quad \dfrac{7h}{7} = \dfrac{154}{7}$ *Use the Division Property of Equality.*

$\quad 1 \times h = 22$ *$7 \div 7 = 1$; $154 \div 7 = 22$*

$\quad\quad\quad h = 22$ *Use the Identity Property of One.*

$\quad\quad 7h = 154$ *Check your solution.*

$7 \times 22 \stackrel{?}{=} 154$ *Replace h with 22.*

$\quad\quad 154 = 154 \checkmark$ *The solution checks.*

So, a koala sleeps an average of 22 hr a day.

Sometimes equations involve fractions.

EXAMPLE 2

Solve and check. $16 = \frac{2}{3}x$

$16 = \frac{2}{3}x$ *Write the equation.*

$\dfrac{16}{\frac{2}{3}} = \dfrac{\frac{2}{3}x}{\frac{2}{3}}$ *Use the Division Property of Equality.*

$16 \div \frac{2}{3} = x$

$16 \cdot \frac{3}{2} = x$ *Multiply by the reciprocal.*

$24 = x$

$16 = \frac{2}{3}x$ *Check your solution.*

$16 \stackrel{?}{=} \frac{2}{3} \cdot 24$ *Replace x with 24.*

$16 = 16$ ✓   *The solution checks.*

So, $x = 24$.

- Solve and check. $\frac{1}{4}x = 22$

TECHNOLOGY LINK

More Practice: Use
***Mighty Math Astro
Algebra,*** Green, Level P.

To solve a division equation, use the inverse operation, multiplication, and the Multiplication Property of Equality.

Multiplication Property of Equality	$10 = 10$
If you multiply both sides of an equation by the same number, the two sides remain equal.	$5 \times 10 = 5 \times 10$ $50 = 50$

EXAMPLE 3

Solve and check. $\frac{a}{5} = {}^-14$

$\frac{a}{5} = {}^-14$ *Write the equation.*

$5 \cdot \frac{a}{5} = 5 \cdot {}^-14$ *Use the Multiplication Property of Equality.*

$\frac{5}{1} \cdot \frac{a}{5} = {}^-70$

$\frac{5a}{5} = {}^-70$

$a = {}^-70$ *$5 \div 5 = 1$ and $1a = a$*

$\frac{a}{5} = {}^-14$ *Check your solution.*

$\frac{{}^-70}{5} \stackrel{?}{=} {}^-14$ *Replace a with $^-70$.*

${}^-14 = {}^-14$ ✓   *The solution checks.*

So, $a = {}^-70$.

- Solve and check. ${}^-9 = \frac{c}{12}$

Think and Discuss ▶ Look back at the lesson to answer the question.

1. Tell how you solve multiplication and division equations and how you solve addition and subtraction equations.

Guided Practice ▶ Solve and check.

2. $\frac{x}{3} = 8$ **3.** $4x = {}^-28$ **4.** $45 = \frac{1}{2}x$ **5.** $1.8 = \frac{y}{4}$

PRACTICE AND PROBLEM SOLVING

Independent Practice ▶ Solve and check.

6. $3x = 6$ **7.** $8k = 48$ **8.** $\frac{y}{5} = 7$ **9.** $\frac{a}{6} = 3$

10. $4 = \frac{m}{3}$ **11.** ${}^-48 = {}^-3p$ **12.** ${}^-15 = \frac{s}{5}$ **13.** $60 = {}^-12n$

14. $\frac{1}{4}d = 60$ **15.** $4.12 = \frac{a}{4}$ **16.** $24 = \frac{4}{5}b$ **17.** $2.1 = \frac{w}{1.3}$

Solve and check.

18. $\frac{8}{15} = \frac{c}{45}$ **19.** $\frac{a}{4} = 2\frac{1}{2}$ **20.** $\frac{3}{5}m = 2\frac{2}{5}$

Problem Solving Applications ▶ **21.** Mike divided his marbles equally among 3 friends. Each friend got 14 marbles. Write and solve an equation to find how many marbles Mike had originally.

22. In the equation $\frac{8.2}{n} = 4$, the variable is in the denominator. Explain how you would solve the equation. Then solve.

23. Write and solve a multiplication equation to find the width of a rectangle with an area of 0.45 cm² and a length of 0.9 cm.

24. ❓ **What's the Error?** Mike's steps to solve the equation $2x = 12$ are shown at the right. Find his error and give the correct solution.

$$2x = 12$$
$$2x - 2 = 12 - 2$$
$$x = 10$$

MIXED REVIEW AND TEST PREP

Solve and check. (p. 290)

25. $x - 5 = 17$ **26.** $6 = y - 10$ **27.** ${}^-15 = z - 3$

28. Evaluate the expression $\sqrt{16} - 3^2 + 12$. (p. 278)

⭐**29. TEST PREP** The low temperatures in Anchorage for five days were 4°F, 6°F, 4°F, 2°F, and 14°F. How much greater is the average low temperature for the five days with the outlier, 14°F, than without the outlier? (p. 109)

A 14°F   **B** 6°F   **C** 4°F   **D** 2°F

Use Formulas

Learn how to use formulas to solve problems.

The Verrazano Narrows Bridge in New York is the longest bridge in the United States. The bridge is 13,700 feet long. Elena walked across the bridge at an average speed of 264 ft per min. How long did it take her to walk across the bridge?

If you know two of the three parts of the formula *distance* = *rate* · *time*, or *d* = *rt*, you can solve for the third part.

EXAMPLE 1

Find how long it took Elena to walk across the Verrazano Narrows Bridge.

$d = rt$	*Write the formula.*
$13{,}700 = 264t$	*Replace d with 13,700 and r with 264.*
$\dfrac{13{,}700}{264} = \dfrac{264t}{264}$	*Solve the equation.*
$51\frac{59}{66} = t$	

So, it took $51\frac{59}{66}$ min, or about 52 min, for Elena to walk across the Verrazano Narrows Bridge.

• Suppose Elena walked at an average speed of 250 ft per min. How long would it take her to cross the bridge?

You can solve for the distance if you know the rate and the time.

EXAMPLE 2

How far would Elena travel if she drove her car at an average speed of 57 mi per hr for $3\frac{1}{2}$ hr?

$d = rt$	*Write the formula.*
$d = 57 \cdot 3\frac{1}{2}$	*Replace r with 57 and t with $3\frac{1}{2}$.*
$d = 199\frac{1}{2}$	*Multiply.*

So, Elena would travel $199\frac{1}{2}$ mi.

You can solve for the rate if you know the distance and the time.

EXAMPLE 3

Suppose Elena took 5 hours to drive 306.5 mi. What was Elena's average speed?

$d = rt$ *Write the formula.*

$306.5 = r \cdot 5$ *Replace d with 306.5 and t with 5.*

$\dfrac{306.5}{5} = \dfrac{5r}{5}$ *Solve the equation.*

$61.3 = r$

So, Elena's average speed was 61.3 mi per hr.

Cleveland is located in the northeast corner of Ohio near Lake Erie. The monthly normal temperature for November in Cleveland is about 10°C. You may be more familiar with the Fahrenheit temperature scale. To convert from degrees Celsius to degrees Fahrenheit, use the formula below.

$$F = \left(\frac{9}{5} \cdot C\right) + 32$$

EXAMPLE 4

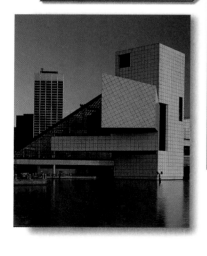

Find the monthly normal temperature for November in Cleveland in degrees Fahrenheit (°F). Write your answer as a decimal.

$F = \left(\frac{9}{5} \cdot C\right) + 32$ *Write the formula.*

$F = \left(\frac{9}{5} \cdot 10\right) + 32$ *Replace C with 10. Operate inside parentheses.*

$F = 18 + 32$ *Add.*

$F = 50$

So, the monthly normal temperature for November is about 50°F.

To convert from degrees Fahrenheit to degrees Celsius, use the formula below.

$$C = \frac{5}{9} \cdot (F - 32)$$

EXAMPLE 5

The monthly normal temperature in Cleveland for June is 78°F. Find the temperature in degrees Celsius (°C). Write your answer as a decimal and round to the nearest tenth of a degree.

$C = \frac{5}{9} \cdot (F - 32)$ *Write the formula.*

$C = \frac{5}{9} \cdot (78 - 32)$ *Replace F with 78. Operate inside parentheses.*

$C = \frac{5}{9} \cdot 46$ *Multiply.*

$C = 25\frac{5}{9} \approx 25.6$

So, the monthly normal temperature for June is about 25.6°C.

Think and ▶ Discuss

Look back at the lesson to answer the questions.

1. **Explain** how you know the unit of time in Example 1 is minutes.

2. **Explain** how you know the unit of length in Example 2 is miles.

3. *REASONING* **Tell** what a comfortable room temperature would be in degrees Celsius.

Guided ▶ Practice

Use the formula $d = rt$ to complete.

4. $d = $ ■ ft	5. $d = 100$ mi	6. $d = 300$ mi
$r = 22$ ft per min	$r = $ ■ mi per hr	$r = 50$ mi per hr
$t = 4$ min	$t = 5$ hr	$t = $ ■ hr

Convert the temperature to degrees Fahrenheit. Write your answer as a decimal.

7. 10°C   8. 20°C   9. 30°C   10. 40°C   11. 5°C

Convert the temperature to degrees Celsius. Write your answer as a decimal and round to the nearest tenth of a degree.

12. 85°F   13. 57°F   14. 212°F   15. 100°F   16. 32°F

PRACTICE AND PROBLEM SOLVING

Independent ▶ Practice

Use the formula $d = rt$ to complete.

17. $d = $ ■ ft	18. $d = 44$ km	19. $d = 162.5$ mi
$r = 8$ ft per sec	$r = $ ■ km per hr	$r = 65$ mi per hr
$t = 4.6$ sec	$t = 2.2$ hr	$t = $ ■ hr

Convert the temperature to degrees Fahrenheit. Write your answer as a decimal.

20. 10°C   21. 37°C   22. 100°C   23. 95°C   24. 32°C

Convert the temperature to degrees Celsius. Write your answer as a decimal and round to the nearest tenth of a degree.

25. 32°F   26. 104°F   27. 47°F   28. 72°F   29. 94°F

Solve.

30. Rita drove for $3\frac{1}{2}$ hours at a rate of 40 mi per hr. How far did Rita travel?

31. A train traveled 350 mi in 4 hours. Find the train's average rate of speed.

32. A hot air balloon rose 4,200 ft in 14 min. Find the balloon's average rate of ascent.

Problem Solving ▶
Applications

33. Science In order to remain in orbit, the space shuttle must reach speeds of about 17,500 mi per hr. Suppose the space shuttle travels 52,650 mi in 3 hr. Is the space shuttle traveling fast enough to remain in orbit? Explain.

Use Data **For 34–36, use the data in the table below.**

34. Space Does Mercury or Venus have the higher extreme temperature? Explain.

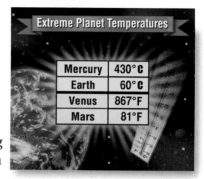

Extreme Planet Temperatures	
Mercury	430°C
Earth	60°C
Venus	867°F
Mars	81°F

35. Does Earth or Mars have the higher extreme temperature? Explain.

36. Graphs Make a bar graph comparing all of the extreme temperatures shown in the table. Use °C.

37. ✎ **Write a problem** where you must use $d = rt$ to solve it.

MIXED REVIEW AND TEST PREP

Solve and check. (p. 297)

38. $3x = 15$

39. $^-42 = 7y$

40. $\frac{z}{8} = 3$

⭐**41. TEST PREP** Which equation shows 18 is 6 more than a number? (p. 284)

A $18 + 6 = x$ **B** $18 + x = 6$ **C** $18 = x + 6$ **D** $18 = x - 6$

⭐**42. TEST PREP** Simplify $2x + 4x - 12$ by combining like terms. Then evaluate the expression for $x = {}^-4$. (p. 272)

F $^-36$ **G** $^-10$ **H** 10 **J** 36

PROBLEM SOLVING [LINKUP to Careers

Mathematician Evelyn Boyd Granville is a mathematician who has needed formulas throughout her career. When the United States space program was brand new, Dr. Granville worked on Project Mercury and Project Apollo, calculating orbits and programming computers. She has also worked with students from kindergarten through college.

• The distance to the moon is approximately 239,000 miles. Suppose a space craft took 6 days to reach the moon. What was its average speed in mph?

Two-Step Equations

MATH LAB

Explore how to use models to solve two-step equations.

You need algebra tiles or paper rectangles and squares.

QUICK REVIEW

Solve.

1. $\frac{x}{3} = 10$ 2. $\frac{k}{3} = 15$

3. $^-6 = \frac{p}{7}$ 4. $8.4 = \frac{m}{12}$

5. $^-36 = \frac{a}{2}$

Sometimes equations involve more than one step. In this lab you will use models to solve two-step equations.

Activity

• Use algebra tiles or paper rectangles and squares to model $2x + 2 = 6$. Use a green rectangle to represent the variable x, and use a yellow square to represent 1.

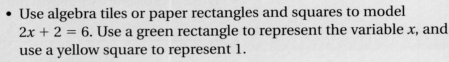

$2x \qquad + 2 \quad = \qquad 6$

• To solve an equation, arrange the model so that a rectangle is alone on one side. First, remove two squares from each side.

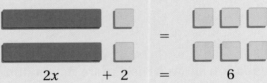

• The model now shows $2x = 4$.

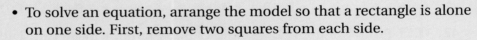

• Separate the model into two equal groups on each side.

• What does each pair of groups model?

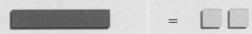

• What is the solution of the equation?

Think and Discuss

- How can you use the original model to check your solution?

- What did you do to the model so that only rectangles were alone on one side? What operation does this model?

- What operation did you model when you separated the model into two equal groups?

- Look at this model. What equation does it represent? Describe how to solve it.

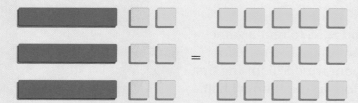

Practice

1. Copy the model below. What equation does it represent? Use the model to solve the equation.

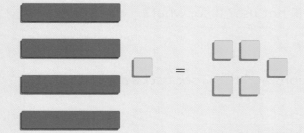

TECHNOLOGY LINK

More Practice: Use E-Lab, *Exploring Two-Step Equations.*
www.harcourtschool.com/elab2002

Use a model to solve each equation.

2. $3x + 1 = 10$

3. $4y + 2 = 10$

4. $2b + 5 = 9$

5. Explain how you would solve the equation $2x - 1 = 5$. Solve the equation.

6. Can you solve $3x + 5 = 2$ by using a model? Explain.

MIXED REVIEW AND TEST PREP

Use the formula $d = rt$ to complete. (p. 300)

7. $d = 165$ mi
   $r = 55$ mi per hr
   $t = \blacksquare$ hr

8. $d = 679.5$ ft
   $r = \blacksquare$ ft per min
   $t = 4.5$ min

9. $d = \blacksquare$ km
   $r = 88$ km per hr
   $t = 4.25$ hr

10. Solve. $x - 7 = {}^-7$ (p. 290)

11. **TEST PREP** Evaluate $(x + 5)^2 - 9 \cdot 12$ for $x = \sqrt{64}$. (p. 278)

   A 1,920   **B** 960   **C** 61   **D** $^-19$

305

PROBLEM SOLVING STRATEGY
Work Backward

Learn how to use the strategy *work backward* to solve problems.

Felipe's parents paid $155 to rent a game room for his birthday party. The rates for renting the game room were $75 for the first hour and then $20 for each additional half hour. For how long did Felipe's parents rent the game room?

Analyze   What are you asked to find?

What information is given?

Is there information you will not use? If so, what?

Choose   What strategy will you use?

You can use inverse operations and the *work backward* strategy to solve the problem.

Solve   How will you solve the problem?

The cost of renting the game room was calculated in this way:

number of additional half hours		cost for additional half hours		cost for first hour		total cost of renting game room
	$\times$		$+$		$=$	
■	$\times$	$20	$+$	$75	$=$	$155

You can work backward by reversing the operations and the order.

($155 $-$ $75) $\div$ $20 $=$ 4

first hour + additional half hours = total time
1 hour + 4 half hours = 3 hours

So, Felipe's parents rented the game room for 3 hours.

Check   How can you check your answer?

What if Felipe's parents had paid $175 to rent the game room? For how long would they have rented the game room?

PROBLEM SOLVING PRACTICE

PROBLEM SOLVING STRATEGIES

Draw a Diagram or Picture

Make a Model

Predict and Test

▶ **Work Backward**

Make an Organized List

Find a Pattern

Make a Table or Graph

Solve a Simpler Problem

Write an Equation

Use Logical Reasoning

Solve the problem by working backward.

1. The local cab company charges $1.25 for the first mile traveled and then $0.35 for each additional mile. Natalie spent $7.20 on a ride in a cab. How many miles did Natalie travel?

2. Abby buys three CDs in a set for $29.98. She saved $6.44 by buying the set instead of buying the individual CDs. If each CD costs the same amount, how much does each of the three CDs cost when purchased separately?

MIXED STRATEGY PRACTICE

For 3–4, use this information.

There are 141 students on the field trip to the Space Museum. It takes 3 full buses to transport the students. The buses are all identical. There will be 51 more students on the next field trip.

3. Which expression could you use to find the number of buses needed for the next field trip if b is the number of students per bus?

   A $141 - (b + 51)$ **C** $(141 + 51) \div b$
   B $51b + 3$ **D** $141 + 51 - b$

4. Which of the equations below could you use to find b, the number of students per bus?

   F $\frac{50}{b} = 3$ **H** $\frac{b}{50} = 141$
   G $50 + b = 141$ **J** $3b = 141$

5. Maria has chorus practice at 5:30 P.M. She needs 25 min to walk there, $\frac{1}{4}$ hr to get ready, and $\frac{1}{2}$ hr to fix and eat a snack before she gets ready. When should she begin to fix the snack?

6. Jana has 89 CDs in her CD collection. She wants to take all of them with her on vacation. A CD carrying case holds 25 CDs. How many carrying cases does Jana need to carry all 89 CDs?

7. Elliot has nickels and dimes in his pocket. He has twice as many dimes as nickels. If the total value of the coins is $1.00, how many of each coin does he have?

8. Amanda is making a small quilt that is 36 in. by 54 in. She wants to add a border of trim. She has 10 feet of the trim. How much more trim does she need?

9. Half of the computer club members use $\frac{3}{8}$ of the computer stations in the technology lab. What is the least number of students that could be in the computer club?

10. **?** **What's the Question?** Sean has a balance of ⁻$183 in his checking account. Last week he wrote a check for $25 and one for $217. The answer is $59.

1. **VOCABULARY** The property which states that if you divide both sides of an equation by the same nonzero number, the two sides remain equal is the __?__. (p. 297)

2. **VOCABULARY** The property which states that if you multiply both sides of an equation by the same number, the two sides remain equal is the __?__. (p. 298)

Solve and check. (pp. 297–299)

3. $6x = 24$

4. $28 = 4a$

5. $\frac{y}{7} = 11$

6. $1.9 = \frac{a}{12}$

7. $1.1n = 55$

8. $\frac{x}{5} = 3$

9. $9y = {}^-72$

10. $^-6 = \frac{x}{3}$

11. $\frac{k}{5} = {}^-12$

12. $\frac{3}{4} = \frac{m}{12}$

13. $2.5x = 42.5$

14. $\frac{1}{3}b = 36$

Use the formula $d = rt$ to complete. (pp. 300–303)

15. $d = 1{,}485$ mi
    $r = 55$ mi per hr
    $t = \blacksquare$ hr

16. $d = 220$ km
    $r = \blacksquare$ km per hr
    $t = \frac{1}{2}$ hr

17. $d = \blacksquare$ ft
    $r = 10$ ft per sec
    $t = 36$ sec

18. $d = 3{,}570$ ft
    $r = 140$ ft per min
    $t = \blacksquare$ min

19. $d = 1{,}625$ mi
    $r = \blacksquare$ mi per hr
    $t = 25$ hr

20. $d = \blacksquare$ m
    $r = 32$ m per sec
    $t = 9.6$ sec

21. $d = 273.42$ km
    $r = 18.6$ km per hr
    $t = \blacksquare$ hr

22. $d = 157.5$ in.
    $r = \blacksquare$ in. per min
    $t = 42$ min

Convert the temperature to degrees Fahrenheit. Write your answer as a decimal. (pp. 300–303)

23. $10°C$

24. $25°C$

25. $2°C$

26. $41°C$

Convert the temperature to degrees Celsius. Write your answer as a decimal and round to the nearest tenth of a degree. (pp. 300–303)

27. $59°F$

28. $89°F$

29. $72°F$

30. $38°F$

Solve.

31. The maximum speed of the X-15A-2 jet is 4,534 mi per hr. How long would it take this jet to travel the 2,787 mi from Los Angeles to New York if it is traveling at its maximum speed? Round your answer to the nearest tenth. (pp. 300–303)

32. André paid $14.35 for a cab ride. The rate for riding in the cab was $1.75 for the first mile and then $0.35 for each additional mile. How many miles long was André's cab ride? (pp. 306–307)

33. Two days ago, Raul had a certain amount of money in his wallet. He then added $2 to his wallet. Yesterday, he was paid for his chores, and the amount of money in his wallet tripled. He now has $18 in his wallet. How much was originally in his wallet? (pp. 306–307)

Decide on a plan.

See item **3.**

Decide what should replace F in the formula. Solve for C and round to the nearest degree.

Also see problem **4**, p. H63.

Choose the best answer.

1. Solve for x.

$$3x = {}^-18$$

A $x = {}^-54$ **C** $x = 6$

B $x = {}^-6$ **D** $x = 54$

2. Solve for x.

$$\frac{x}{3} = {}^-9$$

F $x = {}^-27$ **H** $x = 3$

G $x = {}^-3$ **J** $x = 27$

3. Use the formula $C = \frac{5}{9} \times (F - 32)$ to convert 60°F to degrees Celsius. What is that amount rounded to the nearest degree?

A 14°C   **C** 50°C

B 16°C   **D** 51°C

4. Solve for a.

$$9a = {}^-108$$

F $a = {}^-12$ **H** $a = 12$

G $a = 11$ **J** $a = 972$

5. Solve for x.

$$4x = 32$$

A $x = 128$ **C** $x = 28$

B $x = 36$ **D** $x = 8$

6. It took Mike 3 hours to drive to his college at an average speed of 52 miles per hour. Which equation can be used to find the distance Mike drove?

F $d = 52 + 3$

G $d = 52 - 3$

H $d = 52 \times 3$

J $d = 52 \div 3$

7. Rita had $7\frac{1}{2}$ yards of fabric. She used $3\frac{7}{8}$ yards for a cushion. How much fabric is left?

A $9\frac{3}{8}$ yd   **C** $3\frac{5}{8}$ yd

B $6\frac{3}{8}$ yd   **D** $2\frac{1}{8}$ yd

8. Tim is surveying members of a hockey club. Which of the following methods should he use to produce a systematic sample?

F A survey of the top-scoring champions

G A survey of 2 out of every 9 goalies

H A survey of the 25 oldest club members

J A random survey of 1 out of every 8 club members

Write What You Know

9. Describe how you can use the formula $d = rt$ to find how much time it takes to go somewhere if the distance and rate are known. Then use your method to find how long it would take a train to travel 220 miles at an average speed of 40 miles per hour.

10. Ten bicyclists rode the following numbers of miles: 5, 4, 3, 6, 4, 7, 5, 3, 4, 6. Find the median of the distances. Explain how you found your answer.

MATH DETECTIVE

Puzzle Power

REASONING The number at each triangle vertex is the sum of the lengths of the two sides that form the angle. For example, in Mystery Triangle 1, $a + b = 9$, $a + c = 13$, and $b + c = 12$. Use your knowledge of triangles and the strategy *predict and test* to find a, b, and c.

Mystery Triangle 1

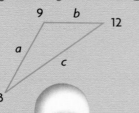

Mystery Triangle 2

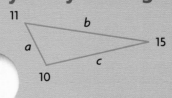

Mystery Triangle 3

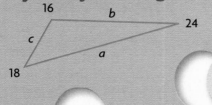

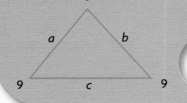

Mystery Triangle 4

Mystery Triangle 5

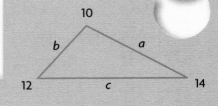

Mystery Triangle 6

Think It Over!

- ✎ **Write About It** Explain how you found the lengths of the sides of Mystery Triangle 1.

- **Stretch Your Thinking** Find a, b, and c shown on the triangle at the right. Then explain why it is actually impossible to draw a triangle with these measurements.

Reflexive, Symmetric, and Transitive Properties

Learn how to recognize and use Reflexive, Symmetric, and Transitive Properties.

Two numbers can relate to each other in many different ways. For example, one number may be *less than* the other. The most familiar relationship between numbers is *equality*. Reflexive, Symmetric, and Transitive Properties describe ways that numbers can relate to each other.

Reflexive Property of Equality

A number is always equal to itself.

$a = a$ $\qquad$ $7 = 7$

Symmetric Property of Equality

Equal numbers remain equal when their order is changed.

If $a = b$, then $b = a$.

If $4 + 9 = 10 + 3$, then $10 + 3 = 4 + 9$.

Transitive Property of Equality

Numbers equal to the same number are equal to each other.

If $a = b$ and $b = c$, then $a = c$.

If $5 + 3 = 8$ and $8 = 10 - 2$, then $5 + 3 = 10 - 2$.

EXAMPLE

Decide whether the Reflexive, Symmetric, or Transitive Property is used.

a. If $2 + 5 = 6 + 1$, then $6 + 1 = 2 + 5$. $\qquad$ *Symmetric*
b. $12 + t = 12 + t$ $\qquad$ *Reflexive*
c. If $4 + 6 = 10$ and $10 = 13 - 3$, then $4 + 6 = 13 - 3$. $\quad$ *Transitive*
d. $15 + 6 = 15 + 6$ $\qquad$ *Reflexive*

TALK ABOUT IT

- In your own words, describe Reflexive, Symmetric, and Transitive Properties.

TRY IT

Identify the property.

1. If $s = 3$ and $3 = d$, then $s = d$. $\qquad$ **2.** $4 + n = 4 + n$

3. If $4 + 6 = 3 + 7$, then $3 + 7 = 4 + 6$.

Decide whether each of the Reflexive, Symmetric, and Transitive Properties holds for the given relationship.

4. "is greater than" $\qquad$ **5.** "is a sister of"

VOCABULARY

1. The parts of an algebraic expression that are separated by an addition or subtraction sign are called ___?___. (p. 273)

2. The terms $3x^2$ and $2x^2$ have the same variable raised to the same power. They are called ___?___ terms. (p. 273)

3. The product of a number and itself is a ___?___. (p. 276)

EXAMPLES

EXERCISES

Chapter 13

- **Write numerical and algebraic expressions.** (pp. 270–271)

  Write an algebraic expression for the word expression.

3.8 less than a	$a - 3.8$
the sum of twice b and 5	$2b + 5$

Write an algebraic expression for the word expression.

4. c divided by 2.6

5. the product of h and $\frac{3}{4}$

6. the difference between b and 45

- **Evaluate numerical and algebraic expressions.** (pp. 272–275, 278–279)

  Evaluate $12 + 8x$ for $x = {}^-4$.

$12 + 8 \times ({}^-4)$	*Replace x with $^-4$.*
$12 + {}^-32$	*Multiply.*
$^-20$	*Add.*

For 7–11, evaluate the expressions for $x = {}^-3$ and $y = 9$.

7. $^-6x - {}^-8$

8. $4x + 2x + 9$

9. $12x - 7x + 15$

10. $4x^2 + xy$

11. $\sqrt{144} - 4^2 + 35y$

12. Evaluate $a - c + b$ for $a = 2.5$, $b = 7.8$, and $c = {}^-12$.

Chapter 14

- **Write word sentences as equations.** (pp. 284–285)

  Forty is fifteen more than a number.

↓	↓	↓	↓	↓
40	=	15	+	n

  Equation: $40 = 15 + n$

Write an equation for the word sentence.

13. $^-9$ is six less than a number.

14. A number divided by 2.5 is fifteen.

15. A number times $\frac{2}{3}$ is eighteen.

- **Solve addition equations.** (pp. 287–288)

$x + 6.4 = 9.8$	*Use the Subtraction Property of Equality.*
$x + 6.4 - 6.4 = 9.8 - 6.4$	
$x = 3.4$	

Solve.

16. $6 + m = {}^-24$

17. $r + 3\frac{2}{3} = 4$

18. $p + 5.8 = 13$

19. $15.1 = w + {}^-3.89$

• **Solve subtraction equations.** (pp. 290–291)

$$x - 17 = {}^-8$$

$$x - 17 + 17 = {}^-8 + 17 \quad \textit{Use the Addition Property of Equality.}$$

$$x + 0 = 9$$

$$x = 9 \quad \textit{Use the Identity Property of Zero.}$$

Solve.

20. $m - 8 = 1$

21. $r - 1\frac{1}{2} = 3\frac{1}{4}$

22. $0.6 = p - 1.4$

23. ${}^-5 = w - 9$

Chapter 15

• **Solve multiplication and division equations.** (pp. 297–299)

$$\frac{m}{6} = 4.8$$

$$6 \cdot \frac{m}{6} = 4.8 \cdot 6 \quad \textit{Use the Multiplication Property of Equality.}$$

$$m = 28.8$$

Solve.

24. $18 = \frac{3}{4}c$

25. ${}^-7.2t = 14.4$

26. ${}^-6z = {}^-96$

27. $\frac{n}{8} = 3.4$

28. $2x = {}^-36$

• **Use formulas.** (pp. 300–303)

Use the formula $d = rt$ to complete.

rate = 54 miles per hour
time = 5 hours   distance = ■ mi

$$d = rt$$

$$d = 54 \times 5 \quad \textit{Replace r with 54 and t with 5.}$$

$$d = 270 \quad \textit{Multiply.}$$

The distance is 270 miles.

Use the formula $A = lw$ to complete.

29. length = 8.6 in., width = 3 in., area = ■

30. Area = 420 in.², length = 70 in., width = ■

Use the formula $P = 2l + 2w$ to complete.

31. length = $9\frac{1}{2}$ ft, width = $4\frac{1}{2}$ ft, perimeter = ■ ft

32. perimeter = 34 cm, length = 12 cm, width = ■ cm

PROBLEM SOLVING APPLICATIONS

Solve. (pp. 306–307)

33. Anna Maria has soccer practice at 4:30 P.M. It takes her 5 min to change her clothes, 10 min to have a snack, and 15 min to get to the field. She wants to spend some time with her grandmother before practice. If she arrives home from school at 3:15 P.M., how much time does Anna Maria have to spend with her grandmother?

34. Andrew is thinking of a number between 1 and 10. If he multiplies the number by 4 and then subtracts 5, the result is 31. What is the number?

TASK A • Horse Sense

Alex wants to buy a saddle for his horse. The saddle costs $300. He has $87.25 saved. In addition to keeping his horse at the stable for free, Alex earns $5.25 an hour grooming horses and cleaning out all the stables. He wants to know how many hours he will have to work before he can buy the saddle.

Alex writes this equation to find the amount of money, *m*, that he needs to buy the saddle.

$$m + \$87.25 = \$300$$

a. Explain how to use the inverse operation to solve for *m*. Then solve.

b. Alex wrote this expression to figure his earnings: $5.25 + *h*. What mistake did he make?

c. How could you find the number of hours Alex needs to work to buy the saddle? How many hours does he need to work?

d. Alex has the chance to work at the stable shop. He can work there 2 hours for every 5 hours he works in the stable. Working in the stable shop pays $9.00 per hour. If Alex decides to do this, how many hours must he work at each job to earn enough to pay for the saddle? Explain your reasoning.

TASK B • On the Road

Materials: maps, mileage charts

Plan a trip to a city in the United States that is at least 200 miles away from where you live.

a. Choose two different routes you can take. Figure out how many miles each route is.

b. Show how you would use the distance formula to find the number of hours of driving it would take to reach the city at 55 miles per hour on each of the routes.

c. How long would it take to drive to that city at a rate of 65 miles per hour?

d. Suppose the speed limit on the longer route is 55 miles per hour and the speed limit on the shorter route is 40 miles per hour. Which route will get you to the city faster?

Data ToolKit

Use a Spreadsheet to Complete Function Tables

The input/output table shown here gives values of the algebraic expression $b \times 7 + 2$. Each output is the value of the expression for a given input value of the variable. You can use a spreadsheet to complete input/output tables.

Tosha orders books online. Each book costs $7. A shipping charge of $2 is added to each book order. How much would it cost to order 1 book? to order 2, 3, 4, 5, or 6 books?

Write an expression. Then use a spreadsheet to evaluate the expression.

$$b \times 7 + 2 \qquad \leftarrow b = \text{number of books}$$

- Enter the number of books into Column A.

- Enter the expression into Column B.
  Type =. Click in cell A1. Type *7 + 2. Press *Enter*.

- Highlight cells B1–B6. Click on *Edit* and choose *Fill Down*.

- The values in Column B are the costs of buying from 1 to 6 books.

So, it would cost $9, $16, $23, $30, $37, and $44 to order 1, 2, 3, 4, 5, and 6 books.

Expression: $b \times 7 + 2$	
Input (*b*)	Output
1	$1 \times 7 + 2 = 9$
2	$2 \times 7 + 2 = 16$
3	$3 \times 7 + 2 = 23$
4	$4 \times 7 + 2 = 30$
5	$5 \times 7 + 2 = 37$
6	$6 \times 7 + 2 = 44$

Practice and Problem Solving

Use a spreadsheet to complete the table for the expression.

1.

$x + 8 \times 2 - 3$	
Input (*x*)	Output
4	
5	
6	
7	

2.

$y \times 3 + 12$	
Input (*y*)	Output
15	
16	
17	
18	

3.

$2 \times z + 3 \times z$	
Input (*z*)	Output
11	
12	
13	
14	

4. **STRETCH YOUR THINKING** Write a rule for a function table. Then use a spreadsheet to complete the function table for 20 values.

Multimedia Math Glossary www.harcourtschool.com/mathglossary

5. **Vocabulary** Look up *square* and *square root* in the Multimedia Math Glossary. Write a description of the relationship between square and square root.

PROBLEM SOLVING ON LOCATION

In Pennsylvania

The First Superhighway

The Pennsylvania Turnpike was the first highway designed for modern high-speed long-distance travel. Completed in 1940, the turnpike crossed the Allegheny Mountains between Harrisburg and Pittsburgh, Pennsylvania. It shortened travel time between those two cities by 3 hours.

1. The turnpike is 132 feet wide on the section called the Southwestern Expansion. That is $1\frac{5}{6}$ times the road's width on the portion called the Northeastern Extension. Write an equation you can use to find the width of the Northeastern Extension. Then solve the equation.

2. Originally 160 miles long, the Pennsylvania Turnpike has since been lengthened to 514 miles. Write an equation you can use to find how much longer the highway is today than it was in 1940. Then solve the equation.

Use Data For 3-5, use the figure, which shows distances on the Pennsylvania Turnpike.

Pittsburgh   Breezewood   Blue Mountain   Valley Forge   Philadelphia

104 miles   40 miles m

It is 100 mi farther from Blue Mountain to Valley Forge than it is from Valley Forge to Philadelphia. (Note: Diagram not drawn to scale.)

3. Use the variable m to write an expression for the distance from Blue Mountain to Valley Forge.

4. Write and simplify an expression for the distance from Pittsburgh to Philadelphia.

5. It is 24 miles from Valley Forge to Philadelphia. How far is it from Pittsburgh to Philadelphia?

Fifteen thousand workers from 18 states labored for 2 years to build the Pennsylvania Turnpike.

The First Oil Well

Pennsylvania was the birthplace not only of America's modern highway system, but of its oil industry too. In 1859, Edwin L. Drake drilled the world's first oil well, beside Oil Creek, near Titusville, Pennsylvania.

At the Drake Well Museum, you can see a replica of the first oil well and learn about the early petroleum industry.

Before Drake drilled his well, small amounts of oil seeped from the ground near Oil Creek. Drake's well produced about 210 times as much oil each day as could be collected above ground.

1. Let *A* represent the amount of oil that could be collected above ground each day. Write an expression for the daily amount produced by Drake's well.

2. About 5 gallons of oil could be collected above ground each day. Use the expression you wrote to find the daily production of Drake's well.

3. There are 42 gallons in 1 barrel of oil. Find the daily production of Drake's well in barrels.

4. In 1859, oil sold for about $\sqrt{400}$ dollars per barrel. Find the value of one day's production of oil at Drake's well.

The surface of the ground at Drake's well is located at an altitude of 1,200 feet above sea level.

5. To reach oil, Drake first drilled through gravel to an altitude of 1,168 ft above sea level. Then he drilled through shale an additional 37.5 ft. What was the total depth below ground level at which Drake struck oil?

6. A few months later, Drake drilled a second well nearby. It was 63 ft deeper than 6 times the depth of the first well. Let *d* represent the depth of the first well. Write an expression for the depth of the second well.

7. How deep was Drake's second well?

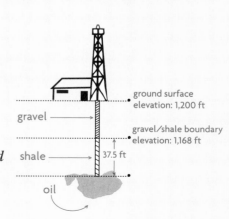

Frank Lloyd Wright greatly influenced modern architecture. He designed many buildings around nature. His buildings often had large walls, terraces, and roof overhangs that reached out into the landscape. Wright built this house between 1936 and 1939. He named it Fallingwater.

PROBLEM SOLVING What types of angles do you see in the decks that hang over the waterfalls?

TYPES OF ANGLES

Name	Figure
acute	
right	
obtuse	
straight	

Check What You Know

Use this page to help you review and remember important skills needed for Chapter 16.

 ## Vocabulary

Choose the best term from the box.

1. Two rays with a common endpoint form a(n) __?__ .

2. An exact location is named by a(n) __?__ .

3. A straight path that extends without end in opposite directions is called a(n) __?__ .

4. A flat surface that extends without end in every direction is called a(n) __?__ .

> acute
> angle
> line
> plane
> point

Classify Angles (For Intervention, see p. H18.)

Classify each angle by stating whether it is *acute, obtuse, right,* or *straight*.

5.

6.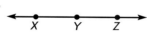

7.

8.

9.

10.

11.

12.

 ## Name Angles (For Intervention, see p. H19.)

Name the angle formed by the blue rays.

13.

14.

15.

16.

17.

18.

19.

20.

> ### LOOK AHEAD
>
> **In Chapter 16 you will**
> - solve problems involving figures in two and three dimensions
> - classify parallel, intersecting, and perpendicular lines
> - name, measure, and draw angles
> - explore relationships among angles

Points, Lines, and Planes

Learn how to describe figures by using the terms of geometry.

Points, lines, and planes are the building blocks of geometry.

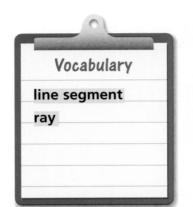

Vocabulary

line segment

ray

A *point* is an exact location. Think of a star in space.	•P Use the letter to point P name the point.
A *line* is a straight path that extends without end in opposite directions. Think of a narrow beam of light flashing across space in both directions.	line AB, $\overleftrightarrow{AB}$, line BA, $\overleftrightarrow{BA}$ Use two points on the line to name the line.
A *plane* is a flat surface that extends without end in every direction. Think of the surface of a large lake.	plane *LMN* Use three points in any order that are not on a line to name the plane.

Line segments and rays are parts of a line. Use points on the line to name line segments and rays.

A **line segment** has two endpoints.	line segment XY, $\overline{XY}$ line segment YX, $\overline{YX}$
A **ray** has one endpoint. From the endpoint, the ray extends forever in one direction only.	ray JK, $\overrightarrow{JK}$

You see models of points, lines, and planes every day. Specks of dust on a window model points. Flagpoles and pencils model line segments. Table tops and ceilings of rooms model planes.

Lines extend without end. So, real-world objects like goal "lines" and boundary "lines" are really just parts of lines. For the same reason, the sides of a skyscraper or of a gemstone are just portions of planes.

CHECK FOR UNDERSTANDING

Think and Discuss ▶ **Look back at the lesson to answer the question.**

1. **Name** the geometric figure modeled by each of the following: a page of a book; a single dot, or "pixel," on a computer screen; the path of a passenger jet across the sky.

Guided Practice ▶ **Name the geometric figure.**

2.

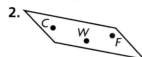

3.

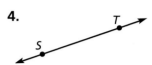

4.

PRACTICE AND PROBLEM SOLVING

Independent Practice ▶ **Name the geometric figure.**

5.

6.

7.

8.

9.

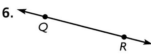

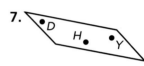

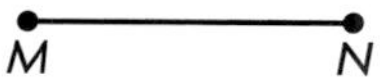

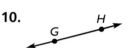

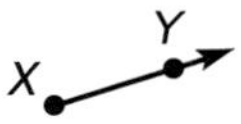

10.

For 11–14, use the figure at the right.

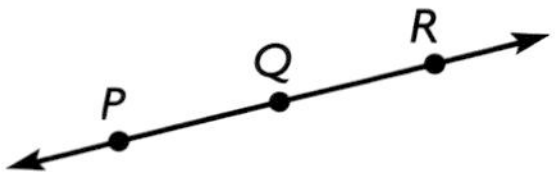

11. Name three different line segments.
12. Name four different rays.
13. Give six names for the line.
14. Give another name for ray *RQ*.

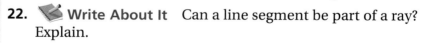

Problem Solving Applications ▶ **Name the geometric figure formed by the intersection of each pair of figures.**

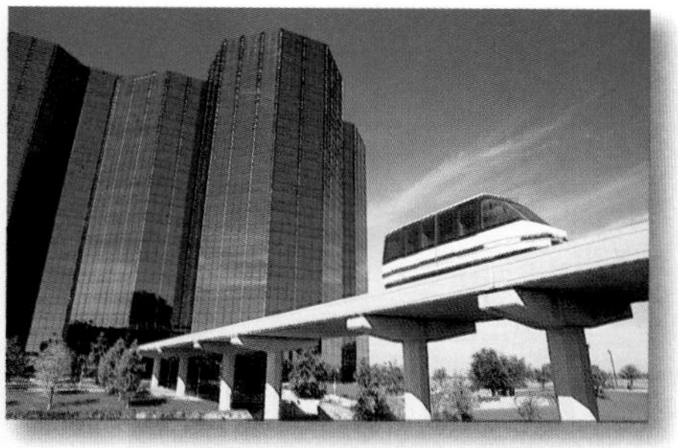

15. two lines
16. two planes
17. line and plane

Give a real-world example of each term.

18. a ray
19. a point
20. a line
21. a line segment
22. **Write About It** Can a line segment be part of a ray? Explain.

MIXED REVIEW AND TEST PREP

23. If $n = pq$, find n when $p = 24$ and $q = 4$. (p. 272)
24. Order from least to greatest: 0.4, 3, 1.19, 0.085 (p. 52)
25. Evaluate $\sqrt{36} - 2^3$ (p. 278)
26. Solve. $5.2 = k - 3.7$ (p. 290)
27. **TEST PREP** Tim has 2 apples that weigh $\frac{3}{16}$ lb and $\frac{1}{4}$ lb. How much less than a pound of apples does Tim have? (p. 182)

A $\frac{1}{6}$ lb **B** $\frac{1}{16}$ lb **C** $\frac{9}{16}$ lb **D** $\frac{1}{4}$ lb

EXTRA PRACTICE page H47, Set A

Angles

Explore how to name, measure, and draw angles.

You need protractor, straightedge.

QUICK REVIEW

1. $5x = {}^-90$ **2.** $6y = 3.6$ **3.** $\frac{1}{4}b = 8$

4. $26 = 4d$ **5.** $\frac{2}{3} = 2k$

Vocabulary

vertex

An angle is formed by two rays with a common endpoint, called the **vertex**. You can name an angle by using the letter of its vertex alone. You can also use the letter of a point on each side, along with the letter of the vertex, to name the angle. The middle letter should always be for the vertex.

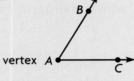

$\angle BAC$, $\angle CAB$, or $\angle A$

An angle is measured in degrees. One degree measures $\frac{1}{360}$ of a circle. The number of degrees determines the type of angle.

An acute angle measures less than 90°.	
A right angle measures 90°.	
An obtuse angle measures more than 90° but less than 180°.	
A straight angle measures 180°.	

Activity 1

You can measure an angle by using a protractor.

- On a sheet of paper, draw an angle that looks approximately like $\angle XYZ$.

- Place the center point of the protractor on the vertex of the angle.

- Place the base of the protractor along ray YZ.

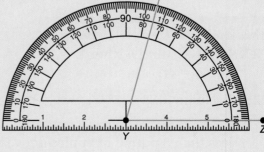

- Read the scale that starts with 0 at ray YZ. The measure of $\angle XYZ$ above is 75°. This can be written as m$\angle XYZ = 75°$.

Activity 2

You can also use a protractor to draw angles of a given measure.

- Draw a ray on a sheet of paper.

- Place the center point of the protractor on the endpoint of the ray. Align the protractor so that the ray passes through 0°.

- Draw an angle of 150° by making a mark on the paper above 150 on the scale you are using.

- Use your straightedge to draw a ray to complete the angle.

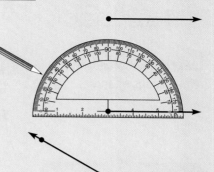

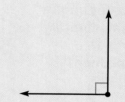

Activity 3

To estimate the size of an angle, you can compare it to an angle whose measure you know.

- Draw an angle on a sheet of paper.

- Compare its measure to the right angle shown here, and estimate how many degrees it contains.

- Measure the angle to check your estimate.

Think and Discuss

- What kind of angle did you draw in Activity 1?

- The protractor has two scales, a top scale and a bottom scale. How can you decide which scale to use?

Practice

TECHNOLOGY LINK

More Practice: Use **Mighty Math Cosmic Geometry,** *Amazing Angles,* Levels A and B.

Trace the angle. Then measure it and tell what kind of angle it is.

1. 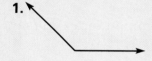   2.    3.

Draw an angle with the given measure.

4. 48°   5. 180°   6. 154°   7. 90°   8. 45°

Estimate the measure. Then trace the angle and use a protractor to check your estimate.

9. 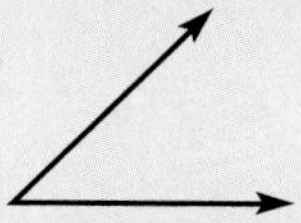   10. 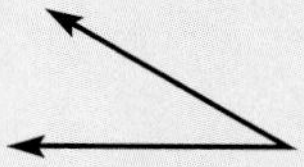   11.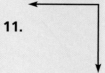

MIXED REVIEW AND TEST PREP

12. Name the geometric figure. (p. 318)

E •———————• F

13. Solve. $96 = 12b$. (p. 297)

14. Find the product. $31.82 × 4 (p. 70)

15. How much greater is $\frac{4}{5} ÷ 3$ than $\frac{4}{5} ÷ 6$? (p. 210)

16. **TEST PREP** Find the mean, median, and mode: 50, 52, 56, 51, 50, 53. (p. 106)

A 51.5, 50, 52   B 50, 52, 51.5   C 52, 51.5, 50   D 51.5, 52, 50

LESSON 16.3

Angle Relationships

Learn how to understand relationships of angles.

QUICK REVIEW

1. 90 − 15   **2.** 90 − 40   **3.** 180 − 50   **4.** 180 − 12

5. What number added to 25 gives 90?

Vocabulary

vertical angles

adjacent angles

complementary angles

supplementary angles

Angle relationships play an important role in many sports and games. Miniature-golf players must understand angles to know where to aim the ball. In the miniature-golf hole shown, m∠1 = m∠2, m∠3 = m∠4, and m∠5 = m∠6.

Certain angle pairs are given special names in geometry.

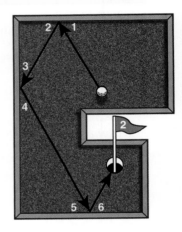

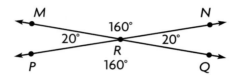

Remember that congruent figures have the same size and shape.

Vertical angles are formed opposite each other when two lines intersect. Vertical angles have the same measure, so they are always congruent.

∠*MRP* and ∠*NRQ* are vertical angles.

∠*MRN* and ∠*PRQ* are vertical angles.

Adjacent angles are side-by-side and have a common vertex and ray.

∠*MRN* and ∠*NRQ* are adjacent angles.

EXAMPLE 1

Identify the type of angle pair shown.

a.

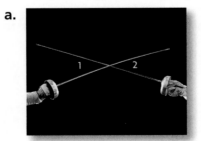

∠1 and ∠2 are opposite each other and are formed by two intersecting lines. They are vertical angles.

b.

∠3 and ∠4 are side-by-side and have a common vertex and ray. They are adjacent angles.

• If ∠1 measures 40° in Example 1, what does ∠2 measure? If ∠4 measures 150°, what does ∠3 measure?

Complementary and Supplementary Angles

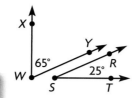

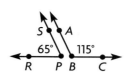

TECHNOLOGY LINK

More Practice: Use
**Mighty Math Cosmic
Geometry,** *Geo Academy*
Level B.

Two angles are complementary if the sum of their measures is 90°.
$65° + 25° = 90°$ $\angle XWY$ and $\angle RST$ are **complementary angles**.

Two angles are supplementary if the sum of their measures is 180°.
$65° + 115° = 180°$ $\angle RPS$ and $\angle ABC$ are **supplementary angles**.

In some cases you can use the definitions of complementary and supplementary angles to find the measures of angles.

EXAMPLE 2

The angles are complementary. Find the unknown measure.

$75° + \blacksquare = 90°$ *The sum of the measures is 90°.*

$90° - 75° = \blacksquare$ *Subtract to find the unknown measure.*

$90° - 75° = 15°$

So, the unknown angle measure is 15°.

EXAMPLE 3

The two adjacent angles are supplementary. Find the unknown measure.

$80° + \blacksquare = 180°$ *The sum of the measures is 180°.*

$180° - 80° = \blacksquare$ *Subtract to find the unknown measure.*

$180° - 80° = 100°$

So, the unknown angle measure is 100°.

When an object bounces off a surface, it always rebounds at the same angle. In the figure, $\angle SWQ$ and $\angle TWR$ have the same measure.

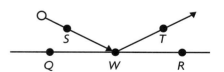

EXAMPLE 4

A hockey puck hits $\overline{BD}$, the wall of an ice rink, at an angle of 35°. $\overline{AC}$ is drawn so that $\angle BCA$ and $\angle DCA$ are right angles. Find the measures of $\angle 1$ and $\angle 3$.
The puck rebounds at the same angle, so the measure of $\angle 3$ is 35°.

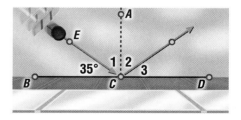

$\angle 1$ and $\angle ECB$ form a right angle, so they are complementary.
$m\angle 1 + 35° = 90°$
$90° - 35° = 55°$, so the measure of $\angle 1$ is 55°.

323

Think and ▶
Discuss

Look back at the lesson to answer each question.

1. **Describe** vertical, adjacent, complementary, and supplementary angles.

2. **Explain** how you can find the measure of an angle complementary to a given angle; an angle supplementary to a given angle.

Guided ▶
Practice

For 3–6, use the figure at the right.

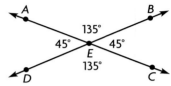

3. Name two pairs of vertical angles.

4. Name two pairs of adjacent angles.

5. Name a pair of complementary angles.

6. Name two pairs of supplementary angles.

Find the unknown angle measure.

7.   8. 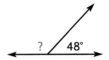   9.

Independent ▶
Practice

For 10–13, use the figure at the right.

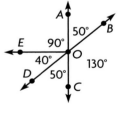

10. Name two angles adjacent to ∠COD.

11. Name the angle vertical to ∠AOB.

12. Name the angle complementary to ∠AOB.

13. Name two angles supplementary to ∠BOC.

Find the unknown angle measure. The angles are complementary.

14.   15.   16.

Find the unknown angle measure. The angles are supplementary.

17.   18. 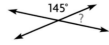   19.

Tell if the angles are *vertical, adjacent, complementary, supplementary,* or *none* of these.

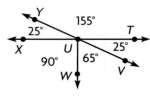

20. ∠TUV and ∠XUY

21. ∠XUY and ∠TUY

22. ∠VUW and ∠TUV

23. ∠XUW and ∠TUV

Problem Solving ▶ Applications

For 24–27, use the figure at the right. Trace the figure. Find the measure of the numbered angles.

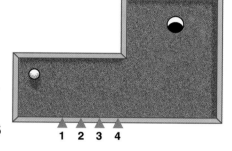

24. Name each pair of vertical angles. Tell how you know the angles are vertical.

25. Name each pair of adjacent angles. Tell how you know the angles are adjacent.

26. Name each pair of complementary angles. Tell how you know the angles are complementary.

27. Name each pair of supplementary angles. Tell how you know the angles are supplementary.

28. Which number should you aim at to hit the golf ball into the hole in one bounce? Use a protractor to decide. Explain your reasoning.

29. **REASONING** Two angles are each complementary to the same angle. What can you conclude about them.

30. **? What's the Question?** Eva drew angles for a project. She drew 6 angles every 2 minutes. The answer is 21 angles.

MIXED REVIEW AND TEST PREP

31. Name the geometric figure. (p. 318)

P   Q

32. Find the distance, d, for $r = 12.6$ ft per sec and $t = 22.5$ sec. (p. 300)

33. $^{-}9 - (^{-}3) = \blacksquare$ (p. 250)

⭐ 34. **TEST PREP** Find $5{,}500 \div 25$. (p. 22)

   A 200   **B** 200 r5   **C** 220   **D** 2,200

⭐ 35. **TEST PREP** John bought 3 sweaters for $49, $29.95, and $36.75. How much change did he get from $120? (p. 66)

   F $4.30   **G** $4.75   **H** $5.15   **J** $6.00

PROBLEM SOLVING | **LiNKÜp** to Art

Design Kuba cloth is made by the Kuba people of Zaire, Africa. The cloth is made from the fiber of the raffia palm. Originally meant to be worn, people now collect and display Kuba cloth because of its artistic beauty.

• Describe how the design shown involves vertical angles.

EXTRA PRACTICE page H47, Set B

325

Classify Lines

Learn how to classify the different types of lines.

Vocabulary

parallel lines

perpendicular lines

The chart shows some of the ways lines can relate to one another.

Parallel lines are lines in a plane that are always the same distance apart. They never intersect and have no common points.	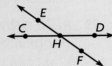	Line *AB* is parallel to line *ML*. $\overleftrightarrow{AB} \parallel \overleftrightarrow{ML}$ (∥ means "is parallel to.")
Intersecting lines cross at exactly one point.	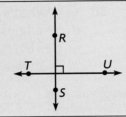	Line *EF* intersects line *CD* at point *H*.
Perpendicular lines intersect to form 90° angles, or right angles.		Line *RS* is perpendicular to line *TU*. $\overleftrightarrow{RS} \perp \overleftrightarrow{TU}$ (⊥ means "is perpendicular to.")

Line segments can also be parallel, intersecting, or perpendicular.

EXAMPLE

The map shows a section of New York City. Describe each relationship as parallel, intersecting, or perpendicular.

a. streets to streets
Streets are parallel to each other.

b. avenues to avenues
Avenues are parallel to each other.

c. streets to avenues
Streets are perpendicular to avenues and intersect avenues.

d. Broadway to streets and avenues
Broadway intersects streets and avenues.

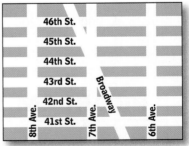

CHECK FOR UNDERSTANDING

Think and ▶ Discuss

Look back at the lesson to answer the question.

1. Draw line 1 parallel to line 2. Draw line 3 perpendicular to line 1. Describe the relationship between line 2 and line 3.

Guided ▶ **Classify the lines.**
Practice

2.   3.   4.   5.

PRACTICE AND PROBLEM SOLVING

Independent ▶ **Classify the lines.**
Practice

6.   7.    8.   9.

The lines in the figure at the right intersect to form a cube.

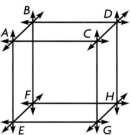

10. Name all the lines that are parallel to $\overleftrightarrow{AC}$.

11. Name all the lines that intersect $\overleftrightarrow{CD}$.

12. Name all the lines that are perpendicular to and intersect $\overleftrightarrow{DB}$.

13. Name all lines that are not parallel to and do not intersect $\overleftrightarrow{AE}$.

Problem Solving ▶ **For 14–15 use this information. Main Street is a straight street. It**
Applications **intersects 3rd, 4th, and 5th Streets, which are parallel to each other. Pine and Oak Streets are perpendicular to 5th Street.**

14. Draw a map showing the six streets.

15. Suppose that Pine and Oak Streets are perpendicular to Main Street rather than to 5th Street. Draw a map showing the streets.

16. ✎ **Write About It**   Can two lines be both parallel and intersecting? Explain.

17. ❓ **What's the Error?**   Aidan said that all perpendicular lines are intersecting lines and all intersecting lines are perpendicular. What is wrong with his statement?

MIXED REVIEW AND TEST PREP

18. Two angles are complementary. If one angle measures 72°, what does the other measure? (p. 322)

19. Suppose it took Cosetta 4.5 hours to drive 266.4 mi. What was Cosetta's average speed? (p. 300)

20. Find the sum. $3\frac{5}{8} + 5\frac{3}{8}$ (p. 186)

21. Which is greater, $\frac{5}{12}$ or $\frac{1}{3}$? (p. 166)

⭐ 22. **TEST PREP**   Protractors sell for $0.09 each. How many more protractors can you buy for $9.00 than for $7.20? (p. 76)

   A 0.01   **B** 10   **C** 20   **D** 30

EXTRA PRACTICE page H47, Set C

1. **VOCABULARY** Part of a line that has two endpoints is a(n) __?__ . (p. 318)

2. **VOCABULARY** __?__ angles are opposite angles formed when two lines intersect. (p. 322)

3. **VOCABULARY** __?__ angles are two angles whose sum is 180°. (p. 323)

4. **VOCABULARY** Lines that intersect to form right angles are called __?__ . (p. 326)

Name the geometric figure. (pp. 318–319)

5. F ●————————● G →

6. C ●———————————————● D

7. ← A ●————————————————● B →

Tell whether the angle is acute, right, obtuse, or straight. (pp. 320-321)

8.

9.

10.

11.

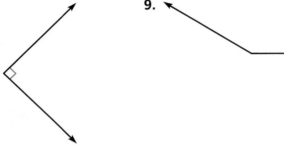

Use the figure to find the measure of each angle. (pp. 322–325)

12. ∠ABF

13. ∠ABE

14. ∠CBD

15. ∠CBF

16. ∠FBE

17. ∠FBD

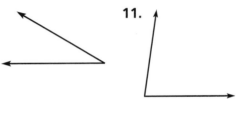

For 18–21, use the figure at the right. (pp. 326–327)

18. $\overleftrightarrow{HJ} \perp$ __?__

19. $\overleftrightarrow{KH}$ intersects __?__

20. $\overleftrightarrow{HI} \perp$ __?__

21. $\overleftrightarrow{HJ} \parallel$ __?__

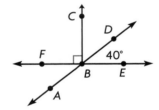

22. Complete: If two angles are complementary, both angles must be __?__ (acute, right, obtuse, straight). (pp. 322–325)

23. Complete: If two lines are perpendicular, then they are also __?__ (parallel, intersecting) (pp. 326–327)

24. Two streets intersect and form a 38° angle. Find the unknown angle measure. (pp. 322–325)

25. Two sections of a picture frame join to form a right angle. Find the unknown angle measure.

(pp. 322–325)

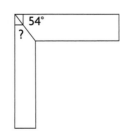

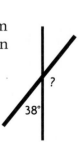

Understand the problem.

See item **5**.

Be sure you know the meaning of each term in the answer choices. Look at the diagram to determine the relationship of $\angle DCA$ and $\angle ACB$.

Also see problem **1**, p. H62.

Choose the best answer.

1. Which term describes an angle that measures 60°?

A Acute   **C** Obtuse

B Right   **D** Straight

For 2–3, use the diagram below.

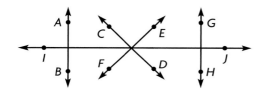

2. Which appears to be parallel to $\overleftrightarrow{AB}$?

F $\overrightarrow{CD}$ **H** $\overleftrightarrow{GH}$

G $\overrightarrow{EF}$ **J** $\overrightarrow{IJ}$

3. Which appears to be perpendicular to $\overleftrightarrow{GH}$?

A $\overleftrightarrow{AB}$ **C** $\overleftrightarrow{EF}$

B $\overrightarrow{CD}$ **D** $\overrightarrow{IJ}$

For 4–5, use the diagram below.

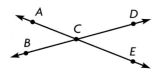

4. Which of these pairs of angles are vertical angles?

F $\angle ACB$ and $\angle ACD$

G $\angle ACB$ and $\angle DCE$

H $\angle DCE$ and $\angle BCE$

J $\angle ACD$ and $\angle DCE$

5. Which term describes the relationship of $\angle DCA$ and $\angle ACB$?

A Complementary angles

B Supplementary angles

C Right angles

D Vertical angles

6. The orchestra rehearsed for $2\frac{1}{2}$ hours on Saturday and $3\frac{1}{2}$ hours on Sunday. Altogether, how many hours did the orchestra rehearse?

F $6\frac{1}{2}$ **H** $5\frac{1}{2}$

G 6   **J** 5

7. Paula made 190 tote bags. She sold 115 of them at the craft fair and made $2,070. Which equation can be used to find p, the price she sold each tote bag for?

A $190 \times 115 = p$

B $190 \times p = 115$

C $190 \times p = \$2,070$

D $115 \times p = \$2,070$

8. What is the value of $y - 54$ for $y = 17$?

F 71   **H** $^-71$

G 37   **J** Not here

Write What You Know

9. Draw two parallelograms and measure the angles. Describe the relationships of the angles in a parallelogram.

10. The supplement of angle A's complement is 134°. Find the measure of angle A. Explain how you found your answer.

CHAPTER 17

Plane Figures

DATA LINK

Computer-aided design, known as CAD, uses computer software to create models that show the geometry and other characteristics of objects.

PROBLEM SOLVING Notice the geometric figures used to design the pool and patio. What geometric figures do you see in the umbrella table?

Check What You Know

Use this page to help you review and remember important skills needed for Chapter 17.

✓ Vocabulary

circle
isosceles
polygon
scalene

Choose the best term from the box.

1. A triangle that has exactly two congruent sides is a(n) __?__ triangle.

2. A closed plane figure formed by three or more line segments is a(n) __?__ .

✓ Identify Polygons (For Intervention, see p. H19.)

Name each polygon.

3.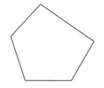

4.

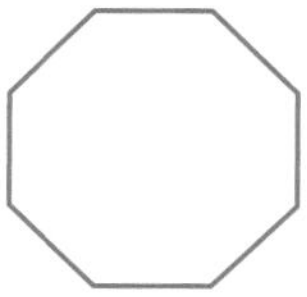

5.

6.

7.

8.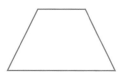

9.

10.

✓ Angles in Polygons (For Intervention, see p. H18.)

Name each angle in the polygon. Tell whether it is *acute*, *obtuse*, *right*, or *straight*.

11.

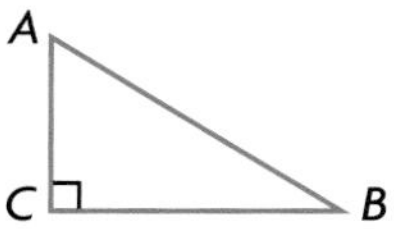

12.

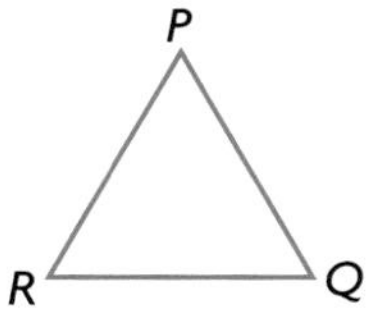

13.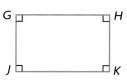

14.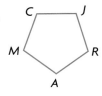

✓ Patterns (For Intervention, see p. H14.)

Describe a possible pattern. Then give the next three numbers in the pattern.

15. 4, 8, 12, 16

16. 2, 6, 18, 54

17. 37, 31, 25, 19

18. 243, 81, 27, 9

19. $1, \frac{1}{2}, \frac{1}{4}, \frac{1}{8}, \frac{1}{16}$

20. 9, 6, 3, 0

LOOK AHEAD

In Chapter 17 you will

- find measures of angles in triangles and solve problems involving quadrilaterals
- use the strategy *find a pattern* to solve problems
- draw two-dimensional figures
- identify parts of circles

LESSON **17.1**

Triangles

Learn how to classify triangles and solve problems involving the angle measures of triangles.

QUICK REVIEW

1. 35 + 45 **2.** 180 − 12 **3.** 54 + 25
4. 180 − 110 **5.** 75 + 32

Vocabulary

acute triangle
obtuse triangle
right triangle

A triangle can be classified by the angles it contains. An **acute triangle** contains only acute angles. An **obtuse triangle** contains one obtuse angle. A **right triangle** contains one right angle.

acute triangle — obtuse triangle — right triangle

In this activity, you will explore the angle measures in a triangle.

Activity (MATH LAB)

You need: protractor, straightedge, scissors

- Use a straightedge to draw a large triangle. Use scissors to cut it out.

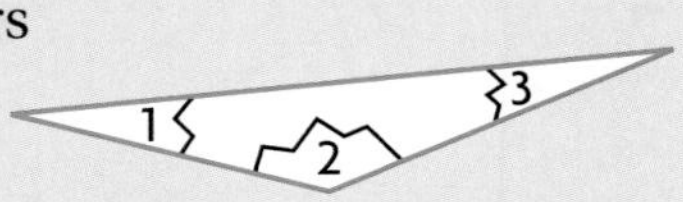

- Tear off the three angles. Arrange them around a point on a line as shown.

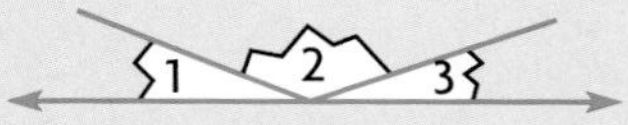

- What is the sum of the measures of the three angles? Explain.
- Compare your results with those of your classmates. What guess can you make?

Math Idea ▶ The sum of the measures of the angles of a triangle is 180°. To decide whether a triangle is acute, obtuse, or right, you need to know the measures of its angles.

EXAMPLE 1

Three stars form a triangle known as "Triangulum," with two angles measuring 20° and 50°. Classify the triangle.

To classify the triangle, find the measure of $\angle B$ in the figure.

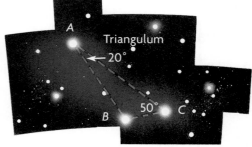

$180° - (20° + 50°) = 180° - 70°$
$= 110°$

Subtract the sum of the known angle measures from 180°.

So, the measure of $\angle B$ is 110°. Because $\triangle ABC$ contains one obtuse angle, Triangulum forms an obtuse triangle.

Remember that an *equilateral* triangle has three congruent sides, an *isosceles* triangle has two congruent sides, and a *scalene* triangle has no congruent sides.

Sometimes you can find the measure of an angle by using what you know about vertical angles.

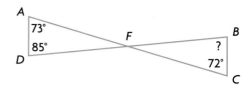

EXAMPLE 2

Find the measure of ∠FBC.

$180° - (73° + 85°) = 180° - 158°$ *Find m ∠AFD.*
$= 22°$

m ∠BFC = 22°   *Vertical angles are congruent.*

$180° - (22° + 72°) = 180° - 94°$ *Find m ∠FBC.*
$= 86°$

So, the measure of ∠FBC is 86°.

Another way to find the measure of a missing angle is to apply what you know about complementary and supplementary angles.

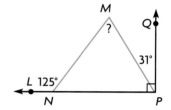

EXAMPLE 3

Remember that the sum of the measures of two *complementary* angles is 90°. The sum of the measures of two *supplementary* angles is 180°.

In the figure at the right, ∠NPQ is a right angle, ∠LNM and ∠MNP are supplementary angles, and ∠NPM and ∠QPM are complementary angles. Find the measure of ∠M.

∠MNP = 180° − 125° = 55°   *The sum of the measures of two supplementary angles is 180°.*

∠MPN = 90° − 31° = 59°   *The sum of the measures of two complementary angles is 90°.*

55° + 59° = 114°   *Find the sum of the known angle measures in △NMP.*

180° − 114° = 66°   *Subtract the sum of the known angle measures for △NMP from 180°.*

So, ∠M measures 66°.

CHECK FOR UNDERSTANDING

Think and ▶ Discuss

Look back at the lesson to answer each question.

1. **Explain** why a triangle cannot have two obtuse angles.

2. **Draw** an example of each: scalene acute triangle; isosceles right triangle; scalene obtuse triangle.

Guided ▶ Practice

Find the measure of ∠B and classify △ ABC.

3.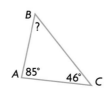

4.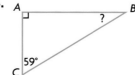

5.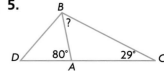

Independent ▶
Practice

Find the measure of ∠B and classify △ABC.

6.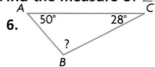

7.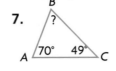

8.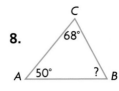

9.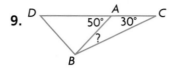

10.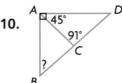

11.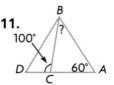

For 12–17, use the figure at the right. Find the measure of each angle.

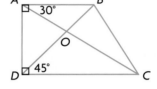

12. ∠DAO

13. ∠ADO

14. ∠ABD

15. ∠AOD

16. ∠AOB

17. ∠DCA

For 18–23, use the figure at the right. Find the measure of each angle.

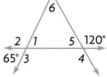

18. ∠1

19. ∠2

20. ∠3

21. ∠4

22. ∠5

23. ∠6

For 24–27, use the figure at the right.

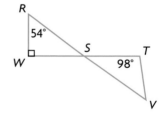

24. What is the relationship between ∠RSW and ∠VST?

25. What is the measure of ∠W?

26. Find the measure of ∠RSW.

27. Find the measure of ∠V.

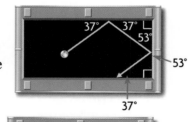

Problem Solving ▶
Applications

28. In a video game, a moving ball always bounces off an edge at the same angle at which it hit the edge. Find the measure of the unknown angle.

a.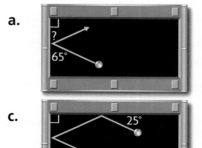

b.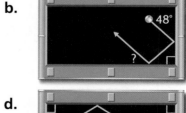

c.

d.

29. The three main stars that are in the constellation known as Leo Minor form an isosceles triangle, with congruent base angles. The third angle in the triangle measures 144°. Find the measure of each base angle.

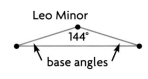
Leo Minor
144°
base angles

30. REASONING How many total degrees are in the angles of a quadrilateral? (**Hint:** Divide it into triangles.)

For 31–32, use the figure at the right.

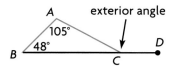
A   exterior angle
105°
48°   D
B   C

31. If you extend a side of a triangle, you form an *exterior angle.* In the figure, $\angle ACD$ is an exterior angle of $\triangle ABC$. Find the measure of $\angle ACD$.

32. ❓ **What's the Question?** The answer is $\angle BCA$ and $\angle ACD$.

MIXED REVIEW AND TEST PREP

33. Classify the lines. (p. 326)

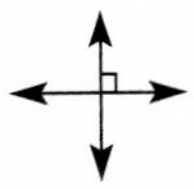

34. Which is greatest, 0.03, 0.009, or 0.107? (p. 52)

35. What is the median of the values 24, 17, 39, 19, and 21? (p. 106)

⭐**36. TEST PREP** Which is the value of x in the expression $\frac{6}{x} = \frac{18}{12}$? (p. 216)

A $x = 2$ **B** $x = 4$ **C** $x = 9$ **D** $x = 24$

⭐**37. TEST PREP** Miguel wants to record a TV movie that begins at 9:40 P.M. and ends at 12:10 A.M. How long does the movie last? (p. 214)

F 9 hr 30 min **G** 3 hr 50 min **H** 2 hr 30 min **J** 2 hr 10 min

---

PROBLEM SOLVING 💡 **Thinker's Corner**

Figure It Out With a partner, draw a triangle, a quadrilateral, a pentagon, and a hexagon. Do not draw the polygons with congruent sides.

For each figure you draw, extend one side at each vertex, as shown for the quadrilateral at the right. The angles you form outside the polygon are exterior angles.

2
1
4
3

For each figure, cut out the exterior angles and arrange them around a common point.

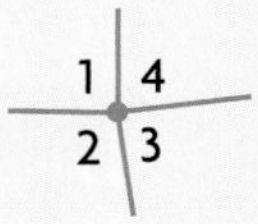

1. Tell the sum of the measures of the exterior angles for the triangle, the quadrilateral, the pentagon, and the hexagon.

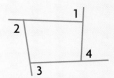
1 4
2 3

2. What can you guess about the sum of the measures of the exterior angles of a polygon?

PROBLEM SOLVING STRATEGY
Find a Pattern

Learn how to use the strategy *find a pattern* to solve problems.

Vocabulary

regular polygon

Mrs. Jones is building a deck in the shape of a regular octagon. A **regular polygon** is a polygon in which all sides are congruent and all angles are congruent. So, a regular octagon has 8 congruent sides and 8 congruent angles. What is the measure of each angle of the octagon?

Analyze

What are you asked to find?

What information is given?

Choose

What strategy will you use?

Look for a pattern in the sums of the measures of the angles of regular polygons that have fewer sides than an octagon.

Solve

How will you solve the problem?

triangle

quadrilateral

pentagon

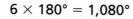

hexagon

Draw regular polygons that have fewer sides than an octagon. Divide each into triangles. Make a table to record your data and help you look for a pattern.

POLYGON	SIDES	TRIANGLES	SUM OF ANGLE MEASURES
Triangle	3	1	180°
Quadrilateral	4	2	2 x 180° = 360°
Pentagon	5	3	3 x 180° = 540°
Hexagon	6	4	4 x 180° = 720°

The number of triangles is always 2 fewer than the number of sides. So, an octagon can be divided into $8 - 2 = 6$ triangles.

$$6 \times 180° = 1,080°$$

The sum of the angle measures of an octagon is 1,080°. To find the measure of each angle in a regular octagon, divide the sum by 8: $1,080° \div 8 = 135°$.

Check

How can you check your answer?

What if Shalana and her mother want to build a deck in the shape of a decagon? What would be the measure of each angle?

PROBLEM SOLVING STRATEGIES

Draw a Diagram or Picture

Make a Model

Predict and Test

Work Backward

Make an Organized List

▶ Find a Pattern

Make a Table or Graph

Solve a Simpler Problem

Write an Equation

Use Logical Reasoning

Solve the problem by finding a possible pattern.

1. Macklin saves 1 dime on Monday, 2 dimes on Tuesday, 4 dimes on Wednesday, 7 dimes on Thursday, and 11 dimes on Friday. If he continues the pattern, how many dimes would you expect him to save on each of the next three days?

2. There were 10 players in the checkers tournament. If each player shook the hand of every other player 1 time, how many handshakes were there?

Steadman is making a design for a quilt using blue and white squares.

3. How many squares will there be in the eighth row?

   A 8   **B** 9   **C** 15   **D** 17

4. After he has 8 rows, how many squares will he have used in all?

   F 8   **G** 16   **H** 36   **J** 64

row 1
row 2
row 3
row 4

MIXED STRATEGY PRACTICE

5. Silvia, David, and Rhoda have blue, green, and brown eyes, though not necessarily in that order. Silvia's eyes are the color of the sky. Rhoda does not have brown eyes. Give the color of each person's eyes.

6. This week Cheryl earned $23 more than last week. Last week she earned $225. She plans to spend $\frac{3}{8}$ of this week's earnings, put half in her checking account, and save the rest. How much will she save?

Use Data   **For 7–8, use the table.**

7. Estimate the amount the Alpha Corporation earns from publishing if its total revenues are $11.6 billion.

8. Draw a circle graph to show the sources of revenue for the Alpha Corporation. Then use the graph to find two revenue sources that together produce more than half of the company's revenues.

SOURCES OF REVENUE FOR ALPHA CORPORATION	
Publishing	35%
Movies and TV	28%
Broadcasting and Cable	20%
Other	17%

9. Julia noted that her car's mileage gauge read 12,445 miles at the start of a trip. At the end of the trip, the gauge read 12,829. If she used 16 gallons of gas on the trip, how many miles per gallon did she get?

10. ✎ **Write a problem** that can be solved by finding a pattern. Exchange with a classmate and solve. Explain your solution.

Quadrilaterals

Learn how to identify, classify, and compare quadrilaterals.

QUICK REVIEW

Find each angle measure.

1. ∠1   2. ∠2   3. ∠3

4. ∠4   5. ∠5

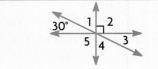

Remember that a quadrilateral is a polygon with four sides and four angles.

These 2 figures are quadrilaterals:

Five types of quadrilaterals are especially important. Their names and properties are listed below. Note: The same marking on two sides or more of a figure indicates that the sides are congruent.

QUADRILATERAL	FIGURE	PROPERTIES
parallelogram		opposite sides parallel and congruent
rectangle		parallelogram with four right angles
rhombus		parallelogram with four congruent sides
square		rectangle with four congruent sides
trapezoid		quadrilateral with exactly two parallel sides

Notice that rectangles and rhombuses are types of parallelograms.

Rectangle:

Parallelogram with **four right angles**

Rhombus:

Parallelogram with **four congruent sides**

A square is both a type of rectangle and a type of rhombus.

Square:

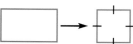

Rectangle with **four congruent sides**

Square:

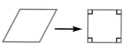

Rhombus with **four right angles**

The figure at the right has four sides, so it is a quadrilateral. A more *exact* name for the figure, however, is *rectangle*, because the name *rectangle* describes the figure's properties more completely than the name *quadrilateral* does.

Quadrilateral:
4 sides
Rectangle:
parallelogram with
4 right angles

EXAMPLE 1

Give the most exact name for the figure.

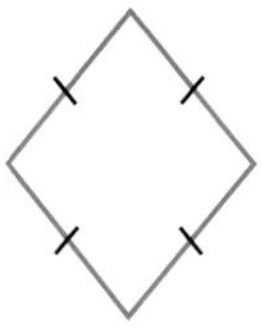

The figure is
a rhombus.

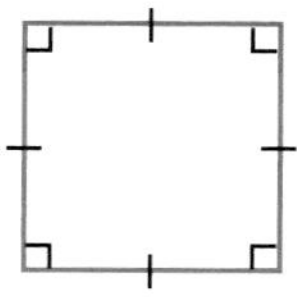

The figure is
a square.

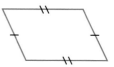

The figure is a
parallelogram.

- Why wouldn't it be more exact to name the figure above at the far right a rhombus or a rectangle?

Statements about geometric figures are often given in "If...then" form.

- *If* a polygon has five sides, *then* it is a pentagon.

- *If* a figure is a triangle, *then* the sum of the measures of its angles is 180.°

You can use what you know about quadrilaterals to write "If...then" statements about quadrilaterals.

EXAMPLE 2

Complete the statement, giving the most exact name for the figure: If a rhombus is also a rectangle, then the rhombus is a ___?___ .

THINK: A rhombus has 4 congruent sides and opposite sides parallel. If it is also a rectangle, then it has 4 right angles, which makes it a square.

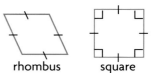

rhombus   square

In the activity below, you will name quadrilaterals by using their properties.

Activity

A   B 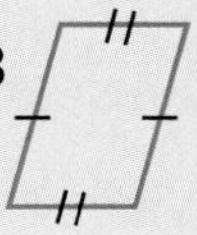   C 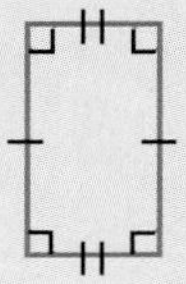   D

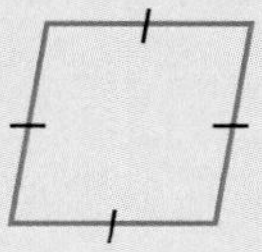

- List the properties of each figure above.

- Give the most exact name for each figure.

**Think and ▶
Discuss**

Look back at the lesson to answer each question.

1. **Compare and contrast** a rectangle and a rhombus.

2. **Explain** how to determine if a quadrilateral is a trapezoid.

**Guided ▶
Practice**

Give the most exact name for the figure.

3.

4.

5.

Complete the statement, giving the most exact name for the figure.

6. If the opposite sides of a quadrilateral are parallel, then the quadrilateral is a ___?___ .

7. If a quadrilateral has four right angles, then it is a ___?___ .

8. If a rectangle has four congruent sides, then it is a ___?___ .

**Independent ▶
Practice**

Give the most exact name for the figure.

9.

10.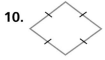

11.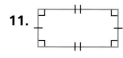

12.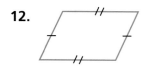

13.

14.

Complete the statement, giving the most exact name for the figure.

15. If a parallelogram has four congruent sides, then it is a ___?___ .

16. If a polygon has five sides, then it is a ___?___ .

17. If a quadrilateral has four congruent sides and four congruent angles, then the quadrilateral is a ___?___ .

18. If exactly two sides of a quadrilateral are parallel, then it is a ___?___ .

19. If a polygon has four sides, then it is a ___?___ .

Problem Solving ▶ Applications

In 20–22, portions of quadrilaterals are hidden from view. Name some possible quadrilaterals each figure could be.

20.
_____

21.
_____

22.
_____

23. A rectangular picture frame is 4 inches wider than it is tall. The total length of the four sides of the frame is 52 inches. What are the dimensions of the frame?

24. Kazumi cut two square frames from a 90-in. length of aluminum. If each side of one frame was 16 in. long, how long was each side of the second frame?

25. **? What's the Error?** Elton says that a quadrilateral with one pair of parallel sides and two right angles is a rectangle. What is wrong with his statement?

26. To raise money for his scout troop, Aldrick sold 16 cans of popcorn for $6.50 each and 18 calendars for $3.75 each. What were Aldrick's total sales?

MIXED REVIEW AND TEST PREP

27. Two angles in a triangle each measure 48°. Find the measure of the third angle. (p. 332)

28. ∠M is complementary to a 75° angle. What is the measure of ∠M? (p. 322)

29. Write 8% as a decimal. (p. 169)

⭐30. **TEST PREP** Which is the greatest common factor of 18 and 24? (p. 150)

    A 2   **B** 3   **C** 6   **D** 72

⭐31. **TEST PREP** Which number is between 17.09 and 17.3? (p. 230)

    F 17.01   **G** 17.08   **H** 17.29   **J** 17.312

PROBLEM SOLVING

 **to Reading**

Strategy • Make Generalizations Sometimes you need to *make generalizations* to solve a problem. When you generalize, you make a statement that is true about a whole group of similiar situations or objects.

Use the Venn diagram to make generalizations about quadrilaterals.

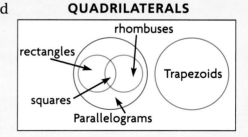

QUADRILATERALS

Complete. Write *always*, *sometimes*, or *never*.

1. A square is __?__ a parallelogram.

2. A trapezoid is __?__ a rectangle.

3. A rhombus is __?__ a trapezoid.

4. A rectangle is __?__ a square.

EXTRA PRACTICE page H48, Set B

341

Draw Two-Dimensional Figures

Learn how to draw two-dimensional geometric figures.

Remember that an isosceles triangle has two congruent sides.

Write *acute, right, obtuse,* or *straight* for each angle measure.

1. 135°   **2.** 180°   **3.** 10°   **4.** 55.7°   **5.** 90°

You can use the properties of geometric figures to draw two-dimensional figures on dot paper.

EXAMPLES

A. Draw a right isosceles triangle.

Use square dot paper to draw figures with right angles.

THINK: *The triangle must have one right angle.*

The triangle also must have two congruent sides.

B. Draw a pentagon with two right angles.

THINK: *The figure must have two right angles.*

The figure also must have five sides.

C. Draw a quadrilateral with four congruent sides and no right angles.

Use isometric dot paper to draw figures with congruent sides and no right angles.

CHECK FOR UNDERSTANDING

Think and ▶ Discuss

Look back at the lesson to answer each question.

1. Describe how you could draw a trapezoid on dot paper.

2. *REASONING* Could you draw an isosceles triangle with two obtuse angles? Explain.

Guided ▶ Practice

Draw the figure. Use square dot paper or isometric dot paper.

3. a triangle with no congruent sides

4. an equilateral triangle

5. a hexagon with no congruent sides

Independent ▶
Practice

Draw the figure. Use square dot paper or isometric dot paper.

6. an acute
isosceles triangle

7. an obtuse
scalene triangle

8. a quadrilateral
with 4 right
angles

9. a quadrilateral
with no
congruent sides
or parallel sides

10. an octagon with
2 right angles

11. a pentagon with
no congruent
sides

Complete the sentence by using *must*, *can*, or *cannot*.

12. A rhombus __?__ have all sides congruent.

13. A right triangle __?__ have two right angles.

14. A trapezoid __?__ have all sides congruent.

15. A scalene triangle __?__ be obtuse.

Problem Solving ▶
Applications

16. Anita drew a quadrilateral. Then she drew a line segment
connecting one pair of opposite corners. She saw that she had
divided the quadrilateral into two right isosceles triangles.
Classify the quadrilateral she began with.

17. On a separate sheet of paper, draw an equilateral triangle with
sides 4 in. long. Cut out the triangle. Then cut from each corner
of the triangle an equilateral triangle with sides 1 in. long. What is
the perimeter of the figure that remains?

18. **(?) What's the Error?** Victor said he drew a triangle with one
acute, one right, and one obtuse angle. Explain his mistake.

19. Ted drew three sides of a quadrilateral as shown
at the right. If he wants a third angle to be 90°,
what will be the measure of the fourth angle?

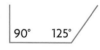

MIXED REVIEW AND TEST PREP

20. Give the most exact name for a
parallelogram with four congruent
sides. (p. 338)

21. Name the figure. (p. 318)

22. Evaluate the expression. (p. 278)
$8^2 + \sqrt{100} - 7$

23. 3.6×0.25 (p. 70)

24. **TEST PREP** A box with a height of $1\frac{9}{16}$ ft is shelved on top of a box
with a height of $2\frac{1}{2}$ ft. The distance between shelves is 5 ft. How far
is it from the top of the upper box to the shelf above? (p. 186)

A $1\frac{15}{16}$ ft   **B** $1\frac{4}{9}$ ft   **C** $1\frac{3}{16}$ ft   **D** $\frac{15}{16}$ ft

Circles

Learn how to identify parts of circles.

radius = 6.25 cm
diameter = 12.5 cm

Vocabulary

radius
diameter
chord
arc
sector

A compact disc (CD) is shaped like a *circle*. A circle is the set of all points a given distance from its center. At the edge of the CD, every point on the circle is 6.25 cm from the center.

A line segment with one endpoint at the center of a circle and the other endpoint on the circle is a **radius** (plural, *radii*). A line segment that passes through the center of a circle and has both endpoints on the circle is a **diameter**. The length of a diameter is always twice the length of a radius.

CD radius = 6.25 cm   CD diameter = $2 \times 6.25 = 12.5$ cm

A circle can be named by its center. Like a polygon, a circle is a plane figure. A circle is not made up of line segments, so it is not a polygon.

EXAMPLE 1

Name the center, the circle, a diameter, and three radii of the circle.

O is the center. The circle is circle *O*. $\overline{AB}$ is a diameter. $\overline{OA}$, $\overline{OB}$, and $\overline{OC}$ are radii.

chord *AB*   arc *AB*, or $\overparen{AB}$

Two points on a circle can indicate a chord, an arc, or a sector. A **chord** is a line segment with its endpoints on a circle. An **arc** is part of a circle. A **sector** is a region enclosed by two radii and the arc joining their endpoints.

sector *AB*

EXAMPLE 2

Music used to be recorded on "long-playing" records (LPs). The shaded sector shows how far an LP turns in one second. Name the radii and arc enclosing the sector.

The radii are $\overline{KR}$ and $\overline{KP}$. The arc is $\overparen{PR}$. Since there are two arcs between *P* and *R*, you can use point *Q* to indicate the longer one: $\overparen{PQR}$.

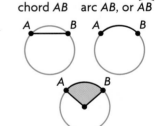

1 second

Think and Discuss ▶ Look back at the lesson to answer each question.

1. **Explain** how you can find the radius of a circle if you know the diameter.

2. **REASONING** What is the longest chord in a circle?

Guided Practice ▶ For 3–6, use the circle at the right. Name the given parts of the circle.

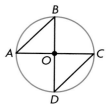

3. center

4. chords

5. radii

6. diameters

PRACTICE AND PROBLEM SOLVING

Independent Practice ▶ For 7–12, use the circle at the right. Name the given parts of the circle.

7. center

8. chords

9. radii

10. diameter

11. a sector

12. two arcs

Problem Solving Applications

13. **REASONING** What happens to the length of a radius as the size of a circle increases?

TECHNOLOGY LINK

More Practice: Use **Mighty Math Cosmic Geometry**, *Geo Academy*, Level G.

Use Data For 14–15, use the table.

14. Find the radius of the LP, the 45, and the 78.

15. How much longer will it take an LP to revolve (spin) 1,000 times than it will take a CD?

16. ✎ **Write About It** The radius of a circle is 6 cm long. Chord AB is 8 cm long. Explain why $\overline{AB}$ cannot be a diameter.

TYPES OF RECORDING DISKS		
Disk	Diameter	Revolutions Per Minute
CD	12.5 cm	200
LP	12 in.	$33\frac{1}{3}$
45	$6\frac{7}{8}$ in.	45
78	10 in.	78

MIXED REVIEW AND TEST PREP

17. Give the most exact name for the figure. (p. 338)

18. Find the unknown angle measure. (p. 322)

19. Evaluate $18 - 9 \div 3 + 5$. (p. 44)

20. What is the LCM of 15 and 20? (p. 150)

⭐21. **TEST PREP** Find $24 \div \frac{3}{4}$. (p. 210)

A 16   **B** 18   **C** 32   **D** 36

1. **VOCABULARY**   A triangle containing one obtuse angle is a(n) ___?___ triangle. (p. 332)

2. **VOCABULARY**   A triangle containing one right angle is a(n) ___?___ triangle. (p. 332)

3. **VOCABULARY**   Any line segment with endpoints on a circle is called a(n) ___?___ . (p. 344)

Find the measure of the angle and classify the triangle. (pp. 332–335)

4.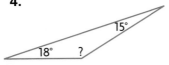

5.

6.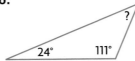

7.

Give the most exact name for the figure. (pp. 338–341)

8.

9.

10.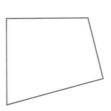

11.

Complete the statement, giving the most exact name for the figure. (pp. 338–341)

12. If the opposite sides of a quadrilateral are parallel, then it is a(n) ___?___ .

13. If a polygon has three sides, then it is a(n) ___?___ .

14. If a parallelogram has four right angles, then it is a(n) ___?___ .

15. If a polygon has eight sides, then it is a(n) ___?___ .

16. If exactly two sides of a quadrilateral are parallel, then it is a(n) ___?___ .

Draw the figure. Use square dot paper or isometric dot paper. (pp. 342–343)

17. a right triangle with two congruent sides

18. a trapezoid with two right angles

19. a parallelogram with four congruent sides

Name the given parts of the circle. (pp. 344–345)

20. center

21. chords

22. radii

23. diameters

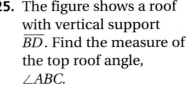

24. Jamie decorated her bike wheels by stringing ribbons between the ends of spokes. Classify △ABO. (O is the center.)

25. The figure shows a roof with vertical support $\overline{BD}$. Find the measure of the top roof angle, ∠ABC.

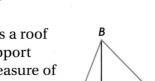

Get the information you need.

See item **1.**

Think about the attributes used to classify polygons. Recall which polygons have all sides the same length.

Also see problem **3**, p. H63.

Choose the best answer.

1. In which type of figure do all the sides always have the same length?

   A Rectangle   **C** Rhombus

   B Trapezoid   **D** Isosceles Triangle

2. What type of triangle has an angle that measures 90°?

   F Equilateral   **H** Obtuse

   G Right   **J** Acute

3. The diameter of a circular garden is 24 meters. What is the radius of the garden?

   A 3.14 m   **C** 48 m

   B 12 m   **D** Not here

4. Lenora drove 156.5 miles in the morning and 204.7 miles in the afternoon. What is the total distance she drove?

   F 361.2 mi   **H** 461.2 mi

   G 389.2 mi   **J** Not here

5. Which term best describes this triangle?

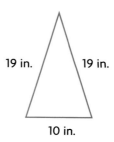

19 in.   19 in.

10 in.

   A Isosceles triangle

   B Equilateral triangle

   C Scalene triangle

   D Right triangle

6. These integers follow a multiplying pattern: $^-1$, $^-2$, $^-4$, . . . Which number might be next?

   F $^-8$ **H** $^-1$

   G $^-6$ **J** $^+8$

7. Kate had $5\frac{1}{4}$ cups of flour. She used $1\frac{3}{8}$ cups in a recipe. How much flour is left?

   A $1\frac{3}{4}$ cups   **C** $3\frac{1}{8}$ cups

   B $2\frac{3}{8}$ cups   **D** $3\frac{7}{8}$ cups

8. Which term best describes this figure?

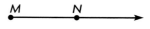

M   *N*

   F Line   **H** Ray

   G Point   **J** Line segment

Write What You Know

9. What is the value of *x* in the figure below? Explain how you found your answer.

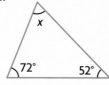

x

72°   52°

10. Tiana drew a rectangle that also was a rhombus. What is the most exact name for the figure? Explain your reasoning.

The skylines of large cities often resemble many different types of solid figures. You can see this in the photograph of New York.

PROBLEM SOLVING The two tallest towers in the center of the photo are part of the World Trade Center. What solid figure does each tower resemble?

DATA LINK

TYPES OF SOLID FIGURES

Rectangular prism

Cylinder

Cone

Check What You Know

Use this page to help you review and remember important skills needed for Chapter 18.

✓ Identify Solid Figures (For Intervention, see p. H20.)

Name each solid figure.

1.

2.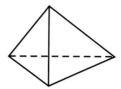

3.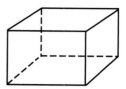

4.

5.

6.

7.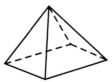

8.

✓ Faces, Vertices, and Edges (For Intervention, see p. H20.)

Copy and complete the table.

	Figure	Number of faces	Number of vertices	Number of edges
9.		▪	▪	▪
10.		▪	▪	▪
11.		▪	▪	▪
12.		▪	▪	▪
13.		▪	▪	▪

LOOK AHEAD

In Chapter 18 you will
- identify solid figures
- draw different views of solids
- use the strategy *solve a simpler problem* to solve problems
- build models of prisms and pyramids

LESSON 18.1

Types of Solid Figures

Learn how to name solid figures.

QUICK REVIEW

Draw each polygon.
1. rectangle
2. triangle
3. pentagon
4. hexagon
5. octagon

You see many solid figures every day.

All solid figures are three-dimensional. In the following activity, you'll explore ways that solid figures can be alike and ways they can be different.

Vocabulary

polyhedron

lateral faces

bases

vertex

Activity

You need: isometric dot paper, colored pencils

• Copy the rectangular prism on isometric dot paper. Use three different colors: one for length, one for width, and one for height.

• Draw a rectangular prism with different dimensions.

• Copy the pyramid. Then draw a rectangular pyramid with different dimensions.

• How are your rectangular prisms alike? How are they different?

• How are your pyramids alike? How are they different?

Math **I**dea ▶ Many common geometric shapes have special names.

A **polyhedron** is a solid figure with flat faces that are polygons.

A prism is a polyhedron with two congruent, parallel bases. Its **lateral faces**, or faces that are not bases, are rectangles. A prism is named for the shape of its **bases**.

Prisms can have any polygon as a base.

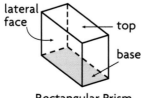

Rectangular Prism

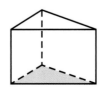

Triangular Prism

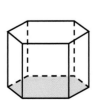

Hexagonal Prism

EXAMPLE 1

Name the figure.

All the faces are flat and are polygons, so the figure is a polyhedron.

The lateral faces are rectangles, so the figure is a prism.

The bases are congruent pentagons.

So, the figure is a pentagonal prism.

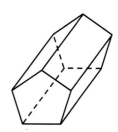

A cylinder is not a polyhedron. It has two flat, parallel, congruent circular bases and a curved lateral surface. Many cans are examples of cylinders.

Remember that a *pyramid* is a solid figure whose one base is a polygon and whose faces are triangles that have a common vertex.

If you connect the single center point on the top base of a cylinder to all the points around the bottom base, you form a cone.

A cone is not a polyhedron. It has one flat circular base, a curved lateral surface, and a **vertex**. The vertex is the top (or bottom) point of this cone.

A pyramid is related to a prism as a cone is related to a cylinder.

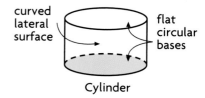

Cylinder

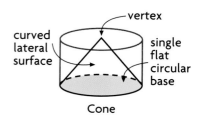

Cone

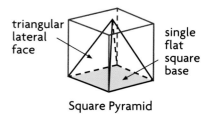

Square Pyramid

EXAMPLE 2

Name the figure.

All the faces are flat and are polygons, so the figure is a polyhedron.

The lateral faces are triangles with a common vertex, so it is a pyramid.

The base is a square.
So, the figure is a square pyramid.

CHECK FOR UNDERSTANDING

Think and ▶ Discuss

Look back at the lesson to answer each question.

1. **Tell** the name of each solid figure: a prism with octagonal bases; a pyramid with a pentagon as a base; a solid figure with flat faces that are polygons.

2. ***REASONING*** **Tell** how a cylinder and a prism are alike. How are they different? Tell how a cone and a cylinder are alike. How are they different?

351

Name the figure. Is it a polyhedron?

3.   4.   5.   6.

PRACTICE AND PROBLEM SOLVING

Independent ▶
Practice

Name the figure. Is it a polyhedron?

7.   8.   9.   10.

Write *true* or *false* for each statement. Rewrite each false statement as a true statement.

11. A cone has no flat surfaces.

12. A cylinder is a polyhedron.

13. A cube is a rectangular prism.

14. The bases of a cylinder are congruent.

15. All pyramids have triangular bases.

16. A hexagonal prism has six faces.

Problem Solving ▶
Applications

17. Niles carved a wooden pyramid with seven faces. What is the shape of the base of the pyramid?

18. Gina made two square pyramids and glued the congruent bases together to make an ornament. How many faces does her ornament have?

19. *REASONING* The lateral faces of a pyramid are equilateral triangles. Is the base a regular polygon? Draw a diagram to explain.

Name the top base of the solid you would obtain by cutting off the top of each figure along the red line.

20.   21.   22.

MIXED REVIEW AND TEST PREP

23. Name the part of a circle that is enclosed by two radii and the arc joining their endpoints. (p. 344)

24. Let *a* represent Alan's age. Write an equation stating that Alan's age 5 years ago was 13. (p. 284)

25. Write $\frac{27}{7}$ as a mixed number. (p. 164)

26. $^-9 + 5$ (p. 244)

27. TEST PREP Maury worked 9.25 hours and earned $7.45 per hour. Which is the total amount he earned to the nearest cent? (p. 70)

    A $16.70   **B** $68.91   **C** $74.50   **D** $689.13

EXTRA PRACTICE page H49, Set A

Different Views of Solid Figures

Learn how to draw and identify different views of a solid.

TECHNOLOGY LINK

More Practice: Use E-Lab, *Different Views of Solid Figures.*
www.harcourtschool.com/ elab2002

Name the base of each solid figure.

1. rectangular prism
2. cone
3. cylinder
4. triangular prism
5. square pyramid

Car designers use computer-aided design (CAD) to view their designs from different angles. These differing views help designers to create cars that are safe, efficient, and attractive.

You can draw different views of a solid.

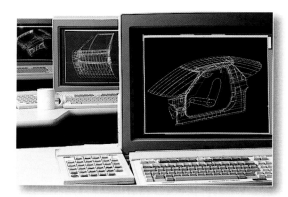

Activity 1

You need: cylinder

- Look at the top of the cylinder. Draw the top view.

- Look at the front of the cylinder. Draw the front view.

- Look at a side of the cylinder. Draw the side view.

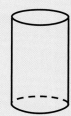

Math **I**dea ▶ You can use different views of a solid figure to identify the figure.

EXAMPLE 1 ▶ Name the solid figure that has the given views.

top view

The top view shows that the base is square and that the faces come together at a point.

front view   side view

The front and side views show that the solid has triangular sides.

So, this solid is a square pyramid.

Activity 2

How would this solid look if you viewed it from the top, the front, and the side? You can build a model to find out.

You need: centimeter cubes, centimeter graph paper

- Use centimeter cubes to build the solid at the right.

- The top view of the solid is shown below:

Draw the front view and the side view on graph paper.

- How many cubes do you see in the top view? the front view? the side view?

- Which views show how tall the solid is?

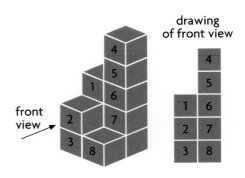

EXAMPLE 2

Draw the front view of the above stack of cubes without building a model.

Think of the front view of each numbered cube. In your drawing, show only the numbered faces, placing them in correct relationship to one another.

drawing of front view

CHECK FOR UNDERSTANDING

Think and Discuss ▶ Look back at the lesson to answer each question.

1. **Explain** why a side view of a solid figure might change if you change your viewpoint from the left side to the right side.

2. **Draw** a top view of the triangular prism.

Guided Practice ▶ Name the solid figure that has the given views.

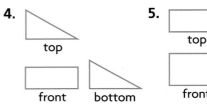

3.
top
front   bottom

4.
top
front   bottom

5.
top
front   side

Independent ▶
Practice

Name the solid figure that has the given views.

6.
top

front   side

7.
top

front   side

8.
top

front   side

Draw the front, top, and bottom views of each solid.

9.

10.

11.

12.

Each solid is made with 10 cubes. On graph paper, draw a top view, a front view, and a side view for each solid.

13. **14.** **15.** **16.**

17. Draw views of the hexagonal prism as seen from points A, back, and B, side.

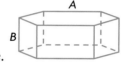

Problem Solving ▶
Applications

18. Pick an object in your classroom. Draw the top, front, and side views.

19. Every side view of a triangular prism shows what shape?

20. ✎ **Write About It**   Name objects in the classroom or at home that have a top view of a rectangle.

MIXED REVIEW AND TEST PREP

21. Name the figure. (p. 350)

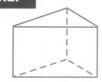

22. Name a parallelogram that has four congruent sides. (p. 338)

23. Evaluate $x^2 + 3x$ for $x = {}^-3$. (p. 272)

24. $\frac{3}{8} + \frac{3}{4}$ (p. 182)

25. TEST PREP   Find the difference in the sums $22.7 + 27.9 + 24.3 + 25.1$ and $140.1 + 36 + 45.6$. (p. 66)

 A 25   **B** 50.3   **C** 121.7   **D** 200.78

Models of Solid Figures

Explore how to build models of prisms and pyramids.

You need 4 in. × 6 in. index card, inch ruler, tape, scissors, protractor, patterns for cylinder and cone, grid paper.

Vocabulary

net

QUICK REVIEW

Tell the number of faces each solid figure has.

1. cube
2. triangular prism
3. square pyramid
4. pentagonal prism
5. triangular pyramid

You can build a solid figure by cutting its faces from paper, taping them together, and then folding them to form the solid.

3 in. 2 in. 1 in.

Activity 1

- Follow these steps to make a pattern for the prism.

Step 1: Draw the faces on the index card.

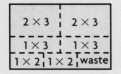

2 × 3   2 × 3
1 × 3   1 × 3
1 × 2 1 × 2 waste

Step 2: Cut out the six rectangles.

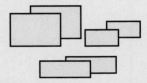

Step 3: Tape the pieces together to form the prism.

Step 4: Remove the tape from some of the edges so that the pattern lies flat.

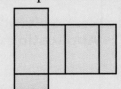

A pattern of polygons that can be folded to form a prism is called a **net** for the prism.

Think and Discuss

- Can you make a different net for the rectangular prism? Draw a sketch to explain.

- How are the nets for a cube and a rectangular prism different?

Practice

On grid paper, draw four nets that can be folded to form a cube. Then draw four nets that cannot form a cube. One of each type is shown at the right.

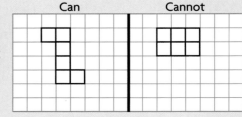

Can   Cannot

- How many triangles make up the pyramid?

- Draw an equilateral triangle on a sheet of paper. Cut out the triangle.

- Use the triangle as a pattern to trace three more triangles. Cut out these triangles.

- Tape the triangles together to form a triangular pyramid.

- Remove the tape from some of the edges to make a net of the triangular pyramid.

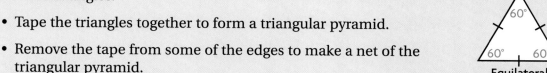

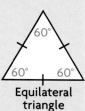

Equilateral triangle

Think and Discuss

- What shapes will always appear in the net of a pyramid? of a prism?

- What solid figure could you make from the net shown at the right? How do you know?

- Explain why the nets shown below cannot be folded to make solid figures.

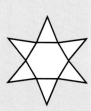

a.

b.

c.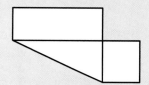

Practice

1. Draw a net for a pentagonal pyramid.

2. Draw a net for a pentagonal prism.

3. Use the cylinder pattern to build a cylinder.

4. Use the cone pattern to build a cone.

MIXED REVIEW AND TEST PREP

5. Draw a top view of a square pyramid. (p. 353)

6. A triangle contains two 48° angles. Find the measure of the third angle. (p. 332)

7. Write an integer to represent a loss of $432. (p. 228)

8. $2\frac{2}{3} \times 1\frac{7}{8}$ (p. 206)

★ **9. TEST PREP** Let t represent today's temperature. If tomorrow's temperature is 8 degrees less than twice today's temperature, which expression represents tomorrow's temperature? (p. 270)

   A $2t - 8$ **B** $t - 16$ **C** $8 - 2t$ **D** $2t + 8$

PROBLEM SOLVING STRATEGY
Solve a Simpler Problem

Learn how to use the strategy *solve a simpler problem* to solve problems.

Find the next number in the pattern.
1. 4, 8, 16, 32
2. 20, 17, 14, 11
3. 3, 6, 9, 12
4. 80, 40, 20, 10
5. 1, 4, 9, 16, 25

Andrew is building models of prisms. He is using balls of clay for the vertices and straws for the edges. How many of each will he need to make a prism with 15-sided bases?

Analyze

What are you asked to find?

What information is given?

Is there information you will not use? If so, what?

Choose

What strategy will you use?

You can use the strategy *solve a simpler problem*. Find the number of vertices and edges for some prisms whose bases have fewer sides. Then use what you learn to solve the harder problem.

Solve

How will you solve the problem?

Find the number of vertices and edges there are on prisms whose bases have 3, 4, and 5 sides. Record your findings in a table.

Sides on base	3	4	5	
Vertices	3 + 3, or 6	4 + 4, or 8	5 + 5, or 10	*top + bottom*
Edges	3 + 3 + 3, or 9	4 + 4 + 4, or 12	5 + 5 + 5, or 15	*top + bottom + sides*

Now, apply the patterns shown in the table to a prism whose bases have 15 sides:

Vertices: 15 + 15 = 30

Edges: 15 + 15 + 15 = 45

So, Andrew will need 30 balls of clay for the vertices and 45 straws for the edges.

Check

What other strategy could you use to solve the problem?

What if the prism had bases with 25 sides each? How many vertices and how many edges would it have?

PROBLEM SOLVING PRACTICE

Solve the problem by first *solving a simpler problem*.

1. Margo wants to make a model of a prism whose bases have 10 sides each. She will use balls of clay for the vertices and straws for the edges. How many balls of clay will she need?

2. Amy wants to make a model of a pyramid whose base has 10 sides. She will use balls of clay for the vertices and toothpicks for the edges. How many balls of clay will she need? How many toothpicks will she need? How many faces will her prism have?

PROBLEM SOLVING STRATEGIES

| Draw a Diagram or Picture |
| Make a Model |
| Predict and Test |
| Work Backward |
| Make an Organized List |
| Find a Pattern |
| Make a Table or Graph |
| ▶ **Solve a Simpler Problem** |
| Write an Equation |
| Use Logical Reasoning |

3. By drawing 1 vertical line and 1 horizontal line, you can divide a sheet of paper into 4 sections. Into how many sections can you divide a sheet of paper using 8 vertical and 8 horizontal lines?

1	2
3	4

   A 10   **B** 16   **C** 64   **D** 81

4. Boxes are stacked with 1 box in the first row, 3 in the second, 5 in the third, and so on. How many boxes are needed altogether to make a 10-row stack?

   F 12   **G** 19   **H** 81   **J** 100

MIXED STRATEGY PRACTICE

Use Data   For 5–7, use the graph.

5. On which day was the difference between the high and low temperatures the greatest? What was the difference?

6. Between which two days was the temperature change the greatest? Did the high or low temperature change more?

7. ✏️ **Write a problem** that can be solved by using the graph.

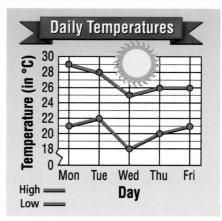

8. Corbin centered a table against a wall that was 13 ft wide. The table was $3\frac{1}{2}$ ft wide. How far was the left end of the table from the left end of the wall?

9. Helmut used 27 toothpicks as edges on a model of a prism. How many sides did the base of the prism have?

10. Keith, Fran, and Lanni divided a basket of tennis balls among themselves. Keith took half of the balls. Fran took half of those that remained. Lanni took the remaining 5 balls. How many balls were in the basket to begin with?

1. **VOCABULARY** A solid figure with flat faces that are polygons is called a(n) ___?___ . (p. 350)

2. **VOCABULARY** A pattern that can be folded to form a solid figure is called a(n) ___?___ . (p. 356)

3. **VOCABULARY** A solid figure with two flat, parallel, congruent circular bases and a curved lateral surface is a(n) ___?___ . (p. 351)

Name the figure. Is it a polyhedron? (pp. 350–352)

4.

5.

6.

7.

Name the solid figure that has the given views. (pp. 353–355)

8.
top

front   side

9.
top

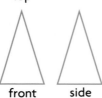
front   side

10.
top

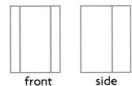
front   side

Draw the front, top, and side views of each solid. (pp. 353–355)

11.

12.

13.

14.

Will the net fold to form a cube? Write *yes* or *no*. (pp. 356–357)

15.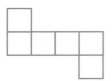

16.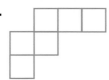

17.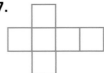

18.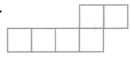

Solve.

19. April made a model of a prism whose bases have 9 sides each. How many vertices and how many edges did her model have? (p. 358)

20. Mark made a model of a pyramid with 16 edges. How many sides did the base of his pyramid have? (p. 358) ·

Check your work.

See item **5.**

Draw a representation of a prism that has 8 vertices. Be sure to account for the edges in all three dimensions.

Also see problem **7**, p. H65.

Choose the best answer.

1. Which solid figure has 4 faces, 4 vertices, and 6 edges?

   A Pentagonal prism

   B Rectangular prism

   C Triangular pyramid

   D Triangular prism

2. What is a name for the solid figure shown below?

   F Pyramid   **H** Cube

   G Cone   **J** Cylinder

3. Which of these does a cone have?

   A One flat, circular base

   B A vertex

   C A curved lateral surface

   D All of the above

4. What model of a solid figure can be made from the net shown below?

   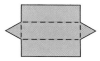

   F Rectangular prism

   G Triangular prism

   H Pentagonal prism

   J Not here

5. Zack is building a model of a prism. He will use balls of clay for the vertices and straws for the edges. How many straws will he use if he uses 8 balls of clay for his model?

   A 4   **B** 8   **C** 12   **D** 16

6. Which solid figure has triangles in all of its views?

   F Triangular pyramid

   G Cone

   H Rectangular prism

   J Triangular prism

7. Solve $x - 34 = 95$.

   A $x = 69$ **C** $x = 129$

   B $x = 121$ **D** Not here

8. What is the sum of the measures of two angles that are complementary?

   F 45°   **G** 90°   **H** 180°   **J** 360°

Write What You Know

9. Explain what a polyhedron is. Give two examples of polyhedrons and explain why a cylinder is not a polyhedron.

10. Show how the order of operations is used to solve the expression below.
    $$30 + (9 - 5)^2 \div 2 - 5$$

OUTLINES OF NESTING DOLLS

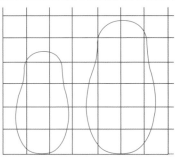

Nesting boxes were first created more than 900 years ago in China. The word nesting refers to the idea of objects fitting inside each other. About 100 years ago, Russian artists started making nesting dolls.

PROBLEM SOLVING Are the dolls shown in the graph congruent? Are they similar?

Check What You Know

Use this page to help you review and remember important skills needed for Chapter 19.

✓ Vocabulary

Choose the best term from the box.

| congruent |
| intersecting |
| line segment |
| parallel |
| perpendicular |

1. Figures that are the same size and shape are __?__.

2. A portion of a line, having two endpoints, is a __?__.

3. Angles that have the same measure are __?__ angles.

4. Lines in a plane that are always the same distance apart, never intersect, and have no common points are __?__.

✓ Congruent Figures (For Intervention, see p. H22.)

Tell whether the figures appear to be congruent or not. Write *yes* or *no.*

5.    6.    7.    8.

9.    10.    11.    12.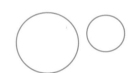

✓ Classify Lines (For Intervention, see p. H14.)

Classify each pair of lines as parallel, perpendicular, or intersecting. There may be two ways to classify a pair.

13.   14.   15.

16. 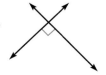   17.    18.

LOOK AHEAD

In Chapter 19 you will
- construct congruent line segments and angles
- bisect line segments and angles
- construct parallel lines
- identify similar and congruent figures

LESSON 19.1

Construct Congruent Segments and Angles

Learn how to construct congruent line segments and angles.

QUICK REVIEW

Draw an angle with the measure indicated.

1. 90° **2.** 120° **3.** 165° **4.** 25° **5.** 47°

Vocabulary

congruent

Line segments that are the same length are said to be **congruent**. Use the symbol ≅ to show that two segments are congruent. In the figure, $\overline{AB}$ and $\overline{CD}$ are 1.5 cm long, so $\overline{AB} \cong \overline{CD}$.

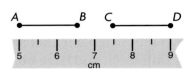

If you have a ruler, you can use it to see whether two line segments are congruent. As you'll see in the following activity, you can also use a compass to test for congruence.

Activity

You Need: compass

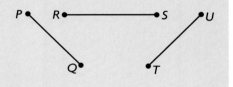

- Trace $\overline{PQ}$, $\overline{RS}$, and $\overline{TU}$ on your paper. Place the compass point on point *P*. Open the compass to the length of $\overline{PQ}$.
- Use the compass to compare the length of $\overline{PQ}$ with the lengths of $\overline{RS}$ and $\overline{TU}$.
- Which segment is congruent to $\overline{PQ}$? Use the symbol ≅ to show the congruence.

Use a compass and a straightedge to construct congruent line segments.

EXAMPLE 1

Trace $\overline{EF}$. Then construct a line segment congruent to $\overline{EF}$.

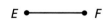

Draw a ray that is longer than $\overline{EF}$. Label the endpoint H.

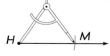

Place the compass point on point E. Open the compass to the length of $\overline{EF}$. Use the same compass opening. Place the compass point on H.Draw an arc that intersects the ray. Label the intersection point M.

So, $\overline{HM} \cong \overline{EF}$.

Congruent Angles

Like line segments, angles can be congruent. Two angles are congruent if they have the same measure in degrees.

When a beam of light hits a mirror, the angle at which it arrives—the *angle of incidence*—is congruent to the angle at which it reflects off the mirror—the *angle of reflection*. In the figure, $\angle ABD \cong \angle DBC$. Both angles measure 28°.

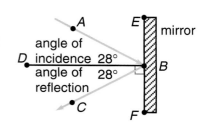

* Do you think $\angle EBA \cong \angle FBC$? Explain.

Use a compass and a straightedge to construct congruent angles.

EXAMPLE 2

Trace $\angle B$. Then construct an angle congruent to $\angle B$.

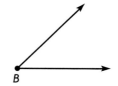

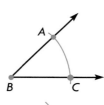

Draw a ray with endpoint M.

Place the compass point on B, and draw an arc through $\angle B$. Label the points of intersection A and C.

Place the compass point on M. Draw the same arc through the ray. Label the point of intersection P.

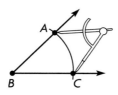

Use the compass to measure the arc in $\angle ABC$.

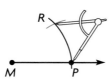

Use the same compass opening to locate point R.

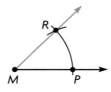

Draw $\overrightarrow{MR}$.

So, $\angle RMP \cong \angle ABC$.

Congruent angles often appear on buildings.

365

CHECK FOR UNDERSTANDING

Think and Discuss ▶ **Look back at the lesson to answer each question.**

1. **Explain** how you know that in the last step in Example 1, *M* is the same distance from *H* as *F* is from *E*.
2. **Give** some examples of congruent line segments in your classroom.

Guided Practice ▶ **Use a compass to decide which two line segments in each group are congruent.**

3.

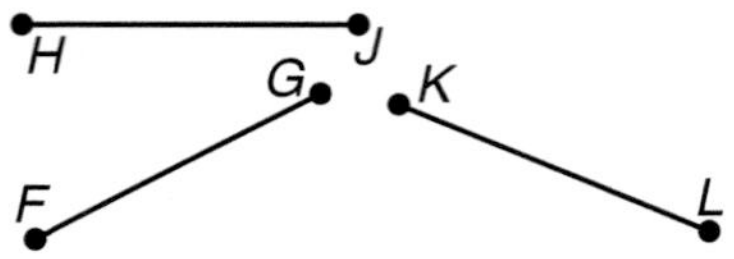

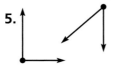

4.

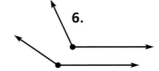

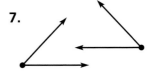

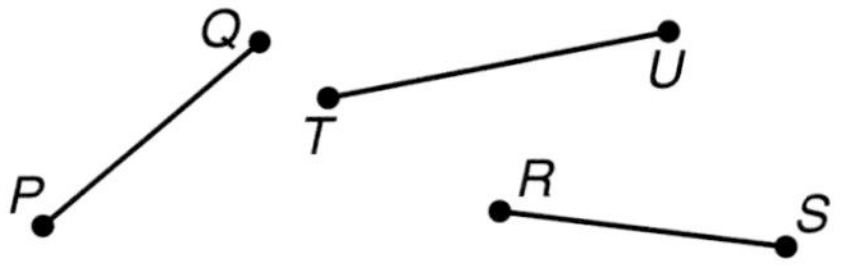

Find the measure of each angle, using a proctractor. Then tell whether the angles in each pair are congruent. Write *yes* or *no.*

5.

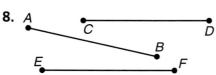

6.

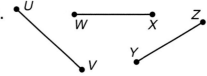

7.

PRACTICE AND PROBLEM SOLVING

Independent Practice ▶ **Use a compass to decide which two line segments in each group are congruent.**

8.

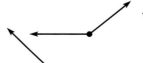

9.

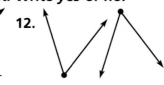

Find the measure of each angle, using a protractor. Then tell whether the angles in each pair are congruent. Write *yes* or *no.*

10.

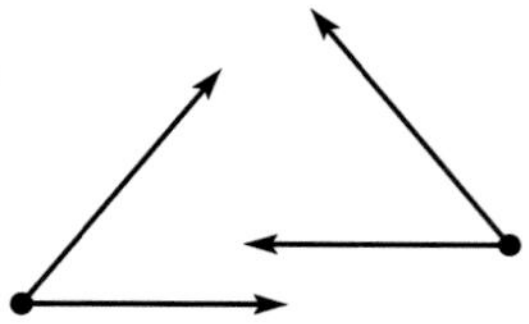

11.

12.

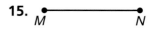

Use a compass and a straightedge to construct a line segment congruent to the given segment.

13.

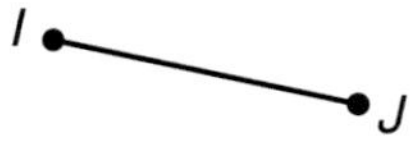

14.

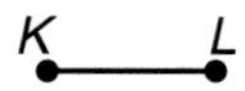

15.

Use a compass and a straightedge to construct an angle congruent to the given angle.

16.

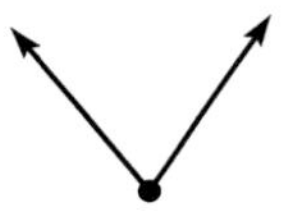

17.

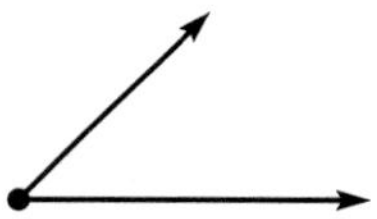

18.

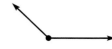

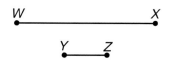

19. Construct an angle with a measure equal to the measure of ∠*ABC* plus the measure of ∠*DEF*.

20. Construct a line segment with a length equal to the length of $\overline{WX}$ minus the length of $\overline{YZ}$.

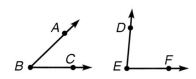

21. A beam of light from point *A* strikes Mirror 1 at point *B*. The angle of incidence is 24°. The beam then strikes Mirror 2 at point *C*. Find the measure of ∠*ECD*.

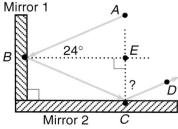

22. ✏️ **Write About It** Explain how you could use a compass to find a point on $\overline{ML}$ that is the same distance from *M* as *P* is from *N*.

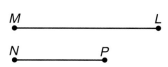

MIXED REVIEW AND TEST PREP

23. $1\frac{1}{4} \div 1\frac{1}{8}$ (p. 198)

24. Complete. $\frac{8}{12} = \frac{\blacksquare}{18}$ (p. 160).

25. Draw the side view of a cylinder. (p. 353)

⭐**26. TEST PREP** Find the sum. $3\frac{2}{3} + 1\frac{5}{6}$ (p. 186)

  A $4\frac{7}{9}$ **B** $5\frac{1}{2}$ **C** $5\frac{8}{12}$ **D** $7\frac{11}{12}$

⭐**27. TEST PREP** Ben's car gets 16 miles per gallon of gasoline. He used 9 gallons of gasoline driving. How far did Ben drive? (p. 22)

  F 25 mi   **G** 100 mi   **H** 144 mi   **J** 154 mi

PROBLEM SOLVING

 ThiNKer's CorNer

GEOMETRIC GEOGRAPHY Find geometric figures in everyday pictures.

Materials: ruler

Most states have borders that twist and turn. But some states have borders that model line segments. Use the map shown at the right to answer the questions.

1. Describe where these four states have borders that appear to model line segments.

2. Does any state have two borders that appear congruent?

3. Why do you think the border between Iowa and Illinois and between Missouri and Illinois twists and turns?

Bisect Line Segments and Angles

Learn how to bisect line segments and angles by using a compass and straightedge.

Vocabulary

bisect

midpoint

When you **bisect** a line segment, you divide it into two congruent parts. The **midpoint** of a line segment is the point halfway between the endpoints of the segment. The midpoint bisects the line segment.

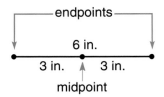

You can use a compass and a straightedge to bisect a segment.

Activity

You Need: compass, straightedge

• Draw a line segment. Label the endpoints R and S.

R •————————• *S*

• Place the compass point on point R. Open the compass to more than half the distance from R to S. Draw an arc through $\overline{RS}$.

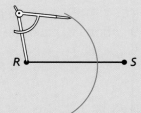

• Keep the same compass opening. Place the compass point on point S. Draw an arc through $\overline{RS}$. Label the points where the arcs intersect as T and U.

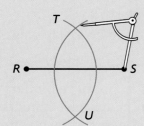

• Use a straightedge to draw a line through T and U. Use P to label the point where $\overline{TU}$ intersects $\overline{RS}$.

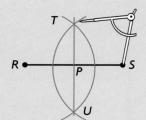

• $\overline{TU}$ bisects $\overline{RS}$. Point P is the midpoint of $\overline{RS}$. Therefore, $\overline{RT} \cong \overline{PS}$.

Bisecting Angles

When you bisect an angle, you divide it into two congruent angles. In the figure, $\overrightarrow{NR}$ bisects $\angle MNP$, so $\angle MNR \cong \angle RNP$.

You can use a compass and a straightedge to bisect an angle.

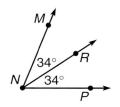

EXAMPLE

Trace $\angle K$. Then bisect the angle.

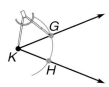

Place the compass point on point K. Draw an arc that intersects both sides of $\angle K$. Label the intersection points of the angle and the arc G and H.

TECHNOLOGY LINK

More Practice: Use E-Lab, *Bisecting Figures.*
www.harcourtschool.com/ elab2002

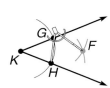

With the compass point on point G, draw an arc in the middle of the angle. Keep the same compass opening. Repeat, placing the compass point on point H. Label the point where the arcs intersect point F.

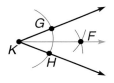

Draw $\overrightarrow{KF}$.

$\overrightarrow{KF}$ is the bisector of $\angle GKH$. So, $\angle GKF \cong \angle FKH$.

CHECK FOR UNDERSTANDING

Think and Discuss ▶ Look back at the lesson to answer each question.

1. **Tell** how many congruent angles are formed when you bisect an angle. How many congruent line segments are formed when you bisect a line segment?

2. **Explain** how bisecting an angle is like bisecting a line segment.

Guided Practice ▶ If a line segment of the given length is bisected, how long will each of the smaller segments be?

3. 38 in.   **4.** 112 cm   **5.** 0.3 m   **6.** 713.6 mm

If an angle of the given measure is bisected, how many degrees will there be in each of the smaller angles that are formed?

7. 52°   **8.** 79°   **9.** 8.1°   **10.** 119.1°

Trace the figure. Then bisect it.

11.

12.

Independent ▶ Practice

If a line segment of the given length is bisected, how long will each of the smaller segments be?

13. 17 ft　　　**14.** 2.01 m　　　**15.** 99.35 cm　　　**16.** 411 yd

If an angle of the given measure is bisected, how many degrees will there be in each of the smaller angles that are formed?

17. 65°　　　**18.** 51°　　　**19.** 142.6°　　　**20.** 179.5°

Trace the figure. Then bisect it.

21.

22.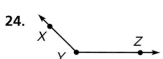

23.

24.

25.

26.

Problem Solving ▶ Applications

27. Trace the square. Use a compass and a straightedge to find the midpoint of each side, then connect them. What figure is formed?

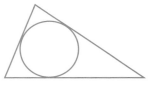

28. With a compass, draw a large circle on a sheet of paper. Use a straightedge to draw a scalene triangle so that each side touches the circle at a single point. Now bisect each angle of the triangle. What is special about the point where the three bisectors intersect?

29. The measures of two angles of a triangle are 61° and 43°. If the third angle is bisected, what is the measure of each of the angles that are formed?

30. ? **What's the Question?** The original angle measures 80°. The answer is 40°.

MIXED REVIEW AND TEST PREP

31. $\angle A \cong \angle R$ and $\angle V \cong \angle R$. If $\angle V$ measures 60°, what is the measure of $\angle A$? (p. 322)

32. A pyramid has 6 faces. What is the shape of the base of the pyramid? (p. 350)

33. Write 0.48 as a fraction in simplest form. (p. 169)

34. 0.82 ÷ 0.4 (p. 76)

★**35. TEST PREP** Ellen has 15 more than twice as many silver dollars as Sheri has. If Ellen has 75 silver dollars, which equation could you use to find the number of silver dollars that Sheri has? (p. 284)

A $b + 15 = 75$　**B** $75 = b - 15$　**C** $2 \times b = 75$　**D** $2b + 15 = 75$

EXTRA PRACTICE page H50, Set B

Construct Parallel Lines

Explore how to construct parallel lines by using a compass and a straightedge.

You need compass, straightedge.

Parallel lines are lines that never intersect, no matter how far they extend. You can use a compass and a straightedge to construct parallel lines.

Activity

• Draw a line and a point P.
Draw a line that goes through point P and intersects the line. Label the point of intersection W.

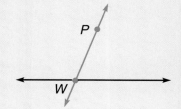

• Construct congruent angles.

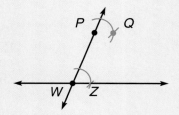

• Draw $\overleftrightarrow{PQ}$. $\overleftrightarrow{PQ}$ is a line parallel to $\overleftrightarrow{WZ}$.

$\overleftrightarrow{PQ} \parallel \overleftrightarrow{YZ}$

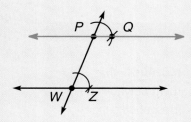

Think and Discuss

• Describe two examples of parallel lines in your classroom.

• What type of figure could you draw with two sets of parallel lines?

Practice

1. Draw a line. Have another student construct a line parallel to it.

2. Use a compass and a straightedge to construct a parallelogram.

Similar and Congruent Figures

Learn how to identify similar and congruent figures.

Vocabulary

similar figures

If a line segment of the given length is bisected, how long will each of the smaller segments be?

1. 10 cm   **2.** 17 in.   **3.** 4 yd   **4.** 5 ft   **5.** 26 in.

Figures that have the same shape are **similar figures** . The red, blue and green triangles all have the same shape, so they are similar.

Congruent figures have the same shape and the same size. The red and green triangles are congruent and similar, because they have the same shape and the same size.

The Venn diagram shows the relationship between pairs of congruent figures and pairs of similar figures. Notice that all congruent pairs are similar. However, some similar pairs are not congruent.

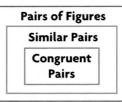

Pairs of Figures

Similar Pairs

Congruent Pairs

Math Idea ▶ Figures can be congruent, similar, both, or neither.

EXAMPLE 1

Remember that congruent polygons have all corresponding sides congruent and all corresponding angles congruent.

Decide whether the figures in each pair appear to be similar, congruent, both, or neither.

A

same shape, not same size
The figures are similar. You can use the symbol ~ to show that figures are similar.

ABCD ~ HJKL   *Read: ABCD is similar to HJKL.*

B

same shape, not same size
The figures are similar.
△*ABC* ~ △*VWX*

C

same shape, same size
The figures are both similar and congruent.

MNPL ≅ YZWX

D

not same shape, not same size
The figures are neither similar nor congruent.

Think and Discuss ▶ Look back at the lesson to answer each question.

1. **Use** the Venn diagram on page 372. Are all pairs of congruent figures similar? Are all pairs of similar figures congruent?

2. **Tell** whether three figures with the same size and the same shape are similar, congruent, or both.

Guided Practice ▶ Tell whether the figures in each pair appear to be *similar, congruent, both,* or *neither.*

3.   4.   5.   6.

PRACTICE AND PROBLEM SOLVING

Independent Practice ▶ Tell whether the figures in each pair appear to be *similar, congruent, both,* or *neither.*

7.    8.   9.

10.   11.   12.

Problem Solving Applications ▶ Write *true* or *false* for each statement.

13. If two figures are the same shape, then they must be congruent.

14. If two figures are congruent, then they must be similar.

15. *REASONING* Would all squares appear to be similar? Would all squares appear to be congruent? Explain.

16. ✎ **Write About It** Explain why all congruent figures are similar but not all similar figures are congruent.

MIXED REVIEW AND TEST PREP

17. If a 73° angle is bisected, what is the measure of each of the smaller angles? (p. 368)

18. Draw the top view of the solid at the right. (p. 353)

19. $3^2 - 2^3$ (p. 250)

20. 30.2×0.75 (p. 70)

★ 21. **TEST PREP** The cafeteria buys packages of peanuts for $0.19 each and sells each package for $0.50. How many packages must be sold to earn at least $10 more than the cost of the peanuts? (p. 22)

A 20   **B** 33   **C** 53   **D** 120

EXTRA PRACTICE page H50, Set C

1. **VOCABULARY** When you divide an angle into two equal parts, you __?__ the angle. (p. 368)

2. **VOCABULARY** The point halfway between the endpoints of a segment is the __?__ of the segment. (p. 368)

3. **VOCABULARY** Figures that have the same shape but not necessarily the same size are __?__. (p. 372)

Use a compass to decide which two line segments in each group are congruent. (pp. 364–367)

4.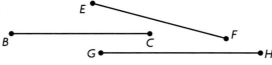

5.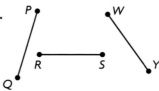

Find the measure of each angle, using a protractor. Then tell whether the angles in each pair are congruent. Write *yes* or *no*. (pp. 364–367)

6.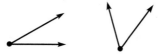

7.

For 8–14, use a compass and a straightedge.

8. Construct two lines that are parallel. (p. 371)

9. Draw a line segment. Construct a line segment congruent to it. (pp. 364–367)

10. Draw an angle. Construct an angle congruent to it. (pp. 364–367)

Trace the figure. Then bisect it. (pp. 368–369)

11.

12.

13.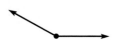

14.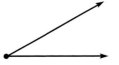

Tell whether the figures in each pair appear to be *similar, congruent, both,* or *neither*. (pp. 372–373)

15.

16.

17.

18.

19. If ∠*DEF* is congruent to ∠*ABC*, and ∠*GHI* is congruent to ∠*DEF*, what can you conclude about ∠*ABC* and ∠*GHI*? (p. 364)

20. The perimeter of a triangle is 54 cm. Two of the sides measure 17 cm and 19 cm. If the third side is bisected, what is the measure of each of the new line segments formed? (p. 368)

Choose the answer.

See item **5.**

If your answer choices are given as pictures, look at each one by itself while you cover the other three. Then choose the answer..

Also, see problem **6**, p. H64.

Choose the best answer.

1. Which expression is equivalent to $(5 \times 4) + 9$?

   A $5 \times 9 + 4 \times 9$
   B $9 + 9$
   C 20×9
   D Not here

2. There are 24 pictures on each roll of film. The PTA wants to take a picture of each of the 720 students for a school mural. How many rolls of film will they need?

   F 36 rolls
   G 30 rolls
   H 24 rolls
   J 12 rolls

3. What is a reasonable estimate for the sum of 2.5, 4.8, 3.1, and 10.2?

   A 200   **C** 23
   B 185   **D** 21

4. Solve for z.

   $$z - 29.3 = 15.8$$

   F $z = 45.1$
   G $z = 14.5$
   H $z = 44.1$
   J $z = 13.5$

5. Which angle appears to be congruent to $\angle ABC$?

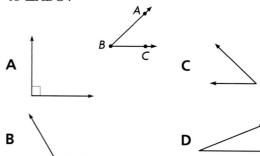

6. What construction do the figures below represent?

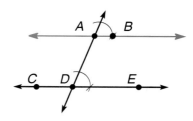

   F parallel lines
   G congruent line segments
   H perpendicular lines
   J angle bisector

7. What is a reasonable estimate for the sum of 2.5, 4.8, 3.1, and 10.2?

   A 200   **C** 33
   B 185   **D** 21

8. Use mental math to find the value of x.

   $$x \div 7 = {}^{-}80$$

   F $x = {}^{-}560$
   G $x = 540$
   H $x = {}^{-}540$
   J $x = 560$

Write What You Know

9. Using a compass and straightedge, construct a line segment that is $1\frac{1}{2}$ times the length of line segment GH. Tell how you did it.

10. Show the outlines for 2 different gardens each with a perimeter of 24 ft. Explain whether the gardens are similar or congruent to one another.

MATH DETECTIVE

Truth or Consequences

REASONING Use your knowledge of quadrilaterals to decide whether each statement is true or false. If a statement is false, use words like *all*, *some*, or *no* to rewrite it to make it true.

Truth or Consequences 1

Every square is a parallelogram.

Truth or Consequences 2

Every rectangle is a square.

Truth or Consequences 3

Every square is a rhombus.

Truth or Consequences 4

Every rhombus is a quadrilateral.

Truth or Consequences 5

A quadrilateral with exactly one pair of parallel sides is a parallelogram.

Truth or Consequences 6

Every rectangle is a trapezoid.

Think It Over!

- **Write About It** Complete this sentence three different ways: "A square is ___?___."

- **Stretch Your Thinking** I'm a quadrilateral. I am symmetrical about each of my diagonals. My diagonals are congruent. What am I?

Explore Perspective

Learn how to make one-point perspective drawings.

Perspective is a method used to make objects drawn on a flat surface appear three-dimensional.

Compare the two drawings on the right. The first drawing looks more flat. The second drawing appears to have depth because it is drawn in perspective.

Parallel lines

Perspective lines

EXAMPLE

Make a one-point perspective drawing of a rectangular prism.
- Draw the front face.
- Draw a point above the face and to one side. This is called a *vanishing point*.
- Connect the four vertices to the vanishing point.
- Draw the top rear edge of the prism. It should be parallel to the top front edge.
- Draw the remaining visible edge of the prism.
- Draw the "hidden" edges, using dashed lines. Erase the segments leading to the vanishing point.

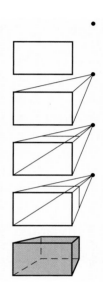

TALK ABOUT IT

- Describe how to draw a triangular prism by using one-point perspective.

TRY IT

Draw each figure. Use one-point perspective.

1. a cube with a vanishing point to the left

2. a cube with a vanishing point to the right

3. a cereal box with a vanishing point in the center

4. The figure on the right shows a rectangular prism drawn using two-point perspective. Draw a rectangular prism by using two-point perspective. Place the two vanishing points in different positions from those shown here.

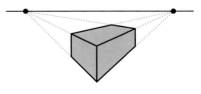

VOCABULARY

1. Two angles whose measures have a sum of 180° are ___?___. (p. 323)

2. A line segment that has both endpoints on a circle and that passes through the center of the circle is a ___?___. (p. 344)

3. A solid figure with flat faces that are polygons is called a ___?___. (p. 350)

4. Figures that have the same shape are called ___?___. (p. 372)

EXAMPLES   **EXERCISES**

Chapter 16

• **Identify and classify angles.** (pp. 320–321)

Tell if the angle *is acute, right,* or *obtuse.*

The angle is acute since its measure is less than 90°.

Tell if each angle is *acute, right,* or *obtuse.*

5.

6.

7.

8.

• **Identify angle relationships.** (pp. 322–325)

Find the measure of ∠*BOF*.

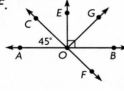

∠*BOF* and ∠*COA* are vertical angles and are congruent. So, the measure of ∠*BOF* is 45°.

Find the measure of each angle.

9. ∠*XTY*

10. ∠*ZTR*

11. ∠*YTZ*

12. ∠*RTV*

13. ∠*VTZ*

14. ∠*XTR*

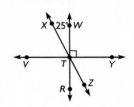

Chapter 17

• **Classify triangles.** (pp. 332–335)

Find the unknown angle measure and classify the triangle.

180° − (60° + 60°) =
180° − 120° = 60°

The measure of the angle is 60°.

The triangle is an acute triangle.

Find the unknown angle measure and classify the triangle.

15.

16.

17.

18.

- **Classify quadrilaterals.** (pp. 342–345)

  Give the most exact
  name for the figure.

  Opposite sides are
  congruent and parallel.
  There are 4 right angles.
  So, the figure is a square.

  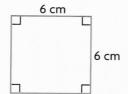

Give the most exact name for the figure.

19.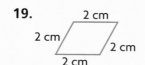

20.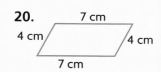

- **Identify parts of a circle.** (pp. 348–349)

  $\overline{AB}$ is a diameter.

  $\overline{OC}$ is a radius.

  $\overline{DE}$ is a chord.

  $\overparen{CD}$ is an arc.

  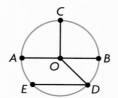

Name the given parts of the circle.

21. chord

22. diameter

23. radius

24. arc

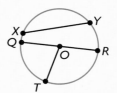

Chapter 18

- **Name solid figures.** (pp. 350–352)

  Name the figure.
  Is it a polyhedron?

  square pyramid
  It is a polyhedron.

Name the figure. Is it a polyhedron? Write *yes* or *no*.

25.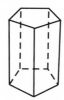

26.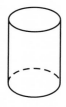

- **Identify nets for solid figures.** (pp. 356–357)

  Name the figure that
  can be formed from
  the net.

  triangular pyramid

  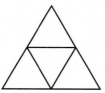

Will the arrangement of squares fold to form a cube? Write *yes* or *no*.

27.

28.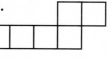

Chapter 19

- **Identify similar and congruent figures.**
  (pp. 372–373)

  Tell if the two figures
  appear to be *similar*,
  congruent, *both*, or
  neither.
  The figures are similar.

Tell whether the figures in each pair appear to be *similar*, *congruent*, *both*, or *neither*.

29.

30.

PROBLEM SOLVING APPLICATIONS

31. Eight players were in a tennis match. Each player had to play every other player. How many matches did they play altogether? (pp. 336–337)

32. Juan made a model of an octagonal prism. How many vertices and how many edges did his model have? (pp. 358–359)

Performance Assessment

TASK A • Monumental Task

The Washington Monument is in Washington, D.C. It is an obelisk, a pillar with a pyramidion (small pyramid) on top. The base is square in shape. The obelisk is wider at the bottom than at the top.

a. Describe how you could estimate the volume of the Washington Monument.

b. The Washington Monument measures 16.8 m on a side at the bottom and 10.5 m on each side at the base of the pyramidion. The monument is 169.3 m high. The pyramidion is 16.8 m high. Show how you would estimate its volume.

c. What additional measurements would you need to be able to get a more accurate estimate of the volume?

TASK B • Lines and Angles

Materials: square sheet of paper, straightedge, protractor, colored pencils

Begin with a square sheet of paper. Label the vertices *A, B, C,* and *D,* as shown in the picture at the right. Then fold the square in half and in half again. Open the square and label the middle points *E, F, G,* and *H.* Label the center *O.* Draw lines from these points to every other point on the square, as shown.

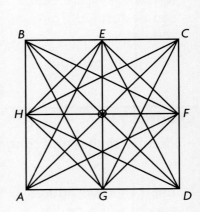

a. Use letters to name a rectangle, a square, and a rhombus that you see in the figure. Add letters to the drawing if needed.

b. Name two vertical angles. Measure your angles to check.

c. Name two complementary angles and two supplementary angles.

d. Use letters to name parallel lines, perpendicular lines, and intersecting lines.

e. Find a quadrilateral and a trapezoid using the lines in the drawing. Color the quadrilateral blue and the trapezoid red. Color an acute triangle green, an obtuse triangle yellow, and a right triangle orange.

Technology Linkup

Mighty Math Cosmic Geometry
Lines, Angles, and Plane Figures

You can use *Geo Movies* to draw and measure line segments, angles, and plane figures.

- Launch Cosmic Geometry.

- Enter the *Geo Movies* building on the far right.

- Click [ ]. Draw 5 line segments to make a pentagon.

- Click [ ]. Point to the figure. Record the perimeter and area of the figure.

- Click [ ]. Point to each angle inside the figure. Record the measurement of each angle.

- Click [ ] to see a movie of your figure being drawn.

- Click [ ] to erase the movie and start over.

Practice and Problem Solving

Use *Geo Movies* to draw the geometric figure. Draw a picture to show your work.

1. line segment
2. obtuse angle
3. acute angle
4. straight angle

5. supplementary angles
6. perpendicular lines
7. intersecting lines
8. parallel lines

9. acute triangle
10. right triangle
11. trapezoid
12. rhombus

13. isosceles triangle
14. obtuse triangle
15. radius of a circle
16. chord of a circle

17. How could you use *Geo Movies* to check whether two figures are congruent?

18. **Write a problem** about a geometric figure that you draw with *Geo Movies*.

Multimedia Math Glossary www.harcourtschool.com/mathglossary

19. **Vocabulary** Look up *plane figure* in the Multimedia Math Glossary. Write a description and then name each of the plane figures shown in the glossary.

PROBLEM SOLVING ON LOCATION

With Geometric Patterns and Structures

Sea Star

Geometry in Nature

The shapes you have studied in geometry are based on the shapes we see around us. A sea urchin's shell is less likely to crack because of its shape. A sea star has tentacles at the ends of its arms to bring food to its mouth.

1. Use geometric terms to describe the shape of a sea urchin.

2. Suppose you drew line segments to connect the ends of the arms of a sea star. What geometric figure would you draw?

Sea Urchin

For 3–5, use the photo of the honeycomb at the right.

3. Bees use geometry when they make honeycombs. What shape do you see at the end of each cell?

4. Suppose a complete honeycomb cell looks like the drawing at the right. What is this figure called?

5. How may faces, edges, and vertices would a sealed honeycomb cell have?

For 6, use the photo of irrigated fields at the right.

6. Sometimes people redesign nature. States such as Nevada and New Mexico have desert regions that can be farmed by using irrigation. What geometric term can you use to describe the distance across one of these irrigated fields?

Geometry and Buildings

The buildings we construct are based on the geometric shapes we learn about in school. Some of the tallest and most interesting buildings in the United States can be found in and around Chicago, Illinois.

The Aon Building stands more than 1,000 feet over Chicago.

Besides its offices, the John Hancock Tower has room for more than 700 apartments.

Use Data   For 1–5, use the photos of the buildings.

1. What shape is each side of the Aon Building?

2. What shape is the base of the Aon Building?

3. What geometric figure describes the shape of the Aon Building?

4. What shape is each side of the John Hancock Building?

5. Does the John Hancock Building have the shape of a pyramid? Explain why or why not.

Around the world, Baha'i Houses are built with 9 entrances. The main floor of the Baha'i House in Wilmette, Illinois is shaped like the polygon at the right.

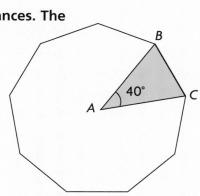

For 6–9, use the figure at the right.

6. How many triangles like △ ABC can be drawn inside the polygon?

7. Find the measure of each angle of △ ABC.

8. What kind of triangle is △ ABC?

9. What is the measure of each angle of the 9-sided regular polygon?

20 Ratio and Proportion

Mini-Europe is an attraction in Belgium. The models in the display are made using the scale 1:25. The tall building is a replica of the Brussels Town Hall in Brussels, Belgium, and is about 4 m tall. The actual building is about 96 m tall.

PROBLEM SOLVING Another model in Mini-Europe is the Eiffel Tower. The actual Eiffel Tower is 320.75 m tall. Using the scale 1:25, about how tall is the model?

DATA LINK

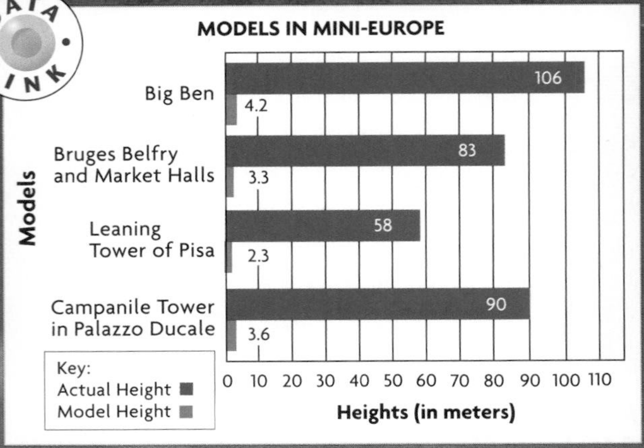

MODELS IN MINI-EUROPE

Models		Heights (in meters)
Big Ben		106 / 4.2
Bruges Belfry and Market Halls		83 / 3.3
Leaning Tower of Pisa		58 / 2.3
Campanile Tower in Palazzo Ducale		90 / 3.6

Key:
Actual Height ■
Model Height ■

0 10 20 30 40 50 60 70 80 90 100 110
Heights (in meters)

Check What You Know

Use this page to help you review and remember important skills needed for Chapter 20.

✓ Vocabulary

Choose the best term from the box.

enlargement
equivalent
polygon
polyhedron
reduction

1. Fractions that name the same number or amount are __?__ fractions.

2. When the size of a figure is increased, the new figure is a(n) __?__ of the original figure.

3. A closed plane figure formed by line segments is a(n) __?__ .

✓ Write Equivalent Fractions (For Intervention, see p. H21.)

Complete each number sentence.

4. $\frac{4}{5} = \frac{\blacksquare}{20}$

5. $\frac{5}{9} = \frac{15}{\blacksquare}$

6. $\frac{\blacksquare}{3} = \frac{28}{21}$

7. $\frac{5}{6} = \frac{\blacksquare}{24}$

8. $\frac{\blacksquare}{9} = \frac{2}{3}$

9. $\frac{5}{\blacksquare} = \frac{25}{30}$

10. $\frac{3}{4} = \frac{12}{\blacksquare}$

11. $\frac{\blacksquare}{6} = \frac{45}{54}$

12. $\frac{1}{2} = \frac{15}{\blacksquare}$

13. $\frac{3}{\blacksquare} = \frac{21}{49}$

14. $\frac{\blacksquare}{25} = \frac{3}{5}$

15. $\frac{24}{64} = \frac{\blacksquare}{16}$

✓ Solve Multiplication Equations (For Intervention, see p. H21.)

Solve.

16. $5x = 20$

17. $2p = 14$

18. $48 = 3k$

19. $45 = 9m$

20. $96 = 8b$

21. $65 = 2.5d$

22. $0.4n = 6$

23. $187 = 11y$

24. $\frac{1}{2}h = 21$

25. $7.2 = 0.03m$

26. $12 = \frac{2}{3}a$

27. $23h = 23$

✓ Congruent and Similar Figures (For Intervention, see p. H22.)

Decide whether the figures in each pair appear to be *similar*, *congruent*, *both*, or *neither*.

28.

29.

30.

31.

32.

33.

LOOK AHEAD

In Chapter 20 you will

- write ratios and rates
- use the strategy *write an equation* to solve problems
- use ratios to identify similar figures
- use proportions to find unknown measures
- read and use scales

Ratios and Rates

Learn how to write ratios and rates and how to find unit rates.

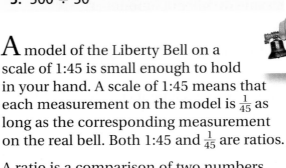

Vocabulary

equivalent ratios

rate

unit rate

A model of the Liberty Bell on a scale of 1:45 is small enough to hold in your hand. A scale of 1:45 means that each measurement on the model is $\frac{1}{45}$ as long as the corresponding measurement on the real bell. Both 1:45 and $\frac{1}{45}$ are ratios.

A ratio is a comparison of two numbers, a and b, and can be written as a fraction $\frac{a}{b}$. You can write a ratio in three ways.

Write: 1 to 45 or 1:45 or $\dfrac{1}{45} \begin{array}{l} \leftarrow \\ \leftarrow \end{array} \dfrac{\text{first term}}{\text{second term}}$ **Read:** one to forty-five

You can write a ratio to compare two amounts—a part to a part, a part to the whole, or the whole to a part.

EXAMPLE 1

The world's largest bells are made from "bell metal." Each 4 pounds of bell metal contains 3 pounds of copper and 1 pound of tin. Write the following ratios.

a. pounds of copper to pounds of tin → $\frac{3}{1}$ *part to part*

b. pounds of copper to total pounds → $\frac{3}{4}$ *part to whole*

c. total pounds to pounds of tin → $\frac{4}{1}$ *whole to part*

Equivalent ratios are ratios that name the same comparisons.

EXAMPLE 2

Write three equivalent ratios to compare the number of stars with the number of stripes.

$\dfrac{\text{number of stars}}{\text{number of stripes}} \rightarrow \dfrac{6}{9}$

$\dfrac{6}{9} \rightarrow \dfrac{6 \div 3}{9 \div 3} = \dfrac{2}{3}$ *Divide both terms by a common factor.*

$\dfrac{6}{9} \rightarrow \dfrac{6 \times 2}{9 \times 2} = \dfrac{12}{18}$ *Multiply both terms by the same number.*

So, $\frac{6}{9}, \frac{2}{3},$ and $\frac{12}{18}$ are equivalent ratios.

Math Idea ▶ A **rate** is a ratio that compares two quantities having different units of measure.

Suppose 24 ounces of cereal costs $4.32.

rate: $\dfrac{\text{price}}{\text{number of ounces}}$ → $\dfrac{\$4.32}{24 \text{ oz}}$ $4.32 for 24 oz

A **unit rate** is a rate that has 1 unit as its second term.

unit rate: $\dfrac{\$4.32}{24} = \dfrac{\$4.32 \div 24}{24 \div 24} = \dfrac{\$0.18}{1}$ $0.18 for 1 oz

The unit rate of the cereal is $0.18 per ounce.

Average speed is average distance traveled in 1 unit of time, such as 1 hour or 1 second. So, average speed is a unit rate.

EXAMPLE 3

The Howards are driving their car from Richmond, VA, to Philadelphia, PA, a distance of 250 mi. They have traveled 200 mi in 4 hr. At this rate, how long will it take them to make the complete trip?

$\dfrac{\text{miles}}{\text{hours}}$ → $\dfrac{200}{4}$ *Write a ratio to compare miles to hours.*

$\dfrac{200}{4} = \dfrac{200 \div 4}{4 \div 4} = \dfrac{50}{1}$ ← miles ← hours   *Find the unit rate, or average speed.*

$\dfrac{50}{1} = \dfrac{50 \times 5}{1 \times 5} = \dfrac{250}{5}$ ← miles ← hours   **THINK:** *50 × 5 = 250. Multiply each term by 5 to find the number of hours for 250 mi.*

So, it will take the Howards 5 hr to travel 250 mi.

CHECK FOR UNDERSTANDING

Think and Discuss ▶ Look back at the lesson to answer each question.

1. **Explain** how to find equivalent ratios.

2. **Tell** which is the better buy, 14 oz of cereal at $2.52 or 20 oz of cereal at $3.20. Explain.

Guided Practice ▶ Write two equivalent ratios.

3. $\dfrac{4}{8}$ 4. $\dfrac{6}{10}$ 5. $\dfrac{9}{12}$ 6. $\dfrac{10}{15}$

Write a ratio in fraction form. Then find the unit rate.

7. 150 points in 10 games   8. $64 in 16 hr

9. 90 words in 2 min   10. $2.56 for 4 pencils

11. $15 for 6 lb   12. 120 mi in 3 hr

Independent ▶
Practice

Write two equivalent ratios.

13. $\frac{2}{4}$ **14.** $\frac{4}{5}$ **15.** $\frac{3}{9}$ **16.** $\frac{15}{21}$

17. $\frac{6}{16}$ **18.** $\frac{21}{30}$ **19.** $\frac{5}{9}$ **20.** $\frac{14}{24}$

Write each ratio in fraction form. Then find the unit rate.

21. $15 for 5 tapes

22. 8 pages in 2.5 hr

23. 60 mi on 3 gal

24. $1.89 for 3 pens

25. $2.10 for 6 fish

26. 900 students for 30 teachers

For 27–28, use the figure at the right.

27. Find the ratio of red sections to blue sections. Then write three equivalent ratios.

28. Find the ratio of blue sections to all the sections. Then write three equivalent ratios.

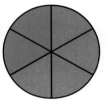

Find the missing term, *m*, that makes the ratios equivalent.

29. 5 to 4, as *m* to 12   **30.** 20:*m*, as 10:8   **31.** $\frac{9}{12}$, as $\frac{45}{m}$

32. Write the following ratio: number of tricycles to number of tricycle wheels.

33. Write the following ratio: number of tires to number of cars.

Problem Solving ▶
Applications

34. Of the 10 members on a team, 4 are girls. What is the ratio of girls to all the members? of boys to girls?

Use Data   For 35–38, use the table.

35. REASONING How can you decide if Brand B or Brand C is the better buy without dividing?

36. Which brand is the best buy?

37. Find the cost of 6 boxes of Brand D.

38.  **What's the Question?** The answer is $0.42 per box.

HOW JUICES COMPARE

Brand	Cost	Number of Boxes
A	$1.17	3
B	$1.23	3
C	$2.52	6
D	$3.15	9

MIXED STRATEGY PRACTICE

39. Describe the pattern. Then find the next term. (p. 336)   1, 4, 9, 16, ■

40. If $k = 4p - 3t$, what is the value of k when $p = 2$ and $t = {}^-2$? (p. 272)

41. Give the prime factorization of 54. (p. 148)

42. Write 45.6% as a decimal. (p. 160)

⭐ **43. TEST PREP**   Solve $n + 3 = {}^-5$. (p. 287)

     A $n = 8$ **B** $n = {}^-2$ **C** $n = {}^-6$ **D** $n = {}^-8$

ALGEBRA
Explore Proportions

Explore how to find equivalent ratios to form a proportion.

You need two-color counters.

Atoms are tiny building blocks of matter. Atoms join in equivalent ratios to form larger building blocks called molecules. For example, the ratio of oxygen atoms to hydrogen atoms in 1 molecule of water is $\frac{1}{2}$. The ratio of oxygen atoms to hydrogen atoms in 2 molecules of water is $\frac{2}{4}$, which is the same ratio as $\frac{1}{2}$.

These ratios can be used to write a proportion. A **proportion** is an equation that shows two equivalent ratios.

$$\frac{1}{2} = \frac{2}{4} \leftarrow \text{proportion}$$

You can use counters to model proportions.

Vocabulary

proportion

Activity

• Make the model shown here. Write the ratio of red counters to yellow counters.

• Separate the red counters into three equal groups. Do the same with the yellow counters. Write the ratio of red to yellow in each group.

• Separate the counters into six equal groups. Write the ratio of red to yellow in each group.

Think and Discuss

• One proportion shown by your model is $\frac{6}{12} = \frac{2}{4}$. What are two other proportions shown by your model?

• Is $\frac{4}{8} = \frac{3}{4}$ a proportion? Use your counters to explain your reasoning.

Practice

TECHNOLOGY LINK

More Practice: Use E-Lab, *Exploring Proportions.*
www.harcourtschool.com/
elab2002

1. How many hydrogen atoms combine with 4 oxygen atoms to form water? How many combine with 6 oxygen atoms? Model the ratios with counters.

Use models to determine whether the ratios form a proportion.

2. $\frac{1}{3}$ and $\frac{4}{12}$ **3.** $\frac{3}{4}$ and $\frac{6}{9}$ **4.** $\frac{2}{4}$ and $\frac{3}{6}$

PROBLEM SOLVING STRATEGY
Write an Equation

Learn how to use the strategy *write an equation* to solve problems.

QUICK REVIEW
1. 24 × 10   2. 30 × 7   3. 50 × 9
4. 400 ÷ 5   5. 300 ÷ 60

Remember that a *proportion* is an equation that shows two equivalent ratios.

The weight of an object depends on the mass and radius of the planet or moon where the object is located. A person who weighs 60 lb on Earth would weigh only 10 lb on the moon. The ratio $\frac{\text{weight on moon}}{\text{weight on Earth}}$ is the same for all objects. Danielle weighs 90 lb. How much would she weigh on the moon?

Analyze   What are you asked to find?

What information is given?

Is there information you will not use? If so, what?

Choose   What strategy will you use?

You can *write an equation* in the form of a proportion. Write two ratios that compare weight on Earth with weight on the moon. Use *n* to represent Danielle's weight on the moon.

Solve   Write and solve a proportion.

$$\frac{\text{weight on moon}}{\text{weight on Earth}} \rightarrow \frac{10}{60} = \frac{n}{90} \leftarrow \frac{\text{Danielle's weight on moon}}{\text{Danielle's weight on Earth}}$$

One Way   Use the Multiplication Property of Equality.

$$\frac{10}{60} = \frac{n}{90}$$ *Multiply both sides by 60 and 90.*

$$\frac{10}{60} \times \frac{\overset{1}{60}}{1} \times \frac{90}{1} = \frac{n}{90} \times \frac{60}{1} \times \frac{\overset{1}{90}}{1}$$ *Simplify.*

$$900 = 60n$$ *Solve the equation.*

$$\frac{900}{60} = \frac{60n}{60}$$

$$n = 15$$

Another Way   Use cross products.

$$\frac{60}{10} \diagdown \!\!=\!\! \diagup \frac{90}{n}$$ *Find the cross products.*

$$60 \times n = 90 \times 10$$

$$60n = 900$$ *Solve the equation.*

$$\frac{60n}{60} = \frac{900}{60}$$

$$n = 15$$

So, Danielle would weigh 15 lb on the moon.

Check   How are the two ways of solving the proportion alike?

What if Danielle weighed 78 lb? What would she weigh on the moon?

PROBLEM SOLVING PRACTICE

Solve the problem by writing an equation.

1. Astronaut Neil Armstrong, the first person to walk on the moon, weighed 165 lb on Earth. How much did he weigh on the moon?

2. Abdul's car uses about 10 gal of gasoline on a 180-mi trip. Predict how much gasoline Abdul will need to travel 540 mi.

3. A map uses a scale of 2 in. for every 9 mi. If the map shows a distance of 18 in., what is the actual distance?

For 4–5, use this information: A machine fills 4 bottles every 6 seconds. The machine fills 12 bottles.

4. Which proportion could you use to find s, the amount of time it takes to fill 12 bottles?

   A $\frac{4}{6} = \frac{s}{12}$ **C** $\frac{4}{6} = \frac{12}{s}$

   B $\frac{6}{4} = \frac{12}{s}$ **D** $\frac{4}{s} = \frac{6}{12}$

5. How long does it take the machine to fill 12 bottles?

   F 8 sec   **H** 16 sec

   G 12 sec   **J** 18 sec

PROBLEM SOLVING STRATEGIES

Draw a Diagram or Picture
Make a Model
Predict and Test
Work Backward
Make an Organized List
Find a Pattern
Make a Table or Graph
Solve a Simpler Problem
▶ **Write an Equation**
Use Logical Reasoning

MIXED STRATEGY PRACTICE

6. At 5:00 P.M. the temperature was 72°F. It fell at a steady rate. At 8:00 P.M. it was 54°F. If the pattern continues, what will the temperature be at 9:00 P.M.?

7. Debra is making 20 invitations. Each invitation is 6 in. long and 5 in. wide. How many square inches of paper does she need?

8. A farmer paid $40,000 for 25 acres of farmland. The farmer wants to buy 15 acres more at the same price per acre. How much will 15 acres cost?

9. Kyle has an average of 88 on four tests. What must he get on his fifth test to have an average of 90?

The histogram shows how many salespeople sold different numbers of cars in the first month of work. Use the histogram for 10–15.

10. How many salespeople sold more than 40 cars in the first month?

11. How many salespeople are represented by the histogram?

12. Did more salespeople sell 0–30 cars or 31–60 cars? How many more?

13. What is the greatest possible number of cars any one salesperson sold?

14. What if no salespeople had sold 31–40 cars? How would the histogram change?

15. ❓ **What's the Error?** Al said that four people sold exactly 21 cars each. What was his mistake?

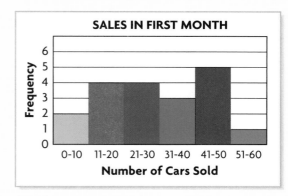

SALES IN FIRST MONTH

389

LESSON 20.4

ALGEBRA
Ratios and Similar Figures

Learn how to use ratios to identify similar figures.

QUICK REVIEW

Simplify.

1. $\frac{6}{8}$ **2.** $\frac{6}{12}$ **3.** $\frac{8}{12}$ **4.** $\frac{15}{18}$ **5.** $\frac{9}{15}$

Vocabulary

corresponding sides

corresponding angles

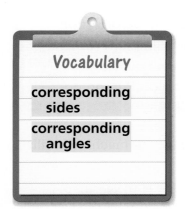

The larger soccer ball and the soccer ball on the key ring have similar black pentagons on them. What other similar polygons can you find on the soccer balls?

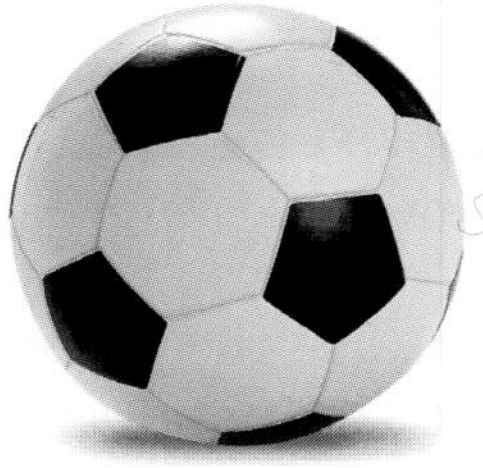

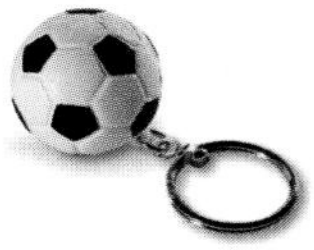

Similar figures have **corresponding sides** and **corresponding angles**. The corresponding sides and angles for similar rectangles *ABCD* and *EFGH* are shown below.

Remember that similar figures are 2 or more figures that have the same shape.

$\overline{AB}$ corresponds to $\overline{EF}$.
$\overline{BC}$ corresponds to $\overline{FG}$.
$\overline{CD}$ corresponds to $\overline{GH}$.
$\overline{DA}$ corresponds to $\overline{HE}$.

$\angle A$ corresponds to $\angle E$.
$\angle B$ corresponds to $\angle F$.
$\angle C$ corresponds to $\angle G$.
$\angle D$ corresponds to $\angle H$.

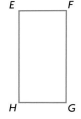

It is usually easier to find the corresponding sides and corresponding angles of similar figures if they are oriented the same way on the page.

Activity

MATH LAB

You need: metric ruler

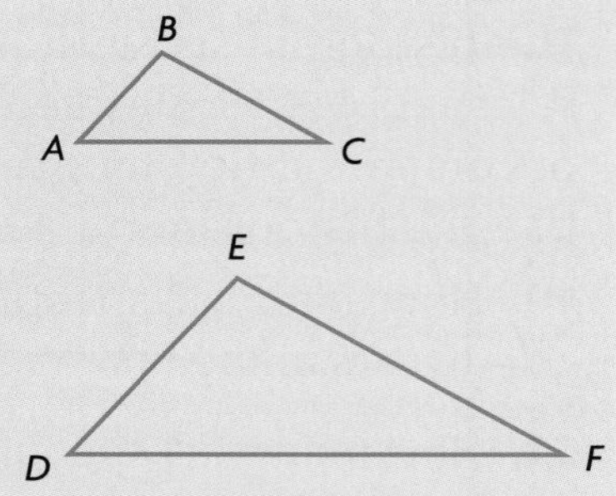

- Do these triangles appear to be similar? Explain.
- Measure and record the lengths of the sides of both triangles in cm.
- Write the ratios *AB* to *DE*, *BC* to *EF*, and *AC* to *DF*.
- Write each of the ratios in simplest form. What do you notice about the ratios?
- Measure and record the angles in both triangles.
- What do you notice about the angle measures?

Math **I**dea ▶ If figures are similar, their corresponding angles are congruent, and the lengths of their corresponding sides have the same ratio.

△MNP and △HJK are similar. Find the measures of ∠M, ∠N, and ∠P and the ratio of the lengths of the corresponding sides.

∠M corresponds to ∠H, so ∠M measures 35°.

∠N corresponds to ∠J, so ∠N measures 23°.

∠P corresponds to ∠K, so ∠P measures 122°.

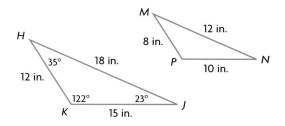

$\overline{MN}$ corresponds to $\overline{HJ}$, $\overline{NP}$ corresponds to $\overline{JK}$, and $\overline{PM}$ corresponds to $\overline{KH}$.

$\frac{MN}{HJ} \rightarrow \frac{12}{18} = \frac{2}{3}$ $\frac{NP}{JK} \rightarrow \frac{10}{15} = \frac{2}{3}$ $\frac{PM}{KH} \rightarrow \frac{8}{12} = \frac{2}{3}$

So, the ratio of the lengths of corresponding sides is $\frac{2}{3}$.

For two polygons to be similar, the ratios of the corresponding sides must be equal, and the corresponding angles must be congruent.

EXAMPLE 2

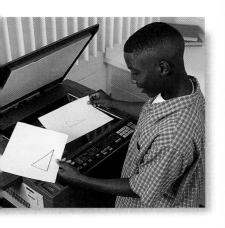

Sean used a copy machine to make an enlargement of △ABC. He labeled his new triangle △DEF. Are the triangles similar?

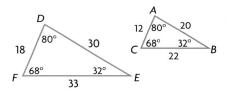

$\frac{DE}{AB} \rightarrow \frac{30}{20} = \frac{3}{2}$ $\frac{EF}{BC} \rightarrow \frac{33}{22} = \frac{3}{2}$ *Write the ratios of the corresponding sides in simplest form.*

$\frac{FD}{CA} \rightarrow \frac{18}{12} = \frac{3}{2}$

$\frac{3}{2} = \frac{3}{2} = \frac{3}{2}$ *Compare the ratios.*

∠D and ∠A are congruent.   *Check whether corresponding*
∠E and ∠B are congruent.   *angles are congruent.*
∠F and ∠C are congruent.

The ratios of the corresponding sides are equal and the corresponding angles are congruent.

So, △ABC is similar to △DEF.

Think and ▶
Discuss

Look back at the lesson to answer each question.

1. **Tell** whether the ratio for corresponding sides in Example 2 could be $\frac{2}{3}$. Explain your decision.

2. **Choose** the one correct statement:
   A All similar figures are also congruent.
   B All congruent figures are also similar.

Guided ▶
Practice

3. Name the corresponding sides and angles. Write the ratio of the corresponding sides in simplest form.

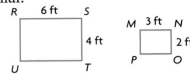

Independent ▶
Practice

Name the corresponding sides and angles. Write the ratio of the corresponding sides in simplest form.

4.

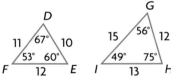

5.

Tell whether the figures in each pair are similar. Write *yes* or *no*. If you write *no*, explain.

6.

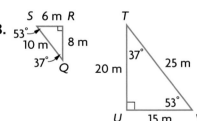

7.

8.

9.

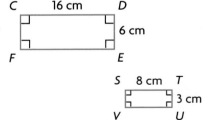

The figures in each pair are similar. Find the unknown measures.

10.

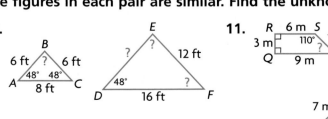

11.

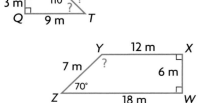

Write _yes_ or _no_. Explain your answers.

12. Are all squares similar?

13. Are all rectangles similar?

14. Are all rhombi similar?

15. Are all right triangles similar?

Problem Solving ▶
Applications

Use the following information for 16–17: The rectangular front of the North End Mall is 60 ft tall and 80 ft wide. The front of the Eastside Mall has sides that are $\frac{3}{4}$ the lengths of the sides of the North End Mall.

16. Are the fronts similar?

17. Find the dimensions of the front of the Eastside Mall.

18. Most basketball courts are 94 ft long and 50 ft wide. A half court is 47 ft by 50 ft. Is a half court similar to a full court? Explain.

19. ✍ **Write About It**
Explain how you know that *ABCD* is similar to *AEFG*.

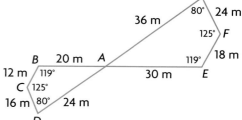

MIXED REVIEW AND TEST PREP

20. What is the unit rate if 5 cans cost $3? (p. 384)

21. If $5y = 15$, what does y equal? (p. 297)

22. How many faces does a rectangular pyramid have? (p. 350)

23. TEST PREP Evaluate $4x + x + 9$ for $x = 3$. (p. 272)
  A 19   **B** 21   **C** 24   **D** 25

24. TEST PREP Evaluate $7^2 - \overline{)49}$. (p. 278)
  F ⁻42   **G** 0   **H** 42   **J** 56

PROBLEM SOLVING LiNKUP to Careers

Artists, photographers, designers, and architects are among the many professionals who use similar figures in their occupations.

1. ❓ **What's the Error?** A photographer took the photo at the right. The photo was then reduced in size. What error did the photographer make in reducing the photo?

2. Suppose that an artist wants to enlarge a drawing that is octagonal in shape. How will the angles of the enlargement be related to those of the original drawing? How will the sides be related?

EXTRA PRACTICE page H51, Set B

393

ALGEBRA
Proportions and Similar Figures

Learn how to use proportions and similar figures to find unknown measures.

Vocabulary

indirect measurement

Arizona's Grand Canyon is more than 1 mile deep and 217 miles long. It has been called one of the Seven Natural Wonders of the World. Each year, tourists snap millions of photos of the canyon. A photo is similar to the scene it depicts.

If two figures are similar and you know the length of one side, you can use a proportion to find the length of another side.

EXAMPLE 1

Remember that a proportion is an equation which states that two ratios are equivalent.

Armando made an enlargement that is similar to a 4-in. × 6-in. photo he took of the Grand Canyon. The enlargement is 12 in. wide. How long is it?

Write the ratios of the corresponding sides. Let x = the length of $\overline{EF}$.

$$\frac{AB}{EF} \to \frac{6}{x} \qquad \frac{AD}{EH} \to \frac{4}{12}$$

$$\frac{6}{x} = \frac{4}{12}$$ *Use the ratios to write a proportion.*

$4 \cdot x = 6 \cdot 12$ *Find the cross products.*

$4x = 72$ *Solve the equation.*

$$\frac{4x}{4} = \frac{72}{4}$$

$$x = 18$$

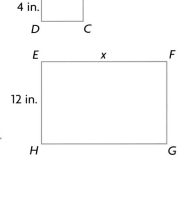

So, the enlargement is 18 in. long.

Math **I**dea ▶ In Example 1, similar figures and a proportion were used to find an unknown length. This method of finding a measurement is called **indirect measurement**. It can be used to find lengths and distances that are too great to be measured with a ruler.

EXAMPLE 2

On a sunny day, one saguaro cactus casts a shadow that is 56 ft long. At the same time, a yardstick casts a shadow that is 4 ft long. Use the similar right triangles shown below to find the height of the cactus.

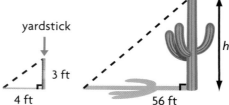

$$\frac{3}{h} = \frac{4}{56} \quad \begin{array}{l} \leftarrow \text{ small triangle} \\ \leftarrow \text{ large triangle} \end{array} \qquad \textit{Write a proportion.}$$

$$4 \times h = 3 \times 56 \qquad\qquad \textit{Find the cross products.}$$
$$4h = 168 \qquad\qquad\quad\;\; \textit{Solve the equation.}$$
$$\frac{4h}{4} = \frac{168}{4}$$
$$h = 42$$

So, the cactus is 42 ft tall.

The saguaro cactus, the world's tallest cactus, is native to Arizona.

CHECK FOR UNDERSTANDING

Think and ▶ Discuss

Look back at the lesson to answer each question.

1. **REASONING** **Explain** why you can use similar figures to measure objects indirectly.

2. **Name** three objects that would be difficult to measure directly.

3. **Determine** how long the picture in Example 1 would be if it were 8 in. wide.

Guided ▶ Practice

The figures are similar. Write a proportion. Then find the unknown length.

4.

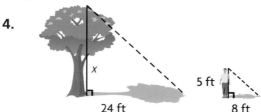

5.

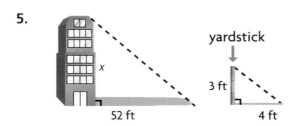

Independent ▶
Practice

The figures are similar. Write a proportion. Then find the unknown length.

6.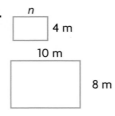

7.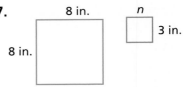

8.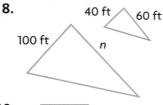

9.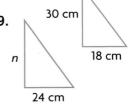

10.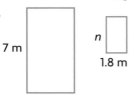

11.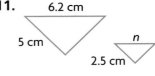

Problem Solving ▶
Applications

12. REASONING Two common envelope sizes are $3\frac{1}{2}$ in. × $6\frac{1}{2}$ in. and 4 in. × $9\frac{1}{2}$ in. Are these envelopes similar? Explain.

13. Art The famous painting known as "Red Interior, Still-life on a Blue Table" was painted by Henri Matisse in 1947. The copy of the painting at the right measures 1 in. tall by $\frac{3}{4}$ in. wide. The height of the real painting is 32 in. Find the width of the real painting.

14. **?** **What's the Error?** The shadow of a 6-ft man measures 5 ft at the same time his son's shadow is 2 ft. Taylor used a proportion to find the son's height as $1\frac{2}{3}$ ft. Explain Taylor's mistake.

MIXED REVIEW AND TEST PREP

15. Triangle A's sides measure 2 cm, 3 cm, and 4 cm. Triangle B's sides measure 6 cm, 7 cm, and 8 cm. Are the triangles similar? Explain. (p. 390)

16. The base of a prism has 20 sides. How many vertices does the prism have? (p. 350)

For 17–18, use the box-and-whisker graph.

17. What is the median? (p. 130)

18. What is the range? (p. 130)

19. TEST PREP Which value for p solves the equation $0.4p = 20$? (p. 297)

    A 5   **B** 19.6   **C** 20.4   **D** 50

EXTRA PRACTICE page H51, Set C

ALGEBRA
Scale Drawings

Learn how to use scale to find the dimensions for a drawing or the actual dimensions of an object.

Vocabulary

scale drawing

scale

1. 490 ÷ 7 **2.** 800 ÷ 10 **3.** 540 ÷ 6 **4.** 3 × 120 **5.** 70 × 11

Floor plans, road maps, and diagrams are examples of scale drawings. A **scale drawing** is a drawing of a real object that is smaller than (a reduction of) or larger than (an enlargement of) the real object. Measurements on a scale drawing are proportional to measurements of the real object.

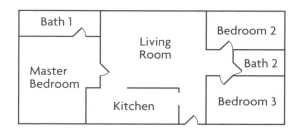

The above floor plan and other types of architectural drawings are called blueprints. A blueprint can be a reduction or an enlargement of an object.

TECHNOLOGY LINK

More Practice: Use **Mighty Math Cosmic Geometry**, *Robot Studio*, Level F.

Activity

You need: centimeter graph paper, metric ruler

• Measure the length and width of the scale drawing above in centimeters.

• On graph paper, make a drawing with measurements that are twice those in the scale drawing.

• Write the ratio of length to width for the original drawing.

• Write the ratio of length to width in your drawing. Are the two ratios equal? Are the drawings similar? Explain.

• Make a drawing with measurements that are half those in the original scale drawing.

• Write the ratio of length to width for your new drawing. Is the ratio equal to the ratio for the original scale drawing? Are the drawings similar? Explain.

• Suppose you made a drawing of the floor plan that was 21 cm long. If your drawing was similar to the floor plan, how wide would it be? Explain.

A **scale** is a ratio between two sets of measurements. The scale 1 in. = 3 ft on this drawing of a bicycle means that 1 in. on the drawing represents 3 ft on the actual bike.

scale: 1 in. = 3 ft

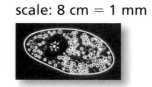

TECHNOLOGY LINK

More Practice: Use E-Lab, *Stretching and Shrinking.* www.harcourtschool.com/ elab2002

EXAMPLE 1

Measure the length of the bike in the scale drawing. Then use the scale to find the actual length of the bike.

scale drawing length: 2 in.

$$\underset{\text{actual (ft)}}{\overset{\text{drawing (in.)}}{}} \rightarrow \overset{\text{scale}}{\frac{1}{3}} = \overset{\text{bike}}{\frac{2}{n}}$$ *Write a proportion. Let n represent the actual length of the bike.*

$1 \times n = 3 \times 2$ *Find the cross products and solve.*

$n = 6$

So, the actual length of the bike is 6 ft.

If an object is very small, a scale drawing might be larger than the object.

EXAMPLE 2

A paramecium is a tiny organism whose entire body consists of a single cell. Use the scale and the scale drawing to find the length of an average-size paramecium.

scale: 8 cm = 1 mm

scale drawing length: 2.4 cm

$$\underset{\text{actual (mm)}}{\overset{\text{drawing (cm)}}{}} \rightarrow \overset{\text{scale}}{\frac{8}{1}} = \overset{\text{paramecium}}{\frac{2.4}{n}}$$ *Write a proportion. Let n represent the actual length of the paramecium.*

$8 \times n = 1 \times 2.4$ *Find the cross products and solve.*

$n = \frac{2.4}{8}$

$n = 0.3$

So, an average paramecium is about 0.3 mm long.

CHECK FOR UNDERSTANDING

Think and Discuss ▶ **Look back at the lesson to answer each question.**

1. **Compare** the actual length of the bike in Example 1 with the length of the bike in the scale drawing.

2. **Make** a scale drawing of a square microchip that has a side of 2 mm. Use a scale of 4 cm:1 mm. How long will the side on your drawing be?

3. **Explain** how the scale drawing of a microchip is different from the scale drawing of the bicycle.

Find the unknown dimension.

4. scale: 1 in.:8 ft
drawing length: 3 in.
actual length: ■ ft

5. scale: 2 in.:7 yd
drawing length: 8 in.
actual length: ■ yd

PRACTICE AND PROBLEM SOLVING

Independent ▶
Practice

Find the unknown dimension.

6. scale: 1 in.:4 ft
drawing length: ■ in.
actual length: 28 ft

7. scale: 1 in.:6 ft
drawing length: 5 in.
actual length: ■ ft

8. scale: 3 in.:4 ft
drawing length: ■ in.
actual length: 14 ft

9. scale: 4 cm:3 mm
drawing length: ■ cm
actual length: 15 mm

10. scale: 2 in.:5 ft
drawing length: $\frac{1}{2}$ in.
actual length: ■ ft

11. scale: 6 cm:3 mm
drawing length: ■ cm
actual length: 2 mm

Problem Solving ▶
Applications

12. A hallway is 15 ft long. How long is it on a floor plan drawn to the scale 2 in. = 3 yd?

13. A model of a storage shed is shown at the right. The model measures 8 in. × 3 in. × 4 in. How many cubical boxes measuring 2 ft on a side can be stacked in the actual shed?

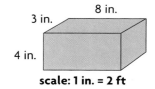

scale: 1 in. = 2 ft

14. Astronomy Gail wanted to make a scale drawing of the solar system. She drew the sun in the center with a diameter of 1 in. The actual diameter of the sun is about 900,000 miles. What problem did Gail run into in drawing the planet Pluto at its average distance of 3.6 billion miles from the sun?

15. ✎ **Write About It** Explain what the scale 1 cm = 2 m on a scale drawing means.

MIXED REVIEW AND TEST PREP

16. Two rectangles measure 9 cm by 12 cm and 12 cm by 16 cm. Are the rectangles similar? (p. 390)

17. True or false? A cone is a polyhedron. (p. 350)

For 18–19, use the stem-and-leaf plot at the right.

18. What is the greatest data value shown? (p. 126)

19. What is the mode of the data? (p. 126)

Stem	Leaves
3	0 0 3 6 9
4	2 3 8
5	3 6 7

⭐**20. TEST PREP** Ted recorded these scores on 10-point history quizzes: 8, 7, 10, 9, 7, 7. What was his median score? (p. 106)

A 3 **B** 7 **C** 7.5 **D** 8

ALGEBRA
Maps

Learn how to read and use map scales.

Mapmakers are called cartographers. Cartographers use ratios, proportions, and scales to convert actual distances on Earth to map distances. You can use the same method to convert map distances into real distances.

EXAMPLE

Use the Maryland map to determine the actual straight-line distance from Washington, DC, to Columbia, MD. The map scale is 1 in. = 20 mi.

Use a ruler to measure the distance from Washington to Columbia.

map distance → $1\frac{1}{4}$, or 1.25 in.

Let n represent the actual distance in miles. Write a proportion.

$$\frac{\text{in.}}{\text{mi}} \rightarrow \frac{1}{20} = \frac{1.25}{n}$$

Find the cross products.

$1 \times n = 1.25 \times 20$

Multiply.

$n = 25$

So, it is 25 mi from Washington to Columbia.

- Philadelphia, PA, is about 130 mi from Washington. If you were a cartographer using the same scale, how far from Washington would you place Philadelphia?

CHECK FOR UNDERSTANDING

Think and Discuss

Look back at the lesson to answer each question.

1. **Find** the actual straight-line distance from Washington, DC, to Baltimore, MD.

Guided Practice

2. **Tell** the map distance that represents 90 mi on the Maryland map.

The map distance is given. Find the actual distance. Use a map scale of 1 in. = 100 mi.

3. $\frac{1}{2}$ in.   4. 5 in.   5. $4\frac{1}{4}$ in.

Independent ▶
Practice

The map distance is given. Find the actual distance. The scale is 1 in. = 80 mi.

6. $1\frac{1}{2}$ in. **7.** 3 in. **8.** $2\frac{1}{2}$ in. **9.** $5\frac{1}{4}$ in.

10. $\frac{1}{2}$ in. **11.** 18 in. **12.** $2\frac{3}{4}$ in. **13.** $4\frac{5}{8}$ in.

The actual distance is given. Find the map distance. The scale is 1 in. = 25 mi.

14. 100 mi **15.** 150 mi **16.** $12\frac{1}{2}$ mi **17.** $62\frac{1}{2}$ mi

Problem Solving ▶
Applications

For 18–20, use the map of Colorado. The scale is $\frac{1}{2}$ in. = 46 mi.

18. Find the straight-line distance in miles from Pueblo to Denver and then from Denver to Steamboat Springs.

19. Stefano drove from Denver to Vail. If he drove 55 mi per hour, about how long did the trip take him?

20. Fran drove for 2.3 hours at 50 mi per hour. If you drew her route as a line on the Colorado map, how long would it be?

21. **REASONING** The auto club map of Iowa has a scale of 1 in. = 20 mi. The tourist map has a scale of 1 in. = 40 mi. On which map does the state of Iowa appear larger?

22. ✎ **Write About It** Explain how you can find an actual distance between two cities if you know the map distance and the scale.

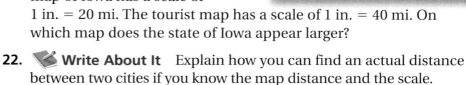

MIXED REVIEW AND TEST PREP

23. Scale: 1 in.:5 ft, actual length: 20 ft, drawing length: ___?___. (p. 397)

24. How many bases does a cone have? (p. 350)

For 25–26, use the line plot at the right. (p. 102)

25. What is the mode of the data?

26. What is the median of the data?

27. TEST PREP Which fraction is greatest? (p. 166)

A $\frac{3}{8}$ **B** $\frac{7}{24}$ **C** $\frac{5}{12}$ **D** $\frac{1}{3}$

1. **VOCABULARY** A comparison of two numbers is a(n) __?__. (p. 384)

2. **VOCABULARY** An equation that shows two equivalent ratios is a(n) __?__. (p. 387)

3. **VOCABULARY** A __?__ is a ratio between two sets of measurements. (p. 398)

Write each ratio in fraction form. (pp. 384–386)

4. white sections to all sections

5. yellow sections to white sections

6. all sections to yellow sections

Write each ratio in fraction form. Then find the unit rate. (pp. 384–386)

7. $0.75 for 3

8. 210 km in 3 hr

9. $4.00 for 10

10. 20 mi in 5 hr

Tell whether the figures in each pair are similar. Write *yes* or *no*. If you write *no*, explain. (pp. 390–393)

11.
15 cm
6 cm
5 cm
3 cm

12.
10 ft 99° 8 ft
34° 14 ft 47°
25 ft 99° 20 ft
34° 35 ft 47°

The figures are similar. Write a proportion. Then find the unknown length. (pp. 394–396)

13.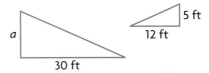
a
30 ft
5 ft
12 ft

14.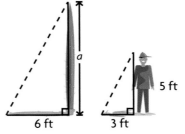
a
6 ft
5 ft
3 ft

Find the unknown dimension. (pp. 397–399)

15. scale: 2 in. = 5 ft
drawing length: 8 in.
actual length: ▪ ft

16. scale: 2 in. = 5 ft
drawing length: ▪ in.
actual length: 40 ft

17. scale: 2 in. = 5 ft
drawing length: 13 in.
actual length: ▪ ft

Use a map scale of 1 in. = 150 mi. Find the actual distance. (pp. 400–401)

18. 5 in.

19. 12 in.

20. $3\frac{1}{2}$ in.

21. $6\frac{3}{4}$ in.

22. $\frac{1}{2}$ in.

23. 0.4 in.

Solve. (pp. 384–386)

24. Beach towels are on sale at 2 for $18. Rich buys 5 of them. How much does he pay?

25. Oranges are selling for $3.00 a dozen. Find the cost of 2 oranges.

Decide on a plan.

See item **6.**

Use the ratios of width to height of the model and width to the unknown height to write a proportion. Find the cross products and solve the equation.

Also see problem **4**, p. H63.

For 1–10, choose the best answer.

1. Tracy drove 400 kilometers in 8 hours. How many kilometers per hour is that?

A 5 km per hr   **C** 392 km per hr
B 50 km per hr   **D** 1,200 km per hr

2. At 6:00 A.M., the temperature was ⁻4°F. By 6:00 P.M., the temperature had risen 17°. What was the temperature at 6:00 P.M.?

F ⁻17°F   **H** 3°F
G ⁻13°F   **J** 13°F

3. Sonny uses 68 strawberry plants to fill 4 rows in his garden. Which proportion could he use to find the number of plants he needs for 3 more of these rows?

A $\frac{4}{68} = \frac{1}{x}$ **C** $\frac{4}{68} = \frac{3}{x}$

B $\frac{4}{3} = \frac{x}{68}$ **D** $\frac{4}{68} = \frac{x}{3}$

4. If 6 pairs of socks cost $4.50, how much will 9 pairs cost?

F $6.75   **H** $9.00
G $7.25   **J** Not here

5. Rectangles *ABCD* and *GHJK* are similar. Which of the following ratios is equivalent to $\frac{AB}{GH}$?

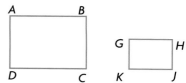

A $\frac{CD}{JK}$ **C** $\frac{AD}{KJ}$

B $\frac{AB}{BC}$ **D** $\frac{JK}{CB}$

6. A model of a building is 5 inches wide and 12 inches high. The actual building is 120 feet wide. How high is the actual building?

F 24 ft   **H** 192 ft
G 50 ft   **J** 288 ft

7. The field goal averages of four players are 0.309, 0.234, 0.312, and 0.289. Which shows these averages in order from least to greatest?

A 0.234, 0.289, 0.312, 0.309
B 0.312, 0.309, 0.289, 0.234
C 0.234, 0.289, 0.309, 0.312
D 0.289, 0.312, 0.234, 0.309

8. What is the range of the data shown below?

PRICES OF DAYPACKS ($)							
21	27	31	26	22	27	20	19
28	31	25	32	20	28	25	20

F 8   **H** 15
G 13   **J** 17

Write What You Know

9. Diane can type 248 words in 4 minutes. Define *unit rate* and find the unit rate for Diane. Then find how long it would take Diane to type 558 words. Describe how you found your answer.

10. The scale on a map is 1 in. = 40 miles. The town you are driving to is 3 inches away on the map. If you average 50 miles per hour, how long will it take you to drive to this town? Explain the steps you used to find your answer.

CHAPTER 21

Percent and Change

Most states, counties, and cities use sales tax and income tax to raise money. Sales tax is given as a percent. Alaska, Delaware, Montana, New Hampshire, and Oregon have no state sales tax. All other states have sales taxes varying from 3% to 7%.

PROBLEM SOLVING Suppose you have $65 and you want to buy a dictionary for $18.75, a map for $6.00, a magazine for $4.50, and a book on tape for $34.00. The sales tax rate is 6.25%. Do you have enough money to buy all of the items? Why or why not?

DATA LINK

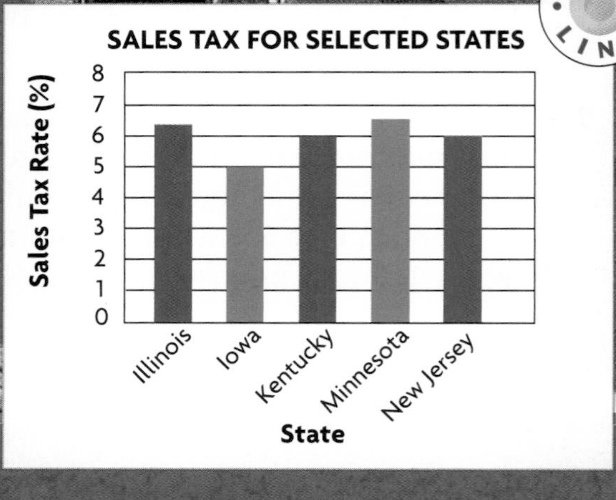

SALES TAX FOR SELECTED STATES

Sales Tax Rate (%)

Illinois Iowa Kentucky Minnesota New Jersey

State

Check What You Know

Use this page to help you review and remember important skills needed for Chapter 21.

 Vocabulary

Choose the best term from the box.

> equivalent
> proportion
> ratio

1. A comparison of two numbers is a(n) __?__ .

2. An equation that shows two equivalent ratios is a(n) __?__ .

 Write Fractions as Decimals (For Intervention, see p. H22.)

Write each fraction as a decimal.

3. $\frac{1}{4}$ 4. $\frac{3}{5}$ 5. $\frac{1}{10}$ 6. $\frac{7}{20}$ 7. $\frac{4}{5}$

8. $\frac{9}{10}$ 9. $\frac{1}{5}$ 10. $\frac{13}{25}$ 11. $\frac{81}{100}$ 12. $\frac{31}{50}$

 Multiply with Decimals and Fractions (For Intervention, see p. H23.)

Find the product.

13. $\begin{array}{r} 3.6 \\ \times 4.8 \\ \hline \end{array}$ 14. $\begin{array}{r} 9.2 \\ \times 4.9 \\ \hline \end{array}$ 15. $\begin{array}{r} 6.5 \\ \times 4.1 \\ \hline \end{array}$ 16. $\begin{array}{r} 2.7 \\ \times 4.3 \\ \hline \end{array}$

17. $\begin{array}{r} 72 \\ \times 0.4 \\ \hline \end{array}$ 18. $\begin{array}{r} 22.4 \\ \times 0.16 \\ \hline \end{array}$ 19. $\begin{array}{r} 19 \\ \times 1.8 \\ \hline \end{array}$ 20. $\begin{array}{r} 0.043 \\ \times\ \ 240 \\ \hline \end{array}$

Find the product. Write the answer in simplest form.

21. $\frac{1}{2} \times \frac{1}{3}$ 22. $\frac{3}{8} \times \frac{3}{5}$ 23. $\frac{2}{3} \times \frac{1}{7}$ 24. $\frac{4}{5} \times \frac{1}{6}$

25. $\frac{3}{4} \times \frac{4}{5}$ 26. $\frac{4}{9} \times \frac{3}{8}$ 27. $\frac{6}{7} \times \frac{8}{9}$ 28. $\frac{5}{12} \times \frac{3}{10}$

 Percents and Decimals (For Intervention, see p. H23.)

Write the percent as a decimal or the decimal as a percent.

29. 45%   30. 0.8

31. 0.14   32. 73%

33. 40%   34. 0.39

35. 53%   36. 125%

37. 100%   38. 1.4

39. 3%   40. 0.007

> ### LOOK AHEAD
>
> **In Chapter 21 you will**
> - write ratios as percents
> - convert among percents, decimals, and fractions
> - find percents
> - construct circle graphs
> - find discounts, sales tax, and interest

Percent

Learn how to write ratios as percents and express parts of a whole area as percents.

Twenty out of 100 teens have a checking account, according to a survey. You can express the ratio 20 out of 100 as a percent.

Math Idea ▶ A percent is the ratio of a number to 100. Percent, %, means "per hundred." So, 20% of teens have a checking account.

EXAMPLE 1

What percent of the squares are shaded?

$\frac{\text{shaded} \rightarrow}{\text{total} \rightarrow} \frac{46}{100}$ *Write the ratio of shaded squares to total squares.*

$\frac{46}{100} = 46\%$ *Write the ratio as a percent.*

So, 46% of the squares are shaded.

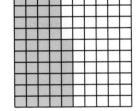

You can write a fraction as a percent.

EXAMPLE 2

Sheri has $25 in her checking account. She writes a check for $9. What percent of the total is the check amount?

$\frac{\text{check amount} \rightarrow}{\text{total amount} \rightarrow} \frac{9}{25}$ *Write a ratio in fraction form.*

$\frac{9}{25} \times \frac{4}{4} = \frac{36}{100} = 36\%$ *Write an equivalent fraction with a denominator of 100. Then write the percent.*

So, the check amount is 36% of the total in the checking account.

You can compare percents just as you compare other numbers.

EXAMPLE 3

The average weekly allowance for 9–11-year-olds in a recent survey was $5. Gina gets 85% of the average. Ricky gets 60% of the average. Compare the percents.

$85 > 60$, so $85\% > 60\%$.

Allowance Amounts, 9–11-year-olds

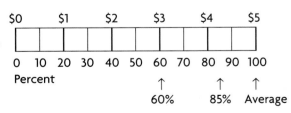

Think and ▶ Discuss

Look back at the lesson to answer the question.

1. **Draw** a model to show 63 out of 100. Then write the ratio as a percent.

Guided ▶ Practice

Write the percent that is shaded.

2.   **3.**   **4.**

Write as a percent.

5. $\frac{31}{100}$ **6.** $\frac{125}{100}$ **7.** $\frac{6}{10}$ **8.** $\frac{43}{50}$

Compare. Write $<$, $>$, or $=$ for each ⬤.

9. 30% ⬤ 25% **10.** 99% ⬤ 100% **11.** 54% ⬤ 54.0%

Independent ▶ Practice

Write the percent that is shaded.

12.   **13.**   **14.**

Write as a percent.

15. $\frac{18}{100}$ **16.** $\frac{150}{100}$ **17.** $\frac{21}{25}$ **18.** $\frac{13}{20}$

Compare. Write $<$, $>$, or $=$ for each ⬤.

19. 6% ⬤ 12% **20.** 50.0% ⬤ 50% **21.** 0.4% ⬤ 0.04%

Problem Solving ▶ Applications

22 Out of 100 students in the cafeteria, 38 wanted chicken and 29 wanted roast beef. What percent did not want either one?

23. The Geckos won 4 out of their 8 games in April, 7 out of 9 in May, and 5 out of 8 in June. What percent of all their games did they win?

24. ✎ **Write a problem** that involves money and can be solved using percents.

MIXED REVIEW AND TEST PREP

25. If a map scale is 1 in. = 60 mi, what map distance represents 150 mi? (p. 400)

26. Write an expression for the phrase "21 decreased by y." (p. 270)

27. $1.5 + 0.51 + 11 + 0.005$ (p. 66)

28. $\frac{9}{16} - \frac{3}{8}$ (p. 182)

29. TEST PREP What is the solution of the equation $2 = 8 + n$? (p. 287)

A $n = \frac{1}{4}$ **B** $n = 4$ **C** $n = 6$ **D** $n = {}^-6$

Percents, Decimals, and Fractions

Learn how to convert among percents, decimals, and fractions.

Every day, data appear in newspapers, on television, and on the Internet. Some data are given in decimal form, some in fraction form, and some in percent form. To fully understand the data you see in your everyday life, you should be able to change from one number form to the others.

Below are two ways to write a decimal as a percent.

One Way   Use place value.

EXAMPLE 1

Only about 0.2 of the people who use the Internet are younger than 18. Write 0.2 as a percent.

$0.2 = \frac{2}{10}$ — *Use place value to express the decimal as a ratio in fraction form.*

$= \frac{2 \times 10}{10 \times 10} = \frac{20}{100}$ — *Write an equivalent fraction with a denominator of 100.*

$= 20\%.$ — *Since percent is the ratio of a number to 100, write the ratio as a percent.*

• Write 0.6 as a percent.

Another Way   Multiply by 100.

EXAMPLE 2

Write the decimal as a percent.

Remember that when multiplying decimal numbers by powers of 10, you move the decimal point one place to the right for each factor of 10.

A. 0.2
   $0.2 = 0.20 = 20\%$ — *Multiply by 100 by moving the decimal point two places to the right.*

B. 0.87
   $0.87 = 0.87 = 87\%$ — *Multiply by 100 by moving the decimal point two places to the right.*

• Write 0.3 and 0.94 as percents.

Write Fractions as Percents

Below are two ways to write a fraction as a percent.

One Way Write an equivalent fraction with a denominator of 100.

EXAMPLE 3

In 1999, about $\frac{2}{5}$ of small businesses in the United States had websites on the Internet. What percent of small businesses had websites?

$$\frac{2}{5} = \frac{2 \times 20}{5 \times 20} = \frac{40}{100}$$ *Write an equivalent fraction with a denominator of 100.*

$$= 40\%$$ *Since percent is the ratio of a number to 100, write the ratio as a percent.*

So, about 40% of small businesses had websites in 1999.

Another Way Use division to write the fraction as a decimal.

$$\frac{2}{5} \rightarrow \begin{array}{r} 0.40 \\ 5\overline{)2.00} \end{array}$$ *Divide the numerator by the denominator.*

$$0.40 = 40\%$$ *Multiply by 100 by moving the decimal point two places to the right.*

You can also write a percent as a fraction.

EXAMPLE 4

Remember that *percent* means "per hundred."

In 1999, about 52% of the world's Internet users lived in the United States. Write 52% as a fraction.

$$52\% = \frac{52}{100}$$ *Write the percent as a fraction with a denominator of 100.*

$$\frac{52}{100} = \frac{52 \div 4}{100 \div 4} = \frac{13}{25}$$ *Write the fraction in simplest form.*

So, 52% written as a fraction is $\frac{13}{25}$.

You can write a percent as an equivalent decimal.

EXAMPLE 5

The number of websites on the Internet in May 1999 was 218% of the websites two years earlier. Write 218% as a decimal.

$$218\% = \frac{218}{100}$$ *Write the percent as a fraction with a denominator of 100.*

$$= 2.18$$ *Write the fraction as a decimal.*

So, 218% written as a decimal is 2.18.

The key sequence below shows how to change a percent to a decimal or a fraction by using a calculator.

70 100

409

Sometimes it takes several steps to rewrite in a different form a percent that is less than 1.

EXAMPLE 6

In 2000, about 30% of people in the United States used the Internet. In China, about 0.8% of the people were Internet users. Write 0.8% as a fraction and as a decimal.

To write 0.8% as a fraction, recall that *percent* means "per hundred."

$$0.8\% = \frac{0.8}{100}$$ *Write the percent as a fraction with a denominator of 100.*

$$\frac{0.8}{100} \times \frac{10}{10} = \frac{8}{1,000}$$ *Multiply numerator and denominator by 10 to remove decimals from the fraction.*

$$= \frac{8 \div 8}{1,000 \div 8}$$ *Simplify.*

$$= \frac{1}{125}$$

So, about $\frac{1}{125}$ of the Chinese people used the Internet.

To write 0.8% as a decimal, write it as a fraction and then as a division problem.

$$0.8\% = \frac{0.8}{100} = 0.8 \div 100$$ *Write the percent as a fraction and then as a division problem.*

$$0.8 \div 100 = 0.008$$ *Divide by 100. This moves the decimal point two places to the left.*

So, about 0.008 of the Chinese people use the Internet.

CHECK FOR UNDERSTANDING

Think and Discuss ▶ Look back at the lesson to answer the question.

1. **Explain** how to move the decimal point to write a percent as a decimal and a decimal as a percent.

Guided Practice ▶ Write as a percent.

2. 0.8 **3.** 0.25 **4.** 1.2 **5.** $\frac{3}{5}$ **6.** $1\frac{1}{2}$

Write each percent as a fraction or mixed number in simplest form.

7. 50% **8.** 40% **9.** 75% **10.** 15% **11.** 180%

Write each percent as a decimal.

12. 33% **13.** 8% **14.** 19% **15.** 0.2% **16.** 240%

PRACTICE AND PROBLEM SOLVING

Independent Practice ▶ Write as a percent.

17. 0.15 **18.** 0.9 **19.** 0.09 **20.** 0.483 **21.** 0.007

22. 1 **23.** 2.5 **24.** $\frac{1}{5}$ **25.** $\frac{1}{8}$ **26.** $\frac{1}{400}$

Write each percent as a fraction or mixed number in simplest form.

27. 88%   **28.** 73%   **29.** 4%   **30.** 700%   **31.** $33\frac{1}{3}$%

Write each percent as a decimal.

32. 58%   **33.** 7%   **34.** 2.5%   **35.** 220%   **36.** 0.03%

Compare. Write $<$, $>$, **or** $=$ **for each** ●.

37. 3 ● 3%   **38.** $\frac{1}{2}$ ● 50%   **39.** 7% ● 0.7   **40.** 12% ● 1.2

Problem Solving ▶ Applications

41. Out of 30 students, 15 prefer dogs as pets and 9 prefer cats. The rest prefer other animals. What percent prefer other animals as pets?

42. Carlo has $10 more than Erin, and Erin has $5 more than Minh. Altogether they have $50. What percent of the money does Erin have?

43. **What's the Question?**   If the answer is "32% of the instruments," what is the question? Use the table.

Central Avenue School Band

Instruments	Number
Brass	28
Woodwinds	16
Percussion	6

MIXED REVIEW AND TEST PREP

44. If 4 of 25 students are absent, what percent are absent? (p. 406)

45. Joe read 78 pages in 3 hours. What was his unit rate? (p. 384)

46. Estimate the quotient: 47.96 ÷ 6.03. (p. 76)

47. TEST PREP   Which is the prime factorization of 45? (p. 148)

A $3 \times 3 \times 5$ **B** 9×5 **C** 45×1 **D** 3×5

48. TEST PREP   Kendra counted cars passing her school each hour. She made this stem-and-leaf plot. What is the median number? (p. 106)

F 55   **G** 45   **H** 44   **J** 18

Number of Cars

Stem	Leaves
3	3 8
4	0 1 7
5	1 5 5

PROBLEM SOLVING   💡 ThinKer's Corner

Math Match   Play this game with a partner. Write each percent shown at the right on an index card, and put the cards down as shown. Then write each fraction and decimal shown below on an index card.

$\frac{1}{20}$ $\frac{1}{2}$ $\frac{3}{5}$ $\frac{4}{100}$ $\frac{25}{100}$ $\frac{8}{50}$ $\frac{10}{200}$ $\frac{5}{10}$ $\frac{6}{10}$ 0.040

0.05   0.50   0.60   0.04   0.25   0.16   0.050   0.5   0.6   0.250

$\frac{1}{25}$ $\frac{1}{4}$ $\frac{4}{25}$ $\frac{5}{100}$ $\frac{5}{10}$ $\frac{60}{100}$ $\frac{2}{50}$ $\frac{4}{16}$ $\frac{12}{75}$ 0.160

Shuffle the fraction and decimal cards. Place them in two equal stacks. Each player draws a card. The first player to place his or her card on the correct equivalent percent card earns 1 point. Continue until all cards have been used. The player with more points is the winner.

50%
60%
5%
4%
25%
16%

EXTRA PRACTICE page H52, Set B

Estimate and Find Percent of a Number

Learn how to estimate and find a percent of a number.

Remember that to change a percent to a ratio in fraction form, write the percent over 100. Then simplify.

$$36\% = \frac{36}{100}$$
$$= \frac{9}{25}$$

QUICK REVIEW

Write each percent as a fraction.

1. 50%   2. 25%   3. 10%
4. $33\frac{1}{3}\%$ 5. 20%

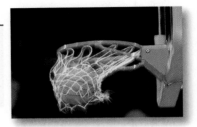

In 1999, basketball star Grant Hill made about 75% of his free throws. At that rate, how many out of 20 free throws would he make?

To find the answer, you need to find 75% of 20.

Activity

You need: two-color counters or colored paper squares

• To find 75% of 20, use 20 yellow counters.

• 75% equals $\frac{3}{4}$, so separate the counters into four groups. Change the color of three groups to red.

• Count the number of red counters. How many free throws would Grant Hill make?

• Use two-color counters to find 50% of 16.

One way to find a percent of a number is to change the percent to a fraction and multiply.

> **EXAMPLE 1**

Due to an injury, Josh played in only 60% of his team's 25 hockey games. In how many games did he play?

Find 60% of 25.

$$60\% = \frac{60}{100}$$ *Write the percent as a ratio in fraction form.*

$$= \frac{3}{5}$$ *Write the ratio in simplest form.*

$$\frac{3}{5} \times \frac{25}{1} = \frac{75}{5} = 15$$ *Multiply the ratio by the number.*

So, Josh played in 15 games.

Sometimes it is easier to change a percent to a decimal.

EXAMPLE 2

In 1998, Randall Cunningham threw 425 passes, and 8% of them resulted in touchdowns. How many touchdown passes did he throw?

Find 8% of 425.

8% = 0.08   *Change the percent to a decimal.*

0.08 × 425 = 34   *Multiply the number by the decimal.*

So, Randall Cunningham threw 34 touchdown passes.

It is customary to leave a tip when you eat in a restaurant.

EXAMPLE 3

Katie's lunch bill came to $13.79. She wants to leave a 15% tip. How much should she leave?

You can estimate the amount of the tip.

Estimate. $13.79 is about $14.
THINK: 15% = 10% + 5%

     10% of $14 is $1.40.
     5% of $14 is half of $1.40, or $0.70.

So, 15% of $13.79 is about $1.40 + $0.70, or $2.10.

To find the amount of the tip to the nearest cent, use a proportion.

Let t = amount of tip.   **THINK:** $15\% = \frac{15}{100}$

So, the tip (t) is to the total bill ($13.79) as 15 is to 100.

$$\frac{t}{13.79} = \frac{15}{100}$$
$$100 \times t = 15 \times 13.79$$
$$100t = 206.85$$
$$t = 206.85 \div 100 = 2.0685$$

To the nearest cent, a 15% tip should be $2.07.

You can use a calculator to find a percent of a number.

EXAMPLE 4

Use a calculator to find 15% of 13.79.

Use the key sequence below.

13.79 [×] 15 [SHIFT] [%]   `2.0685`

So, 2.0685 is 15% of 13.79.

You can use mental math to estimate the percent of a number.

EXAMPLE 5

Estimate 15% of 81.

THINK: 81 is about 80.

15% = 10% + 5%

10% of 80 = 8.
5% of 80 is half of 10% of 80, or half of 8, or 4.

So, 15% of 81 is about 8 + 4, or 12.

• In Example 2, 8% of 425 is given as 34. Use estimation to determine if 34 is reasonable.

CHECK FOR UNDERSTANDING

Think and Discuss ▶ **Look back at the lesson to answer each question.**

1. **Tell** how you can easily find 50% of a number if you know 25% of the number.

2. **Explain** how you can tell whether a percent of a number should be less than or greater than the number.

Guided Practice ▶ **Use a fraction in simplest form to find the percent of the number.**

3. 20% of 30　　**4.** 50% of 80　　**5.** 75% of 48　　**6.** 15% of 20

Use a decimal to find the percent of the number.

7. 52% of 15　　**8.** 40% of 90　　**9.** 28% of 75　　**10.** 76% of 25

PRACTICE AND PROBLEM SOLVING

Independent Practice ▶ **Use a fraction in simplest form to find the percent of the number.**

11. 25% of 80　　**12.** 45% of 10　　**13.** 80% of 55　　**14.** 90% of 120

Use a decimal to find the percent of the number.

15. 12% of 9　　**16.** 28% of 20　　**17.** 71% of 84　　**18.** 95% of 18

Use the method of your choice to find the percent of the number.

19. 21% of 88　　**20.** 35% of 92　　**21.** 40% of 106　　**22.** 2% of 12

23. 51% of 30　　**24.** 99% of 99　　**25.** 82% of 150　　**26.** 63% of 85

27. 5.5% of 70　　**28.** 200% of 100　　**29.** 150% of 38　　**30.** 0.5% of 400

Estimate a 15% tip for each amount.

31. $8.00　　**32.** $12.25　　**33.** $15.75　　**34.** $18.09　　**35.** $32.18

Estimate a 20% tip for each amount.

36. $12.00 **37.** $25.00 **38.** $18.14 **39.** $3.95 **40.** $46.12

ALGEBRA Each proportion shows *n* as a percent of a number. What is the percent? What is *n*?

41. $\frac{25}{100} = \frac{n}{60}$ **42.** $\frac{40}{100} = \frac{n}{280}$ **43.** $\frac{n}{6} = \frac{10}{100}$ **44.** $\frac{n}{240} = \frac{75}{100}$

Problem Solving ▶ Applications

Mille Lacs Lake

45. Geography The surface area of Earth is about 200,000,000 mi². If 70% of the surface is water, about how many square miles of Earth's surface are water?

46. Geography The largest lakes in Minnesota are Red Lake and Mille Lacs Lake. Red Lake is about 218% as large as Mille Lacs Lake. If Mille Lacs Lake covers 207 square miles, how much area do the two lakes cover together?

47. Arnold wants to leave a 15% tip on a bill of $32. He has $35. Does he have enough money for the tip? Explain.

48. *REASONING* A diner left a $7.20 tip on a $40 meal. What percent of the meal price was the tip?

MIXED REVIEW AND TEST PREP

49. Write 0.001 as a percent. (p. 408)

50. $3\frac{2}{5} - 1\frac{7}{10}$ (p. 186)

51. Do the figures appear to be *similar, congruent, both,* or *neither*? (p. 390)

⭐52. TEST PREP An obtuse triangle contains an angle __?__ 90°. (p. 332)

  A greater than **B** less than **C** equal to **D** less than or equal to

⭐53. TEST PREP Two perpendicular lines form an angle which is __?__ . (p. 326)

  F straight **G** right **H** obtuse **J** acute

PROBLEM SOLVING LINKUP to Reading

Strategy • Sequence Sometimes the events in a problem are given in an order which is different from the order in which they occurred. To solve the problem, you must first *sequence* the events in the correct order.

The Brainiac computer was introduced in 1981. Today, a Brainiac sells for 40% of its original price. The price increased an average of $200 per year during the first 6 years. The price topped out at $3,600 in 1987 and has dropped steadily ever since. How much does a Brainiac sell for today?

• Use the words *increased* and *decreased* to sequence the changes in the price of a Brainiac from 1981 till today.

Construct Circle Graphs

Explore how to construct circle graphs.

You need compass, protractor, straightedge.

ANNUAL PASTA SALES	
Shape	**Pounds Sold (millions)**
Spaghetti	310
Elbows	115
Noodles	75

QUICK REVIEW

Write each fraction as a percent.

1. $\frac{1}{2}$ 2. $\frac{1}{4}$ 3. $\frac{1}{10}$ 4. $\frac{1}{5}$ 5. $\frac{27}{100}$

A circle graph shows parts of a whole. If you think of a complete circle as 100%, you can express portions of a circle graph as percents.

- Ms. Shipley's class earned $400 at the school fair. What fraction of the $400 did the class earn at the bake sale?

- What percent of the $400 did the class earn at the bake sale?

MONEY RAISED AT SCHOOL FAIR

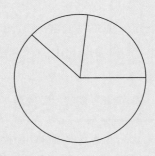

Activity

People in the United States spend over $1.1 billion per year on pasta. The table gives the approximate total weights of the three most popular shapes sold.

- Use a compass to draw a circle on a piece of paper. Mark the center.

- Find the combined weight of the three shapes of pasta.

- Find the percent of the total weight represented by each shape.

  spaghetti: $\dfrac{310}{\text{total weight}} = \blacksquare\%$

  elbows: $\dfrac{115}{\text{total weight}} = \blacksquare\%$

  noodles: $\dfrac{75}{\text{total weight}} = \blacksquare\%$

- Since there are 360° in a circle, multiply each of the percents by 360° to find the degrees in each sector of the circle graph. Round your answers to the nearest degree.

- Use a protractor to draw each angle of 180° or less. Angles greater than 180° remain.

- Label the sectors and write a title.

TOP-SELLING PASTA (in millions of lb)

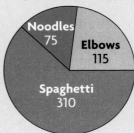

Think and Discuss

- In the circle graph you drew, what does the whole circle represent?

- Suppose the combined weight of the three shapes of pasta was 930 million pounds but the weight of the spaghetti remained the same. Find the number of degrees in the sector representing spaghetti.

Practice

Use Data **Make a circle graph of the data.**

1. The table shows how Shari budgets her earnings from her job.

TECHNOLOGY LINK
More Practice: Use E-Lab, *Exploring Circle Graphs.* www.harcourtschool.com/ elab2002

MONTHLY BUDGET	
Item	**Percent**
Rent	30%
Food	25%
Clothes	12.5%
Other	32.5%

2. The table shows the number of videos rented in one week.

VIDEOS RENTED			
Drama	**Comedy**	**Kids**	**Classics**
250	250	300	200

3. The table records the results of a survey.

FAVORITE SPORTS SURVEY				
Tennis	**Soccer**	**Baseball**	**Golf**	**Other**
50	20	30	80	20

4. Collect data from your classmates about their favorite color. Use the data to make a circle graph with no more than five categories. Compare your graph with those of your classmates.

MIXED REVIEW AND TEST PREP

5. Find 20% of 30. (p. 412)

6. Evaluate $(m - 2m)^2$ for $m = {}^-2$. (p. 272)

7. A rectangle is 16 in. long and 12 in. wide. A similar rectangle is 4 in. long. Find the perimeter of the similar rectangle. (p. 394)

8. Is this survey question biased? "Do you agree that our excellent mayor should be re-elected?" Explain. (p. 98)

9. **TEST PREP** Which shows the value of y for the equation ${}^-3y = 15$? (p. 297)

   A $y = {}^-6$ **B** $y = {}^-5$ **C** $y = 5$ **D** $y = 6$

Discount and Sales Tax

Learn how to solve problems involving discounts and sales tax.

QUICK REVIEW

1. $25 − $10　　**2.** $13 + $8　　**3.** $40 − $11
4. $9 + $51　　**5.** $60 − $9

Vocabulary

discount

sales tax

Anthony wants to buy a backpack for summer camp. He saw the newspaper ad shown below. How much will the pack cost with a 30% discount?

Backpack Sale!
Regular Price $129
NOW 30% OFF!!

Math **I**dea ▶ To find the sale price, you can first find the discount. A **discount** is an amount that is subtracted from the regular price of an item.

discount = regular price × discount rate
　　= $129 × 30%
　　= $129 × 0.30　　*Write the percent as a decimal.*
　　= $38.70

So, the amount of the discount is $38.70.

To find the sale price of the backpack, subtract the discount from the regular price.

Regular Price

Discount

Sale Price

sale price = regular price − discount

　　= $129.00 − $38.70
　　= $90.30

So, the sale price of the backpack is $90.30.

EXAMPLE 1

A volleyball set that regularly sells for $32 is on sale at a 25% discount. Find the amount of the discount and the sale price of the set.

You can use mental math to find the discount if you change 25% to the fraction $\frac{1}{4}$. **THINK:** $\frac{1}{4}$ of $32 is $8.

You can also find the discount by writing the percent as a decimal.

discount $= \$32 \times 25\%$
$\quad\quad\quad = \$32 \times 0.25$
$\quad\quad\quad = \$8$

sale price $=$ regular price $-$ discount
$\quad\quad\quad\quad = \$32 - \$8$
$\quad\quad\quad\quad = \$24$

So, the discount is $8 and the sale price is $24.

In Example 1, you found the sale price when you knew the regular price and the discount rate. Sometimes you may want to find the regular price when you know the sale price and the discount rate.

EXAMPLE 2

Find the regular price of the in-line skates.

The regular price has been discounted 40%. That means the sale price must be 60% of the regular price, since 100% − 40% = 60%.

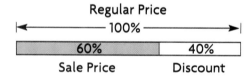

Regular Price

|← ——— 100% ——— →|

60% | 40%

Sale Price | Discount

THINK: sale price $= 60\% \times$ regular price

Let $n =$ the regular price.

$\$72 = 60\% \times n$ *Write an equation relating the sale price and the regular price.*

$\$72 = 0.6 \times n$ *Write the percent as a decimal.*

$\dfrac{\$72}{0.6} = \dfrac{0.6 \times n}{0.6}$ *Divide both sides of the equation by 0.6.*

$\$120 = n$

So, the skates regularly sell for $120.

419

Most states charge a **sales tax** on purchases. A sales tax is calculated as a percent of the cost of an item. To find the amount of the sales tax, multiply the amount of the purchase by the sales tax rate.

EXAMPLE 3

Martina bought a table-tennis table for $119. The sales tax rate was 5%. How much did she pay in sales tax?

Estimate. THINK: 10% of $119 is about $12.00, so 5% of $119 is about $6.

$119 × 5% = $119 × 0.05   *Multiply the price by the percent.*
      = $5.95

So, she paid $5.95 in sales tax. This is close to the estimate of $6, so the answer is reasonable.

You can find the total cost of a purchase that includes sales tax.

EXAMPLE 4

The Carduso family purchased a pool table for $825. The sales tax rate was 7%. What was the total cost of the purchase?

One Way   Add the sales tax to the price.

  $825 × 7% = $825 × 0.07
       = $57.75

$825 + $57.75 = $882.75

Another Way   Multiply the price by 107% since 7% is added to the cost.

$825 × 107% = $825 × 1.07
        = $882.75

So, the total cost of the purchase was $882.75.

CHECK FOR UNDERSTANDING

Think and ▶ Discuss

Look back at the lesson to answer each question.

1. **Explain** how you can estimate the 25% discount on a $96.50 coat.

2. **Describe** an easy way to find the regular price of an item that has been discounted 50%, if you know the discount price.

Guided ▶ Practice

Find the sale price.

3. regular price: $80

4. regular price: $55

5. regular price: $96

6. regular price: $110

 ALGEBRA   **Find the regular price.**

7. sale price: $80
   discount rate: 20%

8. sale price: $150
   discount rate: 40%

Independent ▶ Practice

Find the sale price.

9. regular price: $47.90

DISCOUNT 30%

10. regular price: $120.00

SAVE 40%

11. regular price: $51.80

SALE 10% OFF

12. regular price: $88.50

20% OFF

 ALGEBRA **Find the regular price.**

13. sale price: $28.56
discount rate: 25%

14. sale price: $595.63
discount rate: 30%

Find the sales tax for the given price. Round to the nearest cent.

15. $512.00
tax: 6%

16. $24.95
tax: 7.5%

17. $84.50
tax: 5.5%

18. $260.00
tax: 6.75%

Find the total cost of the purchase. Round to the nearest cent.

19. price:
$44.00
tax: 5%

20. price:
$125.00
tax: 7%

21. price:
$56.95
tax: 8%

22. price:
$899.99
tax: 8.75%

Problem Solving ▶ Applications

Use Data   For 23–24, use the graph.

23. In February, apples cost 12% less per pound than they cost in January. Estimate the cost per pound in February.

24. Maureen bought 15 lb of apples in January and paid 4% sales tax. What was the cost of her purchase?

25. *REASONING*   A refrigerator that regularly sold for $800 was discounted 25%. After the sale, the sale price was raised 25%. Was the new price less than, greater than, or equal to $800? Explain.

PRICE OF APPLES

Price per lb: $0.80, $0.75, $0.70, $0.65, $0.60, $0.55, $0.50, $0

Month: Oct Nov Dec Jan

26. Find 80% of 12.5. (p. 412)

27. Solve for b: $2 = b - 15$. (p. 290)

28. On a scale of 2 in.:3 ft, what is the actual length of a room that is 12 in. long in a scale drawing? (p. 397)

29. TEST PREP   Which is $\frac{37}{8}$ written as a mixed number? (p. 164)

A $3\frac{5}{8}$ **B** $4\frac{3}{8}$ **C** $4\frac{5}{8}$ **D** $8\frac{29}{37}$

30. TEST PREP   The part of a circle shown in red is a(n) __?__ . (p. 344)

F chord   **G** arc   **H** sector   **J** radius

Simple Interest

Learn how to find simple interest.

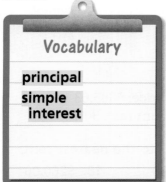

Vocabulary

principal

simple interest

When money is in a savings account, the bank regularly adds money to the account. The original amount you put in the account is called the **principal**. The amount the bank adds is called interest.

Simple interest is a fixed percent of the principal and is paid yearly. Use the formula $I = prt$ to calculate simple interest, where I = interest earned, p = principal, r = interest rate per year, and t = time in years.

EXAMPLE 1

Katrina put $500 in a savings account at a simple interest rate of 6% per year for 2 years. How much interest will she earn?

$I = prt$ *p = $500, r = 6%, t = 2 years*

$I = \$500 \times 0.06 \times 2$ *Multiply.*

$I = \$60$

So, she will earn $60 in interest.

Activity

You need: 12 index cards, as shown

Shuffle and stack the principal cards and turn them face down. Do the same with the interest-rate and time cards.

p	r	t
$50	4%	1 yr
$125	5%	2 yr
$200	8%	3 yr
$300	$6\frac{1}{2}$%	4 yr

- Each player draws one card from each pile and computes the interest earned. The player with the most interest earns points equal to the total interest of all the players.

- Play for five rounds. What is the most interest a player could earn in one round? What is the least?

EXAMPLE 2

Bill borrows $500. He will repay it in 2 years at a simple interest rate of 8%. How much will he have to repay at the end of 2 years?

$I = prt$ *Find the amount of interest.*

$I = \$500 \times 0.08 \times 2 = \$80$

$\$80 + \$500 = \$580$ *Add the interest to the loan.*

So, Bill will have to repay $580.

Think and ▶
Discuss

Look back at the lesson to answer each question.

1. **Explain** what $I = prt$ means.

2. **Tell** how to find the simple interest earned on $500 for 6 months at an annual interest rate of 5%. Then find the simple interest.

Guided ▶
Practice

Find the simple interest.

3. principal: $100
   rate: 4%
   time: 1 year

4. principal: $750
   rate: 8%
   time: 4 years

5. principal: $2,400
   rate: 12%
   time: 6 years

PRACTICE AND PROBLEM SOLVING

Independent ▶
Practice

Find the simple interest.

6. principal: $12,000
   rate: 8%
   time: 2 years

7. principal: $640
   rate: 6.5%
   time: 5 years

8. principal: $21,000
   rate: 5.25%
   time: 18 years

Find the simple interest.

	Principal	Rate	1 Year	5 Years
9.	$70	2%	▦	▦
10.	$250	3.5%	▦	▦
11.	$1,200	6.9%	▦	▦

Problem Solving ▶
Applications

12. Amber put $2,000 in a savings account for 6 years at a simple interest rate of 6%. Haines put $2,400 in an account for 5 years at a rate of 5.5%. Who earned more interest? How much more?

13. Ray needs to borrow $1,000. Bank A offers a 3-year loan with a simple interest rate of 8.5%. Bank B offers a 5-year loan at a simple interest rate of 4%. Which bank should Ray borrow from? Explain.

14. **REASONING** Jake earned $54 in simple interest on a principal of $900 invested for 1 year. What was the interest rate?

15. **(?) What's the Error?** Jeff found simple interest of $360 for a principal of $200, interest rate of 6%, and a time of 3 years. Find his error and the correct amount of simple interest.

MIXED REVIEW AND TEST PREP

16. What is the sale price on a $150 item discounted 20%? (p. 418)

17. Two towns are 125 mi apart. How far apart do they appear on a map with a scale of 2 in. = 50 mi? (p. 400)

18. Solve $4x = 96$. (p. 297)

19. Evaluate $12 + 24 \div 12 - 4$. (p. 44)

⭐20. **TEST PREP** Which is the best estimate of the quotient $618.6 \div 5.914$? (p. 76)

   A 1   **B** 10   **C** 100   **D** 1,000

1. **VOCABULARY** An amount subtracted from the regular price of an item to give the sale price is called the __?__. (pp. 418–421)

2. **VOCABULARY** A fixed percent of the principal, paid yearly, is the __?__. (p. 422)

Write the percent that is shaded. (p. 406)

3.

4.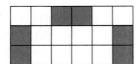

5.

Write as a percent. (pp. 408–411)

6. 0.38

7. $\frac{1}{4}$

8. 0.09

9. $\frac{2}{5}$

Write each percent as a decimal and as a fraction in simplest form. (pp. 408–411)

10. 20%

11. 18%

12. 75%

13. 10%

Find the percent. (pp. 412–415)

14. 40% of 300

15. 25% of 24

16. 1% of 350

17. 100% of 80

18. 200% of 7

19. 15% of 24

20. 11% of 80

21. 90% of 300

22. Use these data to make a circle graph of the Vernon family's monthly budget of $2,000: clothing and entertainment, 25%; food, 10%; rent, 35%; other, 30%. (pp. 416–417)

23. Find the amount spent on food in the Vernon family's budget.

24. Find the amount spent on rent in the Vernon family's budget.

Find the sale price. (pp. 418–421)

25. regular price: $40 SALE 25% OFF

26. regular price: $79 Reduced 10%

Find the regular price. (pp. 418–421)

27. sale price: $36
    20% off

28. sale price: $95
    50% off

29. What is the sales tax on an item that costs $33.50 if the sales tax rate is 6%?

30. What is the total cost of an item that costs $98 if the sales tax rate is 8%?

Find the simple interest. (p. 422)

31. principal: $200
    rate: 6%
    time: 1 year

32. principal: $8,000
    rate: 7.5%
    time: 4 years

33. Julia put $1,500 in a savings account. The yearly simple interest rate was 5%. How much did she have in her account after 3 years?

Eliminate choices.

See item **1.**

Estimate the simple interest at 10% per year for 2 years. Interest at 5% would be half of that amount, so interest at 7% must be more than at 5% and less than at 10%.

Also see problem **5**, p. H64.

Choose the best answer.

1. Tonya put $1,800 in the bank and left it for 2 years. The account earned simple interest at the rate of 7% per year. How much interest did Tonya get?

  A $70   **C** $252

  B $140   **D** $360

2. The Fremont Tennis Club has 120 members. Teenagers make up 85% of the club. How many teens are in the club?

  F 18   **H** 96

  G 24   **J** 102

3. How is 46% written as a decimal?

  A 0.46%   **C** 4.6

  B 0.46   **D** 46

4. Which pair shows two numbers equivalent to $\frac{3}{5}$?

  F 66% and 6.6   **H** 40% and 0.4

  G 0.6 and 60%   **J** 3.5 and 35%

5. What is 18% of 250?

  A 45   **C** 18

  B 36   **D** Not here

6. If 6 pens cost $2.50, how much will 9 pens cost at the same rate?

  F $2.75   **H** $3.25

  G $2.95   **J** $3.75

7. A triangle has angles that measure 20° and 75°. What is the measure of the third angle?

  A 55°   **C** 95°

  B 85°   **D** 180°

8. What is the value of $s - t + v$ for $s = 25$, $t = 12$, and $v = 3$?

  F 16   **H** 10

  G 13   **J** Not here

9. Which group shows an equivalent fraction, decimal, and percent?

  A $\frac{11}{25}$, 0.44, 44%   **C** $\frac{11}{25}$, 4.4, 44%

  B $\frac{11}{25}$, 0.44, 4.4%   **D** $\frac{11}{25}$, 44, 44%

10. $1\frac{3}{4} \times 9\frac{3}{5}$

  F $17\frac{3}{5}$ **H** $18\frac{4}{5}$

  G $18\frac{2}{5}$ **J** Not here

Write What You Know

11. A jacket that regularly sells for $120 is on sale for 30% off. Hannah has a coupon for an additional 10% off the sale price. Find the final price Hannah would pay for this jacket. Explain how you found your answer.

12. Shannon correctly answered 17 out of 20 questions on a math test. Show two different ways to find her score as a percent.

CHAPTER 22

Probability of Simple Events

Many states require bicyclists to wear helmets and have lights and reflectors on their bikes for safety.

PROBLEM SOLVING Suppose your school does random safety checks on the bikes in the bike rack. The graph shows data for the bike rack on a day when you ride your bike to school. What is the probability that your bike will be the one chosen for inspection? Which color has the greatest chance of being chosen?

BICYCLES IN RACK

DATA LINK

Bicycle Colors	
Green	
Pink	
Blue	
Red	
Black	

key: each ◯ = 4 bikes

Check What You Know

Use this page to help you review and remember important skills needed for Chapter 22.

✔ Vocabulary

Choose the best term from the box.

1. A(n) __?__ event will never happen.

2. A(n) __?__ is a specific set of outcomes in a probability experiment.

3. Events that have the same probability of occurring are __?__ .

> equally likely
> event
> impossible
> outcome

✔ Fractions, Decimals, and Percents (For Intervention, see p. H23.)

Write each fraction as a decimal and a percent.

4. $\frac{2}{5}$

5. $\frac{1}{2}$

6. $\frac{7}{10}$

7. $\frac{3}{4}$

8. $\frac{9}{20}$

9. $\frac{1}{10}$

10. $\frac{1}{4}$

11. $\frac{3}{5}$

12. $\frac{1}{5}$

13. $\frac{18}{25}$

14. $\frac{3}{8}$

15. $\frac{3}{25}$

16. $\frac{5}{8}$

17. $\frac{19}{20}$

18. $\frac{3}{50}$

✔ Simplest Form of Fractions (For Intervention, see p. H9.)

Write each fraction in simplest form.

19. $\frac{4}{12}$

20. $\frac{9}{6}$

21. $\frac{32}{64}$

22. $\frac{40}{15}$

23. $\frac{36}{54}$

24. $\frac{9}{75}$

25. $\frac{22}{55}$

26. $\frac{96}{60}$

27. $\frac{49}{56}$

28. $\frac{81}{120}$

29. $\frac{25}{75}$

30. $\frac{72}{60}$

31. $\frac{70}{120}$

32. $\frac{155}{200}$

33. $\frac{200}{800}$

✔ Certain, Impossible, Likely, Unlikely (For Intervention, see p. H24.)

Tell if the event is *certain*, *impossible*, *likely*, or *unlikely*.

34. Having homework this week

35. The sun rising today

36. Spinning an odd number on a spinner that is numbered 1, 3, 5, 6, 7, and 9

37. The month of January having 34 days next year

38. Reading 3 novels in one day

> ### LOOK AHEAD
>
> **In Chapter 22 you will**
> - find the theoretical probability of an event
> - use a simulation to model an experiment
> - find the experimental probability of an event

Theoretical Probability

Learn how to find the theoretical probability of an event.

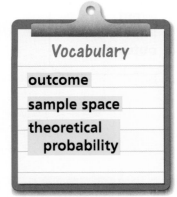

Vocabulary
outcome
sample space
theoretical probability

Each color on the spinner shown represents one equally likely **outcome**, or possible result, of spinning. The **sample space**, or set of all possible outcomes, is white, purple, red, yellow, and green.

Math **I**dea ▶ The **theoretical probability**, P, of an event is a comparison of the number of favorable outcomes to the number of possible, equally likely outcomes. It can be written as a ratio.

$$P(\text{event}) = \frac{\text{number of favorable outcomes}}{\text{number of possible equally likely outcomes}}$$

EXAMPLE 1

In a game students created, if the pointer on the spinner above lands on red, you get an extra turn. What is the probability that you get an extra turn?

1 favorable outcome: red   *List the favorable outcomes.*

5 possible outcomes.   *Count the possible outcomes.*

$P(\text{red}) = \frac{1 \text{ favorable outcome}}{5 \text{ possible outcomes}} = \frac{1}{5}$ *Write the probability as a ratio.*

So, the probability that you get an extra turn in the game is $\frac{1}{5}$.

Probabilities can be expressed as fractions, decimals, and percents.

EXAMPLE 2

Miyoko rolled a number cube labeled 1 to 6. Find each probability. Express your answers as fractions, decimals, and percents.

A. $P(2) = \frac{1}{6}$, $0.1\overline{6}$, or $16\frac{2}{3}\%$ ← 1 choice out of 6

B. $P(2 \text{ or } 3) = \frac{1+1}{6}$ ← 2 choices out of 6

$= \frac{2}{6} = \frac{1}{3}$, $0.\overline{3}$, or $33\frac{1}{3}\%$

C. $P(\text{greater than } 3) = \frac{1+1+1}{6}$ ← 3 choices out of 6

$= \frac{3}{6} = \frac{1}{2}$, 0.50, or 50%

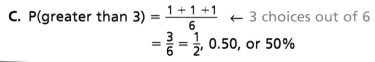

The number line below shows that the probability of an event ranges from 0, or impossible, to 1, or certain. The closer a probability is to 1, the more likely the event is to occur.

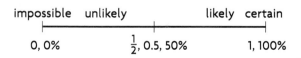

impossible unlikely   likely certain

0, 0% $\frac{1}{2}$, 0.5, 50%   1, 100%

EXAMPLE 3

Each letter of the word *OUTCOMES* is written on a card and placed in a bag. One card is chosen at random. Which outcome is more likely to occur, an *M* or an *O*?

P(*M*) = $\frac{1}{8}$ *Find each probability.*

P(*O*) = $\frac{2}{8}$ = $\frac{1}{4}$

$\frac{1}{4} > \frac{1}{8}$ *Compare the probabilities.*

So, the outcome *O* is more likely to occur than the outcome *M*.

EXAMPLE 4

Use the spinner at the right to find each probability.

A. P(blue) = $\frac{0}{4}$. None of the sections are blue.

B. P(green) = $\frac{1}{4}$

C. P(not green) = $\frac{1 + 1 + 1}{4}$ = $\frac{3}{4}$

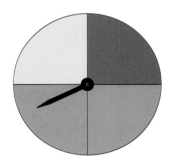

Look at P(green) and P(not green) in Example 4.
The sum of those probabilities is $\frac{1}{4} + \frac{3}{4}$, or 1. If *P* is the probability of an event occurring, $1 - P$ is the probability of an event not occurring.

EXAMPLE 5

Use the spinner above to find each probability. Write each answer as a fraction, a decimal, and a percent.

A. P(not yellow) = 1 − P(yellow)

$= 1 - \frac{1}{4}$

$= \frac{3}{4}$

So, P(not yellow) is $\frac{3}{4}$, 0.75, or 75%.

B. P(not orange) = 1 − P(orange)

$= 1 - \frac{1}{2}$

$= \frac{1}{2}$

So, P(not orange) is $\frac{1}{2}$, 0.50, or 50%.

Think and ▶ Discuss

Look back at the lesson to answer each question.

1. **Write** the sample space for the number cube in Example 2.

2. **Explain** why the probabilities computed in Example 4 are reasonable.

Guided ▶ Practice

Use the spinner at the right to find each probability. Write each answer as a fraction, a decimal, and a percent.

3. P(yellow)

4. P(purple)

5. P(red)

6. P(red or purple)

7. P(not yellow)

8. P(green)

Independent ▶ Practice

Use the spinner at the right to find each probability. Write each answer as a fraction, a decimal, and a percent.

9. P*(L)*

10. P*(T)*

11. P*(B)*

12. P*(E or R)*

13. P*(L, F, or T)*

14. P(vowel)

A number cube is labeled 1, 2, 3, 4, 5, and 6. Find each probability. Write each answer as a fraction.

15. P(2)

16. P(not 5)

17. P(4 or 6)

18. P(less than 4)

19. P(1, 2, or 3)

20. P(even)

21. P(multiple of 3)

22. P(divisible by 6)

23. P(integer)

Cards numbered 1, 1, 2, 2, 3, 4, 5, and 6 are placed in a box. You choose one card without looking. Compare the probabilities. Write <, >, or = for each ●.

24. P(1) ● P(2)

25. P(1) ● P(5 or 6)

26. P(6) ● P(2)

27. P(1 or 2) ● P(3 or 4)

28. P(even) ● P(odd)

29. P(multiple of 3) ● P(greater than 2)

Use P(A), the probability of event A, to find P(not A).

30. $P(A) = \frac{1}{2}$

31. P(A) = 0.4

32. P(A) = 80%

33. $P(A) = \frac{7}{12}$

34. P(A) = 0.61

35. P(A) = 23%

Problem Solving ▶
Applications

36. A rental car agency has 55 blue cars, 32 red cars, and 70 white cars. If a customer selects at random, what is the probability that she will select a green car? Write your answer as a percent.

37. Suppose the probability of an event is $\frac{5}{12}$. Which is greater, the probability that the event will occur or the probability that it will not occur?

38. You are given a choice of three answers to a trivia question. If you don't know the answer, what is the probability of guessing the correct answer?

39. ❓ **What's the Question?** Shalonda has a number cube labeled 1 to 6. The answer is 1.

40. **REASONING** Brenda has a bag with 8 cubes all the same size: 1 red, 4 yellow, and 3 blue. She chooses a blue cube without looking and does not put it back in the bag. What is the probability of choosing a yellow cube after the blue cube has been removed?

41. A box has 24 marbles. If $\frac{3}{8}$ of the marbles are blue, find the number of marbles that are not blue.

MIXED REVIEW AND TEST PREP

42. Tong invested $420 at a simple interest rate of $3\frac{1}{2}\%$. How much interest will he earn in 5 years? (p. 422)

43. Order $7\frac{3}{8}$, $^-7.75$, $7\frac{1}{4}$, and 7.5 from least to greatest. (p. 234)

44. Divide. $135.892 \div 6.41$ (p. 76)

⭐**45. TEST PREP** Evaluate $3p$ for $p = 7.8$.
(p. 82)
    A 2.6   **B** 4.8   **C** 10.8   **D** 23.4

⭐**46. TEST PREP** Find the product.
$^-10 \times {}^-12$ (p. 252)
    F $^-120$ **G** $^-22$ **H** 2   **J** 120

PROBLEM SOLVING | 💡 Thinker's Corner

Geometric Probability

You can use the areas of geometric figures to find probabilities.

In a certain game, students throw a quarter onto a target that looks like the square at the right. To find the probability that the quarter lands on the green section, compare the areas. Assume that the center point of the quarter lands on the target at some random point and that all points are equally likely.

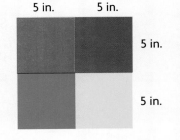

area of green section

5 in. × 5 in. = 25 in.²

area of target

10 in. × 10 in. = 100 in.²

P(green) = $\frac{\text{area of green section}}{\text{area of target}}$ = $\frac{25}{100}$ = $\frac{1}{4}$

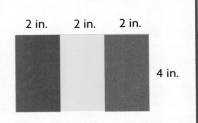

• If the center point of a randomly thrown quarter lands on the target at the right, what is the probability of its being in the blue section?

PROBLEM SOLVING SKILL
Too Much or Too Little Information

Analyze
Choose
Solve
Check

Learn how to use the skill of determining whether there is *too much or too little information* to solve problems.

QUICK REVIEW

Write each fraction in simplest form.

1. $\frac{8}{12}$ **2.** $\frac{6}{9}$ **3.** $\frac{15}{20}$ **4.** $\frac{12}{21}$ **5.** $\frac{35}{100}$

Karyn is waiting in line with 14 other people to ride the bumper cars at a carnival. Each person in line is randomly assigned to one of 15 bumper cars. There are 3 blue, 5 red, 4 yellow, and 3 green bumper cars. What is the probability that Karyn is assigned to a red bumper car?

If a problem has too much information, you must decide what to use to solve the problem.

> **1. Want to know:** P(red bumper car)
>
> **2. Know:** number of red bumper cars, total number of bumper cars
>
> **3. Don't need to know:** number of people waiting in line; number of blue, yellow, and green bumper cars
>
> **4. Missing and need to know:** none
>
> **Decision: Too much information**
>
> **Solve:** 5 red bumper cars, 15 bumper cars total; P(red) $= \frac{5}{15} = \frac{1}{3}$
>
> So, the probability that Karyn is assigned to a red bumper car is $\frac{1}{3}$.

Sometimes there is too little information to solve a problem.

What fraction of the people waiting in line to ride the bumper cars are females?

> **1. Want to know:** fraction of people in line who are females
>
> **2. Know:** 15 people waiting in line
>
> **3. Don't need to know:** number of blue, yellow, green, and red bumper cars; number of bumper cars
>
> **4. Missing and need to know:** number of females in line
>
> **Decision: Too little information**
>
> **Solve:** can't; need more information

Tell if each problem has *too much*, *too little*, or *the right amount* of information. Then solve the problem if possible, or describe the information needed to solve it.

1. A club has 32 female members and 24 male members. The members range in age from 22 years to 51 years. If a club member is randomly selected to be president, what is the probability that a female is chosen?

2. The Jenkins family is trying to decide on a name for their dog. They place the names Rusty, Sporto, and Buddy in a box and randomly select one name. What is the probability that they select the name Rusty or Buddy?

3. G. de Chasseloup-Laubat held the official land speed record in 1898. He drove a vehicle, the Jeantaud, at a speed of 39.24 mi/hr. How many times as fast was the land speed record in 1970?

4. Lina is saving to buy a bicycle and a helmet. The price of the bike is $199. So far, she has $155 saved. How much more does she need to save?

MIXED STRATEGY PRACTICE

Use Data For 5–6, use this table.

5. What is the difference between the average high temperature and the average low temperature for the 4 cities?

   A ⁻81°F **B** ⁻39°F **C** 39°F **D** 81°F

6. **ALGEBRA** Choose the high temperature in Atlanta in °C. Use the formula $C = \frac{5}{9} \times (F - 32)$.

   F ⁻35°C **G** ⁻10°C **H** 10°C **J** 35°C

TEMPERATURES (°F)		
City	Low	High
Atlanta, Ga	19	94
Pittsburgh, PA	6	91
Charleston, SC	22	100
Baltimore, MD	9	95

7. Lydia studies twice as long as Jorge. Jorge and Olga study a total of 7 hr, but Olga studies 3 hr longer than Jorge. How long does Lydia study?

8. A bus travels 55 mi per hour. Misty starts out by bus at 11:45 A.M. to go to the carnival. How many miles does she travel if she arrives at the carnival at 1:15 P.M.?

9. There are 164 steps in the Astoria Column on Coxcomb Hill in Astoria, Oregon. Rick decided to climb to the top once a week for a year. How many steps must he climb up to do that?

10. The Stefan family spent a total of $42.75 at the carnival. They bought food for $12.50, souvenirs for $5.25, and show tickets for $5.00 each. How many tickets did they buy?

11. ✎ **Write a problem** about probability and include too much information. Then write a related problem, giving too little information.

Astoria Column

Simulations

MATH LAB

Explore how to use a simulation to model an experiment.

You need a 5-section spinner and a calculator.

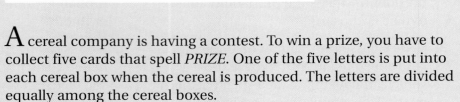

QUICK REVIEW

Find the mean.

1. 4, 5, 4, 6, 6   **2.** 2, 3, 1, 2, 3, 1

3. 10, 9, 17, 10   **4.** 15, 10, 14

5. 250, 200, 150

A cereal company is having a contest. To win a prize, you have to collect five cards that spell *PRIZE*. One of the five letters is put into each cereal box when the cereal is produced. The letters are divided equally among the cereal boxes.

You can conduct an experiment to simulate how many boxes of cereal you have to buy to get all five letters.

Activity 1

- Use a spinner to generate random numbers. Each of the numbers 1 to 5 will represent one of the letters.

- Spin the pointer on the spinner, and tally the numbers you get.

- Continue to spin the pointer until you get every number at least once.

- Repeat the experiment.

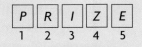

P	R	I	Z	E
1	2	3	4	5

NUMBER	TALLY
1	III
2	III
3	I
4	I
5	III

Think and Discuss

- How many spins did it take in the first experiment to get all five numbers? in the second experiment?

- What is the mean of the spins in your two experiments?

- How many boxes of cereal do you expect you will have to buy to get all five letters? If you bought this many boxes, would you be sure to win? Explain.

Practice

- Repeat the experiment three more times.

- Combine your data with data from four classmates. Find the mean of the spins from all five sets of data.

- How many boxes of cereal do you expect you will have to buy to get all five letters? How does this differ from what you expected after the first two experiments?

You can produce random numbers by using a calculator.

Activity 2

• Use this key sequence on some calculators to produce random numbers from 1 to 15.

   1 15

• Record the number you get each time.

• Continue to press **ENTER** until you have gotten each of the numbers 1 to 15.

• Make a bar graph to show the results.

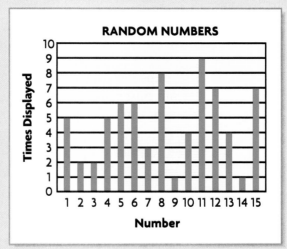

Think and Discuss

• Compare the numbers of times you and your classmates used the key sequence to get all 15 numbers.

• Look at the results in the graph above. What was the last random number produced?

Practice

• Use a calculator to produce random numbers from 1 to 25. Record the number of times you get each number. Continue until you get each number.

MIXED REVIEW AND TEST PREP

1. A spinner has 8 equal sections numbered 1, 2, 1, 3, 2, 4, 1, 5. Find P(1 or 2). (p. 428)

2. The regular price of an item is $140. The discount rate is 25%. Find the amount of discount and the sale price. (p. 418)

3. Evaluate $y^2 + 8$, for $y = 9$. (p. 272)

4. Evaluate $b^2 \div (6 + 4)$, for $b = 8$. (p. 272)

5. TEST PREP How much greater is the product 8.42×13.1 than the product 6.507×14? (p. 70)
   A 19.204 **B** 20.507 **C** 26.028 **D** 71.098

Experimental Probability

Learn how to find the experimental probability of an event.

Vocabulary

experimental probability

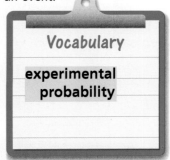

1. $\frac{1}{3} \times 27$ 2. $\frac{1}{5} \times 55$ 3. $\frac{3}{4} \times 24$ 4. $\frac{4}{5} \times 100$ 5. $\frac{2}{10} \times 3{,}000$

Mateo put some red, yellow, and green marbles in a box. He randomly selected a marble and recorded the color. He returned the marble to the box each time. Mateo did this a total of 25 times.

By performing an experiment, Mateo can find the experimental probability of selecting each marble color.

Math Idea ▶ The **experimental probability** of an event is the number of times a certain outcome actually occurs compared with the total number of trials, or the total number of times you do the activity.

experimental probability $= \dfrac{\text{number of times outcome occurs}}{\text{total number of trials}}$

EXAMPLE 1

Mateo recorded his results in a table. Use these results to find the experimental probability of selecting each color. Express your answers as fractions, decimals, and percents.

MARBLE COLOR	Green	Red	Yellow
	2	10	13

P(green) $= \frac{2}{25}$, 0.08, 8%   P(red) $= \frac{10}{25} = \frac{2}{5}$, 0.40, 40%

P(yellow) $= \frac{13}{25}$, 0.52, 52%

You can use the experimental probability to predict future events.

Based upon his experimental results, how many times can Mateo expect to select a red marble in his next 10 selections?

EXAMPLE 2

P(red) $= \frac{2}{5}$ *Use the experimental probability.*

$\frac{2}{5} \times 10 = 4$ *Multiply 10 selections by $\frac{2}{5}$.*

So, he can expect to select a red marble 4 times.

CHECK FOR UNDERSTANDING

Think and ▶ Discuss

Look back at the lesson to answer the question.

1. **What if** Mateo has 10 green, 50 red, and 40 yellow marbles? Tell how the experimental probability of selecting each color compares with the theoretical probability.

Guided ▶ Practice Celia rolled a number cube 30 times. For 2–7, use her results shown in the table below to find the experimental probability of each event. Write the answer in simplest form.

NUMBER	1	2	3	4	5	6
TIMES ROLLED	2	6	7	3	8	4

2. P(1) **3.** P(2) **4.** P(3)

5. P(4) **6.** P(5) **7.** P(6)

8. What is the theoretical probability of rolling each number?

PRACTICE AND PROBLEM SOLVING

Independent ▶ Practice

For 9–15, use the table.
Vincent spun the pointer of this spinner 40 times.

COLOR	Red	Yellow	Blue	Green
SPINS	10	6	16	8

Find the experimental probability.

9. P(red) **10.** P(yellow)

11. P(blue) **12.** P(green)

13. P(yellow or red) **14.** P(not red)

15. Based on his experimental results, how many times can Vincent expect a result of yellow in the next 100 spins?

Problem Solving ▶ Applications

16. Conduct an experiment in which you toss a coin 50 times. Keep a tally of the number of times you toss heads. Combine your results with those of five other students to find the experimental probability of tossing heads. How does it compare to the theoretical probability?

17. ✎ **Write About It** Explain how to find the experimental probability of rolling a 4 using a number cube labeled 1 to 6.

18. Science On a plant called the four o'clock, the flowers may be red, white, or pink. Pink flowers are twice as common as either red or white flowers. If a four o'clock has 140 flowers, how many would you expect to be red? white? pink?

MIXED REVIEW AND TEST PREP

19. For a certain experiment, P(vowel) = $\frac{3}{7}$. Find P(consonant). (p. 428)

Find the percent of the number. (p. 412)

20. 82% of 150 **21.** 7.5% of 130 **22.** 200% of 95

⭐**23. TEST PREP** Which type of graph would you use to show the lengths of the five longest rivers? (p. 120)

A circle **B** bar **C** histogram **D** line

1. **VOCABULARY** The number of times a certain outcome occurs compared with the total number of trials is called the __?__ . (p. 436)

2. **VOCABULARY** A comparison of the number of favorable outcomes to the number of possible equally likely outcomes is called the __?__ . (p. 428)

A bag has 10 new pencils: 3 red, 4 yellow, 1 blue, and 2 green. Find the probability of randomly choosing the given color from the bag. Write each answer as a fraction, a decimal, and a percent. (pp. 428–431)

3. P(green)

4. P(yellow)

5. P(orange)

6. P(not yellow)

7. P(red or green)

8. P(red, yellow, or blue)

A bag has slips of paper numbered 2–10. Find the probability of randomly drawing the number from the bag. Write each answer as a fraction. (pp. 428–431)

9. P(5)

10. P(7 or 9)

11. P(1)

12. P(2, 5, or 9)

13. P(even number)

14. P(not 8)

15. P(5 or 6)

16. P(odd number)

17. P(a number less than 10)

18. P(multiple of 5)

19. P(integer)

For 20–23, use the table. The table shows the results of rolling a number cube 100 times. Find the experimental probability. (436–437)

NUMBER	1	2	3	4	5	6
TIMES ROLLED	8	12	24	13	17	26

20. P(3)

21. P(6)

22. P(2 or 5)

23. Suppose the number cube is rolled 500 times. Based on the experimental results, how many times can you expect to roll a 1 or a 2?

Tell if each problem has *too much, too little,* or *the right amount* of information. Then solve the problem if possible, or describe the information needed to solve it. (pp. 432–433)

Forty-two people have signed up for a race. Sixteen of them are male. Five of the people are wearing red running shoes.

24. What is the probability that the winner of the race will be a female?

25. What is the probability that the winner of the race will be a male wearing red running shoes?

Look for important words.
See item **7**.

Important words are *that is not green*. Use the spinner to determine the probability of spinning any color that is not green. Then choose the expression.

Also see problem **2**, p. H62.

Choose the best answer.

1. Tong has a spinner with 5 equal sections labeled 1 to 5. What is the probability of spinning a number that is not 2?

   A $\frac{1}{5}$ **B** $\frac{2}{5}$ **C** $\frac{4}{5}$ **D** 1

2. A spinner is divided into 8 equal sections. Of the sections, 4 are colored white, 3 are colored blue, and 1 is colored yellow. What is the probability that the pointer will stop on a white section?

   F $\frac{1}{8}$ **G** $\frac{1}{2}$ **H** $\frac{3}{5}$ **J** $\frac{7}{8}$

3. Stan has a number cube labeled 1 to 6. What is the probability that he will roll a number less than 3?

   A $\frac{1}{6}$ **B** $\frac{1}{3}$ **C** $\frac{1}{2}$ **D** $\frac{2}{3}$

4. A bag contains 4 blue marbles, 2 green marbles, 3 yellow marbles, and 6 red marbles. What is the probability of randomly pulling either a blue or a red marble from the bag?

   F $\frac{1}{10}$ **G** $\frac{2}{5}$ **H** $\frac{1}{2}$ **J** $\frac{2}{3}$

5. Alice's class toured a toy factory. At the end of the tour, one person's name was drawn to win a toy. What information do you need to know to determine the probability that Alice won?

   A The kind of toy to be given away

   B The number of toys made at the factory

   C The name of the person drawing

   D The number of names in the drawing

Use the spinner for 6–7.

6. Which is the probability of spinning red?

   F $\frac{1}{2}$ **H** $\frac{1}{4}$

   G $\frac{1}{3}$ **J** $\frac{1}{6}$

7. Which expression shows the probability of spinning a color that is not green?

   A $1 + \frac{1}{6}$ **C** $1 - \frac{1}{3}$

   B $1 - \frac{1}{6}$ **D** $1 - \frac{1}{2}$

8. If two angles are supplementary and the measure of one of the angles is 63°, what is the measure of the second angle?

   F 27°   **H** 137°

   G 117°   **J** Not here

Write What You Know

9. Tell how you would find the probability of spinning yellow on the spinner above. Use your method to find the probability.

10. A store manager decided to survey 1 out of every 10 shoppers. If there were 350 shoppers, how many would be surveyed? Explain how you found your answer.

23 Probability of Compound Events

The yo-yo is the second oldest toy in the world, dating from the ancient Greeks 3,000 years ago. Today, yo-yos are popular toys among collectors.

PROBLEM SOLVING Suppose you are a yo-yo collector. At a flea market, one vendor has grab bags. Each bag contains a yo-yo. You want to get a yo-yo with plastic sides and a metal axle. How can a chart like the one below help you figure out the probability of getting the yo-yo you want?

DATA LINK

GRAB BAG YO-YOS				
Yo-Yo	Sides		Axle	
	Wooden	Plastic	Wooden	Metal
1	X		X	
2		X	X	
3	X		X	
4		X		X
5		X		X
6	X		X	

Check What You Know

Use this page to help you review and remember important skills needed for Chapter 23.

 Vocabulary

Choose the best term from the box.

event
outcome
theoretical
probability
sample space

1. The set of all possible outcomes in a probability experiment is called the __?__ .

2. The ratio of the number of favorable outcomes to the number of possible outcomes is the __?__ .

3. A possible result in a probability experiment is a(n) __?__ .

 Multiply Fractions (For Intervention, see p. H23.)

Multiply. Write your answer in simplest form.

4. $\frac{1}{3} \times \frac{1}{2}$
5. $\frac{1}{2} \times \frac{2}{3}$
6. $\frac{1}{4} \times \frac{1}{8}$
7. $\frac{3}{8} \times \frac{1}{2}$
8. $\frac{1}{3} \times \frac{3}{7}$

9. $\frac{1}{2} \times \frac{4}{5}$
10. $\frac{1}{9} \times \frac{2}{5}$
11. $\frac{3}{4} \times \frac{1}{6}$
12. $\frac{3}{5} \times \frac{1}{5}$
13. $\frac{1}{10} \times \frac{2}{5}$

 Analyze Data (For Intervention, see p. H24.)

Use the line graph for 14–16.

14. During which month did the greatest amount of snow fall?

15. About how many inches of snow fell in all five months?

16. In which two months did 6 inches of snow fall?

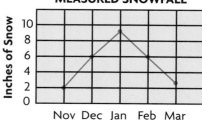

Use the circle graph for 17–20.

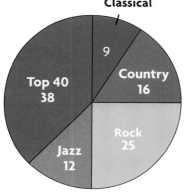

17. How many students were surveyed?

18. What was the most popular music?

19. How many more students preferred rock than preferred classical?

20. How many students preferred rock or jazz?

> **LOOK AHEAD**
>
> **In Chapter 23 you will**
> - find outcomes of compound events
> - identify and find probabilities of independent and dependent events
> - find probabilities and make predictions from sample data

PROBLEM SOLVING STRATEGY
Make an Organized List

Learn how to use the strategy *make an organized list* to solve problems.

QUICK REVIEW

A number cube labeled 1–6 is rolled. Find each probability.

1. P(2)   **2.** P(4)

3. P(1 or 3)   **4.** P(not 5)

5. P(even)

The 5 members of the Montoya family are planning their next vacation. They will travel in June or July to Charleston, SC, Washington, D.C., Hershey, PA, or Chicago, IL. How many different vacations involving one place and one month are possible?

Analyze   What are you asked to find?

What facts are given?

Is there any numerical information you will not use? If so, what?

Choose   What strategy will you use?

You can use the strategy *make an organized list* to show the sample space.

Solve   How will you solve the problem?

List the two months and pair each with the four places the Montoyas might visit.

June, Charleston	July, Charleston
June, Washington, D.C.	July, Washington, D.C.
June, Hershey	July, Hershey
June, Chicago	July, Chicago

So, there are 8 possible choices for the Montoyas' next vacation.

Check   How can you check your answer?

What if the Montoya family adds Kansas City as a choice? How many different vacations are possible?

PROBLEM SOLVING PRACTICE

PROBLEM SOLVING STRATEGIES

Draw a Diagram or Picture

Make a Model

Predict and Test

Work Backward

▶ **Make an Organized List**

Find a Pattern

Make a Table or Graph

Solve a Simpler Problem

Write an Equation

Use Logical Reasoning

Solve by making an organized list.

1. Mrs. Chen is making a dental appointment. It can be on Monday, Tuesday, Wednesday, or Thursday, at 9:30 A.M., 11:00 A.M., or 2:00 P.M. How many choices does Mrs. Chen have?

2. Gilbert is packing for a ski trip. He has 2 jackets, 2 pairs of ski pants, and 2 hats. How many different outfits can Gilbert make if each outfit consists of a jacket, pants, and a hat?

MIXED STRATEGY PRACTICE

For 3–4, use this information. Chad spends $44.39 on two packages of computer disks and a cartridge for his printer.

3. If each package of disks costs $9.95, what equation would you use to find the cost of the cartridge for his printer?
   A $9.95 + c = 44.39$
   B $2(9.95) + c = 44.39$
   C $44.39 - 9.95 = c$
   D $9.95 + 2c = 44.39$

4. Chad's total includes $2.88 in sales tax. What would you do first to find the cost of the items he buys?
   F Add $2.88 to $44.39.
   G Subtract $2.88 from $44.39.
   H Multiply $44.39 by $2.88.
   J Divide $44.39 by $2.88.

5. **Use Data** Use the graph below. How many more visits did Great Smoky Mountains National Park have than Olympic National Park and Yosemite National Park combined? Estimate to determine if your answer is reasonable.

6. Yumi arranged five of her friends in line for a photograph. Alberto stood between Harley and Joshua. Eleni stood between Joshua and Merika. Merika was on the left. Starting from the left, write the order in which Yumi's friends were arranged for the photo.

7. A dance company performs in nonoverlapping groups of 8 and 10. There are 64 dancers in all. There are more groups of 10 than groups of 8. How many groups of each size are there?

8. **?** **What's the Error?** Jay's lunch cost $7.55. He left $0.15 for a 20% tip. Describe his error. How much should a 20% tip be?

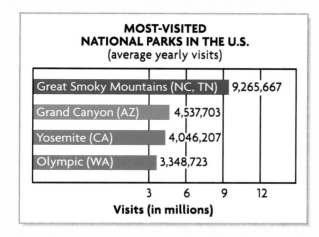

**MOST-VISITED
NATIONAL PARKS IN THE U.S.**
(average yearly visits)

Great Smoky Mountains (NC, TN)	9,265,667
Grand Canyon (AZ)	4,537,703
Yosemite (CA)	4,046,207
Olympic (WA)	3,348,723

Visits (in millions)

Compound Events

Learn how to find all possible outcomes for compound events.

Vocabulary

compound event

tree diagram

Fundamental Counting Principle

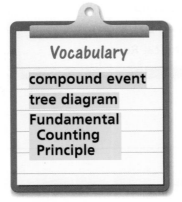

How many choices of a beverage, a sandwich, and a dessert does Carmella have for lunch?

Choosing three different items for lunch is an example of a **compound event**. A compound event includes two or more simple events.

One Way Find the number of possible outcomes for a compound event by drawing a **tree diagram**.

Beverage	Sandwich	Dessert	Outcome
milk (M)	chicken (C)	fruit cup (F) →	M, C, F
		pudding (P) →	M, C, P
	tuna (T)	fruit cup (F) →	M, T, F
		pudding (P) →	M, T, P
	vegetable (V)	fruit cup (F) →	M, V, F
		pudding (P) →	M, V, P
juice (J)	chicken (C)	fruit cup (F) →	J, C, F
		pudding (P) →	J, C, P
	tuna (T)	fruit cup (F) →	J, T, F
		pudding (P) →	J, T, P
	vegetable (V)	fruit cup (F) →	J, V, F
		pudding (P) →	J, V, P

So, Carmella has 12 choices of a beverage, a sandwich, and a dessert.

Another Way Find the number of possible outcomes for 2 or more events by using the **Fundamental Counting Principle**.

Fundamental Counting Principle	If one event has *m* possible outcomes and a second independent event has *n* possible outcomes, then there are *m* × *n* total possible outcomes for the two events together.

EXAMPLE 1

What if the school cafeteria adds 2 more desserts to the menu? Find the total number of beverage-sandwich-dessert selections.

Beverages Sandwiches Desserts
2 × 3 × 4 = 24

So, there are 24 beverage-sandwich-dessert selections.

You can also show the sample space by using a grid to make a table.

EXAMPLE 2

Find the number of possible outcomes if Brianna selects one card at random and spins the pointer on the spinner.

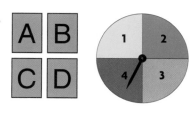

Spinner

Card		1	2	3	4
	A	(A,1)	(A,2)	(A,3)	(A,4)
	B	(B,1)	(B,2)	(B,3)	(B,4)
	C	(C,1)	(C,2)	(C,3)	(C,4)
	D	(D,1)	(D,2)	(D,3)	(D,4)

List the card choices in rows and the spinner choices in columns.

So, there are 16 possible outcomes if Brianna selects one card at random and spins the pointer on the spinner.

CHECK FOR UNDERSTANDING

Think and Discuss ▶ **Look back at the lesson to answer each question.**

1. **What if** Brianna spins the pointer on the spinner in Example 2 and rolls a number cube labeled 1 to 6? Find the number of possible outcomes.

2. **Draw** a tree diagram to show all possible outcomes for tossing three coins.

Guided Practice ▶ **Draw a tree diagram or make a table to find the number of possible outcomes for each situation.**

3. a choice of vanilla, chocolate, or strawberry yogurt in a small, medium, or large cup

4. spinning the pointers on these two spinners

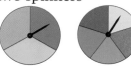

Use the Fundamental Counting Principle to find the number of possible outcomes for each situation.

5. rolling two number cubes labeled 1 to 6

6. a choice of 5 cards, 3 envelopes, and 2 stickers

PRACTICE AND PROBLEM SOLVING

Independent Practice ▶ **Draw a tree diagram or make a table to find the number of possible outcomes for each situation.**

7. a choice of a car or a van with a blue, black, white, or silver exterior, and a black or gray interior

8. tossing a coin and rolling a number cube labeled 1 to 6

Use the Fundamental Counting Principle to find the number of possible outcomes for each situation.

9. a choice of 3 pizza crusts and 6 toppings

10. a choice of 2 ties, 5 shirts, 3 trousers, 2 belts

11. rolling 3 number cubes

12. tossing a coin 4 times

Problem Solving ▶ Applications

13. A cafe offers 12 different sandwiches, 4 different salads, and 8 different soups. If the cafe is open every day of the year, is it possible to have a different combination of sandwich, salad, and soup for lunch every day of the year? Explain.

14. **Write About It** Is it easier to draw a tree diagram, make a table, or use the Fundamental Counting Principle to find the number of possible outcomes for a compound event? Explain.

MIXED REVIEW AND TEST PREP

15. Out of 60 rolls of a number cube, Alicia rolls a six 12 times. What is the experimental probability of rolling a six? (p. 436)

Find the measure of the supplement of each angle. (p. 322)

16. 86°   17. 17°   18. 39°

19. **TEST PREP** Which fraction is greater than the sum of $\frac{3}{16}$ and $\frac{1}{4}$? (p. 166)

   A $\frac{1}{4}$ B $\frac{1}{3}$ C $\frac{3}{8}$ D $\frac{5}{8}$

PROBLEM SOLVING   Thinker's Corner

Explore Combinations Suppose each student is to write a report on 2 of the following careers in the television industry: director, actor, producer, and sound mixer. How many combinations of 2 careers are possible?

A **combination** is a selection of objects or individuals in which the order is not important. You can make an organized list to find the number of combinations.

Choose a career. Pair it with each of the other careers. Do not use any combinations that reverse the pairs already listed. The combination "director, actor" is the same as "actor, director."

director, actor   actor, producer   producer, sound mixer
director, producer   actor, sound mixer
director, sound mixer

So, there are 6 possible combinations of 2 careers.

• How many combinations of 2 careers are possible if another career, makeup artist, is added to the list of report topics?

EXTRA PRACTICE page H54, Set A

Independent and Dependent Events

Learn how to identify and find probabilities of dependent and independent events.

Vocabulary

independent events

dependent events

1. $\frac{1}{4} \times \frac{1}{2}$ 2. $\frac{2}{3} \times \frac{1}{6}$ 3. $\frac{1}{5} \times \frac{1}{4}$ 4. $\frac{1}{2} \times \frac{1}{2}$ 5. $\frac{3}{5} \times \frac{1}{6}$

The table below shows the contents of a bag of marbles. Juan randomly selects a marble from the bag, replaces it, and selects again.

RED	BLUE	GREEN	YELLOW
3	8	5	4

Because Juan replaces the marble after the first selection, the outcome of the second event does not depend on the outcome of the first event. These are **independent events**.

Marbles have been around for at least 4,000 years.

To find the probability of two independent events, multiply their probabilities.

> If A and B are independent events, then
> P(A, B) = P(A) × P(B).

EXAMPLE 1

Remember that to find the theoretical probability of an event with equally likely outcomes, you write the ratio of the number of favorable outcomes to the number of possible outcomes.

What is the probability that Juan first selects a blue marble, replaces it, and then selects a yellow marble?

Find P(blue, yellow).   *First blue, then yellow.*

first selection: P(blue) = $\frac{8}{20}$, or $\frac{2}{5}$ *Find the probability of blue.*

second selection: P(yellow) = $\frac{4}{20}$, or $\frac{1}{5}$ *Find the probability of yellow.*

P(blue, yellow) = $\frac{2}{5} \times \frac{1}{5} = \frac{2}{25}$ *Multiply the probabilities.*

So, the probability is $\frac{2}{25}$, 0.08, or 8%.

EXAMPLE 2

Lisa spins the pointer on the spinner two times.

Find the probability of each event.

A. P(1, 2) = P(1) × P(2) = $\frac{1}{5} \times \frac{1}{5} = \frac{1}{25}$

B. P(not 3, not 4) = P(not 3) × P(not 4) = $\frac{4}{5} \times \frac{4}{5} = \frac{16}{25}$

C. P(1 or 2, 2) = P(1 or 2) × P(2) = $\frac{1+1}{5} \times \frac{1}{5} = \frac{2}{25}$

Dependent Events

Without looking, Isabella selects a magnet from a box, does not replace it, and then passes the box to Jason, who also selects a magnet. Since Isabella does not replace the magnet, Jason's magnet selection depends on the result of Isabella's selection.

If the outcome of the second event depends on the outcome of the first event, the events are called **dependent events**. To find the probability of two dependent events, use the formula below.

> **If A and B are dependent events, then**
> $$P(A, B) = P(A) \times P(B \text{ after } A).$$

EXAMPLE 3

The contents of Isabella's magnet collection are shown in the table. Assuming that Jason and Isabella select without replacement, find the probability that they both select animal magnets.

FLOWERS	ANIMALS	CITIES
4	5	16

Find P(animal, animal).

Isabella's selection:
$P(\text{animal}) = \frac{5}{25} = \frac{1}{5}$

There are 5 animal magnets and 25 equally likely outcomes.

Jason's selection:
$P(\text{animal after animal}) = \frac{4}{24} = \frac{1}{6}$

There are now 4 animal magnets and 24 equally likely outcomes.

$P(\text{animal, animal}) = \frac{1}{5} \times \frac{1}{6} = \frac{1}{30}$

Multiply the probabilities.

So, the probability of Isabella and Jason both selecting animal magnets is $\frac{1}{30}$, $0.0\overline{3}$, or $3\frac{1}{3}\%$.

Math Idea ▶ Compound events consist of either independent events or dependent events. To find the probability of a compound event, multiply the probabilities of the simple events.

CHECK FOR UNDERSTANDING

Think and Discuss ▶ Look back at the lesson to answer each question.

1. **Find** the probability of spinning two even numbers in Example 2. Find the probability of spinning two odd numbers. Express your answers as percents. Are your computed probabilities reasonable? Explain.

2. **Give an example** of two events that are dependent.

Guided Practice ▶ Write *independent* or *dependent* to describe the events.

3. Draw one card from a box, do not replace it, and draw another card.

4. Choose a sandwich and a soup from a menu.

John rolls two number cubes labeled 1 to 6. Find the probability of each event.

5. P(1, 3)   **6.** P(not 2, 4)   **7.** P(7, 3)   **8.** P(1 or 2, 4)

A bag contains five lettered tiles labeled *P, I, Z, Z, A*. Without looking, Kazuhiro selects a tile, does not replace it, and selects another tile. Find the probability of each event.

9. P(*P, A*)   **10.** P(*Z,* not *I*)   **11.** P(*A* or *P, Z*)   **12.** P(vowel, *U*)

PRACTICE AND PROBLEM SOLVING

Independent ▶ Practice

Write *independent* or *dependent* to describe the events.

13. Draw a card from a deck of cards, do not replace it, and draw a second card.

14. Toss a coin three times.

Without looking, you take a card out of the jar and replace it before selecting again. Find the probability of each event. Then find the probability, assuming the card is not replaced after each selection.

15. P(1, 3)   **16.** P(2, 4)   **17.** P(4, 1 or 2)

18. P(4, not 2)   **19.** P(1, 2 or 4)   **20.** P(3, odd)

21. P(2, 2, 2)   **22.** P(2, 1, 2)   **23.** P(4, 1, 1)

Problem Solving ▶ Applications

24. REASONING   Mong has two numbered spinners. The probability of spinning a 5 on both spinners is 0.075. If the probability of spinning a 5 on the first spinner alone is 0.375, what is the probability of spinning a 5 on the second spinner alone?

25. **?** **What's the Question?**   Owen rolls two number cubes labeled 1 to 6. The answer is $\frac{1}{9}$.

26. Donte and his two sisters collect stamps. Donte has twice as many as his older sister, who has 21 stamps, and three times as many as his younger sister. How many stamps do they have in all?

MIXED REVIEW AND TEST PREP

Find the number of possible outcomes of making choices for each situation. (p. 444)

27. 5 shirts, 3 pairs of slacks   **28.** 4 cards, 6 envelopes   **29.** 3 beverages, 6 entrees

30. A bag contains 3 blue, 4 yellow, 5 purple, and 3 red marbles. One marble is chosen at random. Find P(not yellow). (p. 428)

⭐ **31. TEST PREP**   Which figure can always be described as a rectangle? (p. 338)

A rhombus   **B** square   **C** parallelogram   **D** trapezoid

EXTRA PRACTICE page H54, Set B

Make Predictions

Learn how to use sample data to make predictions about a population.

Write in simplest form.

1. $\frac{28}{70}$ 2. $\frac{9}{36}$ 3. $\frac{20}{30}$ 4. $\frac{85}{100}$ 5. $\frac{12}{18}$

Kevin and Shaudra conducted a survey of a random sample of sixth graders at Washington Middle School. The table below shows the results of the survey.

Remember that a population is a group being studied. A sample is a representative part of that group.

FAVORITE PETS	
Pet	**Number of Students**
Cat	30
Dog	25
Hamster	9
Rabbit	5
Bird	3
Other	3

Math Idea ▶ You can use the results from a sample to make predictions about preferences or actions of the population.

EXAMPLE 1

What is the probability that a randomly selected sixth grader at Washington Middle School chooses a cat as his or her favorite pet?

$$P(\text{cat}) = \frac{\text{number of students who prefer cats}}{\text{number of students surveyed}} = \frac{30}{75}, \text{ or } \frac{2}{5}$$

So, the probability that a sixth grader chooses a cat as his or her favorite pet is about $\frac{2}{5}$, 0.40, or 40%.

EXAMPLE 2

There are 210 sixth graders at Washington Middle School. Predict about how many of them would choose cats as their favorite pets.

$$\frac{2}{5} = \frac{n}{210}$$ *Write and solve a proportion.*

$2 \times 210 = 5 \times n$ *Find the cross products.*

$420 = 5n$ *Solve the equation.*

$$\frac{420}{5} = \frac{5n}{5}$$

$84 = n$

So, about 84 sixth graders would choose cats as their favorite pets.

Think and ▶
Discuss

Look back at the lesson to answer each question.

1. **Explain** how to predict how many sixth graders from Washington Middle School would choose rabbits as their favorite pets.

2. **What if** there were 300 sixth graders at Washington Middle School? About how many of them would you predict would choose cats as their favorite pets?

Guided ▶
Practice

The results of a survey of 500 randomly selected teenagers in Iowa indicate that 175 of them use the Internet regularly.

3. What is the probability that a randomly selected teenager in Iowa uses the Internet on a regular basis?

4. Predict about how many Iowa teenagers out of 4,500 use the Internet on a regular basis.

Independent ▶
Practice

Use Data   For 5–6, use the graph.
The graph shows the favorite colors of 100 randomly selected sixth graders from Glenville Middle School.

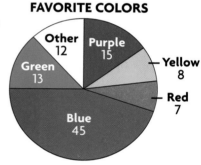

FAVORITE COLORS

Other 12
Purple 15
Yellow 8
Green 13
Red 7
Blue 45

5. If there are 340 sixth graders at Glenville Middle School, about how many more would prefer the color blue than would prefer the color purple?

6. If there are 400 sixth graders at Glenville Middle School, about how many would prefer a color that is not green?

Problem Solving ▶
Applications

7. In a random sample of 400 remote control cars, the quality control department found that 24 were defective. If the company manufactures 10,000 remote control cars, about how many of them would you predict would be defective?

8. ✍ **Write a problem** about making a prediction about a population of 150 sixth graders based on a sample.

There are 6 cards that spell out *DIVIDE*. Suppose you choose one card at random. Find the probability of each event. (p. 428)

9. P(*D*)

10. P(*I* or *D*)

11. P(vowel)

12. Two number cubes labeled 1 to 6 are rolled. Find P(3, not 4). (p. 447)

★ 13. **TEST PREP**   Find $4^3 \div 2^2 + 4$. (p. 278)

   A 6   **B** 8   **C** 20   **D** 64

1. **VOCABULARY** If the outcome of the second event depends on the outcome of the first event, the events are called __?__ events. (p. 448)

2. **VOCABULARY** If the outcome of the second event does not depend on the outcome of the first event, the events are called __?__ events. (p. 447)

Draw a tree diagram or make a table to find the number of possible outcomes for each situation. (pp. 444–446)

3. Miguel tosses a coin and selects 1 marble at random from a jar containing 1 red, 1 blue, 1 yellow, and 1 green marble.

4. To get to work, Lea can walk or ride the subway to Central Station. From there, she takes either a bus, a cab, or a train.

Use the Fundamental Counting Principle to find the number of possible outcomes for each situation. (pp. 444–446)

5. A choice of 4 entrees, 3 drinks, and 2 desserts

6. A choice of 5 ties, 6 shirts, and 2 jackets

For 7–10, suppose you spin the pointer and select a card at random. (pp. 447–449)

7. Find P(green, 1).

8. Find P(red, 2).

9. Find P(red, odd).

10. Find P(blue, 4).

A bag contains 1 green, 2 purple, and 2 white blocks. Without looking in the bag, you choose a block, do not replace it, and then choose another block. For 11–14, find the probability. Write your answer as a fraction, a decimal, and a percent. (pp. 447–449)

11. P(green, white)

12. P(purple, black)

13. P(white or green, purple)

14. P(purple, not white)

Sixty students from Center Middle School were randomly surveyed about their favorite types of movies. The results are shown in the table. For 15–18, find the probability that a student picked at random from Center Middle School prefers each type of movie. (pp. 450–451)

15. drama

16. comedy

17. action

18. horror

Favorite Types of Movies	
Movies	Number of Students
Action	18
Comedy	12
Drama	15
Horror	7
Other	8

19. Suppose there are 300 students at Center Middle School. Predict the number of students who prefer horror movies.

Make an organized list to solve. (pp. 442–443)

20. Armand is making a haircut appointment. It can be on Tuesday, Wednesday, or Friday at 11:00 A.M., 1:00 P.M., 2:30 P.M., or 4:00 P.M. How many possible appointments are there?

Percent of Increase and Decrease

Learn how to find the percent of increase or decrease when given the amount of increase or decrease and the original amount.

The city of Sandville is increasing bus fares. The cost of a monthly bus pass for students will increase from $25 to $26.

Use this formula to find the percent of increase.

$$\% \text{ increase} = \frac{\text{amount of increase}}{\text{original amount}}$$

You divide by the original amount since you are comparing the increase to the original amount.

EXAMPLE 1

Find the percent of increase in a student's bus pass.

$\% \text{ increase} = \dfrac{\text{amount of increase}}{\text{original amount}}$ *Write the formula.*

$\% \text{ increase} = \dfrac{26 - 25}{25} = \dfrac{1}{25}$ *Amount of increase: 26 − 25 = 1 original amount: 25*

$\% \text{ increase} = 0.04, \text{ or } 4\%$ *Write the fraction as a percent.*

So, the percent of increase is 4%.

Use this formula to find the percent of decrease.

$$\% \text{ decrease} = \frac{\text{amount of decrease}}{\text{original amount}}$$

You divide by the original amount since you are comparing the decrease to the original amount.

EXAMPLE 2

Beth observed that 64 elephants drank from a water hole this week. Last week, 80 elephants drank from the water hole. What is the percent of decrease?

$\% \text{ decrease} = \dfrac{\text{amount of decrease}}{\text{original amount}}$ *Write the formula.*

$\% \text{ decrease} = \dfrac{80 - 64}{80} = \dfrac{16}{80}$ *Amount of decrease: 80 − 64 = 16 Original amount: 80*

$\% \text{ decrease} = 0.20, \text{ or } 20\%$ *Write the fraction as a percent.*

So, the percent of decrease is 20%.

TRY IT

Find the percent of increase or decrease.

1. 30 is increased to 39.   **2.** 88 is decreased to 66.

3. $72 is increased to $99.   **4.** 50 is increased to 169.

VOCABULARY

1. A ratio that compares two quantities having different units of measure is a __?__. (p. 385)

2. A fixed percent of the principal, paid yearly, is called __?__. (p. 422)

3. The set of all possible outcomes is called the __?__. (p. 428)

EXAMPLES | **EXERCISES**

Chapter 20

- **Find the unit rate.** (pp. 384–386)

  Apples cost $2.97 for 3 lb. What is the cost per pound?

  $\dfrac{cost}{pounds} = \dfrac{\$2.97}{3}$ *Write a ratio to compare cost to pounds.*

  $\dfrac{2.97 \div 3}{3 \div 3} = \dfrac{0.99}{1}$ *Find the cost per pound.*

  So, the cost is $0.99 per pound.

Find the unit rate.

4. 480 mi in 6 hr
5. $51.96 for 4 CDs
6. 15 books in 5 months
7. 540 mi on 30 gal
8. $1.80 for a dozen rolls
9. 468 tickets for 3 games

- **Use proportions to solve problems using scale drawings.** (pp. 397–399)

  Find the actual length if the drawing length is 6 cm and the scale is 1 cm:5 m.

  $\dfrac{drawing\ (cm)}{actual\ (m)} \rightarrow \dfrac{6}{n} = \dfrac{1}{5}$ *Write a proportion. Let n represent the actual length.*

  $6 \times 5 = n \times 1$ *Find the cross products.*
  $n = 30$

  The actual length is 30 m.

Find the unknown dimension.

10. scale: 3 cm:5 mm
    drawing length: 12 cm
    actual length: ■ mm

11. scale: 1 in.:4 ft
    drawing length: 6 in.
    actual length: ■ ft

12. scale: 2 in.:10 ft
    drawing length: ■ in.
    actual length: 5 ft

13. scale: 4 cm:1 mm
    drawing length: ■ cm
    actual length: 12 mm

Chapter 21

- **Convert among percents, decimals, and fractions.** (pp. 408–411)

  Write 0.4 as a percent.

  $0.4 = 40\%$

  Write $\frac{3}{4}$ as a percent.

  $\frac{3}{4} = \frac{75}{100} = 75\%$

  Move the decimal point two places to the right. Write an equivalent fraction with a denominator of 100.

Write as a percent.

14. 0.9
15. 0.34
16. 2.5
17. $\frac{4}{5}$
18. $\frac{1}{200}$
19. $\frac{5}{8}$

Write as a decimal.

20. 45%
21. $\frac{1}{4}$
22. 135%

Chapter 22

- **Find the theoretical probability of a simple event.** (pp. 428–431)

  A cube is numbered 1 to 6. What is the probability you will roll a number greater than 4?

  $$P(\text{event}) = \frac{\text{number of favorable outcomes}}{\text{number of possible outcomes}}$$

  $$P(\text{number greater than 4}) = \frac{2}{6} = \frac{1}{3}$$

A bag has 10 red marbles, 5 blue marbles, 6 yellow marbles, and 4 green marbles. Find each probability. Express as a fraction, a decimal, and a percent.

23. P(green)

24. P(blue, yellow, or red)

25. P(white)

- **Make predictions based on experimental probabilities.** (pp. 436–437)

COIN	Heads	Tails
TOSS	15	35

  Based upon the experimental results, how many times would you expect to get heads if you tossed the coin 80 times?

  $$P(\text{heads}) = \frac{15}{50} = \frac{3}{10} \qquad \frac{3}{10} \times 80 = 24$$

  You would expect to get heads 24 times.

For 26–27, use the table to find the experimental probability.

26. If you tossed the coin 120 times, how many times would you expect to get tails?

27. If you tossed the coin 30 times, how many heads would you expect to get?

28. A baseball player hit the ball 42 times out of the last 150 times at bat. How many hits can he expect to get during his next 100 times at bat?

Chapter 23

- **Find the number of possible outcomes of a compound event.** (pp. 444–446)

  Find the number of possible outcomes for tossing a coin and selecting card A, B, C, or D.

  Coin: heads or tails—2 choices
  Card: A, B, C, D—4 choices

  $2 \times 4 = 8$; 8 possible outcomes

Find the number of possible outcomes for each situation.

29. picking chocolate or vanilla ice cream and sprinkles, hot fudge, nuts, or whipped cream toppings

30. picking a white, blue, striped, or plaid shirt and black, khaki, or navy pants

---

- **Find the probability of a dependent event.** (pp. 447–449)

  Find the probability of picking an X and then a Y from a bag of cards labeled X, Y, Z, W, Q if you do not replace the cards drawn.

  $$P(X) = \frac{1}{5} \quad P(Y \text{ after } X) = \frac{1}{4}$$

  $$P(X,Y) = \frac{1}{5} \times \frac{1}{4} = \frac{1}{20}$$

Six cards labeled *R, E, V, I, E,* and *W* are in a jar. Suppose you pick a card, do not replace it, and then pick a second card. Find the probability of each event.

31. P(R,V) 32. P(W,E)

33. P(V,E) 34. P(E,E)

PROBLEM SOLVING APPLICATION

35. A drawing of a birdhouse has a scale of 1 in. = 4 in. If the height of the birdhouse on the drawing is 8 in., how high is the actual birdhouse? (pp. 387–389)

Performance Assessment

TASK A • Taxing Decision

James and his friends are shopping at the mall. They are browsing at the sporting goods store and are wondering if they could buy any of these items. Then they remember that there is sales tax on each item. Help James and his friends find the total cost of the items plus tax.

The sales tax on the bicycle is $15.

a. Find the rate of sales tax.

b. At the same rate, find the sales tax on the two other items.

c. If all bikes go on sale at 10% off and shoes are 5% off, how much will two bikes and two pairs of shoes cost, sales tax included?

d. Describe how to use your calculator to find the total cost of the bikes and the shoes.

TASK B • A Hairy Exercise

Collect data from the students in your class. Don't forget to include yourself. Copy and complete the table to show the number of students with each hair color.

Hair Color	Number of Students
Black	
Brown	
Red	
Blond	
Other	

a. Calculate the percent of students with each hair color.

b. Make a circle graph to show the results of your survey.

c. Find out how many students there are in your school. How many students would you expect to have each hair color?

E-Lab • Simulations

Simulations are experiments that are used to predict real-life outcomes.

You can use E-Lab to conduct simulations.

- Click *Simulations*.

- Click the circle.

- Predict how many of 10 random points will fall within the circle.

- Click *10 Random Points*. Do this 8 times.

- Use your results from the 8 trials to predict how many of 10 random points will fall within the circle on the ninth trial.

- Click *10 Random Points*.

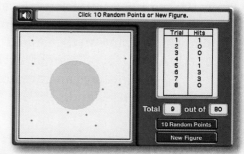

How does the actual outcome compare to your prediction?

Practice and Problem Solving

Use E-Lab. Record the results of each simulation.

1. Click *New Figure*. Click ▪ . Predict how many points will hit the figure. Click *10 Random Points*. Do this 9 times. Record your results. Compare the actual results to your prediction.

2. Click *New Figure*. Click ○ . Predict how many points will hit the figure. Click *10 Random Points*. Do this 9 times. Record your results. Compare the actual results to your prediction.

3. Click *New Figure*. Click ✤ . Predict how many points will hit the figure. Click *10 Random Points*. Do this 9 times. Record your results. Compare the actual results to your prediction.

4. *REASONING* Describe a time when you might use a simulation to predict outcomes.

Multimedia Math Glossary www.harcourtschool.com/mathglossary

5. Vocabulary Visit the Multimedia Math Glossary to locate the terms *independent events* and *dependent events*. Use these terms to describe real-life events.

PROBLEM SOLVING ON LOCATION
In New York City

Ellis Island

From 1892 to 1954, Ellis Island, located in New York harbor, was the first stop for many immigrants to the United States. More than 40 percent of all U.S. citizens alive today have an ancestor who came through Ellis Island.

Where immigrants once entered, New Yorkers and tourists now visit the Ellis Island Museum.

More than 10,700,000 immigrants passed through Ellis Island in its first 40 years of operation. The table shows the number of immigrants from the two countries that sent the greatest number of people through Ellis Island. For 1–4, use the table.

1. About what percent of the immigrants came from Italy?

2. About what percent of the immigrants came from Russia?

Country	Number of Immigrants
Italy	2,500,000
Russia	1,900,000

3. During this period, about 500,000 immigrants came to Ellis Island from Ireland. Suppose you randomly choose the name of an immigrant from this period. Estimate the probability that the immigrant came from Ireland.

4. Find the probability that an immigrant in this 40-year period came from a country other than Italy or Russia.

AVERAGE HIGH/LOW TEMPERATURE (IN °F) FOR NEW YORK												
	Jan	Feb	Mar	Apr	May	Jun	Jul	Aug	Sep	Oct	Nov	Dec
High	38	42	51	65	74	82	84	84	79	70	48	46
Low	26	26	38	40	49	59	67	66	56	46	37	30

Use Data For 5–8, use the table above.

5. What percent of the months have an average high temperature greater than 80°F?

6. What percent of the months have an average low temperature less than 70°F?

7. Write the ratio of the number of months with an average high temperature of 65°F or greater to the number of months with an average high temperature less than 65°.

8. Write the ratio of the number of months with an average low temperature of 50°F or less to the total number of months.

The tablet which the Statue holds in her left hand reads (in Roman numerals) "July 4th, 1776."

Statue of Liberty

As immigrants sailed toward New York harbor with dreams of a new life, their first sight in the distance was the Statue of Liberty. A gift from France in 1886, the Statue of Liberty stands directly across from Ellis Island in New York harbor.

1. When you visit the Statue of Liberty, you can climb 22 stories to the crown. In all, you will walk up 354 steps. Find the number of steps per story. Round your answer to the nearest whole number.

2. To get to the Statute of Liberty, you take a 2-mile ferry ride from Battery Park in New York City. If the ferry ride takes 15 minutes, what is your average speed in miles per hour?

3. The Statue of Liberty weighs 450,000 lb. Sixty percent of its weight is steel. How much of the Statue of Liberty's weight is steel?

Use Data A popular New York City souvenir is a model of the Statue of Liberty. Suppose you made a scale drawing of the statue, using the scale 1 in. = 8 ft, for a school project. For 4–7, use the table.

Statue of Liberty	
Height of base	154 ft
Height of statue	151 ft
Length of hand	$16\frac{1}{2}$ ft
Length of index finger	8 ft

4. Explain what the scale 1 in. = 8 ft means.

5. What would be the length of the index finger in your drawing?

6. What would be the length of the hand in your drawing?

7. How tall would the drawing be from the bottom of the base to the tip of the torch?

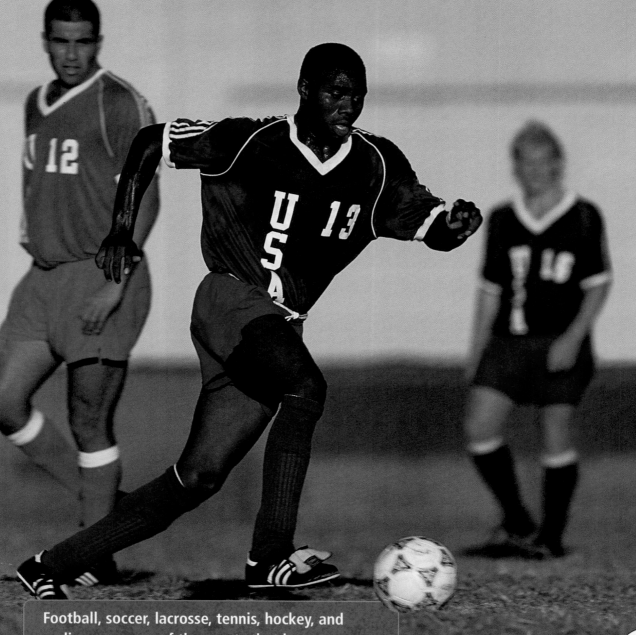

CHAPTER 24 Units of Measure

Football, soccer, lacrosse, tennis, hockey, and curling are some of the sports that have carefully defined rules about the sizes of the courts or fields on which they are played.

PROBLEM SOLVING Lacrosse is played on a field that is 300 ft long. It may be anywhere from 160 to 180 ft wide. Approximately how many times as long is the field as it is wide?

DATA LINK

DIMENSIONS OF PLAYING AREAS		
Sport	**Length (in feet)**	**Width (in feet)**
Tennis		
Singles	78	27
Doubles	78	36
Football	360	160
Soccer	300 to 390	150 to 300
Curling	146	14
Ice Hockey	200	85

Check What You Know

Use this page to help you review and remember important skills needed for Chapter 24.

✓ Vocabulary

Choose the best term from the box.

multiply
divide
feet
yards

1. You can change __?__ to inches by multiplying by 12.

2. To change meters to kilometers, you __?__ by 1,000.

3. To change pints to fluid ounces, you __?__ by 16.

✓ Customary Units (For Intervention, see p. H25.)

Change to the given unit.

4. 1 yd = ■ ft

5. 9 ft = ■ yd

6. 24 in. = ■ ft

7. $4\frac{2}{3}$ ft = ■ in.

8. 6 yd = ■ ft

9. 1 gal = ■ qt

10. 3 gal = ■ qt

11. $1\frac{1}{2}$ lb = ■ oz

12. 5 lb = ■ oz

13. 36 ft = ■ yd

14. 8 qt = ■ gal

15. 60 in. = ■ ft

✓ Metric Units (For Intervention, see p. H25.)

Change to the given unit.

16. 1 m = ■ cm

17. 1 m = ■ mm

18. 2,000 g = ■ kg

19. 1 cm = ■ mm

20. 30 mm = ■ cm

21. 1 km = ■ m

22. 3,000 mL = ■ L

23. 2,000 mm = ■ m

24. 2 cm = ■ mm

25. 4 L = ■ mL

26. 9 kg = ■ g

27. 900 cm = ■ m

✓ Solve Proportions (For Intervention, see p. H26.)

Solve for _n_.

28. $\frac{2}{n} = \frac{6}{84}$

29. $\frac{8}{15} = \frac{24}{n}$

30. $\frac{3}{5} = \frac{30}{n}$

31. $\frac{3}{8} = \frac{n}{24}$

32. $\frac{n}{6} = \frac{8}{12}$

33. $\frac{4}{10} = \frac{n}{25}$

LOOK AHEAD

In Chapter 24 you will

- use and relate customary and metric measurements
- use appropriate tools and units to solve problems

ALGEBRA
Customary Measurements

Learn how to use proportions to convert one customary unit of measurement to another.

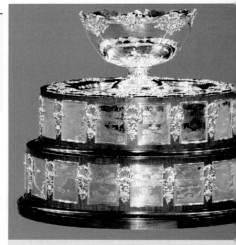

The Davis Cup is 13 in. high and 18 in. across the top.

The Davis Cup is a prized trophy presented to the nation with the highest number of team tennis tournaments won. The trophy was donated in 1900 by the American player Dwight F. Davis. Tournaments are played on a rectangular tennis court that measures 12 yards wide and 26 yards long. How many feet long is the tennis court?

You can write a proportion to convert units of length.

Math Idea ▶ In a proportion, both numerators must have the same units and both denominators must have the same units.

$$\frac{1 \text{ yd}}{3 \text{ ft}} = \frac{26 \text{ yd}}{x \text{ ft}} \rightarrow \frac{1}{3} = \frac{26}{x}$$ *Use the relationship of 1 yard to 3 feet to write a proportion.*

$$1 \cdot x = 3 \cdot 26$$ *Find the cross products.*

$$x = 78$$ *Multiply.*

So, the tennis court is 78 ft long.

• How many feet wide is the tennis court?

You can also use proportions to convert units of weight.

Remember that multiplication problems with variables can be written in different ways.

$7 \times a$ $7(a)$
$7a$ $7 \cdot a$

EXAMPLE

In the powerlifting competition held April 5–6, 1997, Chris Lawton lifted a total of 992,288 lb. How many tons is 992,288 lb?

$$\frac{1 \text{ T}}{2,000 \text{ lb}} = \frac{x \text{ T}}{992,288 \text{ lb}}$$ *Use the relationship of 1 ton to 2,000 pounds to write a proportion.*

$$\frac{1}{2,000} = \frac{x}{992,288}$$ *Find the cross products.*

$$2,000x = 992,288$$ *Solve for x.*

$$\frac{2,000x}{2,000} = \frac{992,288}{2,000}$$

$$x = 496.144$$

So, Chris Lawton lifted 496.144 T or about 496T.

• How many tons is 630,932 lb?

Think and ▶ Discuss

Look back at the lesson to answer each question.

1. **Describe** how you could use a proportion to convert quarts into gallons.

2. **Explain** why sometimes it is important for a proportion to have both numerators with the same units and both denominators with the same units.

Guided ▶ Practice

Use a proportion to convert to the given unit.

3. 2 qt = ■ pt   **4.** 84 in. = ■ ft   **5.** 8 lb = ■ oz

PRACTICE AND PROBLEM SOLVING

Independent ▶ Practice

Use a proportion to convert to the given unit.

6. 10 c = ■ fl oz   **7.** 36 ft = ■ yd   **8.** 8 days = ■ hr

9. 8 oz = ■ lb   **10.** 3.5 T = ■ lb   **11.** 54 in. = ■ yd

12. 20 pt = ■ fl oz   **13.** 15 gal = ■ qt   **14.** 34 pt = ■ qt

15. 36 oz = ■ lb   **16.** 27 c = ■ qt   **17.** 39 in. = ■ ft

Compare. Write <, >, or = for each ●.

18. 31 ft ● 16 yd   **19.** 144 hr ● 6 days   **20.** 77 fl oz ● 9 c

Problem Solving ▶ Applications

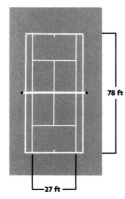

78 ft

27 ft

21. The width of the singles half-court section on a tennis court is $13\frac{1}{2}$ ft. How many yards wide is the singles half-court section?

22. Juan says that each guest will drink $2\frac{1}{2}$ cups of punch. How many gallons of punch will Juan need for 80 guests to each have $2\frac{1}{2}$ cups?

23. **(?) What's the Error?** A sign on a bridge says, "Weight limit 12,000 lb." Eric's truck weighs 12 tons. He says he will be able to drive across the bridge. What's his error? How many pounds does his truck weigh?

MIXED REVIEW AND TEST PREP

24. Write $\frac{4}{5}$ as a percent. (p. 408)

25. $3\frac{2}{5} \div 2\frac{3}{4}$ (p. 210)

26. $\frac{5}{9} - \frac{1}{2}$ (p. 182)

27. $0.218 + 2.143$. (p. 66)

★28. TEST PREP Martha had 132 trading cards. She gave $\frac{1}{4}$ to her sister and $\frac{1}{3}$ to her brother. How many trading cards does she have left? (p. 202)

   A 33   **B** 44   **C** 55   **D** 77

LESSON **24.2**

ALGEBRA
Metric Measurements

Learn how to use proportions to convert one metric unit of measurement to another.

Remember

kilo- = 1,000
hecto- = 100
deka- = 10
deci- = 0.1
centi- = 0.01
milli- = 0.001

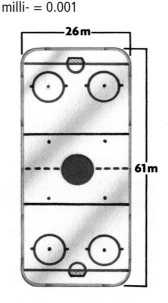

The original Stanley Cup is displayed in the Hockey Hall of Fame in Toronto, Canada.

The Stanley Cup is the oldest trophy competed for by professional ice hockey teams in North America. The trophy was donated in 1892 by Sir Frederick Arthur, Lord Stanley of Preston. Ice hockey games are played on an ice rink 61 meters long and 26 meters wide. How many kilometers long is the ice rink?

You can write a proportion to convert metric units of length.

$\frac{1 \text{ km}}{1,000 \text{ m}} = \frac{x \text{ km}}{61 \text{ m}}$ *Use the relationship of kilometers to meters to write a proportion.*

$\frac{1}{1,000} = \frac{x}{61}$ *Find the cross products.*

$1 \cdot 61 = 1,000x$ *Solve for x.*

$x = 0.061$

So, the ice rink is 0.061 km long.

• How many kilometers wide is the ice rink?

You can also use proportions to convert metric units of liquid.

EXAMPLE

It takes about 30,200 liters of water to make the ice for the ice rink. How many kiloliters of water does it take to make the ice?

$\frac{1 \text{ kL}}{1,000 \text{ L}} = \frac{x \text{ kL}}{30,200 \text{ L}}$ *Use the relationship of kiloliters to liters to write a proportion.*

$\frac{1}{1,000} = \frac{x}{30,200}$ *Find the cross products.*

$1,000x = 30,200$ *Solve for x.*

$\frac{1,000x}{1,000} = \frac{30,200}{1,000}$

$x = 30.2$

So, it takes 30.2 kiloliters of water.

• How many kiloliters are in 1,800 L?

Think and ▶ Discuss

Look back at the lesson to answer each question.

1. **Describe** how you could use a proportion to convert decimeters to meters.

2. **Explain** what you could multiply or divide by to convert one metric measurement to another.

Guided ▶ Practice

Use a proportion to convert to the given unit.

3. 5 m = ■ km **4.** 35 cm = ■ mm **5.** 9 L = ■ mL **6.** 2 g = ■ kg

PRACTICE AND PROBLEM SOLVING

Independent ▶ Practice

Use a proportion to convert to the given unit.

7. 200 kL = ■ L   **8.** 40 dm = ■ cm   **9.** 500 m = ■ km

10. 10 g = ■ kg   **11.** 1,200 m = ■ km   **12.** 12 mm = ■ m

13. 440 g = ■ mg   **14.** 9 L = ■ mL   **15.** 18 kL = ■ L

16. 0.34 m = ■ km   **17.** 425,000 mg = ■ g   **18.** 320 mm = ■ cm

Compare. Write $<$, $>$, or = for each ●.

19. 40 m ● 400 cm   **20.** 200 kL ● 2,000,000 L   **21.** 10 g ● 0.01 kg

Problem Solving ▶ Applications

22. Karen carries her hockey equipment in her gym bag when she goes to practice. When the bag is empty its mass is 2,000 g. What is the mass of the empty bag in kilograms?

23. ❓ **What's the Question?** Marie needs 2,500 mL of soup for a dinner party. The answer is 2.5 L.

24. Jason made a scale drawing of his room. The drawing is 5.5 in. by 6 in. If the scale is 1 in.:2 ft, what are the dimensions of the room?

25. *REASONING* Shawna bought a 3-m length of tissue paper to use for an art project. She used a piece 200 cm long. What fraction of the paper is left over?

MIXED REVIEW AND TEST PREP

26. Solve for n. 8 days = n hours (p. 462)   **27.** Write $\frac{2}{5}$ as a percent. (p. 408)

28. Solve for x. $8.2x = 131.2$ (p. 297)   **29.** What is the opposite of 9? (p. 228)

⭐ **30. TEST PREP** Mina spends $7.60 for $9\frac{1}{2}$ lb of apples. What is the cost per pound? (p. 210)

    A $0.80   **B** $0.84   **C** $0.95   **D** $1.29

ALGEBRA
Relate Customary and Metric

Learn how to convert between customary and metric measurements.

Use a proportion to solve.

1. 10 c = ■ fl oz
2. 5 years = ■ months
3. 60 in. = ■ ft
4. 9 L = ■ kL
5. 4 g = ■ mg

At least 100,000 cubic feet of water passes over the falls every second during daylight hours between April 1 and October 31.

Niagara Falls consists of two waterfalls, the Horseshoe Falls and the American Falls. While Catherine is visiting the Horseshoe Falls, she does some sightseeing in Ontario, Canada. Catherine notices that Canada uses the metric system. She reads that Horseshoe Falls is 792 m wide at its widest point. Catherine wants to find how many feet 792 m is.

Use the table of measures below.

1 in. ≈ 2.54 cm	1 m ≈ 3.28 ft	1 lb ≈ 0.45 kg
1 ft ≈ 30.48 cm	1 mi ≈ 1.61 km	1 qt ≈ 0.95 L
1 yd ≈ 0.91 m	1 oz ≈ 28.35 g	1 gal ≈ 3.79 L

$\dfrac{1\ m}{3.28\ ft} = \dfrac{792\ m}{x\ ft}$ *Use the relationship of meters to feet to write a proportion.*

$\dfrac{1}{3.28} = \dfrac{792}{x}$ *Find the cross products.*

$1 \cdot x = 3.28 \cdot 792$ *Solve for x.*

$x = 2{,}597.76$

So, 792 m ≈ 2,597.76 ft.

EXAMPLES

A. How many kilograms is 365 lb?

THINK: A kilogram is heavier than a pound. So the number of kilograms in the answer should be *less than* 365.

$\dfrac{1\ lb}{0.45\ kg} = \dfrac{365\ lb}{x\ kg}$

$\dfrac{1}{0.45} = \dfrac{365}{x}$

$1 \cdot x = 0.45 \cdot 365$

$x = 164.25$

So, 365 lb ≈ 164.25 kg.

B. How many quarts is 24 L?

THINK: A quart is a bit smaller than a liter. So the number of quarts in the answer should be slightly *greater than* 24.

$\dfrac{1\ qt}{0.95\ L} = \dfrac{x\ qt}{24\ L}$

$\dfrac{1}{0.95} = \dfrac{x}{24}$

$24 = 0.95x$

$\dfrac{24}{0.95} = \dfrac{0.95x}{0.95}$

$25.26 \approx x$

So, 24 L ≈ 25.26 qt.

Think and ▶
Discuss

Look back at the lesson to answer each question.

1. **Tell** if you get a greater or smaller number when you convert miles to kilometers. Explain your reasoning.

2. **Write** the proportion you would use to convert 6 lb to kilograms.

Guided ▶
Practice

Use a proportion to convert to the given unit.

3. 1 in. ≈ ■ mm

4. 4 in. ≈ ■ cm

5. 470 yd ≈ ■ m

6. 70.5 L ≈ ■ gal

7. 25.4 cm ≈ ■ in.

8. 1,000 g ≈ ■ oz

PRACTICE AND PROBLEM SOLVING

Independent ▶
Practice

Use a proportion to convert to the given unit.

9. 2.5 ft ≈ ■ cm

10. 1 lb ≈ ■ g

11. 5 qt ≈ ■ L

12. 10 kg ≈ ■ lb

13. 24 m ≈ ■ yd

14. $\frac{1}{2}$ yd ≈ ■ cm

Compare. Write <, >, or ≈ for each ●.

15. 12 oz ● 737.10 g

16. 33 km ● 20.50 mi

17. 89.85 L ● 15 gal

Problem Solving ▶
Applications

18. Ramona saw a road sign near the Canadian border stating the speed limit was 80 km per hour. To the nearest mile per hour, how many miles per hour is 80 km per hour?

19. Some cameras use film that is 35 mm wide. How many inches wide is 35 mm?

20. ✎ **Write a problem** in which you have to change 137 grams to pounds.

21. More than $\frac{2}{3}$ of Canada has an average January temperature of ⁻18°C. The coastal areas of British Columbia have average January temperatures of 0°C. What is the difference in the average January temperatures?

MIXED REVIEW AND TEST PREP

22. Change 9 L to kL. (p. 464)

23. Write an integer to show a 10 ft decline. (p. 228)

24. Write the ratio one to eight as a fraction. (p. 384)

25. Simplify $\frac{21}{18}$. (p. 160)

⭐26. **TEST PREP** A shirt originally cost $13.65. It was discounted 40% during a sale. What was the sale price of the shirt? (p. 418)

A $5.46   **B** $8.19   **C** $9.10   **D** $10.19

Appropriate Tools and Units

Learn how to measure to a given degree of precision by using the appropriate mathematical tools and units of measure.

Which number is greater?

1. $\frac{1}{8}, \frac{1}{4}$ 2. $\frac{5}{16}, \frac{3}{4}$ 3. $\frac{7}{8}, \frac{6}{40}$ 4. 0.2, 0.02   5. 1.4, 1.41

When Doris runs to train for the track and field competition, she measures her time to the nearest minute. When she runs in the competition, the time is measured to the nearest hundredth of a second.

Doris's time is measured more precisely when seconds are used instead of minutes.

The smaller the unit of measure used, the more precise the measurement is.

Math **I**dea ▶ Metric and customary units are used to measure length, capacity, temperature, and weight (or mass). When you measure quantities, the measurement you get is an approximation.

Activity

You need: 2 different-size bags of rice; quart, pint, and cup measuring tools

Copy the table and record your findings.

	CUP	PINT	QUART
LARGER BAG			
SMALLER BAG			

Work with a partner, and begin with the larger bag of rice.

• Use the quart measure. Record the amount of rice you have to the nearest quart.

• Use the pint measure. Record the amount of rice you have to the nearest pint.

• Use the one-cup measure. Record the amount of rice you have to the nearest cup.

• Repeat the activity with the smaller bag of rice.

Think and Discuss

• Which measurement for each bag of rice is most precise? Why?

• How could you get a more precise measurement?

Measured to the nearest centimeter, this segment is about 4 cm.

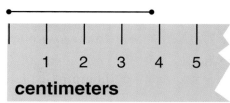

You can measure the length of the same line segment more precisely with a ruler that is divided into millimeters.

A more precise measurement of the length is 38 mm.

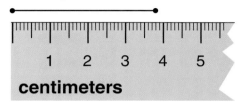

EXAMPLE 1

Measure the segment to the nearest $\frac{1}{4}$ in., $\frac{1}{8}$ in., and $\frac{1}{16}$ in. Which measurement is most precise?

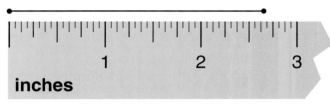

To the nearest $\frac{1}{4}$ in., the line segment is $2\frac{3}{4}$ in. long.

To the nearest $\frac{1}{8}$ in., the line segment is $2\frac{5}{8}$ in. long.

To the nearest $\frac{1}{16}$ in., the line segment is $2\frac{11}{16}$ in. long.

So, $2\frac{11}{16}$ in. is most precise.

When you measure an object, you must decide which unit of measurement is most appropriate. If you wanted to measure the amount of water in a bathtub, for example, you would use liters or gallons. It would not be as appropriate to use milliliters or cups.

EXAMPLE 2

What is an appropriate unit of measure for each item?

A. The mass of two small apples *gram* or *kilogram* Gram is more appropriate.	**B.** The amount of punch per serving *quart* or *cup* Cup is more appropriate.
C. The length of your fingernail *millimeter* or *centimeter* Millimeter is more appropriate.	**D.** The weight of a math book *ounce* or *pound* Pound is more appropriate.

469

Think and
Discuss

Look back at the lesson to answer each question.

1. **Tell** whether the mass of a textbook is more precise measured to the nearest gram or kilogram. Explain your reasoning.

2. **Give** a more precise measurement for a room that is about 12 ft long.

Guided
Practice

Measure the line segment to the given length.

3. nearest centimeter; nearest millimeter

4. nearest inch; nearest half inch

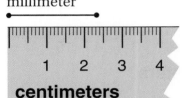

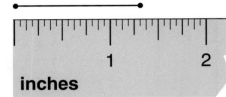

centimeters   **inches**

Tell which measurement is more precise.

5. 7 ft or 85 in.   6. $\frac{1}{2}$ c or $\frac{1}{4}$ c   7. 4 lb or 65 oz   8. 6 cm or 61 mm

Name an appropriate customary or metric unit of measure for each item.

9. length of a classroom

10. liquid in a soft-drink can

11. mass or weight of a new pencil

Independent
Practice

Measure the line segment to the given length.

12. nearest centimeter; nearest millimeter

13. nearest half inch, nearest quarter inch

14. nearest centimeter; nearest millimeter

15. nearest quarter inch; nearest eighth inch

Tell which measurement is more precise.

16. 71 mm or 7 cm   17. 3 yd or 8 ft   18. 3 pt or 5 c   19. $\frac{1}{2}$ lb or 9 oz

Name an appropriate customary or metric unit of measure for each item.

20. weight or mass of a chair

21. thickness of a magazine

22. distance from one city to another

Compare. Write <, >, or = for each ●.

23. 8 qt ● 16 pt   24. 9,000 lb ● 6 T   25. 739 mL ● 0.739 L

Problem Solving ▶ Applications

26. Chris said the county repaired 2,650 yd of highway in her neighborhood. Her brother said the county repaired 1.5 mi of highway. Which measurement is more precise?

27. Which is more precise, measuring to the nearest half centimeter or to the nearest millimeter? Explain.

28. **Write About It** Explain the most precise unit of measurement you might use to measure the length and width of a photograph.

Use Data For 29–30, use the table below.

29. **ALGEBRA** Write an expression to find how much shorter in length the Peace Bridge is than the Brooklyn Bridge and the Golden Gate Bridge combined.

30. What is the length of the Golden Gate Bridge in miles?

BRIDGES	
Bridge	Length in feet
Golden Gate	8,981
Brooklyn Bridge	$1,595\frac{1}{2}$
The Peace Bridge	5,800

MIXED REVIEW AND TEST PREP

31. Change 4.5 gal to L. (p. 466)

32. $^{-}7 - {^{-}12}$ (p. 250)

33. $^{-}12 \times 8$ (p. 252)

34. **TEST PREP** Kim has to allow 45 min to practice her cello before she leaves for soccer practice. She also wants to study for 1 hr 10 min and spend 25 min eating breakfast and getting dressed. If she has to leave for soccer practice at 1 P.M., what time should she get up? (p. 462)

  A 9:40 A.M.   **B** 10:40 A.M.   **C** 11:00 A.M.   **D** 10:40 P.M.

35. **TEST PREP** Which shows 34% of 124? (p. 412)

  F 4.216   **G** 42.16   **H** 421.6   **J** 4,216

PROBLEM SOLVING ⎯ Thinker's Corner

WHAT'S THE UNIT? Choose the appropriate unit of measure from the column on the right. Write the letters in the order shown below the blanks to learn the name of a customary measure that is equal to 4 pecks.

1. Volume of gasoline   **B** mile
2. Length of a football field   **E** yard
3. Distance from Chicago to New York   **H** gram
4. Weight of a hummingbird   **L** gallon
5. Weight of a herd of elephants   **S** pound
6. Weight of a pork roast   **U** ton

    __ __ __ __ __ __
    3   5   6   4   2   1

EXTRA PRACTICE page H55, Set D

PROBLEM SOLVING SKILL
Estimate or Find Exact Answer

Analyze
Choose
Solve
Check

Learn how to decide when to estimate and when to find an exact answer.

Alex is a radio-station disk jockey. He is making a list of songs that should last about a half hour, but no longer. Below is his first list of songs and their playing times. Does he have enough music?

SONG	LENGTH	SONG	LENGTH
Color Me Blue	4.5 min	Smile on Me	5.7 min
Top Dog	3.6 min	Kelso Blues	4.3 min
Hittin' the Road	7.2 min	A Long Time Ago	6.4 min
Stand Up and Shout	2.6 min		

To find out if he has enough music, Alex can estimate the total amount of time for the songs in his list. To estimate, he rounds each amount of time to the nearest minute.

$$5 + 4 + 7 + 3 + 6 + 4 + 6 = 35$$

Since his estimate is 35 min, he has more music than he needs.

Alex also has to play 5.7 min of commercials. How much music time does he need to cut?

Now Alex needs to know the exact amount of time. He adds the lengths of the songs and the lengths of the commercials.

$$4.5 + 3.6 + 7.2 + 2.6 + 5.7 + 4.3 + 6.4 + 5.7 = 40$$

He has 40 min of music and commercials, so he needs to cut the music by 10 min.

Math **I**dea ▶ Sometimes an estimate is all you need to answer a problem and sometimes you need to find an exact answer.

Talk About It ▶ • Which songs could Alex remove from his list so he has 30 min of music and commercials? Is an estimate or exact answer needed?

Decide whether you need an estimate or an exact answer. Solve.

1. Eric is having a party. He is planning for each of his six guests to have two 0.25-lb meat patties. He buys four packages of meat weighing 6 oz, 9 oz, 4.5 oz, and 11.5 oz. Does he have enough meat?

2. Sally uses cereal, $\frac{1}{2}$ c of nuts, $\frac{1}{4}$ c of coconut, and 1 c of raisins to make trail mix. She wants to make a total of 1 qt of mix. How many cups of cereal does she need to add?

3. Sherry took $20.00 to spend at the mall. She paid $8.95 for a blouse and $5.25 for lunch at the food court. Will she be able to spend $10.00 at another store?

4. Ronny takes $50.00 to the music store to buy three books of music. He spends $13.95, $18.99, and $15.75 on the music. How much money does Ronny have left?

MIXED APPLICATIONS

Karen has a checking account at her bank. The bank charges $4 per month plus $0.25 for each check she writes.

5. Which of the following expressions can be used to determine the amount Karen pays each month? Let c = the number of checks.

   A $4 \times 0.25c$ **C** $4 + 0.25c$

   B $4c + 0.25$ **D** $4 - 0.25c$

6. Karen is charged $6.00 by the bank during the month of September. How many checks did she write in September?

   F 2   **H** 8

   G 6   **J** 24

Use Data For 7–9, use the table at the right.

7. How much more rain fell during the month with the most rainfall than the month with the least rainfall?

8. Jon kept records of the rainfall at his home for six months. Estimate the total amount of precipitation from November through April.

Precipitation In Inches

Nov.	Dec.	Jan.	Feb.	Mar.	Apr.
$5\frac{2}{5}$	$6\frac{3}{4}$	$8\frac{1}{2}$	$10\frac{1}{4}$	$5\frac{5}{8}$	$1\frac{1}{10}$

9. The record rainfall for the months November through April is 40 inches. How much above or below the record is this year's rainfall?

10. The month of April has 30 days. What percent of the dates in April are prime numbers?

11. The sum of Maris's and Ramona's ages is 40 yr. Maris is 6 yr older than Ramona. How old is Maris?

12. Thirty-six students try out for baseball and basketball. If 20 students try out only for baseball and 8 try out for both sports, how many try out only for basketball?

13. ✎ **Write a problem** about measurement for which an estimate would be appropriate as an answer. Then write a related problem for which an exact answer is needed.

Use a proportion to convert to the given unit. (pp. 462–463)

1. 5 ft = ▪ in.

2. 4 yd = ▪ in.

3. 8 qt = ▪ gal

4. 40 oz = ▪ lb

5. 6 c = ▪ pt

6. 2.3 T = ▪ lb

Use a proportion to convert to the given unit. (pp. 464–465)

7. 50 mm = ▪ cm

8. 3,000 mg = ▪ g

9. 100 g = ▪ kg

10. 3.7 L = ▪ mL

11. 1.25 km = ▪ m

12. 4,700 mm = ▪ m

Use a proportion to convert to the given unit. Use the table on page 466. Round to the nearest hundredth if necessary. (pp. 466–467)

13. 2 ft ≈ ▪ cm

14. 4 oz ≈ ▪ g

15. 3.2 gal ≈ ▪ L

16. 70 cm ≈ ▪ in.

17. 200 g ≈ ▪ oz

18. 6 L ≈ ▪ qt

19. 7 in. ≈ ▪ mm

20. 190 m ≈ ▪ yd

21. 6 km ≈ ▪ mi

Measure the line segment to the given length. (pp. 468–471)

22. nearest centimeter; nearest millimeter

23. nearest half inch; nearest quarter inch

24. nearest centimeter; nearest millimeter

25. nearest quarter inch; nearest eighth inch

26. nearest centimeter; nearest millimeter

27. nearest inch; nearest half inch

Tell which measurement is more precise. (pp. 468–471)

28. 2 cm or 21 mm

29. 4 ft or 49 in.

30. 27 lb or 427 oz

31. 4,210 g or 4 kg

32. 4,827 mL or 5 L

33. 785 mg or 1 g

Name an appropriate customary or metric unit of measure for each item. (pp. 468–471)

34. weight of a person

35. thickness of a large phone book

36. height of a 1-story home

37. Mike is having a party. He wants each of his guests to have at least 3 cups of punch. If there are 40 guests, how many gallons of punch are needed? (pp. 462–463)

38. The newspaper reported that the new courthouse was 97 ft tall. A television reporter said the courthouse was 32 yd tall. Which measurement is more precise? (pp. 468–471)

Tell whether you need an estimate or an exact answer. Solve. (pp. 472–473)

39. Marla has 5 fl oz of lime juice, 16 fl oz of orange juice, 12 fl oz of white grape juice, and 6 c of pineapple juice for a punch. Does she have enough for twenty 5-fl oz servings?

40. Amy and 3 friends are working on a craft project. Each person needs 39 cm of ribbon. They have 1.25 m of ribbon. How much more do they need?

Decide on a plan.

See item **5.**

Use the ratio of 1 inch to the given number of centimeters. Then write a proportion using the ratio for 6 inches to an unknown number of centimeters.

Also see problem **4**, p. H63.

Choose the best answer.

1. How many feet are in 240 inches?

   A 2 ft   **C** 20 ft

   B 12 ft   **D** 24 ft

2. How many meters are in 12.5 kilometers?

   F 12,500 m

   G 1.25 m

   H 0.125 m

   J 0.0125 m

3. Which measurement is the most precise?

   A $\frac{1}{2}$ ft

   B 6 in.

   C $6\frac{1}{2}$ in.

   D $6\frac{1}{4}$ in.

4. Sue bought 3 pounds of cheddar cheese and 12 ounces of Swiss cheese. Altogether, how many ounces of cheese did she buy?

   F 48 oz   **H** 68 oz

   G 60 oz   **J** Not here

5. Which proportion could be used to determine the approximate number of centimeters in 6 inches?
   (Use 1 in. $\approx$ 2.54 cm.)

   A $\frac{1}{6} = \frac{x}{2.54}$

   B $\frac{x}{6} = \frac{1}{2.54}$

   C $\frac{1}{2.54} = \frac{6}{x}$

   D $\frac{1}{x} = \frac{6}{2.54}$

6. Which term best describes the shape of a cereal box?

   F Rectangular prism

   G Triangular prism

   H Cone

   J Cylinder

7. Which symbol makes this a true number sentence?

   $$^-6 \bullet {}^+1$$

   A $<$ **C** $=$

   B $>$ **D** $-$

8. On a map, the distance between Harding and Gilford is 8 inches. In the scale, 1 inch represents 40 miles. What is the actual distance?

   F 5 mi

   G 128 mi

   H 240 mi

   J 320 mi

Write What You Know

9. Describe how you can convert 18 kg to grams.

10. Find a relationship between fluid ounces and milliliters by using these facts: 1,000 mL = 1 L; 32 fl oz = 1 qt; and 1 qt $\approx$ 0.95 L. Explain how you found your answer.

CHAPTER 25 Length and Perimeter

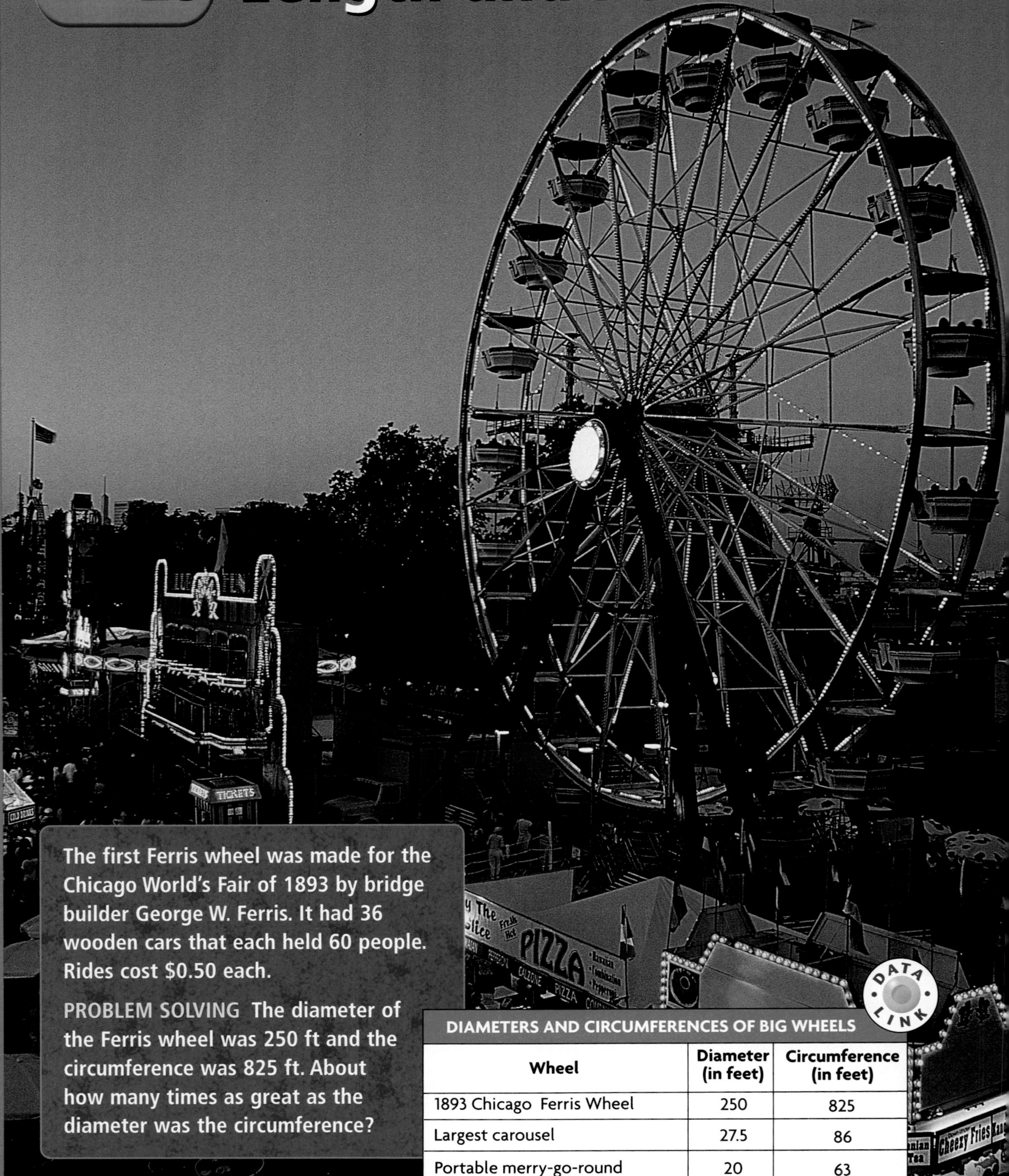

The first Ferris wheel was made for the Chicago World's Fair of 1893 by bridge builder George W. Ferris. It had 36 wooden cars that each held 60 people. Rides cost $0.50 each.

PROBLEM SOLVING The diameter of the Ferris wheel was 250 ft and the circumference was 825 ft. About how many times as great as the diameter was the circumference?

DATA LINK

DIAMETERS AND CIRCUMFERENCES OF BIG WHEELS

Wheel	Diameter (in feet)	Circumference (in feet)
1893 Chicago Ferris Wheel	250	825
Largest carousel	27.5	86
Portable merry-go-round	20	63
2000 London Millennium Wheel	443	1,392

Check What You Know

Use this page to help you review and remember important skills needed for Chapter 25.

✅ Vocabulary

Choose the best term from the box.

area
perimeter
radius

1. The distance around a figure is the ___?___ .

2. A line segment with one endpoint at the center of a circle and the other endpoint on the circle is a ___?___ .

✅ Perimeter (For Intervention, see p. H26.)

Find the perimeter of the figure.

3.
8 ft
8 ft

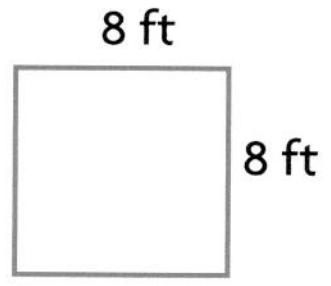

4.
12.5 m 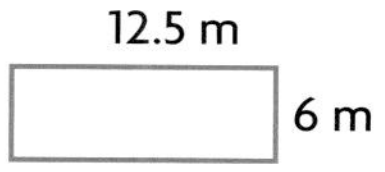
6 m

5.
6 cm   5 cm 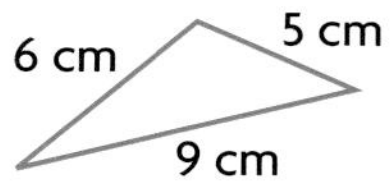
9 cm

6.
0.5 m   1.3 m
1.2 m

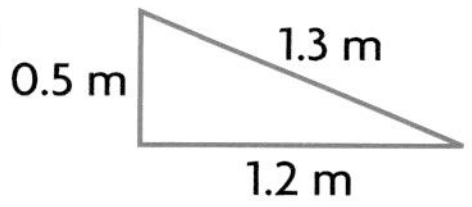

7.
26 in.
26 in.

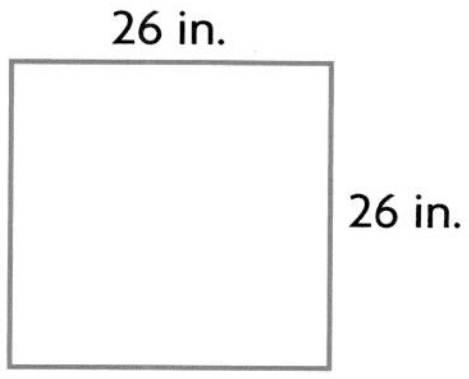

8.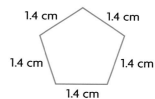
1.4 cm   1.4 cm
1.4 cm   1.4 cm
1.4 cm

✅ Change Units (For Intervention, see p. H27.)

Change to the given unit.

9. 48 in. = ▦ ft

10. 4 yd = ▦ ft

11. 12 yd = ▦ ft

12. 3 mi = ▦ yd

13. 40 mm = ▦ cm

14. 72 in. = ▦ yd

15. 3 m = ▦ mm

16. 7 m = ▦ cm

17. 5 km = ▦ m

18. 4.9 cm = ▦ mm

19. 0.09 m = ▦ km

20. 8.2 cm = ▦ m

✅ Multiply with Fractions and Decimals (For Intervention, see p. H23.)

21. $3\frac{1}{7} \times 28$

22. $2\frac{2}{3} \times 112$

23. $12\frac{1}{2} \times 210$

24. 3.7×9

25. 21.06×25

26. 3.14×8

27. $3\frac{3}{4} \times 16$

28. 5.25×14

29. 0.8×92

30. $15 \times 9\frac{4}{5}$

> ### LOOK AHEAD
>
> **In Chapter 25 you will**
> - find the perimeter of polygons
> - find the circumference of circles

Estimate Perimeter

MATH LAB

Explore how to estimate perimeter.

You need metric ruler, string, perimeter worksheet.

Mrs. Johnson is working with the Jackson Middle School student council to clean up the shores of Pine Lake. She gives all the members drawings of the lake and asks them to estimate the perimeter. Mrs. Johnson can then assign teams of students to clean portions of the lake's shore.

You can estimate the perimeter, or distance around a figure, by using string and a ruler.

Activity

- Lay a piece of string around the perimeter of the drawing of the lake. Mark the string where it meets itself.

- Decide on a metric unit of measure. Use the ruler to measure the string from its beginning to the mark you made on the string.

- Compare your measurement with those of the other students.

Pine Lake

Think and Discuss

- **Explain** why there are different estimates for the perimeter of the drawing of the lake.

- ***REASONING*** Suppose the scale of the drawing is 1 cm = 100 m. How many meters would your estimate represent? How many kilometers?

Practice

Use string and a ruler to estimate the perimeter of each.

1. the outline of your hand

2. the outline of your shoe sole

Perimeter

Learn how to find the perimeter of a polygon.

Activity

You need: metric ruler

• Use the rectangle below. Measure the length of each side to the nearest centimeter.

Think and Discuss

• What is the perimeter to the nearest centimeter?

• How could you get a more precise measurement for the perimeter? What would the perimeter be then?

• What is another way to find the perimeter of a rectangle?

Math Idea ▶ The perimeter, P, of a polygon is the distance around it. To find the perimeter of any polygon, you can use a formula.

EXAMPLE 1

Tony is building a brick patio with a wooden frame. The lengths of the sides are $8\frac{1}{3}$ ft, $7\frac{1}{2}$ ft, $5\frac{3}{4}$ ft, $4\frac{2}{3}$ ft, and $6\frac{1}{4}$ ft. Tony needs to find the perimeter so he will know how much wood he needs for the frame. What is the perimeter?

$P = a + b + c + d + e$ 　　　　*Write a formula.*

$P = 8\frac{1}{3} + 7\frac{1}{2} + 5\frac{3}{4} + 4\frac{2}{3} + 6\frac{1}{4}$ 　*Replace the variables with the lengths.*

$P = \left(8\frac{1}{3} + 4\frac{2}{3}\right) + \left(5\frac{3}{4} + 6\frac{1}{4}\right) + 7\frac{1}{2}$ 　*Use the Commutative and Associative Properties.*

$P = 13 + 12 + 7\frac{1}{2}$ 　　　　*Use mental math to add.*

$P = 32\frac{1}{2}$

So, the perimeter is $32\frac{1}{2}$ ft.

Since the opposite sides of a rectangle are equal in length, you can find the perimeter by finding 2 × length and 2 × width and then adding the products. You can write a formula to find the perimeter of a rectangle. The formula is $P = 2l + 2w$.

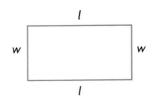

EXAMPLE 2

Find the perimeter of Sara's rectangular yard.

$P = 2l + 2w$ *Write the formula.*
$P = (2 \times 28) + (2 \times 15)$ *Replace l with 28 and w with 15.*
$P = 56 + 30$ *Add the products.*
$P = 86$

So, the perimeter is 86 yd.

15 yd

28 yd

Sometimes you know the perimeter of a figure, but the length of one side is unknown.

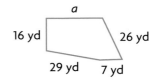

EXAMPLE 3

The polygon at the right has a perimeter of 105 yd. Find the unknown length.

$P = a + b + c + d + e$ *Write the formula.*
$105 = a + 16 + 29 + 7 + 26$ *Use the values you know.*
$105 = a + 78$ *Add the known lengths.*
$105 - 78 = a + 78 - 78$ *Solve for a.*
$27 = a$

So, the unknown length is 27 yd.

EXAMPLE 4

The length of a rectangle is 1 cm more than 3 times the width. What is the perimeter of the rectangle if the width is 32 cm?

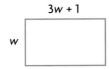

First find the length.
$l = 3w + 1$ *Length is 1 more than 3 times the width.*
$l = (3 \cdot 32) + 1$ *Replace w with 32.*
$l = 96 + 1$ *Multiply.*
$l = 97$ *Add.*

Find the perimeter.
$P = 2l + 2w$ *Write the formula.*
$P = 2(97) + 2(32)$ *Replace w with 32 and l with 97.*
$P = 194 + 64$
$P = 258$

So, the perimeter is 258 cm.

CHECK FOR UNDERSTANDING

Think and Discuss ▶ Look back at the lesson to answer each question.

1. **Explain** how to find the unknown length of one side of a triangle when you know the perimeter and the lengths of the other two sides.

2. **Tell** what formula you could write to find the perimeter of a regular pentagon.

Guided Practice ▶ Find the perimeter.

3. 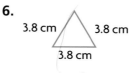 5.56 m, 3.8 m, 4.21 m

4. 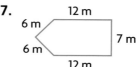 $4\frac{3}{4}$ in. $4\frac{3}{4}$ in. $4\frac{3}{4}$ in. $4\frac{3}{4}$ in. $4\frac{3}{4}$ in.

5. 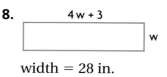 3 ft, 18 in.

PRACTICE AND PROBLEM SOLVING

Independent Practice ▶ Find the perimeter.

6. 3.8 cm, 3.8 cm, 3.8 cm

7. 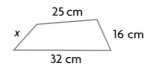 12 m, 6 m, 7 m, 6 m, 12 m

8. 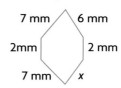 $4w + 3$, w, width = 28 in.

The perimeter is given. Find the unknown length.

9. 25 cm, x, 16 cm, 32 cm

   perimeter = 86.5 cm

10. 7 mm, 6 mm, 2mm, 2 mm, 7 mm, x

    perimeter = 30 mm

Problem Solving Applications ▶

11. 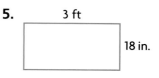 **ALGEBRA** Write the formula to find the perimeter of the square at the right. If $x = 31$ ft, what is the perimeter? x ft

12. **Write a problem** that can be solved by finding perimeter.

MIXED REVIEW AND TEST PREP

13. Solve. 13.7 cm ≈ ■ in. (p. 466)

14. Name the figure. ▱ (p. 338)

15. Solve. $316 = \frac{a}{4}$ (p. 297)

16. Write an expression for 43.8 times a number. (p. 82)

☆ 17. **TEST PREP** Mario earned $45 one week and $75 the next. He saved $30 and spent the rest. What percent of the money he earned did Mario spend? (p. 412)

   A 25% **B** $33\frac{1}{3}$% **C** 50% **D** 75%

EXTRA PRACTICE page H56, Set A

481

PROBLEM SOLVING STRATEGY
Draw a Diagram

Learn how to use the strategy *draw a diagram* to find perimeter.

1. $(2 \times 12) + (2 \times 7)$ **2.** $4\frac{3}{4} + 2\frac{1}{4} + 7\frac{1}{4}$

3. 4×9.8 **4.** $23 + 47 + 42$

5. $62.03 + 12.7 + 5.19 + 35.28$

Roy has a rectangular piece of land that he wants to put a barbed wire fence around. He will place a post every 9 ft along the perimeter. Each post is 5 ft tall. The land is 63 ft long and 45 ft wide. How many posts is Roy going to need?

Analyze

What are you asked to find?

What information are you given?

Is there information you will not use? If so, what?

Choose

What strategy will you use?

Use the strategy draw a diagram to find the number of posts needed to go around the perimeter of Roy's land.

Solve

How will you solve the problem?

Draw a rectangle that is similar to the shape of Roy's land. Place marks along the perimeter of the rectangle to represent the posts.

Count the number of marks you placed around the rectangle. Each corner should have only one mark. So, Roy will need 24 posts.

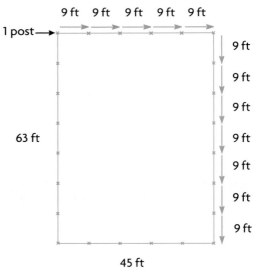

Check

How can you check your answer?

What if Roy wanted to place a post every 8 ft along a rectangular perimeter 64 ft long and 48 ft wide? How many posts would he need?

PROBLEM SOLVING PRACTICE

PROBLEM SOLVING STRATEGIES

▶ **Draw a Diagram or Picture**
Make a Model
Predict and Test
Work Backward
Make an Organized List
Find a Pattern
Make a Table or Graph
Solve a Simpler Problem
Write an Equation
Use Logical Reasoning

Solve the problem by drawing a diagram.

1. Each hexagon on a quilt is made up of six congruent equilateral triangles with 11-in. sides. Each triangle shares two sides with other triangles. What is the perimeter of one hexagon?

2. Darin's pool is 4 yd wide and 10 yd long. He wants to put a cement deck around the pool. It will be $5\frac{1}{2}$ ft wide on all sides. What will be the perimeter of the deck?

For 3–4, use this information.
A window is in the shape of a square with an isosceles triangle attached at the top. The side of the triangle that is attached to the square is the same length as one side of the square.

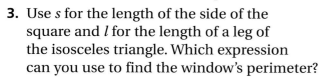

3. Use s for the length of the side of the square and l for the length of a leg of the isosceles triangle. Which expression can you use to find the window's perimeter?

   A $2s + 2l$ **C** $4s + 2l$
   B $3s + 2l$ **D** $4s + 3l$

4. If the square is 2.5 m on a side and each leg of the isosceles triangle is 1.8 m, how much molding will the builder have to buy to go around the window?

   F 6.8 m   **H** 11.1 m
   G 8.6 m   **J** 13.6 m

MIXED STRATEGY PRACTICE

Use Data **For 5–7, use the circle graph.**

5. Which three categories make up about 50% of the expenses? Explain.

6. Sharon spent a total of $98 this year and a total of $81 last year for her fall party. She spent the same percent for pumpkins this year as last year. How much more did she spend on pumpkins this year?

7. What percent of the expenses was spent on decorations, entertainment, and costumes?

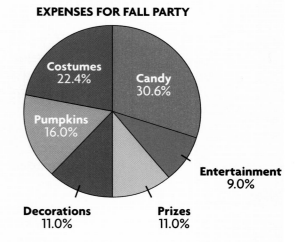

EXPENSES FOR FALL PARTY

Costumes 22.4%
Candy 30.6%
Pumpkins 16.0%
Entertainment 9.0%
Decorations 11.0%
Prizes 11.0%

8. There are 90 ft between bases on major league baseball diamonds and 60 ft between bases on Little League diamonds. How much farther does a major league player have to run if he hits a home run?

9. Kathy wants to work on her car. She needs a 25-mm wrench but her wrenches are all in customary units. To the nearest inch, what size of a customary wrench should she use?

10. Ali and Walt are taking a 50-question test. Ali answers every sixth question incorrectly and Walt answers every eighth question incorrectly. Which questions do both Ali and Walt answer incorrectly?

11. ❓ **What's the Question?** May, Kevin, and Sean chopped wood. May chopped $\frac{2}{5}$ of the wood, Kevin chopped $\frac{1}{10}$ of it, and Sean chopped the rest. The answer is $\frac{1}{2}$.

Circumference

1. 3.14×8 **2.** 5.5×2 **3.** $14 \div 2$ **4.** $2 \times 3 \times 3.14$ **5.** $2 \times 5 \times 3.14$

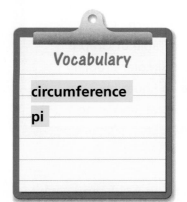

Vocabulary

circumference

pi

Find the **circumference**, or distance around a circle, by using a compass, string, and a ruler.

> ### Activity 1
>
> **You need:** compass, string, ruler, calculator
>
> Find the circumference of a circle with a radius of 4 in.
>
> - Open the compass to a width of 4 in. Use the compass to construct a circle with radius of 4 in.
>
> - Lay the string around the circle. Mark the string where it meets itself.
>
> - Use the ruler to measure the string from its beginning to the mark you made.
>
> - What is the diameter of the circle? What is the circumference of the circle?
>
> - Divide the circumference by the diameter. About how many times as long is the circumference?
>
> - Share your results with those of other students. What do you notice about the relationship between the diameter and the circumference of circles with different diameters?

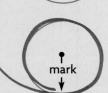

4 in.

mark

The ratio of the circumference to the diameter, $\frac{C}{d}$, is the same for any circle. This ratio is called **pi**, or π. The value of π is usually approximated as 3.14 or $\frac{22}{7}$.

EXAMPLE 1

The Space Needle is about 605 ft high.

The circular restaurant at the top of the Space Needle in Seattle has a diameter of 94.5 ft and makes one revolution every hour. To the nearest foot, how far will a person seated at the edge of the restaurant travel in one revolution?

You need to find the circumference of the restaurant. When you know the diameter of a circle, you can use the formula $C = \pi d$.

$C = \pi d$	*Write the formula.*
$C \approx 3.14 \times 94.5$	*Replace π with 3.14 and d with 94.5*
$C \approx 296.73$	
$C \approx 297$	*Round to the nearest foot.*

So, a person would travel about 297 ft.

EXAMPLE 2

The Vienna Giant Ferris wheel was completed in 1897. It has a radius of 30.48 m. What is the circumference of the Ferris wheel? Round to the nearest hundredth.

$C = 2\pi r$ *Write the formula.*

$C \approx 2 \times 3.14 \times 30.48$ *Replace π with 3.14 and r with 30.48.*

$C \approx 191.4144$

$C \approx 191.41$ *Round to the nearest hundredth.*

So, the circumference is about 191.41 m.

Sometimes it makes sense to use the approximation $\frac{22}{7}$ for π.

EXAMPLE 3

TECHNOLOGY LINK

To learn more about *Circumference,* watch the **Harcourt Math Newsroom Video,** *Lighthouse Repair.*

Sean has a circular outdoor wooden planter with a radius of $3\frac{1}{2}$ in. He wants to put trim around the top. What is the length of the trim he will need?

$C = 2\pi r$ *Write the formula.*

$C \approx \frac{2}{1} \times \frac{22}{7} \times 3\frac{1}{2}$ *Replace π with $\frac{22}{7}$ and r with $3\frac{1}{2}$.*

$C \approx \frac{2}{1} \times \frac{\overset{11}{\cancel{22}}}{\underset{1}{\cancel{7}}} \times \frac{\overset{1}{\cancel{7}}}{\underset{1}{\cancel{2}}}$ *Write 2 and $3\frac{1}{2}$ as fractions. Simplify.*

$C \approx \frac{2}{1} \times \frac{11}{1} \times \frac{1}{1}$ *Multiply.*

$C \approx \frac{22}{1}$

$C \approx 22$

So, Sean will need about 22 in. of trim.

You can also use the π key on a calculator. If you use a calculator, you may have to round your answer.

EXAMPLE 4

Use a calculator with a π key. Find the circumference of a circle with diameter 12.7 m. Round to the nearest tenth.

Use this key sequence.

| SHIFT | π | × | 12.7 | = | *39.8982267* |

$C \approx 39.8982267$

$C \approx 39.9$ *Round to the nearest tenth.*

So, the circumference is about 39.9 m.

485

Think and ▶
Discuss

Look back at the lesson to answer each question.

1. **Compare and contrast** the two formulas you can use to find the circumference of a circle.

2. **Express** $\frac{22}{7}$ as a decimal rounded to the nearest hundredth. How does it compare to 3.14?

Guided ▶
Practice

Find the circumference of the circle.
Use 3.14 or $\frac{22}{7}$ for π. Round to the nearest whole number.

3.
5 m

4.
$4\frac{3}{8}$ ft

5.
1.9 cm

6. $r = 12.9$ mm

7. $d = 90$ yd

8. $r = 14$ mi

9. $d = 8.6$ in.

10. $r = 9.9$ m

11. $d = 10.6$ cm

Independent ▶
Practice

Find the circumference of the circle.
Use 3.14 or $\frac{22}{7}$ for π. Round to the nearest whole number.

12.
14 m

13.
$6\frac{1}{2}$ in.

14.
7.6 cm

15.
$1\frac{2}{5}$ yd

16.
$2\frac{1}{3}$ in.

17.
35 cm

18. $r = 29.62$ km

19. $d = 100.5$ in.

20. $r = 6\frac{2}{3}$ yd

21. $d = 7$ ft

22. $r = 3.5$ m

23. $d = 5.1$ cm

24. Look at the figure of a circle inside a square. The length of each side of the square is 3.5 ft. What is the circumference of the circle?

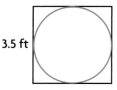
3.5 ft

Find the radius for each circumference.

25. $C \approx 28.26$ ft

26. $C \approx 47.1$ m

27. $C \approx 58.404$ cm

28. A Ferris wheel at the local amusement park has a circumference of 800 ft. This is the same distance as one complete revolution. Find the radius of the Ferris wheel.

29. **REASONING** How does the circumference of two circles compare if the diameter of one is twice the diameter of the other? Give an example to explain your reasoning.

30. **❓ What's the Error?** Rhonda and Caren each used 3.14 to find the circumference of a circle with a radius of 34 mm. Rhonda's answer is 213.52 mm and Caren's is 106.76. Which girl made an error? What is the error?

31. George Ferris created the first Ferris wheel in 1893. The wheel had a maximum height of 264 ft. How many yards is this?

32. **ALGEBRA** The cost of four pizzas was shared equally among six friends. Let c represent the total cost of the pizzas. Write an algebraic expression that shows the amount each friend paid.

MIXED REVIEW AND TEST PREP

33. Find the perimeter of a rectangle that is 46.3 m long and 28.2 m wide. (p. 479)

34. Is 131 prime or composite? (p. 148)

35. **TEST PREP** $12\frac{2}{3} \div 5\frac{3}{5}$ (p. 210)

  A $2\frac{11}{42}$ **B** $3\frac{11}{42}$ **C** $7\frac{1}{2}$ **D** $17\frac{6}{15}$

36. **TEST PREP** At 8:00 A.M. the temperature was 9°F. At 4:00 P.M. the temperature had dropped 12°F. At 11:00 P.M. the temperature was ¯7°F. How much greater was the temperature at 4:00 P.M. than at 11:00 P.M.? (p. 244)

  A ¯4°F **B** ¯3°F **C** 3°F **D** 4°F

PROBLEM SOLVING

LiNKÜP to Social Studies

Making Money The United States uses the dollar as the basic unit of money. France uses the franc and Mexico uses the peso. Artists play a big part in producing money. A new coin begins with an artist's design. The artist makes a clay model of the coin, and a mold is made from that model. A plaster cast is made from the mold. The artist then carves the details of the coin into the plaster.

1. Use a metric ruler to measure the diameter of a penny, nickel, dime, and quarter to the nearest millimeter. Find the circumference of each coin.

2. Research different countries to find the sizes of the coins they use. Compare the sizes of their coins with the ones used in the United States.

The Pythagorean Theorem

Explore how to use the Pythagorean theorem and its converse.

You need red and blue construction paper, graph paper, scissors.

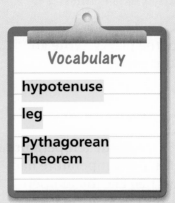

Vocabulary

hypotenuse

leg

Pythagorean Theorem

1. 8^2 **2.** 13^2 **3.** 2.3^2 **4.** 30^2

5. Find the circumference of a circle whose radius is 4.8 cm.

An important and famous relationship in mathematics involves the three sides of a right triangle. One special type of right triangle is an isosceles right triangle.

Activity 1

- Use construction paper to make the squares and isosceles triangle shown at the right.

- Cut apart the two red squares and fit them on top of the blue square.

- What relationship did you find among the areas of the squares?

- Try other isosceles right triangles. Is the relationship true for them?

Activity 2

- Make a graph-paper diagram with squares like the ones at the right.

- What relationship is there between the areas of the red squares and the area of the blue square?

- What are the lengths of the sides of the triangle?

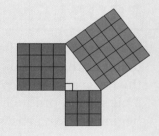

The sides of a right triangle have special names. The side opposite the right angle is the **hypotenuse.** The other two sides of the right triangle are the **legs.**

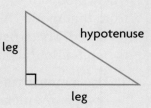

The relationship you discovered is called the **Pythagorean Theorem**. This theorem says that if a and b are the lengths of the legs of a right triangle and c is the length of the hypotenuse, then $a^2 + b^2 = c^2$.

- When you know the lengths of two sides of a right triangle, you can use the Pythagorean Theorem to find the third side. Study this problem.

The legs of a right triangle are 5 ft and 12 ft. Find the length of the hypotenuse.

$a^2 + b^2 = c^2$
$5^2 + 12^2 = c^2$ *Replace the variables.*
$25 + 144 = c^2$ *Solve for c.*
$169 = c^2$
$\sqrt{169} = c$
$13 = c$ So, the hypotenuse has length 13.

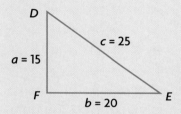

- You can use the Pythagorean Theorem to tell if a triangle is a right triangle. Study this problem.

Is $\triangle DEF$ a right triangle?

$a^2 + b^2 = c^2$
$15^2 + 20^2 \overset{?}{=} 25^2$ *Replace the variables.*
$225 + 400 \overset{?}{=} 625$ *Perform the operations.*
$625 = 625$

Since $a^2 + b^2 = c^2$, $\triangle DEF$ is a right triangle.

Think and Discuss

- **REASONING** What could you conclude if $a^2 + b^2 \neq c^2$?

Practice

Find the length of the hypotenuse of the right triangle.

1.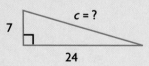

2.

3.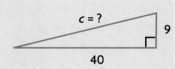

Determine whether each triangle is a right triangle. Write *yes* or *no*.

4. $a = 10$, $b = 24$, $c = 26$

5. $a = 5$, $b = 12$, $c = 19$ ft

6. $a = 4$, $b = 5$, $c = 6$

7. $a = 12$, $b = 35$, $c = 37$

MIXED REVIEW AND TEST PREP

Find the circumference of the circle. Use 3.14 for π. (p. 484)

8. $r = 6$ ft

9. $r = 17.5$ mm

10. $d = 9$ m

11. $d = 100$ in.

⭐ **12. TEST PREP** A quadrilateral has a perimeter of 19.5 yd. Three of the sides have lengths 5 yd, 7 yd, and 2.25 yd. Find the length of the fourth side. (p. 479)

1. VOCABULARY The distance around a polygon is the ___?___ .
(pp. 479–481)

2. VOCABULARY The distance around a circle is called the ___?___ .
(pp. 484–487)

3. VOCABULARY In a right triangle, the side opposite the right angle is the ___?___ .
(pp. 488–489)

Find the perimeter. (pp. 479–481)

4.
35 yd
19 yd

5.
$9\frac{1}{4}$ ft
5 ft
$12\frac{1}{2}$ ft

6.
8.2 cm   8.2 cm
14.7 cm   14.7 cm
10.5 cm

7.
19 in.
6 in.   14 in.
24 in.

8.
5.8 m
5.8 m   5.8 m
5.8 m   5.8 m
5.8 m

9.
$4\frac{1}{10}$ ft $5\frac{1}{5}$ ft
$6\frac{7}{10}$ ft $8\frac{1}{2}$ ft

The perimeter is given. Find the unknown length. (pp. 479–481)

10.
13.4 m   11.9 m
11.06 m   11.06 m
x

$P = 65.7$ m

11.
13.8 cm   13.8 cm
22.2 cm   22.2 cm
x

$P = 84$ cm

12.
13 mi
35 mi
r
29 mi

$P = 135$ mi

Find the circumference. Use 3.14 or $\frac{22}{7}$ for π. Round to the nearest whole number. (pp. 484–487)

13.
9 in.

14.
35 ft

15.
4.3 m

16.
12.5 in.

17.
15.6 m

18.
60.8 km

Solve. (pp. 482–483)

19. An official tournament-size dartboard has a radius of 9 in. Find its circumference to the nearest whole number.

20. Juli has a piece of fabric 32 in. long and 25 in. wide. What is its perimeter?

Get the information you need.

See item **5.**

You need to remember the formula for finding circumference when you know the radius of a circle. Since the choices have decimals, you should use 3.14 for pi.

Also see problem **3**, p. H63.

Choose the best answer.

1. What is the perimeter of this figure?

8 ft

12 ft

A 10 ft   **C** 40 ft
B 20 ft   **D** 100 ft

2. The circumference of a circle measures about 38 inches. Which is a reasonable estimate of the radius?

F 6 in.   **H** 20 in.
G 12 in.   **J** 120 in.

3. The diagram shows the dimensions of a lot. The perimeter of the lot is 211 feet. What is the unknown length?

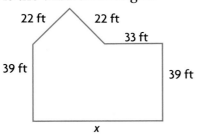

22 ft   22 ft

33 ft

39 ft   39 ft

x

A 23 ft   **C** 37 ft
B 35 ft   **D** Not here

4. One side of a regular pentagon measures 6.25 centimeters. What is the perimeter of the pentagon?

F 12.5 cm   **H** 30 cm
G 25 cm   **J** 31.25 cm

5. Which is the best approximation of the circumference of a circle with a 3.5 cm radius?

A 10.9 cm   **C** 22.0 cm
B 12.3 cm   **D** 38.5 cm

6. The principal of Lincoln Middle School wants to display data about the number of absent students for each day during the past three months. Which type of graph would be best to use if she wants to identify the median number quickly?

F Histogram
G Stem-and-leaf plot
H Circle graph
J Multiple-line graph

7. Gil walks dogs to earn money. On Monday he spent $1\frac{2}{3}$ hours walking dogs. On Thursday he spent $2\frac{1}{2}$ hours walking dogs. What is the total time he spent walking dogs?

A $3\frac{1}{2}$ hr   **C** $4\frac{1}{6}$ hr
B $3\frac{3}{5}$ hr   **D** $4\frac{1}{3}$ hr

8. A survey showed that 1 out of every 25 students at Hillside High takes German. Which of the following decimals represents this ratio?

F 0.04   **H** 0.25
G 0.12   **J** 0.4

Write What You Know

9. A rectangle has a perimeter of 24 ft. The width of the rectangle is 3 ft. Find the length of the rectangle. Explain how you found your answer.

10. Draw and label a rectangle using *l* for length and *w* for width. Then explain what the formula $P = 2l + 2w$ means and why it makes sense.

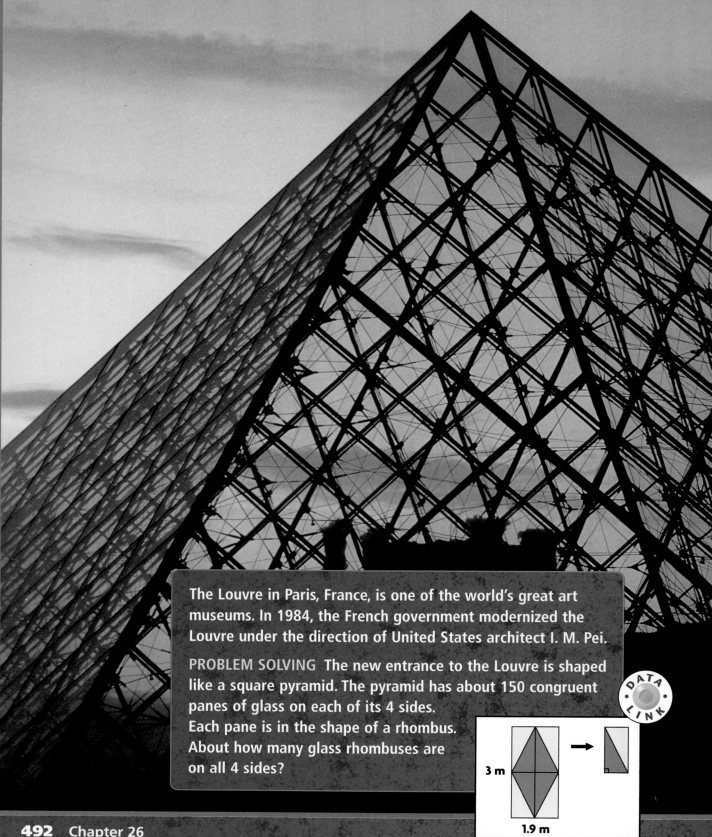

The Louvre in Paris, France, is one of the world's great art museums. In 1984, the French government modernized the Louvre under the direction of United States architect I. M. Pei.

PROBLEM SOLVING The new entrance to the Louvre is shaped like a square pyramid. The pyramid has about 150 congruent panes of glass on each of its 4 sides. Each pane is in the shape of a rhombus. About how many glass rhombuses are on all 4 sides?

3 m

1.9 m

Check What You Know

Use this page to help you review and remember important skills needed for Chapter 26.

 Vocabulary

Choose the best term from the box.

1. To multiply a number by itself is to ___?___ that number.

2. A ___?___ is a symbol that represents a number.

3. The polygons that form the sides of prisms and pyramids are called ___?___ .

> cube
> faces
> square
> variable

 Find the Square of a Number (For Intervention, see p. H27.)

Find the square of each number.

4. 8	**5.** 12	**6.** 15	**7.** 20
8. 50	**9.** 9	**10.** 1.3	**11.** 2.4
12. $1\frac{1}{2}$	**13.** 49	**14.** 83	**15.** 129

 Evaluate Expressions (For Intervention, see p. H18.)

Evaluate each expression.

16. lw for $l = 6$ and $w = 9$

17. $\frac{1}{2}bh$ for $b = 5$ and $h = 10$

18. $2ab$ for $a = 4$ and $b = 12$

19. $3c^2$ for $c = 8$

20. $3(w + 5)$ for $w = 7$

21. $\frac{1}{2}s^2 + 2s$ for $s = 4$

 Faces of Prisms and Pyramids (For Intervention, see p. H20.)

Tell the number of faces of each figure.

22.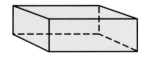

23.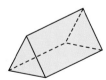

24.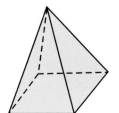

25.

LOOK AHEAD

In Chapter 26 you will

- estimate and find the areas of rectangles, triangles, parallelograms, trapezoids, and circles
- find the surface areas of prisms and pyramids

Estimate and Find Area

Learn how to estimate area and find the area of rectangles and triangles.

QUICK REVIEW

1. 12×6 2. 9×40 3. 3×96
4. 50×20 5. 7×42

Elena wants to put a layer of black netting on her garden to keep the weeds from growing through the mulch. She needs to know the area of her garden so she will know how much netting to buy.

It is not easy for Elena to find the area of her garden since it is not a polygon. She made a drawing of her garden on grid paper to estimate the area. Each square on the grid represents 1 ft².

The **area** of a figure is the number of square units needed to cover it. Use this drawing of Elena's garden to estimate the area.

Remember that area is measured in square units such as square feet (ft²) and square meters (m²).

There are 12 full red squares.

There are 6 almost-full blue squares.

There are 4 half-full orange squares. Combine the half-full squares to make 2 full squares. Do not count the squares that are less than half full.

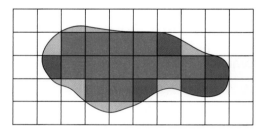

Add the number of squares you counted.

$12 + 6 + 4 \times \frac{1}{2} = 12 + 6 + 2 = 20$

So, the area of Elena's garden is about 20 ft².

You can also use a grid to estimate the area of polygons. In the rectangle at the right, there are 28 full squares and 4 almost-full squares. There are 0 half-full squares.

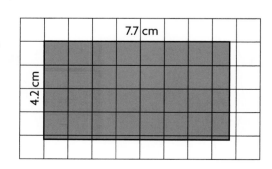

$28 + 4 = 32$

So, the area is about 32 cm².

You can use the formula $A = lw$ to find the area of a rectangle and the formula $A = s^2$ to find the area of a square.

EXAMPLE 1

Find the area of the rectangle shown at the right.

$A = lw$ *Write the formula.*
$A = 16 \times 13$ *Replace l with 16 and w with 13.*
$A = 208$ *Multiply.*

So, the area of the rectangle is 208 in.^{2}

13 in. (width)
16 in. (length)

EXAMPLE 2

Cinzia is going to wallpaper the wall shown at the right. How many square feet of wallpaper will she need?

Think of the wall as a rectangle that is 8 ft × 6 ft and a square that is 4 ft × 4 ft.

Find the area of each figure and then find the sum.

Rectangle	Square
$A = lw$	$A = s^2$
$A = 8 \times 6$	$A = 4^2 = 4 \times 4$
$A = 48$	$A = 16$

Total area: $48 + 16 = 64$

So, Cinzia will need 64 ft^2 of wallpaper.

6 ft
8 ft   4 ft
     4 ft
10 ft

6 ft
8 ft   4 ft
     4 ft
10 ft

If you draw the diagonal of the rectangle at the right, you divide it into two congruent right triangles. The area of each triangle is half that of the rectangle. The area of each of the right triangles is

$\frac{1}{2}lw$, or $\frac{1}{2} \times 16 \times 13 = 104$.

16 in.
(height or width) 13 in.
13 in.
16 in. (base or length)

In triangles, the dimensions are called the base (b) and the height (h). So, the formula for the area of a triangle is written as $A = \frac{1}{2}bh$.

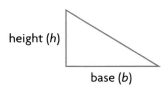

height (h)
base (b)

EXAMPLE 3

Use the formula $A = \frac{1}{2}bh$ to find the area of the triangle at the right.

$A = \frac{1}{2}bh$ *Write the formula.*

$A = \frac{1}{2} \times 8 \times 5$ *Replace b with 8 and h with 5.*

$A = \frac{1}{2} \times 40$ *Multiply.*

$A = 20$ So, the area is 20 cm^2.

5 cm
8 cm

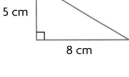

495

Look at triangle ABC at the right below.

Area of triangle ABD: $A = \frac{1}{2}bh = \frac{1}{2} \times 4 \times 6$.

Area of triangle CBD: $A = \frac{1}{2}bh = \frac{1}{2} \times 7 \times 6$.

To find the area of triangle ABC, find the sum of those areas.

$A = \left(\frac{1}{2} \times 4 \times 6\right) + \left(\frac{1}{2} \times 7 \times 6\right)$

$A = \left(\frac{1}{2} \times 24\right) + \left(\frac{1}{2} \times 42\right)$ *Multiply.*

$A = 12 + 21$ *Add.*

$A = 33$ So, the total area of triangle ABC is 33 in.²

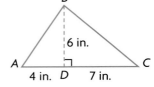

- Find the area of triangle ABC by using 4 in. + 7 in., or 11 in., as the base and 6 in. as the height. Compare your answer to 33 in.²

Sometimes the height is shown outside the triangle. The height of this triangle is 1.8 m.

$$A = \frac{1}{2} \times 2.8 \times 1.8 = 2.52$$

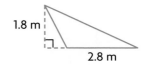

CHECK FOR UNDERSTANDING

Think and ▶ Discuss

Look back at the lesson to answer the question.

1. **Explain** how the areas of a rectangle and a triangle with the same base and height are related.

Guided ▶ Practice

Estimate the area of the figure. Each square on the grid is 1 m².

2.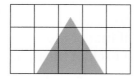

3.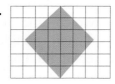

4.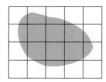

Find the area.

5.

6.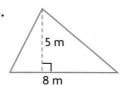

7.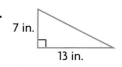

PRACTICE AND PROBLEM SOLVING

Independent ▶ Practice

Estimate the area of the figure. Each square on the grid is 1 m².

8.

9.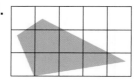

10.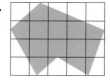

Find the area.

11.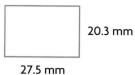

20.3 mm

27.5 mm

12.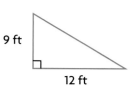

9 ft

12 ft

13.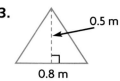

0.5 m

0.8 m

14.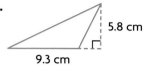

5.8 cm

9.3 cm

15.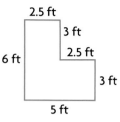

2.5 ft

3 ft

2.5 ft

6 ft

3 ft

5 ft

16.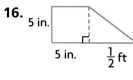

5 in.

5 in.

$\frac{1}{2}$ ft

Problem Solving ▶
Applications

17. The diagram shows a backyard. It will be covered with grass. How many square yards of grass are needed?

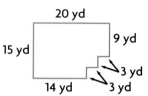

20 yd

15 yd

9 yd

3 yd

14 yd

3 yd

18. *REASONING* The length of this rectangle is twice the width.
   a. Find the area of the rectangle.

   b. Find the perimeter of the rectangle.

2w

w

MIXED REVIEW AND TEST PREP

19. Evaluate $\frac{1}{2} \times c$ for $c = 24$. (p. 216)

20. Evaluate $2d - 6$ for $d = {}^-6$. (p. 272)

21. Write $\frac{3}{5}$ as a percent. (p. 169)

22. TEST PREP If Rod is twice as tall as Cathy, and Rod's height is x, what expression represents Cathy's height? (p. 216)

   A $2x$ **B** x **C** $\frac{1}{2}x$ **D** $x - 2$

23. TEST PREP Which shows the circumference of a circle with a radius of 13 cm? (p. 486)

   F 23.55 cm   **G** 40.82 cm   **H** 81.64 cm   **J** 93.01 cm

---

PROBLEM SOLVING ✦ Thinker's Corner

ALGEBRA How is the area of a rectangle changed when its length and its width are each doubled?

For Rectangle 1, Area = $l \times w$.

For Rectangle 2, Area = $(2l) \times (2w)$.

$$= (2 \times 2) \times (l \times w)$$

$$= 4 \times (l \times w)$$

Rectangle 1 w

l

Rectangle 2 2w

2l

So, the area of a rectangle is four times as great when its length and width are each doubled.

• Use algebra to show how the area of a triangle changes when the base and height are each doubled.

EXTRA PRACTICE page H57, Set A

ALGEBRA
Areas of Parallelograms and Trapezoids

Learn how to find the areas of parallelograms and trapezoids.

QUICK REVIEW

1. $(12 \times 9) \div 2$
2. $(4.2 \times 6) \div 2$
3. $(9\frac{1}{2} \times 3) \div 2$
4. $(240 \times 4) \div 2$
5. $(83.5 \times 24.2) \div 2$

A contractor has to put glass tiles on the side of an escalator. The side has the shape of a parallelogram. How many square feet of tiles will the contractor need?

Activity

You need: scissors, graph paper

• The parallelogram at the right is a drawing of the side of the escalator. Draw the parallelogram on a piece of graph paper and cut it out.

height (*h*) 3 ft
base (*b*) 22 ft

• Cut along the dotted line. Move the triangle to the right side to form a rectangle.

length (*l*) 22 ft
width (*w*) 3 ft

• What is the area of the rectangle? What is the area of the parallelogram?

• How are the dimensions and area of the parallelogram related to those of the rectangle?

• What formula can you write for the area of a parallelogram?

You can use the formula for the area of a rectangle to write a formula for the area of a parallelogram.

$A = lw$ *The length of the rectangle is the base of the parallelogram.*
↓ *The width of the rectangle is the height of the parallelogram.*
$A = bh$

Use $A = bh$ to find the area of the side of the escalator.

$A = bh$
$A = 22 \times 3$ *Replace b with 22 and h with 3.*
$A = 66$

So, the area of the side of the escalator is 66 ft².

> **EXAMPLE 1**

Find the area of the parallelogram at the right.

$A = bh$ *Write the formula.*

$A = 1.5 \times 2.6$ *Replace b with 1.5 and h with 2.6.*

$A = 3.9$ *Multiply.*

So, the area is 3.9 m².

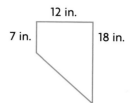
2.6 m
1.5 m

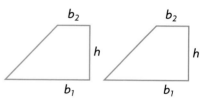
12 in.
7 in.　18 in.

The trapezoid at the right is the shape of the glass tiles used by the contractor. You can use what you know about finding the area of a rectangle to find the area of each glass tile.

If you place two of these trapezoids end to end, you form a rectangle.

The length of the rectangle is $b_1 + b_2$, where b_1 and b_2 are the lengths of the two bases of the trapezoid. The width of the rectangle is the height of the trapezoid.

The area of the rectangle is $(b_1 + b_2) \times h$.

Since each trapezoid is half of the rectangle, you can use $\frac{1}{2} \times (b_1 + b_2) \times h$, or $\frac{1}{2}h(b_1 + b_2)$ to find the area.

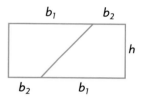
b_2 b_2
h h
b_1 b_1

b_1 b_2
h
b_2 b_1

> **EXAMPLE 2**

Find the area of each glass tile. Use the bases of 7 in. and 18 in. and the height of 12 in.

$A = \frac{1}{2}h(b_1 + b_2)$ *Write the formula.*

$A = \frac{1}{2} \times 12 \times (18 + 7)$ *Replace h with 12, b_1 with 18, and b_2 with 7.*

$A = \frac{1}{2} \times 12 \times 25$ *Multiply.*

$A = \frac{1}{2} \times 300$

$A = 150$

So, the area is 150 in.²

CHECK FOR UNDERSTANDING

Think and ▶ Discuss

Look back at the lesson to answer each question.

1. **Tell** how to find the area of the trapezoid at the right.

2. **Explain** which formula you would use to find the area of a rhombus.

4.2 m
4 m
6.5 m

Guided ▶ Practice

Find the area of each figure.

3.
2 ft
5 ft

4.
3 in.
10 in.
12 in.

5.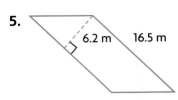
6.2 m   16.5 m

PRACTICE AND PROBLEM SOLVING

Independent ▶ Practice

Find the area of each figure.

6.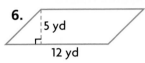
5 yd
12 yd

7.
16.5 m
8.3 m

8.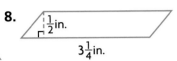
$\frac{1}{2}$ in.
$3\frac{1}{4}$ in.

9.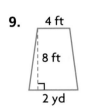
4 ft
8 ft
2 yd

10.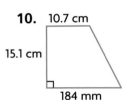
10.7 cm
15.1 cm
184 mm

11.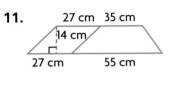
27 cm   35 cm
14 cm
27 cm   55 cm

Problem Solving ▶ Applications

12. When triangular flaps are raised on opposite sides of the square table at the right, it has the shape of a parallelogram. The base of each triangle is 2 ft. What is the area of the table top?

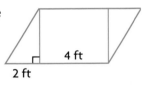
4 ft
2 ft

13. REASONING Copy the parallelogram in Exercise 6 and the trapezoid in Exercise 9. Show how you can divide each figure into 2 triangles to find the area. Then find the areas and compare your answers to the answers you originally got for Exercises 6 and 9.

14. Four months ago the JB Paper Company had 1,480 employees. Three months ago it had 1,590 employees and two months ago it had 1,700. Describe the pattern. Then find the number of employees one month ago if the pattern continued.

MIXED REVIEW AND TEST PREP

15. Find the area of a triangle where $b = 4$ m and $h = 12$ m. (p. 494)

16. $7\frac{1}{6} \times \frac{3}{4}$ (p. 206)   **17.** $3 \text{ g} = \blacksquare \text{ kg}$ (p. 464)   **18.** $^-1,125 \div {^-15}$ (p. 252)

★ **19. TEST PREP** A bakery has a 40 lb bag of flour. Six pounds of flour are used to make cakes. Half of what is left is used to make cookies. How much flour remains in the bag after the cookies are made? (p. 22)

A 17 lb   **B** 18 lb   **C** 19 lb   **D** 20 lb

EXTRA PRACTICE page H57, Set B

Area of a Circle

M A T H LAB

Explore how to find the area of a circle.

You need compass, scissors.

TECHNOLOGY LINK

More Practice: Use E-Lab, *Area of a Circle.*
www.harcourtschool.com/ elab2002

Remember that the formula for the circumference of a circle is $C = \pi d$ or $C = 2\pi r$.

π is approximately 3.14, or $\frac{22}{7}$.

Activity

To see how the area of a circle and its radius are related, you can rearrange the circle to approximate a parallelogram.

- Use the compass to construct a circle on a piece of paper.

- Cut out the circle. Fold it three times as shown.

- Unfold it, and trace the folds. Shade one half of the circle as shown.

- Cut along the folds. Fit the pieces together to make a figure that approximates a parallelogram.

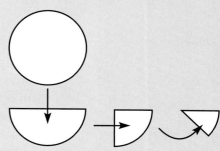

Think and Discuss

Think of the figure as a parallelogram. The base and the height of the parallelogram relate to the parts of the circle.

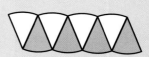

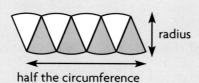

radius

half the circumference

base (b) = $\frac{1}{2}$ the circumference of the circle, or πr

height (h) = the radius of the circle, or r

- What is the formula for the area of a parallelogram?

- Use the formula for the area of a parallelogram to write a formula for the area of a circle. Use πr for the base and r for the height.

- Use your formula to find the area of a circle with a radius of 7 yd to the nearest whole number.

Practice

Find the area of each circle with the given radius. Use your formula for the area of a circle, and round to the nearest meter.

1. $r = 1$ m   **2.** $r = 3$ m   **3.** $r = 7$ m   **4.** $r = 4$ m

ALGEBRA
Areas of Circles

Learn how to use a formula to find the area of a circle.

Jackie is on the girls' wrestling team at her high school. The photo at the right shows a wrestling mat used for practice and competition. Square mats used for high school students have sides measuring 38 ft long. The circle inside the square is the wrestling circle.

EXAMPLE 1

Find the area of the wrestling circle for a radius of 14 ft. Use the formula for the area of a circle, $A = \pi r^2$. Use $\frac{22}{7}$ for π.

$A = \pi r^2$ *Write the formula.*

$A \approx \frac{22}{7} \times (14)^2$ *Replace π with $\frac{22}{7}$ and r with 14.*

$A \approx \frac{22}{7} \times 196$

$A \approx \frac{22}{7} \times \frac{\overset{28}{196}}{1}$ *Simplify.*

$A \approx \frac{22}{1} \times \frac{28}{1}$ *Multiply.*

$A \approx \frac{616}{1}$, or 616

So, the area of the wrestling circle is about 616 ft^2.

• **What if** you use 3.14 for π? What do you get as the area of the wrestling circle, to the nearest whole number?

Sometimes you are given the diameter of a circle.

EXAMPLE 2

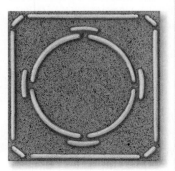

The wrestling circle for sumo, a Japanese form of wrestling, has a diameter of 4.6 m. What is the area of the circle, to the nearest whole number? Use 3.14 for π.

$4.6 \div 2 = 2.3$ *Find the radius.*

$A = \pi r^2$ *Write the formula.*
$A \approx 3.14 \times (2.3)^2$ *Replace π with 3.14 and r with 2.3.*
$A \approx 3.14 \times 5.29$
$A \approx 16.6106$

So, the area of the circle is about 17 m^2.

Think and ▶
Discuss

Look back at the lesson to answer each question.

1. **Explain** how to find the area of a circular dartboard with a diameter of $1\frac{1}{2}$ ft.

2. **Tell** which has the greater area: a circle with a radius of 5 m or a circle with a radius of 50 cm.

Guided ▶
Practice

Find the area of each circle to the nearest whole number.

3.
4.
5.
6.

PRACTICE AND PROBLEM SOLVING

Independent ▶
Practice

Find the area of each circle to the nearest whole number.

7.
8.
9.
10.

11. $r = 4$ mm   12. $r = 0.9$ m   13. $d = 66$ in.   14. $d = 100$ ft

Find the area of the semicircle or quarter circle to the nearest whole number. Use 3.14 for π.

15.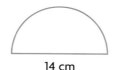

14 cm

16.

8 cm

Problem Solving ▶
Applications

17. College wrestling rules require that the mat have a wrestling circle with a minimum diameter of 32 ft. What is the minimum area of the wrestling circle to the nearest whole number?

18. **? What's the Error?** A circle has a diameter of 18 m. Sean says the area is 28 m² to the nearest meter. What error did Sean make? What is the area?

MIXED REVIEW AND TEST PREP

19. Find the area of a parallelogram with a base of 12.5 ft and height of 8 ft. (p. 498)

20. $^-15 \times ^-5$ (p. 252)

21. Complete. $\frac{16}{24} = \frac{2}{\blacksquare}$ (p. 160)

22. Find the LCM of 5, 8, and 20 (p. 150)

☆23. **TEST PREP** Rectangle *A* is 12 ft by 18 ft. Rectangle *B* is 9 ft by 13 ft. How much greater is the perimeter of *A* than the perimeter of *B*? (p. 479)

A 16 ft   **B** 44 ft   **C** 60 ft   **D** 99 ft

ALGEBRA
Surface Areas of Prisms and Pyramids

Learn how to find the surface areas of prisms and pyramids.

Vocabulary

surface area

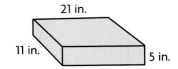

Sam wants to paint the box above. He wants to find the total area of the outside of the box. How many square inches will Sam have to paint?

You can use the formula for the area of a rectangle to find the surface area of a rectangular prism. The **surface area** is the sum of the areas of the faces of a solid figure.

EXAMPLE 1

Use a net to find the surface area. Use the formula $A = lw$.

face A: $A = 11 \times 5 = 55$

face B: $A = 21 \times 11 = 231$

face C: $A = 21 \times 5 = 105$

face D: $A = 21 \times 11 = 231$

face E: $A = 21 \times 5 = 105$

face F: $A = 11 \times 5 = 55$

$55 + 231 + 105 + 231 + 105 + 55 = 782$ *Find the sum.*

So, Sam will paint 782 in.^{2}

Use the formula for the area of a rectangle to find the area of each face.

Another way to find the surface area, *S*, of a prism is to remember that opposite faces have the same area.

EXAMPLE 2

Find the surface area of the prism at the right. Use the formula $A = lw$.

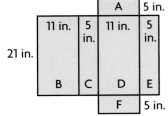

front and back: $(6 \times 8) \times 2 = 96$

top and bottom: $(6 \times 4) \times 2 = 48$

left and right sides: $(4 \times 8) \times 2 = 64$

$S = 96 + 48 + 64 = 208$

So, the surface area is 208 cm^2.

Multiply by 2 to include opposite faces.

Find the sum.

Remember that a pyramid is named by the shape of its base.

To find the surface area of a pyramid, think about its net.

The surface area of a pyramid equals the sum of the areas of the triangular faces and the area of the base. This pyramid has 4 triangular faces and a square base.

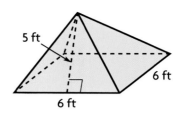

EXAMPLE 3

Find the surface area of the pyramid.

S = area of square + 4 × (area of triangle)

$S = s^2 + 4 \times \left(\frac{1}{2}bh\right)$ *Replace s with 6, b with 6, and h with 5.*

$S = 6^2 + 4 \times \left(\frac{1}{2} \times 6 \times 5\right)$

$S = 36 + 4 \times 15$

$S = 36 + 60$

$S = 96$

So, the surface area is 96 ft².

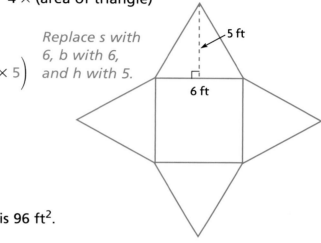

EXAMPLE 4

Find the surface area of the square pyramid at the right.

S = area of square base + 4 × (area of each triangular face)

$S = s^2 + 4 \times \left(\frac{1}{2} bh\right)$

$S = 17^2 + 4 \times \left(\frac{1}{2} \times 17 \times 12\right)$ *Replace s with 17, b with 17, and h with 12.*

$S = 289 + 4 \times 102$

$S = 289 + 408$

$S = 697$

So, the surface area is 697 yd².

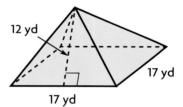

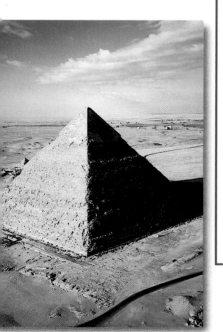

Think and ▶ Discuss

Look back at the lesson to answer each question.

1. **Describe** how you would find the surface area of a pentagonal prism.

2. **Describe** how to find the surface area of a square pyramid.

Guided ▶ Practice

Find the surface area.

3.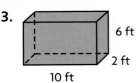
6 ft
2 ft
10 ft

4.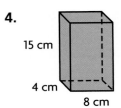
15 cm
4 cm
8 cm

5.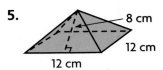
8 cm
12 cm
12 cm

Independent ▶ Practice

Find the surface area.

6.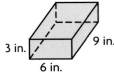
3 in.   9 in.
6 in.

7.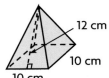
12 cm
10 cm
10 cm

8.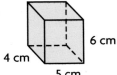
6 cm
4 cm
5 cm

9.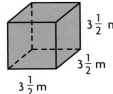
$3\frac{1}{2}$ m
$3\frac{1}{2}$ m
$3\frac{1}{2}$ m

10.
12 cm
14 cm   14 cm

11.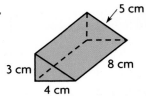
5 cm
3 cm   8 cm
4 cm

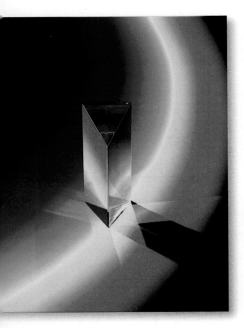

Find the surface area of each cube with the given side length, *s*.

12. $s = 1\frac{1}{2}$ m

13. $s = 3.4$ cm

14. $s = 28$ ft

15. Find the length, width, and height of each rectangular prism. Then find the surface area of each.

   a. The length is twice the width. The height is twice the length. The width is 3 meters.

   b. The width is half the length. The height is twice the width. The length is 6 feet.

   c. The height is three times the length. The length is half the width. The width is 10 inches.

   d. The length is four times the height. The width is one fourth the height. The height is 8 centimeters.

Problem Solving ▶ Applications

For 16–17, use the figure at the right.

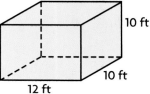

16. Tamara is painting a 12 ft × 10 ft × 10 ft room. She will not paint the ceiling or the floor. How much surface area is she painting?

17. One can of paint covers 350 ft². How many cans of paint will Tamara need to paint the 4 walls?

Use Data For **18–19**, use the graph at the right.

18. About how many more square inches of cardboard are used for Box C than Box A?

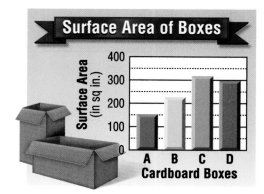

19. If Box A is a cube, what is the length of each side?

20. (?) **What's the Question?** The surface area of a cube is 96 m². The answer is 4 m.

MIXED REVIEW AND TEST PREP

21. Find the area of the circle at the right. (p. 502)

22. Evaluate $^{-}3\frac{1}{3} \times {}^{-}4\frac{1}{2}$. (p. 256)

23. Which is greater, $\frac{-1}{4}$ or 0.25? (p. 234)

12 ft

⭐24. TEST PREP Lydia wants to put a ribbon around the edge of a circular tablecloth with a radius $1\frac{1}{2}$ yards. The ribbon costs $0.45 per yard. What is the cost of the ribbon to the nearest cent? (p. 484)

  A $3.20   **B** $4.24   **C** $4.71   **D** $12.20

⭐25. TEST PREP Which shows the perimeter of a hexagon with sides $9\frac{1}{2}$ ft long? (p. 479)

  F 38 ft   **G** $47\frac{1}{2}$ ft   **H** 57 ft   **J** $66\frac{1}{2}$ ft

---

PROBLEM SOLVING ⟩ 💡 Thinker's Corner

REASONING Look at the large cube at the right. It is made up of 27 unit cubes. Each unit cube is 1 in. long, 1 in. wide, and 1 in. high.

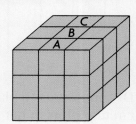

1. What is the surface area of the large cube?

2. What would happen to the surface area if you removed one cube from each of the eight corners? Explain.

3. What would happen to the surface area if you removed the cubes labeled A, B, and C? Explain.

Find the area. (pp. 494–500)

1.
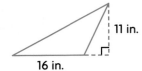
11 in.
16 in.

2.
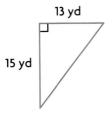
13 yd
15 yd

3.
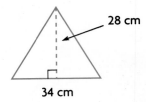
28 cm
34 cm

4.
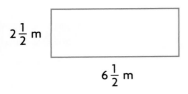
$2\frac{1}{2}$ m
$6\frac{1}{2}$ m

5.
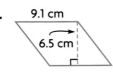
9.1 cm
6.5 cm

6.

2 ft
20 in.

7.
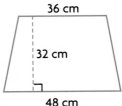
36 cm
32 cm
48 cm

8.
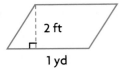
2 ft
1 yd

9.
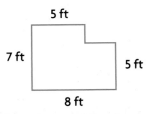
5 ft
7 ft
5 ft
8 ft

Find the area of each circle. Round to the nearest tenth. Use 3.14 or $\frac{22}{7}$ for π. (pp. 502–503)

10. $r = 6$ m

11. $r = 12$ ft

12. $d = 18$ cm

13. $d = 10.5$ m

14. $r = 1\frac{1}{2}$ yd

15. $d = 15$ in.

16. $r = 2.9$ cm

17. $d = 8.4$ m

Find the surface area. (pp. 504–507)

18.
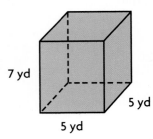
7 yd
5 yd
5 yd

19.
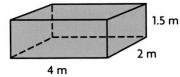
1.5 m
2 m
4 m

20.
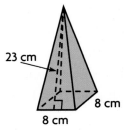
23 cm
8 cm
8 cm

21.
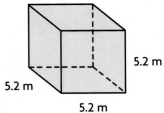
5.2 m
5.2 m
5.2 m

22.
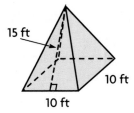
15 ft
10 ft
10 ft

23.
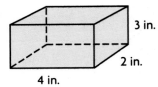
3 in.
2 in.
4 in.

24. Deanna wants to plant grass in three circular plots in her garden. The three plots have diameters of 9 ft, 8 ft, and 3 ft. To the nearest square foot, how much grass must she buy? Use 3.14 for π. (pp. 502–503)

25. The area of the base of a square prism is 25 ft². Each of the 4 faces has a length of 6 ft. What is the surface area of the prism? Explain. (pp. 504–507)

Look for important words.
See item **5.**

Important words are *rounded to the nearest tenth.* You need to remember to round your answer.

Also see problem **2,** p. H62.

Choose the best answer.

1. The area of a rectangular closet is 42 square feet. The width of the closet is 6 feet. What is the length of the closet?

   A 7 ft   **C** 14 ft

   B 12 ft   **D** 36 ft

2. What is the area of a triangle with a base of 3.2 meters and a height of 1.5 meters?

   F 1.8 m²   **H** 2.75 m²

   G 2.4 m²   **J** 3.2 m²

For 3–4, use this diagram.

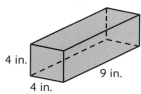

3. Which numerical expression can be used to find the surface area of the figure?

   A $9 \times 4 \times 4$

   B $4(9 \times 4) + 2(4 \times 4)$

   C $2(9 \times 4) + 4(4 \times 4)$

   D $9(2 \times 4) + (9 \times 4) + (4 \times 4)$

4. What is the surface area of the solid?

   F 36 in.²   **H** 144 in.²

   G 72 in.²   **J** 176 in.²

5. What is the area of the trapezoid, rounded to the nearest tenth?

   A 26.7 cm²

   B 67.7 cm²

   C 85.3 cm²

   D 670.3 cm²

6. The simple interest rate paid on a principal of $600 is 6%. How much interest will be earned in 1 year?

   F $3.60   **H** $360

   G $36   **J** Not here

7. The figures are similar. What is the unknown length?

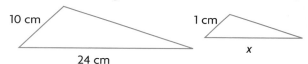

   A $x = 2.4$ cm   **C** $x = 20$ cm

   B $x = 12$ cm   **D** $x = 24$ cm

8. What is the value of $4m - 3 \times 2$ for $m = 3$?

   F 6   **H** 18

   G 9   **J** Not here

Write What You Know

9. The large circle has a radius of 4 ft. The small circle has a radius of 2 ft. Find the area of the unshaded region between the two circles. Explain how you found your answer.

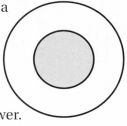

10. The figure shown is a trapezoid. Show how to find the area of the shaded triangle inside it.

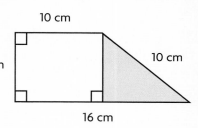

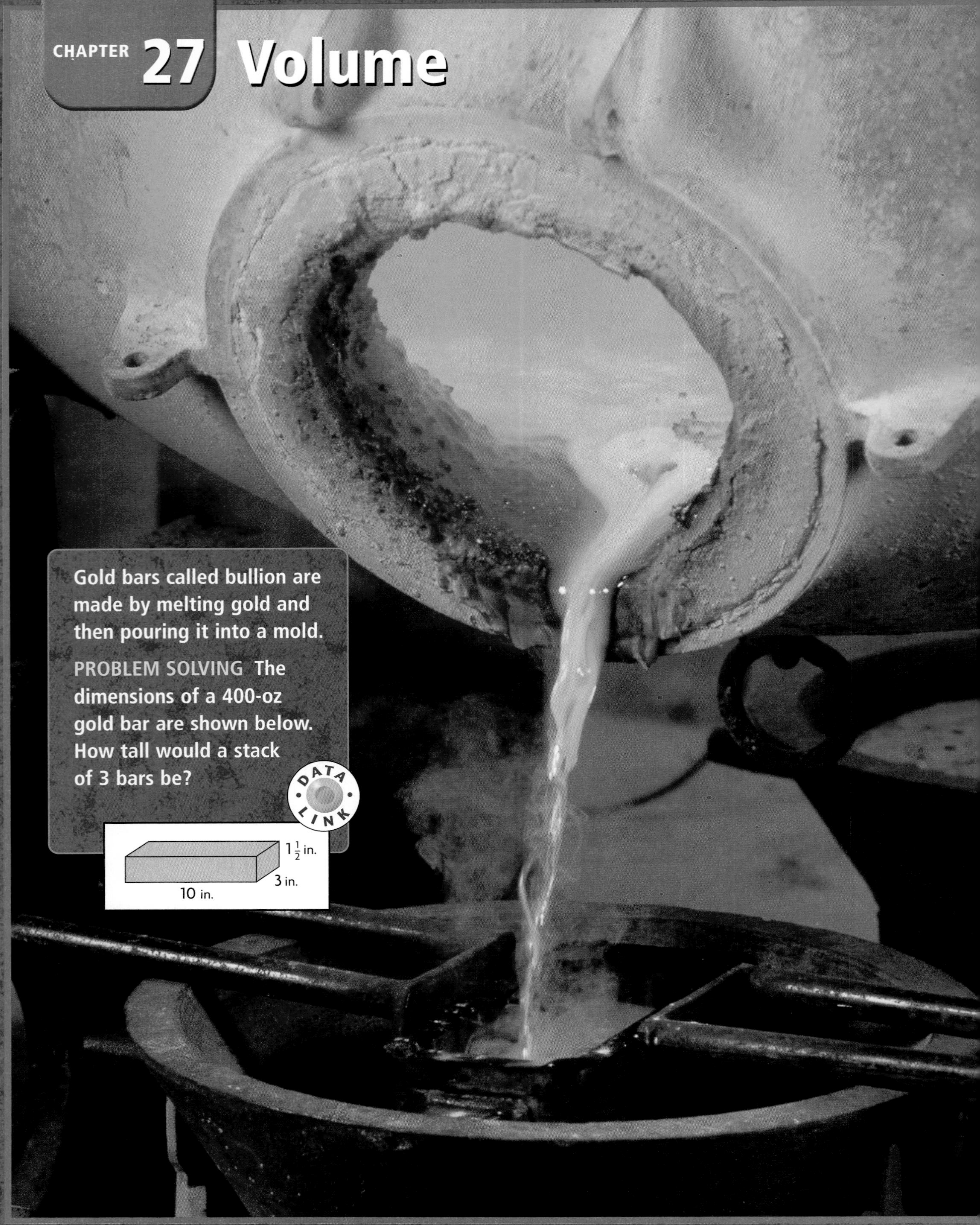

Gold bars called bullion are made by melting gold and then pouring it into a mold.

PROBLEM SOLVING The dimensions of a 400-oz gold bar are shown below. How tall would a stack of 3 bars be?

DATA LINK

$1\frac{1}{2}$ in.

3 in.

10 in.

Check What You Know

Use this page to help you review and remember important skills needed for Chapter 27.

✓ Vocabulary

Choose the best term from the box.

1. The length and width of a rectangle are the ___?___ .

2. Area is measured in ___?___ units.

3. A rectangular prism has length, width, and ___?___ .

> base
> cube
> dimensions
> height
> square

✓ Find the Cube of a Number (For Intervention, see p. H27.)

Find the value.

4. 2^3

5. 4^3

6. 5^3

7. 3^3

8. 8^3

9. 6^3

10. 1.1^3

11. 0.2^3

12. 10^3

13. 12^3

✓ Areas of Squares, Rectangles, and Triangles (For Intervention, see p. H28.)

Find the area of each figure.

14.
9 cm
6 cm

15.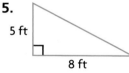
5 ft
8 ft

16.
15 mi
15 mi

17.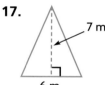
7 m
6 m

18.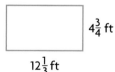
60.3 mm
3.62 cm

19.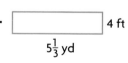
$4\frac{3}{4}$ ft
$12\frac{1}{3}$ ft

20.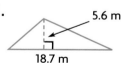
4 ft
$5\frac{1}{3}$ yd

21.
5.6 m
18.7 m

✓ Areas of Circles (For Intervention, see p. H28.)

Find the area to the nearest centimeter.

22.
4 cm

23.
6.7 cm

24.
13 cm

25.
21 cm

LOOK AHEAD

In Chapter 27 you will
- estimate and find the volumes of prisms, pyramids, and cylinders

Estimate and Find Volume

Learn how to estimate and find the volumes of rectangular prisms and triangular prisms.

Vocabulary

volume

Jean wants to estimate how much sand will fill a sandbox. To estimate, Jean needs to think of the volume of the sandbox.

The **volume** is the number of cubic units needed to occupy a given space.

Jean will use a net to make an open box to estimate the volume.

Activity

You need: box net, scissors, tape, 30 centimeter cubes

- Cut out the net. Fold along the dashed lines, and tape the sides to make an open box.

- Estimate how many cubes will fit in the box. Put as many cubes as you can in the box.

- Is your estimate greater than or less than the actual number of cubes you put in the box?

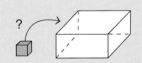

You can visualize how many cubes will fill a prism.

EXAMPLE 1

Use the centimeter cube to estimate the volume.

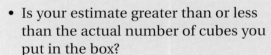

About 5 cubes fit along the length and about 3 cubes along the width. The bottom layer has about 15 cubes. There are about 4 layers of about 15 cubes each.

4 layers × 15 cubes = 60 cubes

So, the volume is about 60 cubic cm, or 60 cm³.

A layer of centimeter cubes has been placed on the base of the rectangular prism below. It takes 10, or 5 × 2, centimeter cubes to make the bottom layer.

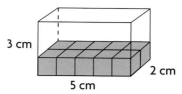

The picture below shows the prism filled with centimeter cubes.

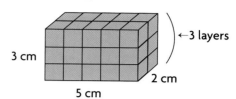

There are 3 layers of 10 cubes each. It takes 30, or 5 × 2 × 3, cubes to fill the prism.

- Look at the table below. What relationship do you see among the length, width, height, and volume?

LENGTH	WIDTH	HEIGHT	VOLUME
5	2	3	30
3	3	6	54
7	4	3	84

The relationship between the dimensions and the volume of each prism can be written as Volume = length × width × height, or $V = lwh$.

The formula $V = Bh$ can also be used to find the volume of a rectangular prism. In this formula, B is equal to $l × w$, which is the area of the base of the prism, and h is the height of the prism.

EXAMPLE 2

Find the volume of the prism at the right.

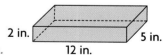

$V = Bh$ where $B = l × w$ *Write the formula.*

$V = (12 × 5) × 2$ *Replace B with 12 × 5 and h with 2.*

$V = 120$ *Multiply.*

So, the volume of the rectangular prism is 120 in.³

- **What if** you cut the rectangular prism into two congruent triangular prisms? What would be the volume of one triangular prism?

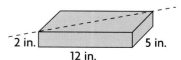

Remember that a triangular prism has bases that are congruent triangles.

The volume of a triangular prism is one half the volume of a rectangular prism with the same length, width, and height. To find the volume of a right triangular prism, you can use the formula $V = \frac{1}{2}lwh$. To find the volume of any triangular prism, you can use the formula $V = Bh$ where B is the area of the triangular base.

EXAMPLE 3

Find the volume of the prism at the right.

One Way

$V = \frac{1}{2}lwh$

$V = \frac{1}{2} \times 2.8 \times 4.2 \times 5$

$V = 29.4$

Another Way

$V = Bh$

$V = \left(\frac{1}{2} \times 2.8 \times 4.2\right) \times 5$

$V = 5.88 \times 5$

$V = 29.4$

So, the volume is 29.4 m³.

• **What if** the dimensions of the prism were tripled? What would the volume be?

CHECK FOR UNDERSTANDING

Think and Discuss

Look back at the lesson to answer each question.

1. **Tell** how to find the volume of a cereal box that measures 26 cubes by 3 cubes by 18 cubes.

2. **Explain** how to find the height of a rectangular prism if you know the length, width, and volume of the prism.

Guided Practice

Find the volume.

3. 4 in. / 3 in. / 2 in.

4. 6 in. / 10 in. / 6 in.

5. 8 cm / 4 cm / 3 cm

PRACTICE AND PROBLEM SOLVING

Independent Practice

Find the volume.

6. 5 cm / 2 cm / 3 cm

7. 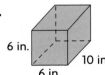 3 ft / 9 ft / 12 ft

8. 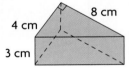 10 m / 8 m / 18 m

9. 3 ft / 7 ft / 14 ft

10. 7.2 cm / 4.5 cm / 3.7 cm

11. 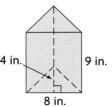 4 in. / 9 in. / 8 in.

Find the unknown length.

12.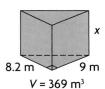
9 cm
5 cm
x
V = 900 cm³

13.
x
8.2 m   9 m
V = 369 m³

14.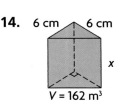
6 cm   6 cm
x
V = 162 m³

Problem Solving ▶
Applications

15. Bobby orders sand to fill the long-jump pit. The pit is 14 ft long, 9 ft wide, and $1\frac{1}{2}$ ft deep. How much sand does Bobby need to order?

16. *REASONING* How many different rectangular prisms with whole-number dimensions have a volume of 8 ft³? Write the dimensions of each prism.

17. A model of a rectangular prism is made using the scale 1 in. = 3 ft. The model measures 4 in. by 3 in. by 2 in. What is the volume of the actual prism?

MIXED REVIEW AND TEST PREP

18. Find the surface area of a rectangular prism 8.2 ft long, 6.4 ft wide, and 4.5 ft high. (p. 504)

19. Evaluate $3y^2$ for $y = 3.9$. (p. 278)

20. Solve $8.1t = 49.41$. (p. 82)

21. **TEST PREP** Which type of angle is formed by two perpendicular lines? (p. 326)

   A straight   **B** obtuse   **C** right   **D** acute

22. **TEST PREP** The price of a $400 stereo is increased 20%. Then the price is discounted 20%. What is the final price? (p. 418)

   F $360   **G** $384   **H** $400   **J** $420

PROBLEM SOLVING LiNKUP to Reading

Analyze Information When you analyze details in a problem, look for information that is needed to solve the problem.

Calvin is mailing a model car to a friend. To make sure the model box does not move around in the mailing box, he must fill the empty space with packing material. How many cubic inches of packing material will he need?

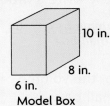
10 in.
8 in.
6 in.
Model Box

12 in.
12 in.
10 in.
Mailing Box

1. What information do you need to solve the problem?

2. Solve the problem. Explain how you solved it.

3. **What if** Calvin uses a mailing box that is 1 ft by $\frac{3}{4}$ ft by $\frac{3}{4}$ ft? Will he have more or less space to fill? How much more or less?

PROBLEM SOLVING STRATEGY
Make a Model

Learn how to use the strategy *make a model* to solve problems.

A rice company uses different-sized boxes. A small box of rice is 2 in. wide, 4 in. long, and 6 in. high. How does the volume of the small box change when the length, width, and height are halved to make a sample box?

Analyze

What are you asked to find?

What information are you given?

Is there information you will not use? If so, what?

Choose

What strategy will you use?

You can use the strategy *make a model*.

Solve

How will you solve the problem?

Make a model of each box. Use a ratio to compare the volumes.

Use cubes to make a model of each box. Count the cubes to find the volume of each box.

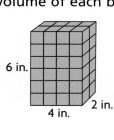
6 in.
4 in. 2 in.

The volume of the small box is 48 in.³
The volume of the sample box is 6 in.³

3 in.
2 in. 1 in.

Now write a ratio to compare the volumes.

$$\frac{sample\ box}{small\ box} = \frac{6}{48} = \frac{1}{8}$$

So, the volume of the sample box is 6 in.³, or $\frac{1}{8}$ the volume of the small box.

Check

How can you check your answer?

What if the dimensions of the small box are doubled to make a larger box? How will the volume of the box change?

PROBLEM SOLVING PRACTICE

Solve the problem by making a model.

1. The volume of the box at the right is 240 in.³ Double the height of the box. How does the volume of the box change?

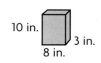
10 in. 3 in. 8 in.

2. The dimensions of the large box at the right are halved to make a smaller box. How does the volume change?

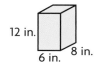
12 in. 6 in. 8 in.

PROBLEM SOLVING STRATEGIES

Draw a Diagram or Picture

▶ **Make a Model**

Predict and Test

Work Backward

Make an Organized List

Find a Pattern

Make a Table or Graph

Solve a Simpler Problem

Write an Equation

Use Logical Reasoning

For 3–4, a rectangular box is 6 cm wide, 8 cm long, and 13 cm high.

3. The length and width of the rectangular box are halved. The height remains the same. How does the volume of the smaller box compare to the volume of the larger box?

A $\frac{1}{8}$ of the volume   **C** $\frac{1}{3}$ of the volume

B $\frac{1}{4}$ of the volume   **D** $\frac{1}{2}$ of the volume

4. The length, width, and height of the rectangular box are doubled. How does the volume of the larger box compare to the volume of the smaller box?

F equal to the volume

G twice the volume

H four times the volume

J eight times the volume

MIXED STRATEGY PRACTICE

Use Data   **For 5–6, use the line graph at the right.**

5. About how much did Howie make in sales during the six months shown on the graph?

6. At what point were sales and costs the same?

7. Grander's Clothes Shop sells shirts for $12 and shorts for $15. Terance will spend $120. How many different combinations of shirts and shorts can Terance buy?

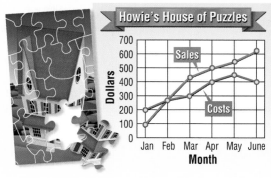

8. Ana makes a circular pattern of stars on a piece of paper. Successive stars are the same distance apart. The sixth star is directly opposite the eighteenth star. How many stars form the circle?

9. Jasmine, Hugh, Sharon, and Mike are standing in line. A girl is not first or last in line. Mike is before Hugh. Jasmine is directly in front of Hugh. Who is first, second, third, and fourth in line?

10. A side of square A is 6 times the length of a side of square B. How many times as great is the area of square A than the area of square B?

11. **❓ What's the Question?** The volume of a cube is 216 cm³. The length of each side is halved. The answer is 27 cm³.

ALGEBRA
Volumes of Pyramids

Learn how to use a formula to find the volumes of pyramids.

When it was built, the Great Pyramid of Khufu was 145.75 m (481 ft) high. Each side measured 229 m (751 ft) in length.

Quang wants to use what he knows about the volume of a prism to find the volume of his model of the Great Pyramid of Khufu.

Activity

You need: prism and pyramid nets, tape, rice

- Cut out the nets for the prism and the pyramid. Fold along the dashed lines and use tape to make an open prism and an open pyramid.

- Compare the height of the prism with the height of the pyramid.

- Compare the base of the prism with the base of the pyramid.

- Fill the pyramid with rice. Pour the rice into the prism. Repeat until the prism is full.

- What is the relationship between the heights and bases of the prism and pyramid?

- What is the relationship between the volume of the pyramid and the volume of the prism?

To find the volume of a pyramid, use the formula $V = \frac{1}{3}Bh$.

EXAMPLE

Find the volume of Quang's pyramid model at the right to the nearest tenth.

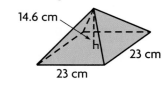
14.6 cm, h, 23 cm, 23 cm

$V = \frac{1}{3}Bh$, *where* $B = l \times w$ *Write the formula.*

$V = \frac{1}{3}(23 \times 23) \times 14.6$ *Replace B with 23 × 23 and*

$V = \frac{1}{3} \times 529 \times 14.6$ *h with 14.6. Multiply.*

$V = \frac{1}{3} \times 7{,}723.4$

$V = 2{,}574.5$ So, the volume is 2,574.5 cm³.

CHECK FOR UNDERSTANDING

Think and ▶ **Look back at the lesson to answer the question.**
Discuss
1. **Compare** the formulas for the volume of a prism and the volume of a pyramid.

Guided ▶ **Find the volume.**
Practice

2.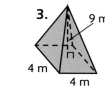
6 in.
3 in.
5 in.

3.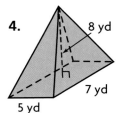
9 m
4 m
4 m

4.
8 yd
7 yd
5 yd

PRACTICE AND PROBLEM SOLVING

Independent ▶ **Find the volume.**
Practice

5.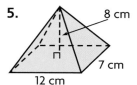
8 cm
7 cm
12 cm

6.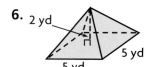
2 yd
5 yd
5 yd

7.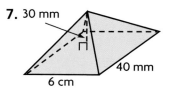
30 mm
40 mm
6 cm

8. square pyramid:
$l = 12$ cm, $w = 12$ cm,
$h = 25$ cm

9. rectangular pyramid:
$l = 18$ ft, $w = 25$ ft,
$h = 30$ ft

10. rectangular pyramid:
$l = 10$ dm, $w = 4$ dm,
$h = 7.5$ dm

11. rectangular pyramid: $l = 9$ ft,
$w = 5$ yd, $h = 4\frac{4}{5}$ yd

Problem Solving ▶
Applications

12. The weight of a cubic foot of steel is 490 lb. What is the weight of a solid steel rectangular pyramid that is 2 ft by 3 ft by 18 in.?

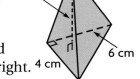
7 cm
6 cm
4 cm

13. ✎ **Write About It** Tell how you would find the volume for the triangular pyramid at the right.

14. Arlena has $8.40 in nickels, dimes, and quarters. She has the same number of each coin. How many of each coin does she have?

MIXED REVIEW AND TEST PREP

15. $13\frac{1}{5} \times 8\frac{1}{3}$ (p. 206)

16. $4\frac{1}{2} + 3\frac{4}{5}$ (p. 186)

17. Write 79% as a decimal. (p. 60)

18. Volume = ■ (p. 514)

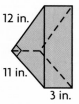
12 in.
11 in.
3 in.

⭐ 19. **TEST PREP** Which list shows $\frac{3}{4}$, $\frac{2}{3}$, and $\frac{5}{8}$ in order from least to greatest? (p. 166)

A $\frac{2}{3}, \frac{5}{8}, \frac{3}{4}$　　B $\frac{5}{8}, \frac{3}{4}, \frac{2}{3}$　　C $\frac{5}{8}, \frac{2}{3}, \frac{3}{4}$　　D $\frac{2}{3}, \frac{3}{4}, \frac{5}{8}$

EXTRA PRACTICE page H58, Set B

Volume of a Cylinder

MATH LAB

In geometry, a cylinder is a solid figure that has two congruent, parallel circles as its bases. In the activity below, you will find the volume of a cylinder.

Activity

Explore how to use manipulatives to find the volume of a cylinder.

You need 2 pieces of centimeter graph paper, scissors, tape.

- Use one piece of graph paper to cut out a rectangle 20 cm long and 8 cm wide. Without overlapping, roll the rectangle the short way and tape it to form a cylinder.

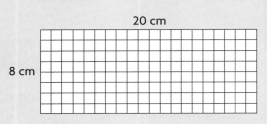
20 cm

8 cm

- Stand the cylinder on the second piece of graph paper. Trace around the base of the cylinder. Count the whole centimeter squares and the parts of centimeter squares inside the tracing.

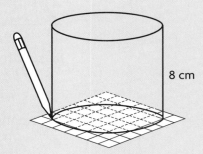
8 cm

Think and Discuss

- About how many cubes would fit on the bottom layer of the cylinder?

- About how many layers of cubes can fit inside the cylinder?

- What is the approximate volume of the cylinder?

Practice

TECHNOLOGY LINK

More Practice: Use E-Lab, *Exploring the Volume of a Cylinder.*
www.harcourtschool.com/ elab2002

Draw the following rectangles on centimeter graph paper. Cut out each one, roll it, and tape it to make a cylinder. Find the volume of each cylinder.

1.
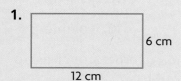
6 cm

12 cm

2.
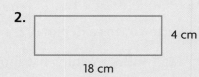
4 cm

18 cm

Volumes of Cylinders

Learn how to find the volume of a cylinder.

You can find the volume of a rectangular prism by multiplying the area of the base by the height. You can use the same method to find the volume of a cylinder.

Math **I**dea ▶ Volume of a cylinder = area of base × height

$V = Bh$

$V = \pi r^2 h$ πr^2 is the area of the base, which is a circle.

EXAMPLE 1

Larisa has a cylindrical flower vase like the one shown at the right. To the nearest cubic inch, what is the volume of water the vase will hold?

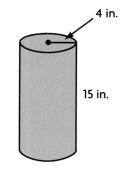

4 in.

15 in.

$V = \pi r^2 h$	*Write the formula.*
$V \approx 3.14 \times 4^2 \times 15$	*Replace π with 3.14 or $\frac{22}{7}$,*
$V \approx 3.14 \times 16 \times 15$	*r with 4, and h with 15.*
$V \approx 753.6$	*Multiply.*

So, the volume of water is about 754 in.³

The volume of a cylinder can also be found if you know the diameter and the height of the cylinder.

EXAMPLE 2

Find the volume to the nearest cubic centimeter.

5 cm

15 cm

$5 \div 2 = 2.5$	*Find the radius.*
$V = \pi r^2 h$	*Write the formula.*
$V \approx 3.14 \times (2.5)^2 \times 15$	*Replace π with 3.14 or $\frac{22}{7}$,*
$V \approx 3.14 \times 6.25 \times 15$	*r with 2.5, and h with 15.*
$V \approx 294.375$	*Multiply.*

So, the volume is about 294 cm³.

• **What if** you have a cylinder with a diameter of 6.5 cm and a height of 12.5 cm? What is the volume of the cylinder?

Sometimes you may have to find the volume of part of a cylinder.

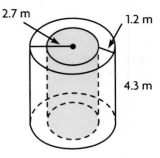

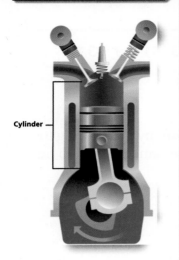

Cylinder

EXAMPLE 3

Look at the water tank at the right. The wall is 1.2 m thick. To the nearest meter, how much water can the inside cylinder hold?

Find the radius of the inside cylinder.

$2.7 - 1.2 = 1.5$ ← radius

Use the radius to find the volume of the inside cylinder.

$V \approx 3.14 \times (1.5)^2 \times 4.3$

$V \approx 3.14 \times 2.25 \times 4.3$

$V \approx 30.3795$

Replace π with 3.14, r with 1.5, and h with 4.3. Multiply.

So, the water tank can hold about 30 m³ of water.

CHECK FOR UNDERSTANDING

Think and Discuss ▶ **Look back at the lesson to answer each question.**

1. **What if** the dimensions of Larisa's vase are doubled? What is the volume of the vase to the nearest cubic inch?

2. **Explain** how the formula for the volume of a cylinder is similar to the formula for the volume of a rectangular prism.

3. **Explain** in your own words which parts of a cylinder are represented by πr^2 and h in the formula $V = \pi r^2 h$.

Guided Practice ▶ **Find the volume. Round to the nearest whole number.**

4.
4 cm
7 cm

5.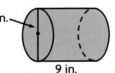
7 in.
9 in.

6.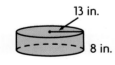
13 in.
8 in.

PRACTICE AND PROBLEM SOLVING

Independent Practice ▶ **Find the volume. Round to the nearest whole number.**

7.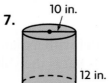
10 in.
12 in.

8.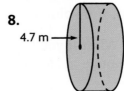
4.7 m
1.6 m

9.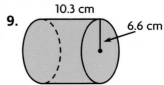
10.3 cm
6.6 cm

Find the volume. Round to the nearest whole number.

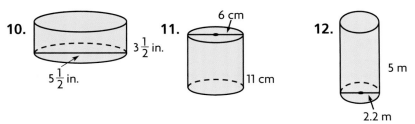

10. $5\frac{1}{2}$ in. $3\frac{1}{2}$ in.

11. 6 cm   11 cm

12. 5 m   2.2 m

Find the volume of the inside cylinder to the nearest whole number.

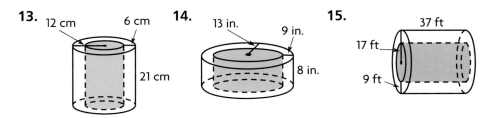

13. 12 cm   6 cm   21 cm

14. 13 in.   9 in.   8 in.

15. 37 ft   17 ft   9 ft

Problem Solving ▶
Applications

16. Nicole bought a decorative outdoor planter with a diameter of 21 in. and a height of 28 in. She wants to fill it with potting soil. How much potting soil will Nicole need to the nearest cubic inch?

17. How does the volume of a cylinder with a radius of 10 ft and a height of 5 ft change when the dimensions are doubled?

18. **? What's the Error?** A circular swimming pool has a diameter of 20 ft and a height of 6 ft. Mr. Alvarez determined that it would take 7,536 cubic feet of water to fill the pool. What did he do wrong? What is the solution?

19. **✏ Write a problem** about finding the volume of a real-life object shaped like a cylinder. Share with your classmates.

20. A large bag of potting soil weighs 40 lb and costs $3.79. Julio has $30 to buy as many of the large bags of potting soil as he can. How many pounds of potting soil can he buy?

MIXED REVIEW AND TEST PREP

21. Find the volume to the nearest foot of a pyramid 1.2 ft long, 2.8 ft wide, and 3.6 ft high. (p. 518)

22. Write the algebraic expression for 6.9 more than 5 times d. (p. 270)

23. Which has the greater value, $\frac{1}{10} \div \frac{1}{2}$ or $\frac{1}{2} \div \frac{1}{10}$? (p. 210)

24. $^-50 \times {}^-11$ (p. 252)

⭐ 25. **TEST PREP** Karl made a model, using the scale 1 in. = 4 ft. The actual length is 200 ft. Which is the length of the model? (p. 387)

A 30 in.   **B** 40 in.   **C** 50 in.   **D** 500 in.

EXTRA PRACTICE page H58, Set C

Find the volume. (pp. 512–515)

1.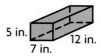
5 in. 12 in. 7 in.

2.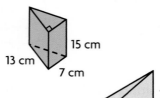
15 cm 13 cm 7 cm

3.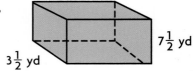
$7\frac{1}{2}$ yd $3\frac{1}{2}$ yd

4.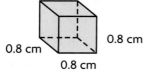
0.8 cm 0.8 cm 0.8 cm

5.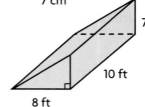
7 ft 10 ft 8 ft

6.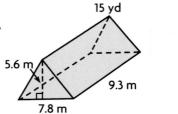
15 yd 5.6 m 9.3 m 7.8 m

For 7–12, find the volume of the pyramid. (pp. 518–519)

7.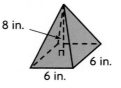
8 in. 6 in. 6 in.

8.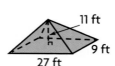
11 ft 9 ft 27 ft

9.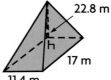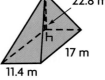
22.8 m 17 m 11.4 m

10. length = 5 m
width = 9 m
height = 10 m

11. base = 25.5 ft^2
height = 12 ft

12. base = 216 m^2
height = 24 m

Find the volume. Round to the nearest whole number. (pp. 521–523)

13.
4 m 9.6 m

14.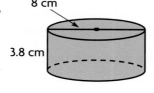
8 cm 3.8 cm

15.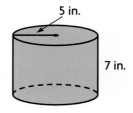
5 in. 7 in.

Find the volume of the inside cylinder to the nearest whole number. (pp. 521–523)

16.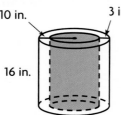
10 in. 3 in. 16 in.

17.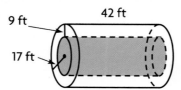
9 ft 42 ft 17 ft

18.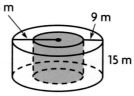
18 m 9 m 15 m

Solve. (pp. 516–517)

19. A box is 8 in. high, 2 in. wide, and 4 in. long. The length and the width of the box are doubled to make a larger box. How does the volume of the larger box compare to the volume of the smaller box?

20. A box is 12 cm long, 8 cm wide, and 9 cm tall. Each dimension is halved to make a smaller box. How does the volume of the smaller box compare to the volume of the larger box?

Choose the answer.
See item **8.**

You need to find the sale price, but first you need to find the amount of the discount. Be sure that the answer you choose answers the question asked.

Also see problem **6**, p. H64.

Choose the best answer.

1. Seth needs a box that has a volume of $6,000 \text{ cm}^3$. Which dimensions could the box have?

A $l = 25$ cm, $w = 20$ cm, $h = 12$ cm

B $l = 20$ cm, $w = 30$ cm, $h = 5$ cm

C $l = 6$ cm, $w = 5$ cm, $h = 20$ cm

D $l = 3$ cm, $w = 2$ cm, $h = 10$ cm

2. Ian's age is 3 years greater than twice Barb's age. Ian is 27 years old. Which of the following equations can be used to find b, Barb's age?

F $b + 5 = 27$ **H** $3 + b = 27$

G $2b = 27$ **J** $2b + 3 = 27$

3. Which is the volume of the figure?

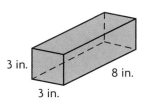

3 in. 8 in. 3 in.

A 9 in.^3 **C** 72 in.^3

B 48 in.^3 **D** 144 in.^3

4. What is the volume of the prism shown?

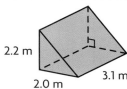

2.2 m 2.0 m 3.1 m

F 6.82 m^3 **H** 7.4 m^3

G 7.3 m^3 **J** 8.36 m^3

5. Which is a reasonable estimate for the volume of the cylinder?

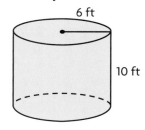

6 ft 10 ft

A $10,000 \text{ ft}^3$ **C** 600 ft^3

B $1,000 \text{ ft}^3$ **D** 60 ft^3

6. What is the median of the data?
30, 25, 40, 35, 25

F 25 **H** 31

G 30 **J** 40

7. Which unit of measure below is the only one that could be used for the volume of a box?

A in. **C** cm^2

B cm **D** ft^3

8. The regular price of a jacket is $56. If it is discounted 30%, what is the sale price?

F $16.80 **H** $34.50

G $22.60 **J** Not here

Write What You Know

9. A rectangular prism has a volume of 64 m^3. Its length is 10 m and its height is 2 m. Find its width. Explain how you found your answer.

10. Which is the better price: a 50% discount on a $50 jacket or a 25% discount on the same jacket followed by another 25% discount. Explain your reasoning.

MATH DETECTIVE

Missing Pieces

Solve each mystery by finding the missing pieces that satisfy the given conditions.

Mystery Number 1

The volume of a rectangular prism is 60 in.^{3} Give two different sets of dimensions for the length, width, and height of the figure.

LENGTH	WIDTH	HEIGHT	VOLUME
▪	▪	▪	60 in.^{3}
▪	▪	▪	60 in.^{3}

Mystery Number 2

The volume of a cube is 64 m^3. What is the surface area of the cube?

LENGTH	WIDTH	HEIGHT	VOLUME	SURFACE AREA
▪	▪	▪	64 m^3	▪

Mystery Number 3

The surface area of a cube is 150 m^2. What is the volume of the cube?

LENGTH	WIDTH	HEIGHT	VOLUME	SURFACE AREA
▪	▪	▪	▪	150 m^2

Think It Over!

Write About It   Explain what happens to the volume if you double all the dimensions of either of the rectangular prisms in Mystery Number 1.

Stretch Your Thinking
A rectangular prism has three whole-number dimensions, two of which are the same. If the volume of the rectangular prism is 36 cm^3, what is the greatest surface area the prism can have?

Challenge Networks

Learn how to use a network to find all possible routes and to find the shortest route.

A **network** is a graph whose edges are labeled with numbers. In the network at the right, the cities are the vertices, and the connecting segments between them are the edges. The estimated distances between the four cities are given in miles.

You can use a network to help you find the shortest route between locations.

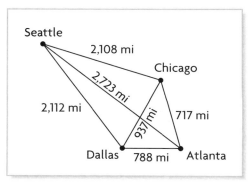

Seattle
2,108 mi
Chicago
2,723 mi
2,112 mi
937 mi
717 mi
Dallas 788 mi Atlanta

EXAMPLE

Use the distances shown on the network. Find the shortest route starting in Seattle that includes all the cities.

Let S = Seattle, A = Atlanta, C = Chicago, and D = Dallas.

SACD = 2,723 + 717 + 937 = 4,377   *List the different routes*

SADC = 2,723 + 788 + 937 = 4,448   *and the number of miles for each one.*

SCAD = 2,108 + 717 + 788 = 3,613

SCDA = 2,108 + 937 + 788 = 3,833

SDAC = 2,112 + 788 + 717 = 3,617

SDCA = 2,112 + 937 + 717 = 3,766   *Compare the distances.*

The shortest route, SCAD, is 3,613 mi.

TALK ABOUT IT

How can you use a network to find the shortest route? Explain.

TRY IT

Find all the possible routes starting from A that include every vertex. Find the distance for each route. Distances are in kilometers. Figures are not drawn to scale.

1.

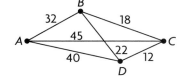

B
32   18
A 45
40   22   12
D C

2.

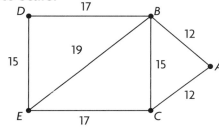

D 17 B
        12
15   19   15 A
        12
E 17 C

3. In Exercise 1, what is the shortest route? the longest route?

4. In Exercise 2, what is the shortest route? the longest route?

VOCABULARY

1. The sum of the areas of the faces of a solid figure is called the __?__. (p. 504)

2. The __?__ is the number of cubic units needed to occupy a given space. (p. 512)

EXAMPLES | **EXERCISES**

Chapter 24

• Convert units of measurement. (pp. 462–465)

Change 20 ounces to pounds.

$\frac{1}{16} = \frac{x}{20}$ *Use the relationship 1 lb = 16 oz to write a proportion.*

$16x = 20$ *Find the cross products. Solve for x.*

$x = 1.25$, or $1\frac{1}{4}$ So, 20 oz = $1\frac{1}{4}$ lb.

Use a proportion to convert to the given unit.

3. 32 qt = ▧ gal
4. 15,000 lb = ▧ T
5. 15 ft = ▧ in.
6. 42 m = ▧ cm
7. 56 oz = ▧ lb
8. 50 m = ▧ km
9. 0.9 kg = ▧ g
10. 60 km = ▧ m

• Convert customary and metric measurements. (pp. 466–467)

Change 6 inches to centimeters.

$\frac{1}{2.54} = \frac{6}{x}$ *Use the relationship 1 in. ≈ 2.54 cm to write a proportion.*

$x = 6 \times 2.54$ *Find the cross products.*

$x = 15.24$ cm   *Solve for x.*

So, 6 in. ≈ 15.24 cm.

Use a proportion to convert to the given unit. Use the table.

11. 4 gal ≈ ▧ L
12. 20 kg ≈ ▧ lb
13. 5 ft ≈ ▧ cm
14. 8 mi ≈ ▧ km
15. 15 m ≈ ▧ yd

1 in. ≈ 2.54 cm
1 ft ≈ 30.48 cm
1 yd ≈ 0.91 m
1 mi ≈ 1.61 km
1 lb ≈ 0.454 kg
1 gal ≈ 3.79 L

Chapter 25

• Find the perimeter of polygons. (pp. 479–481)

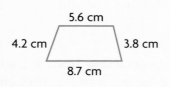

Add the lengths.

$5.6 + 3.8 + 8.7 + 4.2 = 22.3$ cm

Find the perimeter.

16.

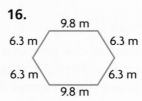

17. a regular pentagon with sides of $12\frac{1}{4}$ in.

• Find the circumference of circles. (pp. 486–489)

Use the formula $C = 2\pi r$.

$C \approx 2 \times 3.14 \times 4.6$

$C \approx 28.888$ *Round to the nearest whole*

$C \approx 29$ cm   *number.*

Find the circumference to the nearest whole number. Use $\pi = 3.14$ or $\pi = \frac{22}{7}$.

18.

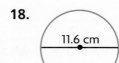

19. $r = 42$ ft

20. $d = 125$ m

Chapter 26

- **Find the area of polygons.**

  (pp. 494–500)

15 cm

13 cm

$A = l \times w$ *Write the formula.*

$A = 15 \times 13$ *Replace l with 15 and w with 13.*

$A = 195$ *Multiply.*

So, the area is 195 cm².

Find the area.

21.
5 cm
8.3 cm

22.
$6\frac{1}{2}$ in.
8 in.

23.
18.3 m
12 m
24.7 m

24.
40 ft
32 ft   15 ft
23 ft

- **Find the area of circles.**

  (pp. 502–503)

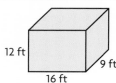

2.5 cm

Use the formula $A = \pi r^2$.

$A \approx 3.14 \times 2.5 \times 2.5$
$A \approx 19.625$
$A \approx 20$ *Round to the nearest whole number.*

So, the area is about 20 cm².

Find the area to the nearest whole number.
Use $\pi = 3.14$ or $\pi = \frac{22}{7}$.

25.
7 cm

26.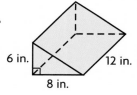
28 m

27. $r = 26$ in.

28. $d = 42$ m

- **Find the surface area of prisms and pyramids.** (pp. 504–507)

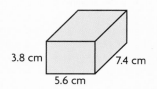

12 ft   9 ft
16 ft

$2 \times 12 \times 9 = 216$ *area of sides*
$2 \times 12 \times 16 = 384$ *area of front and back*
$2 \times 9 \times 16 = 288$ *area of top and bottom*
$216 + 384 + 288 = 888$

So, the surface area is 888 ft².

Find the surface area.

29.
6 cm
11 cm   11 cm

30.
6 in.   12 in.
8 in.

Chapter 27

- **Find the volume of prisms, pyramids, and cylinders.**

  (pp. 512–523)

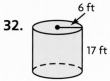

3.8 cm   7.4 cm
5.6 cm

$V = B \times h$ *Base is a rectangle.*
$V = l \times w \times h$ *Area of a rectangle is l × w.*
$V = 5.6 \times 7.4 \times 3.8$
$V = 157.472$

So, the volume is 157.472 cm³.

Find the volume.

31.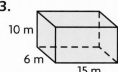
9 mm
8 mm
15 mm

32.
6 ft
17 ft

33.
10 m
6 m
15 m

34.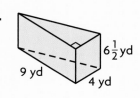
$6\frac{1}{2}$ yd
9 yd   4 yd

PROBLEM SOLVING APPLICATIONS

35. Kwan framed a rectangular picture that measured 15 in. ×12 in. The framing strips were $2\frac{1}{2}$ in. wide. What was the perimeter of the frame? (pp. 482–483)

36. A box is 8 m × 6 m × 9 m. If each dimension is doubled, how does the volume change? (pp. 516–517)

Performance Assessment

TASK A • Adding the Trim

Materials: calculator

Kate has decorated a box in which to store photos. She wants to trim the top edge of the box with ribbon. To do this, Kate needs to find the perimeter of a rectangle with length 8 in. and width 5 in.

- Write the formula for the perimeter of a rectangle.

- Show the steps needed to find the perimeter on a calculator by using the formula.

- Use the Distributive Property to write the perimeter formula in another form.

- Show the calculator steps needed to find the perimeter by using this formula.

- Write a sentence comparing the way you use the two formulas on your calculator.

TASK B • Measuring Up

Materials: inch ruler, metric ruler, plain paper

Draw a composite figure using squares, rectangles, triangles, other polygons, or circles. Draw the figure large enough so you can measure all the dimensions.

- Describe how you formed your figure.

- Measure the dimensions of your figure to the nearest inch. Determine the perimeter of your figure.

- Convert the perimeter measurement to centimeters.

- Measure the dimensions of your figure to the nearest half inch, to the nearest centimeter, and to the nearest millimeter. Copy and complete the table.

Unit	in.	cm (converted)	$\frac{1}{2}$ in.	cm (measured)	mm
Perimeter					

Technology Linkup

Use a Spreadsheet to Find
the Perimeter and Area of Rectangles

Vernetta is making a stained-glass window. She uses a spreadsheet to compute the perimeter and area of each rectangular piece of glass.

LENGTH AND WITH OF GLASS				
Rectangle	Length (*l*)	Width (*w*)	Perimeter (*P*)	Area (*A*)
A	12.25 cm	6.5 cm		
B	19.6 cm	12.75 cm		
C	12.4 cm	18.5 cm		
D	16.7 cm	11.25 cm		
E	6.21 cm	4.11 cm		

- Enter the column headings. Then enter the rectangle letters and their lengths and widths.

- To find the perimeter, type = 2*B2 + 2*C2 in cell D2.

- Use the *Fill Down* command to enter the formula in cells D3 through D6.

- To find the area, type = B2*C2 in cell E2.

- Use the *Fill Down* command to enter the formula in cells E3 through E6.

All	A	B	C	D	E
1	Rectangle	Length (l)	Width (w)	Perimeter (P)	Area (A)
2	A	12.25	6.5		
3	B	19.6	12.75		
4	C	12.4	18.5		
5	D	16.7	11.25		
6	E	6.21	4.11		

All	A	B	C	D	E
1	Rectangle	Length (l)	Width (w)	Perimeter (P)	Area (A)
2	A	12.25	6.5	37.5	79.625
3	B	19.6	12.75	64.7	249.9
4	C	12.4	18.5	61.8	229.4
5	D	16.7	11.25	55.9	187.875
6	E	6.21	4.11	20.64	25.5231

Practice and Problem Solving

Make a spreadsheet to find the perimeter and area of each rectangle.

1. length: 4.35 in.
   width: 9 in.

2. length: 22.7 in.
   width: 35.6 in.

3. length: 7.49 in.
   width: 3.5 in.

4. length: 55.3 cm
   width: 3.8 cm

5. length: 5.51 in.
   width: 5.62 in.

6. length: 43.8 cm
   width: 37.5 cm

7. length: 4.1 in.
   width: 1.9 in.

8. length: 83.3 cm
   width: 15.2 cm

9. **STRETCH YOUR THINKING** Explain how you could use a spreadsheet to compute the volume of a rectangular prism.

Multimedia Math Glossary www.harcourtschool.com/mathglossary

10. **Vocabulary** Look up *surface area* in the Multimedia Math Glossary. Using the picture of the rectangular prism as a model draw a picture of a square pyramid and its net. Label the dimensions of the square pyramid and find the surface area.

Fish, Whales, and More

The Chicago Bears play football at Chicago's Soldier Field. But the city's turtles, sharks, and whales play at Shedd Aquarium in Grant Park. Since the aquarium opened in 1930, more than 100 million visitors have walked through its doors to view the Shedd's spectacular sealife exhibits.

Use Data For 1–6, use the data in the caption for the photo of the beluga whale. Find each measurement and round to the nearest hundredth.

Adult beluga whales are 13 to 16 ft long and weigh 1,500 to 3,000 lb. Babies are about 5 ft long and weigh 80 to 150 lb.

1. Weight range of baby in ounces

2. Length range of adult in yards

3. Weight range of baby in kilograms

4. Weight range of adult in tons

5. Length of baby in centimeters

6. Length of baby in meters

7. The length of the Florida pygmy seahorse ranges from 2 cm to $4\frac{1}{2}$ cm. Write the range in millimeters and in inches.

8. Nurse sharks can be found in coral reefs at a depth of 70 meters. About how many feet is that? Round to the nearest foot.

9. In 1977, a hawksbill sea turtle was taken from Shedd Aquarium. It was later found in a suitcase at a Chicago airport. At the time the turtle weighed 1 pound. It now weighs 170 pounds. By how many kilograms did its weight increase?

10. **Stretch Your Thinking** A harbor seal can swim at speeds up to 11 miles per hour. About how many feet per second is that?

Caribbean Reef

The Shedd Aquarium's popular Caribbean Reef exhibit contains 90,000 gallons of water. Inside the tank are some 250 sea creatures representing 70 different species.

1. How many liters of water does the Caribbean Reef tank hold?

2. Water passes through the Caribbean Reef tank at a rate of 1,400 gallons per minute. How long does it take to circulate all the water in the exhibit?

Use Data   For 3–7, use the drawing, which shows a typical aquarium window.

3. What is the perimeter of the viewing surface?

4. What is the area of the viewing surface?

5. Classify the solid figure formed by the window.

6. What is the volume of the window in cubic feet? Explain.

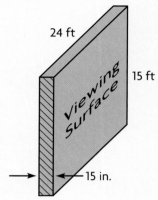

7. 📓 **Write About It** Explain how you could find the volume of the window in cubic inches. Then find it.

8. What if the Caribbean Reef exhibit were shaped like a square prism with a base 30 feet on a side and a height of 16 feet? How many cubic inches of water could the tank hold?

9. What if the Caribbean Reef exhibit were shaped like a cylinder with a diameter of 30 feet and a height of 16 feet? How many cubic inches of water could the tank hold?

10. A gallon of water occupies 231 cubic inches of volume. Would either tank in Exercises 8 or 9 be large enough to hold the water in the Caribbean Reef exhibit? Explain.

The Shedd Aquarium is a great place to learn about marine animals up close.

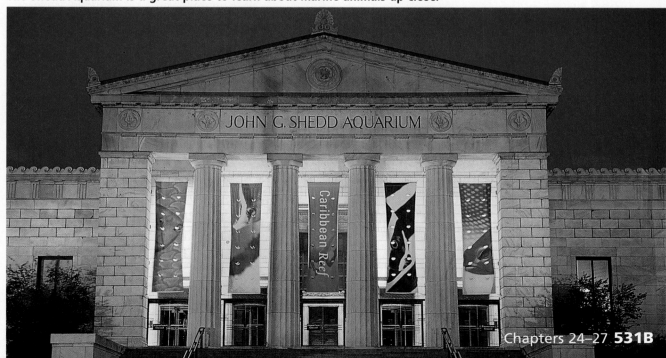

28 Algebra: Patterns

In 1992, 100 women from 20 countries set the women's skydiving record for the largest free-fall formation. Formation skydivers must plan what they want to do ahead of time. Some sketch out patterns. Others use tables to plan how many people join up at a time.

PROBLEM SOLVING Suppose 30 skydivers want to form linked rings. Each ring has 4 skydivers, one of whom is also part of the next ring. Complete the pattern shown in the table. Then show the pattern graphically with dots. How would you extend this pattern?

DATA LINK

SKYDIVING LINKED RINGS						
Ring	1	2	3	4	5	6
Number of Skydivers	4	7	10	13	■	■

Check What You Know

Use this page to help you review and remember important skills needed for Chapter 28.

✓ Vocabulary

Choose the best term from the box.

> equation
> expression
> variable

1. An algebraic or numerical sentence showing that two quantities are equal is a(n) __?__ .

2. A mathematical phrase that combines operations, numerals, and variables to name a number is a(n) __?__ .

✓ Compare Numbers (For Intervention, see p. H29.)

Compare. Write < or > for each .

3. 1.01 ● 1.1
4. $1\frac{1}{4}$ ● $1\frac{1}{2}$
5. ‾8 ● ‾2
6. 576 ● 479

7. 0.7 ● 0.35
8. 9.23 ● 9.31
9. 0 ● ‾2
10. 2,108 ● 2,110

11. ‾3.1 ● ‾2.9
12. $\frac{2}{3}$ ● $\frac{3}{4}$
13. 10.1 ● 9.09
14. 0.5 ● 0.06

15. $2\frac{4}{5}$ ● $3\frac{1}{5}$
16. 3 ● ‾4.56
17. 0.8 ● 0
18. $\frac{‾3}{4}$ ● $\frac{‾3}{8}$

✓ Input/Output Tables (For Intervention, see p. H29.)

Copy and complete each table.

19.

INPUT	OUTPUT
x	$x - 6$
16	▨
27	▨

20.

INPUT	OUTPUT
y	$y + 14.7$
12	▨
12.5	▨

21.

INPUT	OUTPUT
w	$w \div 4$
100	▨
102	▨

22.

INPUT	OUTPUT
a	$a \times 15$
10	▨
23	▨

✓ Evaluate Expressions (For Intervention, see p. H18.)

Evaluate each expression.

23. $x + 10$ for $x = 5$

24. $24 - y$ for $y = 2$

25. $20 \div z$ for $z = {}^-4$

26. $8.9a$ for $a = {}^-3$

27. $9s$ for $s = 7$

28. $w - 8$ for $w = 40$

29. $x^2 - {}^-4$ for $x = {}^-5$

30. $(w - 7) \div 4$ for $w = {}^-9$

31. $m \cdot \frac{1}{3}$ for $m = \frac{9}{14}$

32. $42 + k - 8$ for $k = 18$

33. $(w - 8) \div 2$ for $w = 14.2$

> **LOOK AHEAD**
>
> **In Chapter 28 you will**
> - identify, extend, and make number patterns in sequences
> - identify, extend, and make number patterns and write a rule for the pattern
> - identify, extend, and make patterns of geometric figures

PROBLEM SOLVING STRATEGY
Find a Pattern

Learn to solve problems by using the strategy *find a pattern*.

The members of a skydiving team make formations in the air by holding on to each other. After one second, one person begins a formation. At three seconds, three others join with the first person. At five seconds, five more people join. At seven seconds, seven more join the group. If this pattern continues, how many people will be in the group after eleven seconds?

Analyze

What are you asked to find?

What information are you given?

Is there information you will not use? If so, what?

Choose

What strategy will you use?

You can use the strategy *find a pattern*.

Solve

How will you solve the problem?

Find patterns for skydivers added and total skydivers.

Skydivers added	Total skydivers
1, 3, 5, 7, . . .	1, 4, 9, 16, . . .
Pattern: consecutive odd numbers	Pattern: squares of consecutive numbers

TECHNOLOGY LINK
More Practice: Use E-Lab, *Geometry and Number Patterns.*
www.harcourtschool.com/ elab2002

Show the patterns in a table to find the number of skydivers in the formation after eleven seconds.

Time	1 sec	3 sec	5 sec	7 sec	9 sec	11 sec
Skydivers added	1	3	5	7	9	11
Total skydivers	1	4	9	16	25	36

So, there are 36 skydivers at 11 seconds.

Check

How can you check your answer?

What if there are 81 skydivers? At how many seconds are there 81?

PROBLEM SOLVING STRATEGIES

| Draw a Diagram or Picture |
| Make a Model |
| Predict and Test |
| Work Backward |
| Make an Organized List |
| ▶ **Find a Pattern** |
| Make a Table or Graph |
| Solve a Simpler Problem |
| Write an Equation |
| Use Logical Reasoning |

Solve the problem by finding a pattern.

1. Alex is beginning a new physical training program. He will do 7 push-ups the first day, 12 push-ups the second day, 17 push-ups the third day, and will continue until he does 42 push-ups in one day. If this pattern continues, on which day will Alex do 42 push-ups?

2. The school choir learns songs for a competition to be held in 7 weeks. The choir knows 5 songs after the first week, 8 after the second week, and 11 after the third week. If this pattern continues, how many songs will the choir know after 7 weeks?

For 3–4, use this information.

The air temperature decreases 8°C for every 1,500-m gain in altitude.

3. Which algebraic expression can be used to find the temperatures if the starting temperature is 20°C and n is the number of 1,500-m gains in altitude?

   A $20 - 8n$ **C** $20 + n$

   B $20 + 8n$ **D** $20 - n$

4. It was 20°C when Rachel began her ascent in a hot-air balloon. How many meters had the balloon gone up when the temperature was ⁻4°C?

   F 3 m   **H** 4,500 m

   G 3,000 m   **J** 6,000 m

MIXED STRATEGY PRACTICE

5. Paul bought previously viewed videos for $119.50. He paid $12.50 each for some and $8.00 each for others. He bought fewer than 8 videos at each price. How many videos did Paul buy for $12.50?

6. A side of square C is 8 times the length of a side of square A. How many times as great as the area of square A is the area of square C?

7. Light poles along a highway are placed at regular intervals. The distance from the first pole to the sixth pole is 300 ft. What is the distance from the twentieth pole to the thirty-first pole?

8. A card collection was divided among Michelle and her three brothers. Armand received $\frac{1}{3}$ of the cards, Ben received $\frac{1}{4}$ of the cards, and Denis received $\frac{1}{6}$ of the cards. Michelle received 345 cards. How many cards were in the collection before it was divided?

9. Marco has a garden that is 5 ft wide and 8 ft long. He decides to double the width. What is the difference in area of the gardens?

10. **? What's the Error?** To solve the proportion $\frac{15}{30} = \frac{x}{12}$, Sean writes the equation $15x = 12 \times 30$. What error did he make? What is the correct equation?

Patterns in Sequences

Learn how to recognize, describe, and extend patterns in sequences.

Vocabulary

triangular number

sequence

term

Mr. Fabano designed patterns for the members of his marching band to make when they perform. His patterns, shown below, are triangular arrays.

A number that can be represented by a triangular array is a **triangular number**.

triangular arrays →

triangular numbers →   1   3   6   10

The pattern of triangular numbers above can be written in a number sequence. A **sequence** is an ordered set of numbers. Each number in the sequence is called a **term**.

To get the next term in the triangular number sequence, you add 1 more than was added previously. Use this rule to find the next term in the sequence.

sequence →   1   3   6   10   ■
                      +2 +3 +4 +5

Since 10 + 5 = 15, the next triangular number is 15.

Math Idea ▶ A sequence can have a repeating pattern that uses addition, subtraction, multiplication, division, or any combination of these operations.

EXAMPLE 1

Identify a pattern in the sequence 1, $5\frac{1}{2}$, 10, $14\frac{1}{2}$, 19, . . . and write the rule. Using your rule, find the next three terms in the sequence.

1          $5\frac{1}{2}$ 10 $14\frac{1}{2}$ 19   *Look for a pattern. Compare each term with the next.*

+$4\frac{1}{2}$ +$4\frac{1}{2}$ +$4\frac{1}{2}$ +$4\frac{1}{2}$

A rule is to add $4\frac{1}{2}$ to each term to get the next term.

$19 + 4\frac{1}{2} = 23\frac{1}{2}$ $23\frac{1}{2} + 4\frac{1}{2} = 28$ $28 + 4\frac{1}{2} = 32\frac{1}{2}$

So, $23\frac{1}{2}$, 28, and $32\frac{1}{2}$ are the next three terms.

Sequences that show a decreasing pattern often involve subtraction or division.

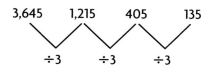 **EXAMPLE 2**

Find the next three possible terms in the sequence.

3,645; 1,215; 405; 135; . . .

3,645   1,215   405   135

÷3   ÷3   ÷3

Look for a pattern. Compare each term with the next.

A rule is to divide each term by 3 to get the next term.

135 ÷ 3 = 45   *Start with 135 and divide by 3.*

45 ÷ 3 = 15

15 ÷ 3 = 5

So, 45, 15, and 5 are the next three terms.

You can use a rule to write a sequence.

EXAMPLE 3

James has $20.00. He spends $2.20 every day on his school lunch. Write a sequence to show how James will spend his $20.00. How much money will James have left after three days?

Start: $20.00

Rule: Subtract $2.20 from each term.

Day 1: $20.00 − $2.20 = $17.80

Day 2: $17.80 − $2.20 = $15.60

Day 3: $15.60 − $2.20 = $13.40

Subtract $2.20 from each term to find the next term.

$20.00, $17.80, $15.60, $13.40, . . .   *Write the terms as a sequence.*

So, James will have $13.40 left after three days.

• **What if** James continues this pattern? How many more school lunches can he buy with the money left after three days?

CHECK FOR UNDERSTANDING

Think and Discuss ▶ **Look back at the lesson to answer each question.**

1. **Write** a rule to make the following sequence. Then find the next three terms. 8, 32, 128, 512, . . .

2. **Tell** whether the sequence 1,200; 240; 48 . . . is increasing or decreasing.

Guided Practice ▶ **Write a rule for each sequence.**

3. 5, 20, 35, 50, . . .   4. 0.001, 0.01, 0.1, 1, . . .

Find the next three possible terms in each sequence.

5. $^-1, 10, 21, 32, \ldots$

6. $0.2, 2, 20, 200, \ldots$

PRACTICE AND PROBLEM SOLVING

Independent ▶
Practice

Write a rule for each sequence.

7. $405, 135, 45, 15, \ldots$

8. $17.5, 16.3, 15.1, 13.9, \ldots$

9. $7, 7.89, 8.78, 9.67, \ldots$

10. $12, ^-30, 75, ^-187\frac{1}{2}, \ldots$

Find the next three possible terms in each sequence.

11. $35, 65, 125, 215, \ldots$

12. $400, 200, 100, 50, \ldots$

13. $91, 90, 88, 85, \ldots$

14. $^-4, 40, ^-400, 4,000, \ldots$

Use the rule to write a sequence.

15. Start with 9; add 3.7.

16. Start with 5; multiply by $^-6$.

Problem Solving ▶
Applications

17. Mr. Amano bought a collector's gold coin for $138. The value of the coin increased the same amount each year he owned it. The sequence $192, $246, $300, $354, . . . shows the pattern of the increasing value. Write the rule for the pattern and find what the coin is worth the sixth year.

18. *REASONING* The sequence 1, 1, 2, 3, 5, 8, 13, . . . is called the Fibonacci Sequence. Write a rule for the pattern, and list the first 15 terms in the sequence. As the sequence continues, what other patterns do you find?

19. ✎ **Write a problem** involving a number sequence that starts with $\frac{1}{2}$ and uses multiplication to get the next number. Explain your rule.

20. The length of a rectangle is 4 more than 3 times the width. What is the perimeter of the rectangle if the width is 2 feet?

MIXED REVIEW AND TEST PREP

21. Find the volume of the cylinder to the nearest whole number. (p. 521)

22. What is the measure of $\angle BEC$ at the right? (p. 322)

23. Evaluate $3 \times (7 + 4) \times 2^3$. (p. 44)

24. Solve. $\frac{h}{1.5} = 30$ (p. 82)

⭐**25.** **TEST PREP** Jeff has $1,000 to landscape his lawn. He plans to spend 35% of it on new trees. Each tree costs $55. How many trees can he buy? (p. 412)

A 6   **B** 7   **C** 8   **D** 9

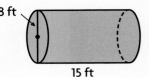

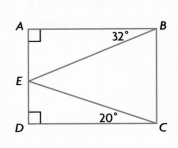

EXTRA PRACTICE page H59, Set A

LESSON 28.3

Number Patterns and Functions

Learn how to write an equation to represent a function.

Vocabulary

function

QUICK REVIEW

Evaluate each expression.

1. $a + 3.2$ for $a = 5.4$ **2.** $5.6 - r$ for $r = 1.2$

3. $0.25 \times m$ for $m = 4$ **4.** $q \div 0.3$ for $q = 9$

5. $y - 16$ for $y = 25.2$

Delia earns $5.25 per hour. She wants to buy a camera that costs $125. How many hours will Delia have to work to earn $125?

Use the pattern below.

HOURS WORKED	MONEY EARNED
1	$5.25
2	$10.50
3	$15.75
4	$21.00
5	$26.25

The pattern shows that for the first five hours, as the number of hours increases by one, the money earned increases by $5.25.

Delia used the pattern to write an equation in which h equals the number of hours worked and m equals the amount of money earned. She can use the equation to find how many hours she needs to work to earn $125.

TECHNOLOGY LINK

To learn more about number patterns and functions, watch the **Harcourt Math Newsroom Video** *British Code Breaker.*

$m = 5.25h$

$125 = 5.25h$ *Replace m with 125.*

$\dfrac{125}{5.25} = \dfrac{5.25\,h}{5.25}$ *Solve the equation.*

$23.81 = h$

So, Delia needs to work about 24 hr.

Math **I**dea ▶ A **function** is a relationship between two quantities in which one quantity depends on the other. Every input has exactly one output. Delia's equation is a function because the amount of money she earns, m, depends on the number of hours, h, that she works.

539

You can use a function table or an equation to represent some functions.

EXAMPLE 1

Mrs. Nien told her students that she used a pattern to determine the different sizes of paper to cut for them to use in class. The sizes, in inches, are shown below. Look for a pattern. Then write an equation to find the length when the width is 10 in.

width, w	3	4	5	6	7
length, l	7	9	11	13	15

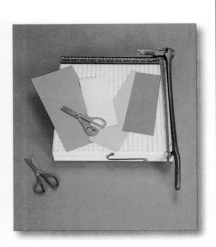

Compare the length and the width.
The length is 1 in. more than 2 times the width.   ← pattern

$l = 2w + 1$ *Write the equation.*

Use the equation to find the length when the width is 10 in.

$l = (2 \times 10) + 1$ *Replace w with 10.*
$l = 20 + 1$
$l = 21$

So, the length is 21 in. when the width is 10 in.

• In Example 1, the value of l depends on what other value?

You can use an equation to find missing terms in a function table.

EXAMPLE 2

Write an equation to describe the function. Then use it to find the missing term.

x	1	2	3	4	5	6
y	1	4	9	16	25	■

THINK: Since each y-value is greater than the corresponding x-value, use either addition or multiplication. What number is each x-value multiplied by to get the y-value?

Each x-value is multiplied by itself.   ← pattern

The equation is $y = x^2$.

$y = x^2$
$y = 6^2$ *Replace x with 6.*
$y = 36$ ← $6^2 = 6 \times 6$

So, the missing term is 36.

• Find y when $x = 13$.

Think and ▶ **Look back at the lesson to answer each question.**
Discuss

1. **Explain** how, in Example 1, you would find the value of w if you knew $l = 21$.

2. **Write** an equation for this function: the amount earned is $6 times the number of hours worked.

Guided ▶ **Write an equation to represent the function.**
Practice

3.

a	10	9	8	7	6
b	4	3	2	1	0

4.

x	0	1	2	3	4
y	0	12	24	36	48

PRACTICE AND PROBLEM SOLVING

Independent ▶ **Write an equation to represent the function.**
Practice

5.

c	1	2	3	4	5
d	2.1	3.1	4.1	5.1	6.1

6.

m	1	4	9	16	25
n	1	2	3	4	5

7.

w	1.2	1.3	1.4	1.5	1.6
l	4.8	5.2	5.6	6.0	6.4

8.

x	5	4	3	2	1
y	9	7	5	3	1

TECHNOLOGY LINK

More Practice: Use E-Lab, *Geometry and Number Patterns*.

www.harcourtschool.com/elab2002

Write an equation to represent the function. Then use it to find the missing term.

9.

k	24	34	44	54
g	12	17	22	▪

10.

s	11	14	17	20
r	22	▪	28	31

11.

c	16	24	32	36
d	-4	-6	▪	-9

12.

a	42	48	52	56
b	14	▪	$17\frac{1}{3}$	$18\frac{2}{3}$

Problem Solving ▶
Applications

13. Write an equation for each function. Tell what each variable you use represents.

   a. The length of a rectangle is 5 times its width.

   b. Distance is the product of the rate of travel and the time traveled.

   c. The cost of shipping a package is $1.25 per pound.

   d. The cost of riding in a cab is $3.00 plus $0.75 per mile.

14. Felicity wants to purchase a CD player for $115.99. She earns $5.25 per hour baby-sitting. Write an equation and find how many hours Felicity needs to baby-sit to buy the CD player.

15. ***REASONING*** In a function, every input has exactly one output. The equation $y = x^2$ is a function. The equation $y^2 = x$ is not a function if x is any real number. Explain.

16. Different sizes of signs are shown below. Look for a pattern. Then write an equation to find the height when the width is 13 ft.

width, w	1	3	5	7	9
height, h	5	11	17	23	29

17. Tasha sells homemade bookmarks for 75¢ each. Her sales for each of seven weeks are shown below. Write an equation to represent the function and complete the table.

n	8	15	22	29	36	43	50
t	$6	$11.25	$16.50	$21.75	$27	$32.25	■

Use Data   For 18–19, use the graph of Vincent's budget.

18. Vincent saved $380 to spend on his vacation. How much did he plan to pay for meals and entertainment?

VACATION BUDGET

Meals and entertainment 45%
Entrance fees 25%
Souvenirs 30%

19. How much money did Vincent plan to spend on souvenirs and entrance fees?

MIXED REVIEW AND TEST PREP

20. Using a rule, name the next three terms. $^-0.82, ^-0.75, ^-0.68, ^-0.61, \ldots$ (p. 536)

21. $\frac{12}{13} + 1\frac{1}{8}$ (p. 182)

22. $(^-324 \div 12) + 4^3$ (p. 256)

23. **TEST PREP**   Oranges are 5 for $1.80. How much will 2 dozen oranges cost? (p. 22)

A $1.92   **B** $3.20   **C** $4.32   **D** $8.64

24. **TEST PREP**   Which shows $\sqrt{25} + ^-13$? (p. 278)

F $^-38$ **G** $^-8$ **H** 12   **J** 18

PROBLEM SOLVING THINKER'S CORNER

Algebra   The binary, or base-two, number system uses only the digits 0 and 1.

DECIMAL	0	1	2	3	4	5	6	7	8	9	10
BINARY	0	1	10	11	100	101	110	111	1000	1001	1010

In the decimal system, each place value is ten times the place value to the right. In the binary system, each place value is twice the place to the right. You can use powers of 2 to find the decimal equivalent of a binary number.

$$10100_{two} = (1 \times 2^4) + (0 \times 2^3) + (1 \times 2^2) + (0 \times 2^1) + (0 \times 2^0)$$

$$= (1 \times 16) + (0 \times 8) + (1 \times 4) + (0 \times 2) + (0 \times 1)$$

$$= 16 + 0 + 4 + 0 + 0$$

$$= 20$$

Find the decimal equivalent of each binary number.

1. 1111_{two} **2.** 10001_{two} **3.** 11000_{two} **4.** 11111_{two}

EXTRA PRACTICE page H59, Set B

Geometric Patterns

Learn how to recognize, describe, and extend patterns of geometric figures.

What is the next possible term?

1. 5, 10, 15, . . .   **2.** ⁻2, ⁻4, ⁻6, . . .

3. 10, 1, 0.1, . . .   **4.** 3, ⁻9, 27, . . .

5. 10, 12, 15, 19, . . .

Vocabulary

fractal

Ayita is making a wall hanging using a geometric figure. Patterns of geometric figures are based on the shape, color, size, position, or number of the figures. The pattern below is what Ayita started with for her wall hanging.

  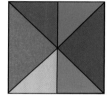

Notice that the size of the figure increases twice before the pattern repeats. The next figure in the pattern will be the large figure.

EXAMPLE 1

Ayita used this pattern for another craft project she made. Look for the pattern. Draw the next three figures.

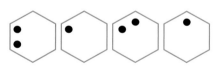

Look for a possible pattern. The circles, alternating between two and one, move clockwise around the inside of the figure.

So, the next three figures might be: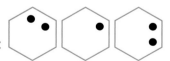

EXAMPLE 2

Look for a possible pattern. Draw the next figure.

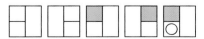

Look for a pattern.

The figures are flipped horizontally about a vertical axis. The top left square and then the bottom left square are changed.

So, the next figure might be:

543

You can also find patterns in three-dimensional figures.

EXAMPLE 3

Mr. Gallo is displaying boxes in his store. Look for the pattern. Draw what the next two displays will look like.

Look for a pattern. The number of boxes in the new bottom row increases by one.

So, these are the next two displays:

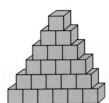

Some geometric figures contain a repeating pattern of smaller and smaller parts. This can be described as a fractal. A **fractal** has a repeating pattern containing shapes that are like the whole but of different sizes throughout.

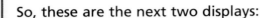

You can build fractals from two-dimensional figures by repeating a process over and over again. This is called iterating. An iteration is a step in the process of iterating.

EXAMPLE 4

Build a fractal from a square by repeating this iteration process two times. Find the number of shaded squares that would be in Stage 3.

Draw a shaded square. → Reduce the entire figure to one-half in length and width. Place one copy in the upper left-hand corner and one in the lower right-hand corner.

Stage 0   Stage 1   Stage 2

Look for a pattern.

Stage 0 has 1 shaded square, Stage 1 has 2, and Stage 2 has 4.

The pattern of shaded squares is 1, 2, 4, ...

So, 8 smaller squares would be shaded in Stage 3.

CHECK FOR UNDERSTANDING

Think and ▶ Discuss

Look back at the lesson to answer the question.

1. **Draw** a geometric pattern that uses a triangle.

Guided ▶ Practice

Draw the next three figures in the pattern.

2. 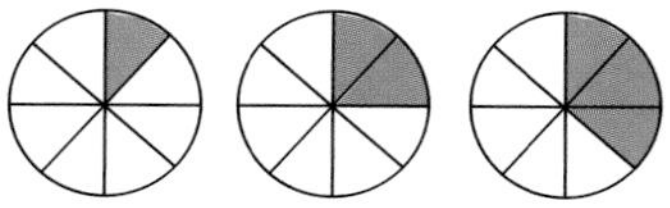 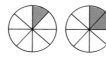   **3.**

PRACTICE AND PROBLEM SOLVING

Independent ▶ Practice

Draw the next three figures in the pattern.

4. 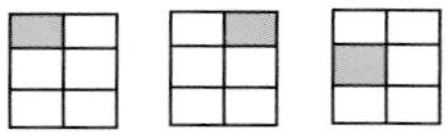 ⊞ ⊞ ⊞   **5.** ⊠ ⊠ ⊠

6.    **7.**

For 8–9, use the figures at the right.

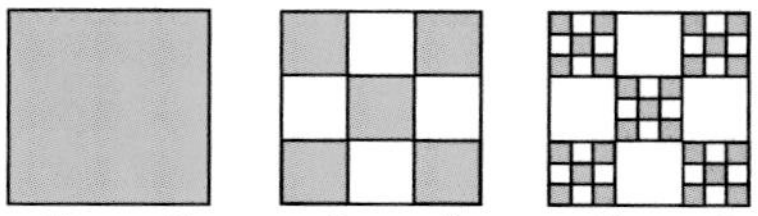

8. The iterative process is (1) reduce the figure to one-third in length and width and (2) place five copies in the corners and center of the original square. How many yellow squares would be in Stage 3?

Stage 0   Stage 1   Stage 2

9. Use exponents to describe the pattern for the number of yellow squares at each stage.

Problem Solving ▶ Applications

10. Ayita saves part of the money made by selling her wall hangings. She saves $55 of the $450 she gets for the first hanging. She will increase the amount saved by $4 each time she sells a hanging. Write the pattern of amounts. For which hanging will she save $75?

11. The figure at the right was made with rectangular prisms. The top level has 1 prism, the next level has 4 prisms, and the bottom has 9 prisms. How many prisms would be in the next two levels added at the bottom?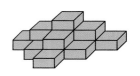

12. ✎ **Write a problem** using a geometric pattern in your school. Describe it in words and with a drawing.

MIXED REVIEW AND TEST PREP

13. Solve $y = 6x + 3$ for $x = {}^-2, {}^-1, 0, 1,$ and 2. (p. 539)

14. Compare. Write $<$ or $>$ for ●.
$\frac{-1}{2}$ ● 0.50 (p. 234)

15. Write the prime factorization for 65. (p. 148)

16. Evaluate $3x + 9x - 36$ for $x = {}^-5$. (p. 272)

☆17. TEST PREP Which triangle has one right angle and no congruent sides? (p. 332)
A an obtuse scalene triangle
B a right isosceles triangle
C a right scalene triangle
D an acute scalene triangle

1. VOCABULARY A relationship in which one quantity depends on another is a(n) __?__ . (p. 539)

2. VOCABULARY An ordered set of numbers is called a(n) __?__ . (p. 536)

3. VOCABULARY Each number in a sequence is called a(n) __?__ . (p. 536)

Write a rule for each sequence. (pp. 536–538)

4. 220, 22, 2.2, . . .

5. $^-$3, 9, $^-$27, . . .

6. 100, 99.1, 98.2, . . .

7. 4, $3\frac{2}{3}$, $3\frac{1}{3}$, . . .

8. 243, 81, 27, . . .

9. $\frac{7}{8}$, $3\frac{1}{8}$, $5\frac{3}{8}$, . . .

Find the next three possible terms in each sequence. (pp. 536–538)

10. 7, $5\frac{3}{4}$, $4\frac{1}{2}$, . . .

11. 90, 79, 68, . . .

12. 9, 15, 21, . . .

13. $^-$4, $^-$2, 0, . . .

14. 32, 16, 8, . . .

15. 5, 17, 29, . . .

Write an equation to represent the function. (pp. 539–542)

16.

x	9	8	7	6	5
y	0	-1	-2	-3	-4

17.

x	10	11	12	13	14
y	17	18	19	20	21

18.

x	-3	-2	-1	0	1
y	-30	-20	-10	0	10

19.

x	64	56	48	40	32
y	8	7	6	5	4

Draw the next three figures in the pattern. (pp. 543–545)

20.

21.

Draw the next two solids in the pattern. (pp. 543–545)

22.

23.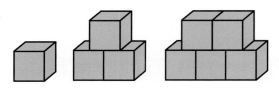

Solve. (pp. 536–538)

24. Ms. Wong received 110 orders the first month, 125 the second month, and 140 the third month. If this pattern continues, how many orders will she receive the sixth month?

25. The temperature was 21°F at 6 P.M. At 7 P.M. it was 18°F, and at 8 P.M. it was 15°F. If this pattern continues, what will the temperature be at midnight?

Check your work.

See item **2.**

You could re-write each term as a decimal to check your work. Your decimal answer should be equivalent to the fraction answer you chose.

Also see problem **7**, p. H65.

Choose the best answer.

1. What is the output when the input is 7?

Input s	Algebraic Expression $s - 3$	Output
4	4 − 3	1
5	5 − 3	2
6	6 − 3	3
7	7 − 3	■

A 4　　　　　　**C** 6

B 5　　　　　　**D** Not here

2. What is the next possible term in the sequence?

$$3\tfrac{1}{4}, 3\tfrac{7}{8}, 4\tfrac{1}{2}, 5\tfrac{1}{8}$$

F $5\tfrac{1}{4}$　　**G** $5\tfrac{3}{4}$　　**H** 6　　**J** $6\tfrac{1}{4}$

3. Alice works $7\tfrac{1}{2}$ hours each day. If she is paid d dollars per hour, which expression shows her earnings for 1 day?

A $d - 7\tfrac{1}{2}$　　　**C** $7\tfrac{1}{2} - d$

B $7\tfrac{1}{2} + d$　　　**D** $7\tfrac{1}{2} \times d$

4. Marisa jogs 400 meters on day 1, then 475 meters on day 2, and 550 meters on day 3. If she continues this pattern, how far will she jog on day 7?

F 700 m　**G** 775 m　**H** 850 m　**J** 925 m

5. Which figure comes next in the pattern?

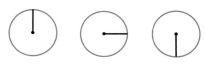

A

C

B

D

6. Mrs. Crain filled her car with gas and then drove 261 miles. When she filled the car with gas again, she determined that the car had averaged 29 miles per gallon. Which equation could be used to find g, the number of gallons of gas it took to fill the car the second time?

F $29 \div g = 261$　**H** $29 + g = 261$

G $261 \times g = 29$　**J** $261 \div g = 29$

7. The area of a rectangular rug is 96 square feet. The rug is 8 feet wide. How long is the rug?

A 6 ft　　　　**C** 24 ft

B 12 ft　　　　**D** Not here

8. What is the LCM for 8 and 12?

F 2　　　　　**H** 24

G 4　　　　　**J** 96

Write What You Know

9. Identify a rule to find the next three possible terms in the sequence below. Tell how you used your rule to find the next three possible terms.

89, 83.3, 77.6, 71.9,…

10. Find an equation to represent the function below. Explain how you know that the equation works for the function.

x	−2	−1	0	1	2
y	−8	−4	0	4	8

29 Geometry and Motion

Many geometric shapes have symmetry. People and many animals also have symmetry. Many starfish, for example, have line symmetry.

PROBLEM SOLVING Look for an object in your classroom that has line symmetry. Does it have one line of symmetry or more? Sketch the object and show its line(s) of symmetry.

DATA
LINK

**NUMBER OF LINES OF SYMMETRY
FOR REGULAR POLYGONS**

Check What You Know

Use this page to help you review and remember important skills needed for Chapter 29.

✓ Vocabulary

Choose the best term from the box.

1. Figures that are the same size and shape are ___?___ .

2. The movement of a geometric figure to a new position without turning or flipping it is a ___?___ .

> equivalent
> congruent
> slide

✓ Slides, Flips, and Turns (For Intervention, see p. H30.)

Identify the transformation as a slide, flip, or turn.

3.

4.

5.

6.

7.

8.

✓ Line Symmetry (For Intervention, see p. H30.)

Trace the figure. Draw the lines of symmetry.

9.

10.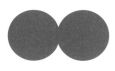

11.

12.

13.

14.

15.

16.

✓ Measure Angles (For Intervention, see p. H31.)

Use a protractor to measure each angle.

17.

18.

19.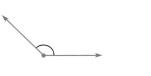

20.

> ### LOOK AHEAD
>
> In Chapter 29 you will
> - make tessellations
> - perform transformations of plane and solid figures
> - examine symmetry in plane figures

Transformations of Plane Figures

Learn how to use translations, rotations, and reflections to transform geometric shapes.

Estimate the measure of each angle.

1.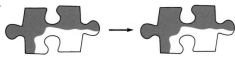

2.

3.

4.

5.

Vocabulary

transformation
translation
rotation
reflection

A movement of a figure without changing the size or shape of the figure is a rigid **transformation**. Since the size and shape do not change, the original figure and the transformation are always congruent.

A **translation** is the movement of a figure along a straight line.

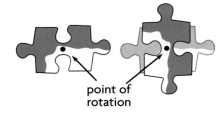

Only the location of the figure changes with translation.

Turning a figure around a point is called a **rotation**.

Both the position and the location of the figure can change.

A point of rotation can be on or outside a figure.

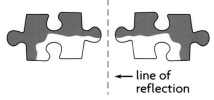

point of rotation

Flipping a figure over a line is called a **reflection** about that line.

Both the position and the location of the figure change with reflection.

line of reflection

EXAMPLE 1

Draw a 90° clockwise rotation of the figure around the given point of rotation.

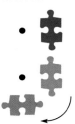

Trace the figure and the point of rotation.

Place your pencil on the point of rotation.

Rotate the figure clockwise 90°.

Trace the figure in its new location.

• How would the rotation look if you turned the figure 180°?

EXAMPLE 2

Draw a horizontal reflection and a vertical reflection of the shape of the state of California.

Trace the figure and the vertical line of reflection.

Reflect the figure horizontally over the line. This is a horizontal reflection.

Trace the figure in its new location.

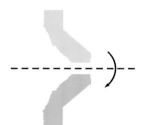

Use the same process for a horizontal line of reflection.

Reflect the figure vertically over the line. This is a vertical reflection.

Trace the figure in its new location.

- Rotate the original state shape 90° clockwise about a point in its upper right corner.

CHECK FOR UNDERSTANDING

Think and Discuss ▶ Look back at the lesson to answer each question.

1. **Explain** why a transformation is always congruent to the original figure.

2. **Describe** how the new position of a figure that has been rotated 360° compares with its original position.

Guided Practice ▶ Tell which type or types of transformation the second figure is of the first figure. Write *translation, rotation,* or *reflection*.

3.   4.  5.  6.

Tell what moves were made to transform each figure into its next positions.

7.   8.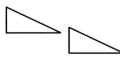

551

Independent ▶
Practice

Tell which type or types of transformation the second figure is of the first figure. Write *translation*, *rotation*, or *reflection*.

9.

10.

11.

12.

Tell what moves were made to transform each figure into its next position.

13.

14.

Trace each figure. Then draw a 90° clockwise rotation around the given point followed by a horizontal reflection.

15. **16.** **17.** **18.**

Problem Solving ▶
Applications

19. Kathy is working on a design for a logo. She drew a profile of an eagle and wants to have two eagles facing each other. What transformation should she do to the profile?

20. The mirror image of the word AMBULANCE is written on the front of ambulances so that it can be read in a car's rearview mirror. Identify the transformation and write the word as it appears on the front of the ambulance.

21. **REASONING** Draw a figure that looks the same when translated by rotation and by reflection.

22. The lowest temperature recorded in a certain month was 48°, and the highest temperature was $1\frac{1}{2}$ times as great as the lowest. What was the range of temperatures for the month?

MIXED REVIEW AND TEST PREP

23. Draw the next three figures in the geometric pattern at the right. (p. 543)

24. Find the area of a circle with a radius of 4 in. (p. 502)

25. Find a rational number between $\frac{3}{8}$ and $\frac{3}{4}$. (p. 230)

26. Write the decimal for 45%. (p. 60)

⭐**27.** **TEST PREP** A spinner is divided into five sections, labeled 1, 2, 3, 4, and 5. If each number is equally likely to occur, what is the probability of the pointer's landing on an odd number? (p. 428)

A $\frac{1}{5}$ **B** $\frac{2}{5}$ **C** $\frac{1}{2}$ **D** $\frac{3}{5}$

EXTRA PRACTICE page H60, Set A

LESSON 29.2

Tessellations

Learn how to use polygons to make tessellations and how to make figures for tessellations.

Vocabulary

tessellation

QUICK REVIEW

Tell which type of transformation the second figure is of the first figure.

1. A ^{A} 2. A A   3. A◁   4. $\underset{A}{\forall}$ 5. $\overset{A}{A}$

A repeating arrangement of shapes that completely covers a plane, with no gaps and no overlaps, is called a **tessellation** of the plane. Although most tessellations are produced by humans, a few occur in nature.

Honeycombs are naturally occurring tessellations using hexagons.

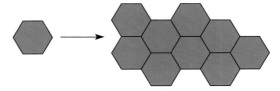

• What other natural tessellations can you think of?

In addition to being made from polygons, tessellations can be made from figures that are not polygons.

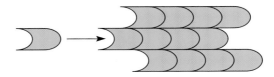

Activity 1

You need: pattern blocks, colored pencils or markers

Make a tessellation.

• Choose a pattern block to use for your tessellation shape.

• Design your tessellation. Remember that the shapes must fit together without overlapping or leaving gaps.

• Record your tessellation. Color it to make a pleasing design.

553

Math Idea ▶ Using regular polygons that can form tessellations, you can make other shapes that tessellate the plane.

Activity 2

You need: paper, scissors, and tape

Make a tessellation shape.

- Cut out a square that is 2 in. × 2 in.

- Use scissors to cut out a part of the square on one side.

- Translate the cutout part of the square to the opposite side of the square. Then tape it.

- Trace the new shape to form at least two rows of a tessellation. You will need to rotate, translate, or reflect the shape.

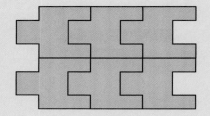

- **What if,** after you cut out part of a square, you didn't tape the cutout to the opposite side? Could the shape tessellate the plane? Explain.

TECHNOLOGY LINK

More Practice: Use **Mighty Math Cosmic Geometry,** *Tessellation Creation Station.*

CHECK FOR UNDERSTANDING

Think and Discuss ▶ **Look back at the lesson to answer each question.**

1. **Explain** how you know when a pattern of shapes forms a tessellation.

2. **Give an example** of a shape that does not form a tessellation.

Guided Practice ▶ **Trace and cut out several of each polygon. Tell whether the polygon can be used repeatedly to tessellate the plane. Write *yes* or *no*.**

3.   4.   5.   6.

Make the tessellation shape described by each pattern. Then form two rows of a tessellation.

7.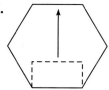

8.

9.

PRACTICE AND PROBLEM SOLVING

Independent ▶ Practice

Trace and cut out several of each shape. Tell whether the shape can be used to form a tessellation. Write *yes* or *no*.

10.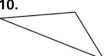

11.

12.

13.

Make the tessellation shape described by each pattern. Then form two rows of a tessellation.

14.

15.

16.

17.

Problem Solving ▶ Applications

18. Cut out a hexagon, and then change it by cutting out a part on one side. Translate the cutout part to the opposite side of the hexagon. Can this new shape form a tessellation?

19. Fergus wants to lay floor tiles in a tessellation pattern. The floor measures 24 yd by 12 yd. What is the total area to be covered?

20. 📖 **Write About It** Explain why you can translate a cutout from the edge of a regular hexagon to the opposite side when you want to make a tessellation pattern.

MIXED REVIEW AND TEST PREP

21. Tell which type of transformation the second figure is of the first figure. (p. 550)

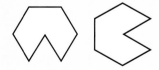

22. Write the ratio four to seven in three ways. (p. 384)

23. $\frac{3}{4} + \frac{1}{6}$ (p. 182)   24. Solve. $x - 8 = 40$ (p. 290)

⭐25. **TEST PREP**   Susan has run for five days. The distances of her runs were 4.5 mi, 3 mi, 4 mi, 3.5 mi, and 5 mi. How many miles will she have to run the sixth day for her mean distance to be 4.5 miles? (p. 106)

A 3.5 mi   **B** 4 mi   **C** 7 mi   **D** 27 mi

EXTRA PRACTICE page H60, Set B

PROBLEM SOLVING STRATEGY
Make a Model

Learn how to solve problems that involve tessellations by making a model.

QUICK REVIEW

QUICK REVIEW

Draw a horizontal reflection for each.

1.  2. 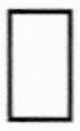 3.

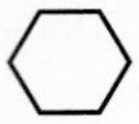

4. 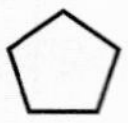 5.

Delaney is designing a tile mosaic for the top of her table. The shape Delaney uses must tessellate a plane. She wants to use the shape shown above. Can she use this shape for her design?

Analyze

What are you asked to find?

What information are you given?

Is there information you will not use?

Choose

What strategy will you use?
You can use the strategy *make a model*.

Solve

How will you solve the problem?
Trace and cut out several copies of the shape. Use the copies to see whether the shape will tessellate a plane.

Begin by moving the shapes around to see if they fit together. The pieces must not overlap or leave any gaps when you place them together.

You may also check to see if the sum of the angles where the vertices meet is equal to 360°. If it is, the shape will tessellate a plane.

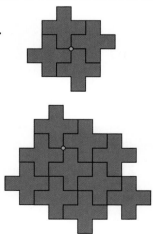

Check

How can you check your answer?

What if Delaney wants to use the figure at the right to make her mosaic? Will this figure tessellate a plane? Make a model to test your ideas.

Solve the problem by making a model.

1. Robert wants to make a design using a combination of two of the three shapes below. He wants the design to tessellate a plane. Which two-shape combinations can he choose from?

2. Will this shape tessellate a plane? Explain.

Mr. Foster wants to use triangular tiles in this size and shape to cover the floor of a hallway that measures 4 ft by 12 ft.

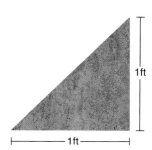

1ft

1ft

3. How many tiles does he need to cover the floor completely?

   A 24   **C** 96

   B 48   **D** 144

4. How many tiles would he need if the dimensions of each tile were half of those shown?

   F 48   **H** 192

   G 96   **J** 384

MIXED STRATEGY PRACTICE

Use Data   For 5–6, use the table.

5. Find the median temperature.

6. Make a stem-and-leaf plot of the temperature data.

HIGH TEMPERATURES IN SAN FRANCISCO						
Mon	Tue	Wed	Thurs	Fri	Sat	Sun
61	63	65	59	59	64	70

7. Shelley wants to make a quilt out of tessellating shapes. Which of the following shapes can she use?

   a.   b.

8. Ed had $40 when he went to the mall. He spent $24.88 on a shirt and $8.42 on a CD. Estimate how much he had left for lunch.

9. Bernie, Nash, Lisa, and Gerta are standing in the lunch line. Bernie is not first in line. Lisa has at least two people ahead of her in line. Nash is third. Give the order of the four in line.

10. A palindrome is a number that is the same backward and forward, such as 1,221. Find a three-digit palindrome in which the product of the digits is divisible by 35.

Transformations of Solid Figures

Learn how to identify and perform transformations of solid figures.

Name the shapes of the faces that make up each solid.

1.   2.   3.   4.   5.

You perform a transformation of a solid figure when you move your math book from your backpack to your desk.

Activity

- Place your textbook in the upper left corner of your desk, and imagine vertical and horizontal lines which divide the desk into fourths.

F = Front

- Perform a 180° rotation of the textbook around the desk's vertical center line.

B = Back

Remember that a transformation can be a rotation, a reflection, or a translation. A solid figure and its transformation are always congruent because the size and shape do not change.

- Perform a 180° rotation around the desk's horizontal center line.

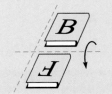

- Translate the textbook from the lower right to the lower left of the desk.

- Perform a 180° rotation around an axis that runs through the middle of the book, perpendicular to the desktop.

EXAMPLE

How many ways can you place your textbook on a piece of paper of the same size so that the paper is completely covered?

For a rectangular textbook, four positions are possible.

Think and Discuss ▶ Look back at the lesson to answer the question.

1. **Explain** whether a solid figure and its transformation are congruent.

Guided Practice ▶ Tell how many ways you can place the solid figure on the outline.

2.    3.    4.

PRACTICE AND PROBLEM SOLVING

Independent Practice ▶ Tell how many ways you can place the solid figure on the outline.

5.   6.   7.

8.    9. 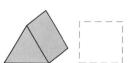   10.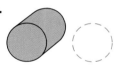

Problem Solving Applications ▶

11. Allison hung a sign that says "Allison's Room" upside-down on her door. What transformation should she do to the sign so that it is hung correctly?

12. **Write About It** Think of transformations of solid figures that happen in your everyday life. Give several examples.

MIXED REVIEW AND TEST PREP

13. Tell which type of transformation the second figure is of the first figure. (p. 550)

14. $25 \div 5 + (7 - 3)^2$ (p. 44)

15. What is the perimeter of a square with sides 6 cm long? (p. 479)

16. Write the prime factorization of 36. (p. 148)

17. **TEST PREP** A computer which normally sells for $800 is on sale at 20% off. What is the total cost if the sales tax rate is 5%? (p. 418)

   A $780   **B** $672   **C** $640   **D** $168

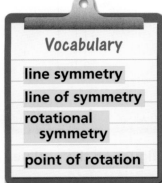

LESSON 29.5

Symmetry

Learn how to identify line symmetry and rotational symmetry.

Vocabulary

line symmetry

line of symmetry

rotational symmetry

point of rotation

QUICK REVIEW

Give the degree equivalent of each fractional portion of a full rotation.

1. $\frac{1}{4}$ 2. $\frac{1}{2}$ 3. $\frac{1}{8}$ 4. $\frac{3}{5}$ 5. $\frac{2}{3}$

Symmetry can be found all around us, both in nature and in manufactured objects. This Oregon Swallowtail butterfly has symmetry because each wing looks like a reflection of the other.

A figure has **line symmetry** if it can be folded or reflected so that the two parts of the figure match, or are congruent. The line across which the figure is symmetric is known as the **line of symmetry**.

Activity 1

You need: paper, scissors, dark crayon

- Fold the paper in half. Use the crayon to write your name in cursive along the fold line.

- Fold the paper along the fold line so your name is inside. Use the handle of the scissors to make a rubbing of your name.

- Unfold the paper. Your name appears on the other half of the paper. Your design has line symmetry.

- Where is the line of symmetry in the design you made from your name?

- How many lines of symmetry does the design have?

Some figures have several lines of symmetry.

EXAMPLE 1

Remember that regular polygons have all sides congruent and all angles congruent.

Find all the lines of symmetry in the regular hexagon.

Trace the figure and cut it out.

Fold it in half in different ways.

If the halves match, then the fold is a line of symmetry.
Count the different lines of symmetry.

The figure has six lines of symmetry.

- Do all hexagons have six lines of symmetry? Explain.

560 Chapter 29

Letters and whole words can have lines of symmetry.

EXAMPLE 2

Do the words DECIDED and MUM have line symmetry?

Copy each word.

Draw the dashed lines as shown.

The two halves of each word are the same size and shape.

Since the halves formed by each line are congruent, the words have line symmetry.

A figure has **rotational symmetry** if it can be rotated less than 360° around a central point, or **point of rotation**, and still look exactly the same as the original figure.

MATH LAB

Activity 2

You need: 4-in.-square paper, scissors

- Fold the square of paper in half and then in half again.

- Draw a petal shape on the folded paper, and cut it out so that the center is connected.

- Trace your flower on a piece of paper.

- Place the point of your pencil at the center of the cutout flower. Match it with its tracing.

- Rotate the flower, using the point of the pencil as the point of rotation, until it matches the tracing again.

The figure you made has rotational symmetry.

EXAMPLE 3

Does the figure of the starfish below have rotational symmetry? If it does, what are the fraction and the angle measure of each turn?

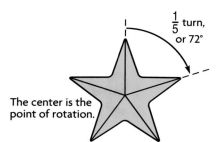

$\frac{1}{5}$ turn, or 72°

The center is the point of rotation.

Trace the figure.

Place your pencil point at the point of rotation.

Rotate the figure until it looks like the original figure.

Use a protractor to measure the degree and fraction of rotation.

The figure has $\frac{1}{5}$-turn, or 72°, rotational symmetry.

- Does the starfish figure above also have line symmetry?

Think and Discuss ▶ Look back at the lesson to answer each question.

    1. Explain the difference between line symmetry and rotational symmetry.

    2. Draw a pentagon that has fewer than five lines of symmetry.

Guided Practice ▶ Trace the figure. Draw the lines of symmetry.

  3.   **4.**   **5.**   **6.**

Trace the figure. Complete the other half of the figure across the line of symmetry.

  7.   **8.**   **9.**   **10.**

Tell whether each figure has rotational symmetry, and, if so, identify the symmetry as a fraction of a turn and in degrees.

  11.   **12.**   **13.**   **14.**

Independent Practice ▶ Trace the figure. Draw the lines of symmetry.

  15.   **16.**   **17.**   **18.**

Trace the figure. Complete the other half of the figure across the line of symmetry.

  19.    **20.**   **21.**   **22.**

  23.   **24.**   **25.**   **26.**

Tell whether each figure has rotational symmetry, and, if so, identify the symmetry as a fraction of a turn and in degrees.

27.   **28.**   **29.**   **30.**

Problem Solving ▶ Applications

31. Philip has a square cake. He cuts it along all the lines of symmetry of the square. How many pieces does he have? Are they all the same size?

32. *REASONING* Does a circle have line symmetry? rotational symmetry? Explain.

33. **(?) What's the Error?** Nick says that all figures have rotational symmetry, because when you rotate any figure 360° it matches up. What's his error?

MIXED REVIEW AND TEST PREP

34. Tell which type of transformation the second figure is of the first figure. (p. 550)

35. Find the volume of a cylinder with a diameter of 4 in. and a height of $\frac{5}{6}$ ft. (p. 521)

36. $\frac{2}{3} \times \frac{3}{8}$ (p. 202)

37. TEST PREP Which shows the number of inches in 5 ft? (p. 462)

A 50 **B** 60 **C** 75 **D** 180

38. TEST PREP Andy drove 900 miles in 18 hours. At what rate was he traveling? (p.384)

F 45 mi per hr **G** 50 mi per hr **H** 55 mi per hr **J** 60 mi per hr

---

PROBLEM SOLVING **LINKUP** to Reading

Strategy • Choose Important Information A word problem may contain more information than is needed. You must decide what information you need to solve the problem.

Susan visited her cousin in Guatemala and brought back fabric to make a quilt that will be 3 ft long and 2 ft wide. She has cotton pieces that are 4 in. on a side, cotton-blend pieces that are 3 in. on a side, and wool pieces that are 6 in. on a side. How many squares will she need for a wool quilt?

• What does the problem ask you to find?

• What information is needed to solve the problem?

• Tell the important information in this problem. Then solve.
Pauline will fold two pieces of paper along each of their lines of symmetry one time. One is a square measuring 3 in. on a side, and one is a rectangle measuring 4 in. by 8 in. How many folds must she make in all?

1. **VOCABULARY**   A movement of a plane or solid figure without changing the figure's size or shape is a ___?___ . (p. 550)

2. **VOCABULARY**   A repeating arrangement of shapes that completely covers a plane with no overlaps or gaps is called a ___?___ . (p. 553)

3. **VOCABULARY**   A line across which a figure is symmetric is known as a ___?___ . (p. 560)

Tell which type or types of transformation the second figure is of the first figure. Write *translation*, *rotation*, or *reflection*. (pp. 550–552)

4.

5.

6.

7.

Tell whether each polygon can be used to form a tessellation. Write *yes* or *no*. (pp. 553–555)

8.

9.

10.

11.

Tell how many ways you can place the solid figure on the outline. (pp. 558–559)

12.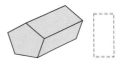

13.

14.

15.

Trace the figure. Draw the lines of symmetry. (pp. 560–563)

16.

17.

18.

19.

Tell whether each figure has rotational symmetry, and, if so, identify the symmetry as a fraction of a turn and in degrees. (pp. 560–563)

20.

21.

22.

23.

24. Richard builds a wall 6 full bricks wide, starting and ending every other layer with a half of a brick. The figure shows the first 3 layers of the wall. How many full bricks and how many half bricks does he need for 8 layers? (pp. 556–557)

25. There are 6 shelves that are 18 inches apart, and the bottom shelf is 24 inches from the floor. How far from the floor is the top shelf? (pp. 556–557)

TIP!

Understand the problem.
See item **2.**

Think about the hexagon pattern block and how 6 of the triangle blocks make the same shape when you put them together. The angle measure of the rotation is related to the measure of the angle in the triangle.

Also see problem **1**, p. H62.

Choose the best answer.

1. Which type of transformation of the first figure is shown by the second figure below?

   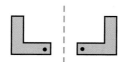

   A Translation   C Reflection
   B Rotation   D Not here

2. A regular hexagon has rotational symmetry. What is the angle measure of each turn?

   F 30°   H 60°
   G 45°   J 90°

3. What is the relationship of the second figure to the first?

   A Reflection   C Translation
   B Rotation   D Not here

4. How many lines of symmetry does a regular pentagon have?

   F 0   H 10
   G 5   J 15

5. What is the measure of each of the angles that surround the circled vertex?

   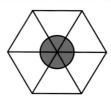

   A 30°   C 50°
   B 45°   D 60°

6. Dana's CD collection is $\frac{4}{9}$ rock and $\frac{2}{9}$ country. What part of her collection is neither rock nor country?

   F $\frac{1}{3}$ H $\frac{1}{2}$
   G $\frac{2}{5}$ J $\frac{2}{3}$

7. Meg sold 47 flowers. Each one was blue or white. She sold 13 more blue than white flowers. How many blue flowers did she sell?

   A 17   C 33
   B 30   D 34

8. The length of a rectangular prism is 8 centimeters, the width is 4 centimeters, and the height is 2 centimeters. What is the volume?

   F 32 cm^3 H 112 cm^3
   G 64 cm^3 J 144 cm^3

Write What You Know

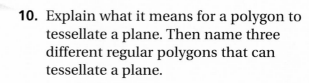

9. Describe how you can tell whether a line drawn on a figure is a line of symmetry. Then draw a line of symmetry for the figure below.

10. Explain what it means for a polygon to tessellate a plane. Then name three different regular polygons that can tessellate a plane.

30 Algebra: Graph Relationships

Before any airline flight, a flight plan is drawn on a map. The plan shows the route the airplane will fly. East-west lines of latitude and north-south lines of longitude form a coordinate grid system on the map. Pilots use these to identify places along the route.

PROBLEM SOLVING Suppose a flight is departing from Charlotte, N.C. The approximate coordinates for this city are (81,36). What are the approximate coordinates of nearby Columbia, S.C.?

Check What You Know

Use this page to help you review and remember important skills needed for Chapter 30.

Vocabulary

Choose the best term from the box.

1. Two numbers used to locate a point on a coordinate plane are called a(n) ? .

2. A movement that does not change the size or shape of a figure is called a(n) ? .

3. Lines that intersect at right angles to each other are ? lines.

> ordered pair
> origin
> parallel
> perpendicular
> transformation

Compare Numbers (For Intervention, see p. H29.)

Compare. Write <, >, or = for each ●.

4. $^-6$ ● $^-7$

5. $^-\frac{1}{2}$ ● $^-0.5$

6. $^-2.44$ ● 2.44

7. 3.10 ● 3.01

8. $3\frac{1}{2}$ ● $^-3.6$

9. $\frac{2}{5}$ ● 0.45

10. 0.75 ● 75%

11. $\frac{^-4}{5}$ ● $\frac{^-5}{6}$

Ordered Pairs (For Intervention, see p. H31.)

Write the ordered pair for each point.

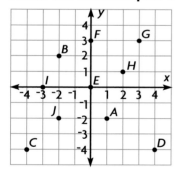

12. point A

13. point B

14. point D

15. point E

16. point G

17. point H

18. point C

19. point F

20. point I

21. point J

Transformations (For Intervention, see p. H30.)

Draw the reflection of the letter after it is flipped across the line.

22.

23. N

24.

25.

LOOK AHEAD

In Chapter 30 you will

- solve algebraic inequalities
- use ordered pairs to describe locations on the coordinate plane and to graph functions
- solve problems by using the skill *make generalizations*
- explore linear and nonlinear relationships
- graph transformations

Inequalities on a Number Line

Learn how to use inequality symbols and solve inequalities.

QUICK REVIEW

Use < or > to compare.

1. $6 + 9 \bullet 13$ **2.** $30 \div 3 \bullet 13$

3. $2 \times 6 \bullet 9$ **4.** $15 \bullet 25 - 9$

5. $68 \div 4 \bullet 16$

An **inequality** is an algebraic sentence that contains the symbol $<, >, \leq, \geq,$ or $\neq$.

These are inequalities:

$$n < 5 \qquad y > {}^-7 \qquad k \leq 3.8 + 2.7 \qquad b \geq 12 \div 4$$

If w represents the total weight the elevator can hold, then $w \leq 3,500$. Since w can equal any positive value less than or equal to 3,500, the inequality has many possible solutions.

ELEVATOR
─────────────
MAXIMUM LOAD:
3,500 POUNDS

Math **I**dea ▶ You can use a number line to show all the solutions of an inequality.

EXAMPLE 1

A. Graph the solutions of the inequality $p > 3$.

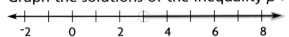

The blue line shows that all values greater than 3 are included in the solutions. The blue arrow means the solutions continue to the right of 8 on the number line. The circle at 3 is not filled in because 3 is *not* a solution of the inequality.

Remember that you can use these symbols when you compare numbers:

$=$ is equal to
$\neq$ is not equal to
$<$ is less than
$>$ is greater than
$\leq$ is less than or equal to
$\geq$ is greater than or equal to

B. Graph the solutions of the inequality $m \leq 4$.

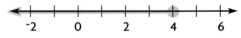

All values less than 4 are included in the solutions. The circle at 4 is filled in because 4 *is* a solution of the inequality.

Sometimes you can solve an inequality the same way you solve an equation.

EXAMPLE 2

Solve the inequality and graph the solutions on a number line.

$x - 2 < 3$
$x - 2 + 2 < 3 + 2$ *Add 2 to both sides.*
$x < 5$ *Simplify.*

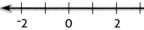

Think and ▶
Discuss

Look back at the lesson to answer the question.

1. **Explain** whether or not the circle at $^-3$ should be filled in on the number line showing the graph of the solutions of $x \geq {}^-3$.

Guided ▶
Practice

Graph the solutions of the inequality.

2. $x < 8$ 3. $x \geq {}^-7$ 4. $a \leq {}^-12$ 5. $y > 100$

PRACTICE AND PROBLEM SOLVING

Independent ▶
Practice

Graph the solutions of the inequality.

6. $x \leq 0$ 7. $x \leq 8$ 8. $x \geq {}^-10$ 9. $x < 30$

Solve the inequality and graph the solutions.

10. $2 + x > 8$ 11. $a + 5 \leq 8$ 12. $y - 2 > {}^-4$ 13. $k + 3 \geq 3$

14. $b - 10 \leq 5$ 15. $n - 1 < {}^-1$ 16. $3x < 9$ 17. $\frac{c}{2} > 1$

TECHNOLOGY LINK

More Practice: Use
Mighty Math Astro
Algebra, Green, Level U.

For 18–19, write an inequality for the word sentence.

18. The value of c is less than or equal to two.

19. The value of p is greater than negative eleven.

Problem Solving ▶
Applications

20. **Measurement** Trout less than 10 in. long must be thrown back. Write an inequality that represents the lengths that may be kept.

Use Data For 21–22, use the data in the table.

21. Write an inequality relating a, the altitude of any U.S. location, to Mount McKinley's altitude.

UNITED STATES SUPERLATIVES	
Highest Point	**Lowest Point**
Mount McKinley, AK	Death Valley, CA
Altitude: 20,320 ft	Altitude: $^-282$ ft

22. Write an inequality relating a, the altitude of any U.S. location, to Death Valley's altitude.

23. ❓ **What's the Error?** Tyra wrote the solutions of the inequality $\frac{a}{4} < 8$ as $a < 2$. What did she do? What is the correct answer?

MIXED REVIEW AND TEST PREP

24. How many lines of symmetry does the figure at the right have? (p. 560)

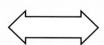

Write the value in the requested form. (p. 408)

25. $\frac{3}{8}$ as a percent   26. 125% as a decimal   27. 60% as a fraction

⭐ 28. **TEST PREP** Keith tossed a number cube labeled 1 to 6. He wanted to throw a 1 or a 2. How much greater is the probability that he will not succeed than that he will succeed? (p. 428)

A $\frac{1}{2}$ B $\frac{1}{3}$ C $\frac{1}{4}$ D $\frac{1}{6}$

EXTRA PRACTICE page H61, Set A

Graph on the Coordinate Plane

Learn how to use an ordered pair to describe a location on the coordinate plane.

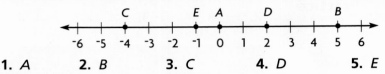

QUICK REVIEW

Identify the number named by the point on the number line.

1. *A*   2. *B*   3. *C*   4. *D*   5. *E*

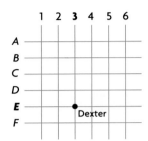

Points on a map are often located using pairs of coordinates. On the map grid shown, the town of Dexter is located at point (3, *E*). The coordinates of the location are 3 and *E*.

Use the same method to locate points on a **coordinate plane** like the one below.

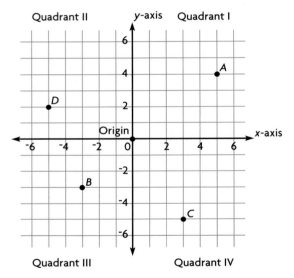

The coordinate plane is divided by a horizontal number line called the **x-axis** and a vertical number line called the **y-axis**. The two **axes** divide the plane into 4 pieces called **quadrants**. The point (0,0), where the axes intersect, is called the **origin**.

Math Idea ▶ Using an **ordered pair** of coordinates like (5,⁻2), you can locate any point on the coordinate plane. The first number of an ordered pair tells you how far to move right or left from the origin. The second number tells you how far to move up or down.

On the coordinate plane above, the coordinates of point *A* are (5,4), of point *B* are (⁻3,⁻3), of point *C* are (3,⁻5), and of point *D* are (⁻5,2). Points can be located at the origin, on the *x*- or *y*- axis, or in one of the four quadrants.

EXAMPLE 1

Describe how to locate the point represented by the ordered pair (6,⁻5) on the coordinate plane.

From the origin, move 6 units to the right, since the 6 is positive. Then move 5 units down, since the 5 is negative.

- Is the point (3,4) the same as the point (4,3)? Explain.

EXAMPLE 2

Name the ordered pair and the quadrant where each point on the coordinate plane below is located.

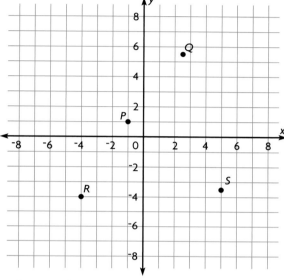

To find the ordered pair, first move right or left from the origin. Then move up or down.

Point *P* is located at (⁻1,1), in Quadrant II.

Point *Q* is located at about $(2\frac{1}{2}, 5\frac{1}{2})$, in Quadrant I.

Point *R* is located at (⁻4,⁻4), in Quadrant III.

Point *S* is located at about (5,⁻3.5), in Quadrant IV.

- Where would you find the point (0,6) on the coordinate plane? Where would you find (⁻5,0)?

You can use what you know about ordered pairs to plot points on a coordinate plane.

EXAMPLE 3

Sketch a coordinate plane, and plot the points *A*(4.5,2) and *B*(⁻3,3).

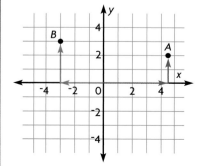

To locate point A, start at the origin and move 4.5 units to the right and 2 units up.

To locate point B, start at the origin and move 3 units to the left and 3 units up.

Think and ▶ **Look back at the lesson to answer each question.**
Discuss

1. **Explain** how the point (5,4) is different from the point (⁻5,⁻4).

2. **Decide** in which quadrant a point with two negative coordinates would appear. Explain your answer.

Guided ▶ **Write the ordered pair for each point on the coordinate plane.**
Practice

3. point A   4. point B

5. point C   6. point D

7. point E   8. point F

9. point G   10. point H

11. point I   12. point J

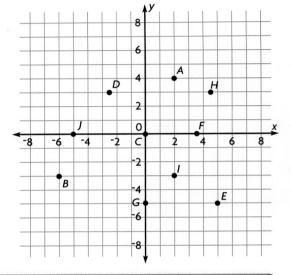

Use the coordinate plane at the right. Identify the points located in the given quadrant.

13. I   14. II

15. III   16. IV

Independent ▶ **Write the ordered pair for each point on the coordinate plane.**
Practice

17. point A   18. point B

19. point C   20. point D

21. point E   22. point F

23. point G   24. point H

25. point I   26. point J

27. point K   28. point L

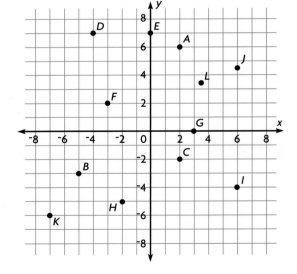

Use the coordinate plane at the right. Identify the points located in the given quadrant.

29. I   30. II

31. III   32. IV

Sketch a coordinate plane and plot the points.

33. $F(^-7,3)$ 34. $L(^-4,6)$ 35. $Y(^-1,3)$ 36. $A(^-4,^-4)$

37. $K(^-2.5,^-6)$ 38. $I(0,^-7)$ 39. $T\left(2\frac{1}{2},^-6\frac{1}{2}\right)$ 40. $E(6,^-3)$

Problem Solving ▶
Applications

41. Geometry Locate and connect the points (5,5), (5,⁻3), and (⁻2,⁻3) on a coordinate plane. What kind of geometric figure do you have? What is the area of the figure?

42. Geometry Locate the points (2,3), (2,⁻3), and (⁻4,3) on a coordinate plane. What ordered pair tells the location of the fourth point that would make the figure a square? What is the area of the square? What is the perimeter?

43. When a point lies on the *x*-axis, what can be said about the *y*-coordinate? What is the *x*-coordinate when a point lies on the *y*-axis?

44. ✎ **Write About It** What do all of the points in Quadrant II have in common? What do all of the points in Quadrant III have in common?

MIXED REVIEW AND TEST PREP

45. Solve. $5 + y \geq ^-2$ (p. 568)

46. A field is 249 ft long. Write the field's length in yards. (p. 462)

47. Find the circumference of a circle with radius 25 cm. (p. 484)

48. TEST PREP How much more is the sum of $6\frac{1}{2}$ and $3\frac{4}{5}$ than the sum of $2\frac{3}{4}$ and $3\frac{1}{5}$? (p. 186)

A $10\frac{3}{10}$ **B** $5\frac{19}{20}$ **C** $5\frac{13}{20}$ **D** $4\frac{7}{20}$

49. TEST PREP Which figure is a reflection of the figure at the right? (p. 558)

   F   **G**   **H**   **J**

PROBLEM SOLVING

 **to Science**

Astronomy Astronomers locate stars in the sky by using a coordinate grid system. Movements to the left and right are measured in hours (h). Movements up and down are measured in degrees (°).

Use the sky map of the Boötes constellation at the right.

1. Name the location of Arcturus, the brightest star in the constellation.

2. Give the letter of the star located at about $(14\frac{1}{2}$ h, 38°).

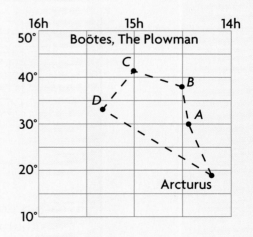

(**EXTRA PRACTICE**) page H61, Set B)

Graph Functions

Learn how to represent functions with ordered pairs, graphs, and equations.

When you see a flash of lightning, count the seconds until you hear thunder. A count of 5 seconds means you are 1 mile from the lightning.

ELAPSED TIME (sec)	5	10	15	20	25	30
DISTANCE TO LIGHTNING (mi)	1	2	3	4	5	6

The values in the table can also be given by using the ordered pairs (5,1), (10,2), (15,3), (20,4), (25,5), and (30,6).

The ordered pairs represent a function. You can use the ordered pairs to show the function on a graph.

HOW FAR AWAY IS THE LIGHTNING?

You can also use ordered pairs to graph data from a function table. Use the x and y values to form the ordered pairs.

EXAMPLE

Graph the data from the function table on a coordinate plane. Then write an equation relating y to x.

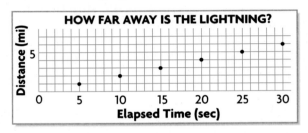

x	-3	-2	-1	0	1	2	3
y	-2	-1	0	1	2	3	4

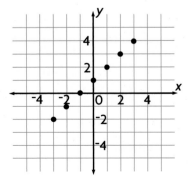

Write the data in the table as ordered pairs:

(−3,−2), (−2,−1), (−1,0), (0,1), (1,2), (2,3), (3,4).

Then plot the points.

In each ordered pair, the y-value is 1 more than the x-value. This means that x and y are related by the equation $y = x + 1$.

Think and ▶ **Look back at the lesson to answer the question.**
Discuss

1. **Name** two ways to represent functions.

Guided ▶
Practice

2. Copy and complete the function table. Then graph the data on a coordinate plane, and write an equation relating y to x.

x	2	3	4	5
y	4	6	8	■

PRACTICE AND PROBLEM SOLVING

Independent ▶ **Copy and complete the function table. Then graph the**
Practice **data on a coordinate plane, and write an equation relating**
y to x.

3.

x	5	6	7	8	9
y	3	4	5	■	■

4.

x	2	1	0	⁻1	⁻2
y	6	3	0	■	■

5. Use the equation $y = 5x$ to make a function table. Use the integers from ⁻2 to 2 as values of x.

Problem Solving ▶ **For 6–7, Sue earns three coupons for each win in an**
Applications **arcade game.**

6. Let w represent the number of wins. Use ordered pairs to show the number of coupons earned, c, for 7, 8, 9, and 10 wins. What equation relates c to w?

7. Let c represent the number of coupons earned. Use ordered pairs to show the number of wins, w, for 45, 48, 51, and 54 coupons. What equation relates w to c?

8. **Write a problem** involving a function. Make a function table for 5 values of x. Write the ordered pairs, and locate them on a coordinate plane.

MIXED REVIEW AND TEST PREP

9. Name the quadrant containing the ordered pairs (4,3), (2,1), and (6,5). (p. 570)

10. Tell which type of transformation the second figure is of the first figure. (p. 558)

11. A rectangle measures 8 cm by 5 cm. What is its area? (p. 494)

12. Greg ran a 5-km race. How many meters did he run? (p. 464)

⭐**13. TEST PREP** What is the value of the expression $k^2 - 24 \div 3$ where k equals the perimeter of a square with sides of length 3? (p. 272)

A 40 **B** 120 **C** 136 **D** 144

PROBLEM SOLVING SKILL
Make Generalizations

Analyze
Choose
Solve
Check

Learn how to solve problems by making generalizations.

For a science fair project, Cara is making a display showing how popular foods such as raisins and maple syrup are made from natural ingredients. She wants to show how to calculate the amount of each type of food that can be produced from given amounts of ingredients.

It takes about 4 lb of grapes to make 1 lb of raisins. How many pounds of raisins can be made from 160 lb of grapes?

Pounds of Grapes, x	4	8	12	16	20
Pounds of Raisins, y	1	2	3	4	5

Notice that each y-value is $\frac{1}{4}$ of the corresponding x-value. So, the equation $y = \frac{1}{4}x$ describes this situation. Since the equation can be used to find the amount of raisins for any amount of grapes, it is called a generalization.

For 160 lb of grapes, $x = 160$. Then $y = \frac{1}{4} \times 160 = 40$.

So, about 40 lb of raisins can be made from 160 lb of grapes.

On average it takes about 40 gal of sap from maple trees to make 1 gal of maple syrup. How many gallons of sap does it take to make 25 gal of syrup?

Gallons of Syrup, x	1	2	3	4	5
Gallons of Sap, y	40	80	120	160	200

Notice that each y-value is 40 times the corresponding x-value. So, the equation $y = 40x$ is a generalization.

For 25 gal of syrup, $x = 25$. Then $y = 40 \times 25 = 1,000$.

So, it takes about 1,000 gal of maple sap to make 25 gal of maple syrup.

Talk About It ▶

• **What if** Cara shows in her display that about 1 qt of apple juice can be made from 60 apples? What equation could she use to find the number of quarts that can be made from a given number of apples?

Solve by making a generalization.

Rod spent $6 on supplies to advertise his mowing business. The table shows his profit, *y*, for several income amounts, *x*.

x	$12	$13	$14	$15	$16
y	$6	$7	$8	$9	$10

1. Which equation can be used to show Rod's profit?

   A $y = 2x$ **C** $y = x - 6$

   B $y = \frac{1}{2}x$ **D** $y = x + 6$

2. How much profit did Rod make if he earned $80 mowing lawns?

   F $86   **H** $74

   G $83   **J** $40

Liz charges $9 for each lawn that she mows.

3. What equation can Liz use to show the amount that she earns, *y*, when she mows *x* lawns?

4. How much money will Liz earn if she mows 27 lawns?

MIXED APPLICATIONS

5. Eduardo is driving from Pittsburgh to San Francisco, a distance of 2,578 mi. He drove 440 mi the first day, 380 mi the second day, and 425 mi the third day. How many miles does he have left to drive?

6. To celebrate school spirit week, Roosevelt School plans to paint every sixth locker blue and every tenth locker orange. If there are 300 lockers in the school, how many will be painted both blue and orange?

7. Mr. Wills needs to catch a flight that leaves at 1:00 P.M. It takes 45 min to get to the airport, and he wants to be there 1 hr 15 min early. At what time should Mr. Wills leave home?

8. Sandra used 60 ft of fencing to fence her garden. The garden is a rectangle that is twice as long as it is wide. What are the dimensions of the garden?

9. Use Data Use the graph at the right. The length of the main span of the Verrazano Narrows Bridge is 4,260 ft. The lengths of several other suspension bridges are shown in the graph. About how much longer than the George Washington Bridge is the Verrazano Narrows Bridge?

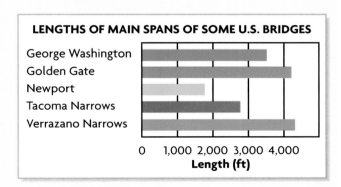

10. **?** **What's the Question?** Sally's Gifts sells an average of 9 beanbag animals each day. The answer is about 126 beanbag animals.

Explore Linear and Nonlinear Relationships

Explore linear and nonlinear relationships.

You need square tiles or paper squares, graph paper.

You can learn about linear relationships by looking at patterns.

Activity 1

The model at the right shows Stage 1 to Stage 3 of a pattern.

- Use square tiles or graph paper to build Stages 4, 5, and 6 of the pattern.

- Record each stage and the perimeter of each figure in a table.

- Graph the ordered pairs (x,y) from the table on a coordinate plane.

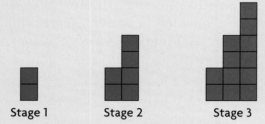

Stage 1   Stage 2   Stage 3

STAGE, x	PERIMETER, y
1	6
2	12
3	▪

Think and Discuss

- Describe the pattern of the points on the graph you drew.

- Write an equation to show the relationship between each stage and its perimeter.

Practice

- Make a table that shows the stage, x, and the perimeter, y, of each figure.

- Write an equation to find the perimeter. Graph the ordered pairs (x, y).

- What pattern do you see in the graph?

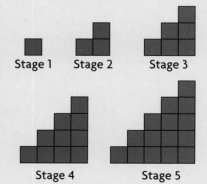

Stage 1   Stage 2   Stage 3

Stage 4   Stage 5

You can also use patterns to learn about nonlinear relationships, which have graphs that do not form straight lines.

Activity 2

You need: square tiles or paper squares, graph paper

The model at the right shows Stage 1 to Stage 3 of a pattern.

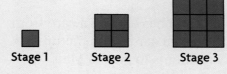

Stage 1 Stage 2 Stage 3

- Use square tiles or paper squares to build Stages 4, 5, and 6 of the pattern.

- Record each stage and the area of each figure in a table.

STAGE, x	AREA, y
1	1
2	4
3	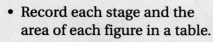

- Graph the ordered pairs (x,y) from the table on a coordinate plane.

Think and Discuss
- What pattern do you see in the y-values of the ordered pairs?

- Describe the pattern of the points on the graph you drew.

- The equation for this relationship is $y = x^2$. Use the equation to find the ordered pair for Stage 8.

Practice
Use the x-values 1, 2, 3, and 4 to find ordered pairs for each equation. Then graph the equation.

1. $y = 2x^2$ **2.** $y = {}^-2x^2$

3. $y = 3x^2$ **4.** $y = \dfrac{12}{x}$

MIXED REVIEW AND TEST PREP

5. Use the table to write an equation that relates x and y. (p. 574)

x	-2	-1	0	1	2
y	6	3	0	-3	-6

Evaluate the expression for $x = 1, 2,$ and 3. (p. 272)

6. ${}^-15 - 3x$ **7.** $0.25x + 6$ **8.** $\dfrac{{}^-18}{x} + 6$

⭐ **9. TEST PREP** A recipe for pumpkin cookies requires $1\frac{3}{4}$ c of pumpkin. Shana wants to triple the recipe. If Shana has 5 c of pumpkin, how much more does she need? (p. 206)

   A $\frac{1}{4}$ c   **B** $\frac{1}{2}$ c   **C** $1\frac{1}{4}$ c   **D** $5\frac{1}{4}$ c

Graph Transformations

Learn how to use transformations to change positions of figures on a coordinate plane.

Remember that a rigid *transformation* is a movement that doesn't change the size or shape of a figure.

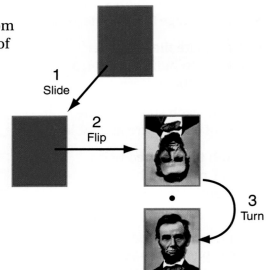

1
Slide

2
Flip

3
Turn

Marla has chosen a card from a pack of cards with pictures of Presidents to give to Matt.

First, she *slides* the card toward Matt, face down. This is a *translation*.

Next, she *flips* the card face up. This is a *reflection*.

Finally, she *turns* the card so that Matt can see it. This is a *rotation*.

Translations, reflections, and rotations are three types of transformations. You can use what you know about the coordinate plane to explore transformations on graph paper.

Activity

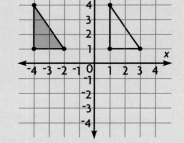

You need: graph paper, scissors

- Draw and number a coordinate plane as shown.

- Plot the points (1,4), (1,1), and (3,1) and connect them to form a triangle.

- Trace the triangle on a sheet of paper. Cut it out.

- Place your cutout triangle on the triangle in the coordinate plane. Then translate it 5 units to the left, keeping the base parallel to the *x*-axis.

- Name the coordinates for the vertices of the new triangle.

- Again, place your cutout triangle on the original triangle in the coordinate plane. Translate the cutout 4 units down, keeping the left side parallel to the *y*-axis.

- Name the coordinates for the vertices of the new triangle.

You can translate a figure on a coordinate plane by sliding it horizontally, vertically, or both.

EXAMPLE 1

Translate square *ABCD* 6 units right and 5 units down. Find the coordinates of the new square, *A'B'C'D'*.

To find the coordinates of the new square, add 6 to each *x*-coordinate and subtract 5 from each *y*-coordinate of the original square.

ABCD		*A'B'C'D'*
$A(^-4,3) \rightarrow$	$A'(^-4 + 6, 3 - 5) \rightarrow$	$A'(2,^-2)$
$B(^-2,3) \rightarrow$	$B'(^-2 + 6, 3 - 5) \rightarrow$	$B'(4,^-2)$
$C(^-2,1) \rightarrow$	$C'(^-2 + 6, 1 - 5) \rightarrow$	$C'(4,^-4)$
$D(^-4,1) \rightarrow$	$D'(^-4 + 6, 1 - 5) \rightarrow$	$D'(2,^-4)$

original

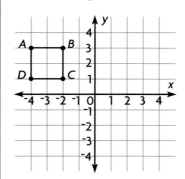

translation

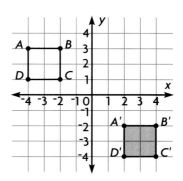

You can also draw the reflection of a figure on a coordinate plane. When you reflect a figure, you flip it across the *x*-axis or the *y*-axis.

EXAMPLE 2

Reflect △*ABC* across the *x*-axis. Find the coordinates of the new figure, △*A'B'C'*.

original

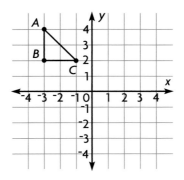

reflection

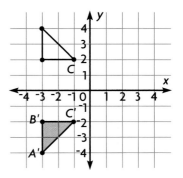

TECHNOLOGY LINK

More Practice: Use E-Lab, *Transformations and Tangrams.*
www.harcourtschool.com/elab2002

△*ABC*		△*A'B'C'*
$A(^-3,4)$	$\rightarrow$	$A'(^-3,^-4)$
$B(^-3,2)$	$\rightarrow$	$B'(^-3,^-2)$
$C(^-1,2)$	$\rightarrow$	$C'(^-1,^-2)$

581

You can draw a rotation of a figure about the origin on a coordinate plane.

EXAMPLE 3

Rotate trapezoid *ABCD* clockwise 90° about the origin. Find the coordinates of the new figure, *A'B'C'D'*.

original

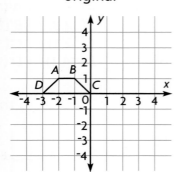

rotation

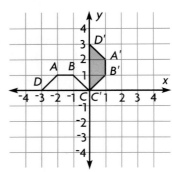

ABCD   A'B'C'D'

A(⁻2,1) → A'(1,2)
B(⁻1,1) → B'(1,1)
C(0,0) → C'(0,0)
D(⁻3,0) → D'(0,3)

• **What if** you had rotated the trapezoid *counterclockwise* 90° about the origin? What would the coordinates of the new figure be?

CHECK FOR UNDERSTANDING

Think and ▶ Discuss

Look back at the lesson to answer each question.

1. **Compare** the *x*- and *y*-coordinates of the original figure and the new figure in Example 2. How are they related?

2. **Explain** why the coordinates of vertex *C* in Example 3 above did not change from the original figure to the new figure.

Guided ▶ Practice

Copy the figure onto a coordinate grid. Transform the figure according to the directions given. Name the new coordinates.

3. translate 2 units left

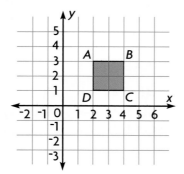

4. rotate 90° clockwise about (0,0)

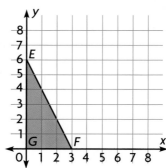

Independent ▶ Practice

Copy the figure onto a coordinate grid. Transform the figure according to the directions given. Name the new coordinates.

5.

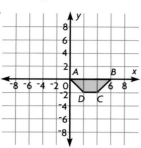

rotate 180° counterclockwise about (0,0)

6.

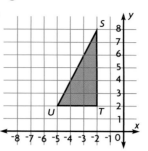

reflect across the *y*-axis

7.

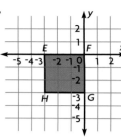

translate 3 units right and 1 unit down

Graph rectangle *EFGH* with coordinates (1,⁻2), (4,⁻2), (4,⁻4), and (1,⁻4).

8. Translate the rectangle 5 units up. What are the new coordinates of the rectangle?

9. Reflect the new rectangle in Exercise 8 across the *x*-axis. What are the new coordinates of the rectangle?

Problem Solving ▶ Applications

For 10–11, use the figure at the right. Triangle *A* has been transformed into triangle *B* by a 180° rotation about (0,0).

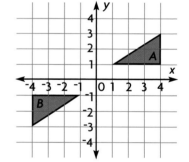

10. Could you complete the same transformation using only reflections? If so, how?

11. Could you complete the same transformation using only translations? Explain.

12. 🌀 **What's the Question?** The coordinates of △*ABC* are *A*(1,⁻3), *B*(1,⁻1), and *C*(3,⁻1). The answer is that the new coordinates are *A*′(1,3), *B*′(1,1), and *C*′(3,1).

MIXED REVIEW AND TEST PREP

13. Use the table to write an equation that relates *x* and *y*. (p. 574)

x	0	1	2	3
y	0	7	14	21

14. Complete: 33, 24, 15, 6, ■ (p. 539)

15. $\frac{1}{2} \times \frac{2}{3} \times \frac{3}{4}$ (p. 202)

16. TEST PREP What is the value of the expression $5v + 32 - 7v$ where *v* equals the volume of a rectangular prism with dimensions 2, 3, and 4? (p. 512)

A 152   **B** 24   **C** ⁻16   **D** ⁻24

17. TEST PREP Find the greatest common factor of 35, 56, and 63. (p. 150)

F 6   **G** 7   **H** 8   **J** 9

1. **VOCABULARY** The point where the x-axis and y-axis intersect is called the __?__ . (p. 570)

2. **VOCABULARY** The x-axis and the y-axis divide the coordinate plane into four __?__ . (p. 570)

Graph the solutions of the inequality. (pp. 568–569)

3. $x < 8$ 4. $x \geq {}^-2$ 5. $x \leq {}^-4$ 6. $x > {}^-1$

Solve the inequality. (pp. 568–569)

7. $n + 5 < 8$ 8. $t - 6 > {}^-2$ 9. $5c \geq 35$

Describe how to locate the point for the ordered pair on the coordinate plane. (pp. 570–573)

10. $(2,4)$ 11. $(3,{}^-2)$ 12. $({}^-1,{}^-6)$

13. List the ordered pairs from the table. (pp. 574–575)

x	1	2	3	4
y	4	5	6	7

14. For the function table in Exercise 13, write an equation that relates y to x. (pp. 574–575)

15. For the values in the function table, write an equation that relates b to a. (pp. 574–575)

a	1	2	3	4
b	3	6	9	12

Solve. (pp. 580–583)

16. Trapezoid *DEFG* has coordinates $D(3,3)$, $E(4,3)$, $F(5,1)$, and $G(2,1)$. If it is reflected across the x-axis, what are the new coordinates?

17. Triangle *ABC* has coordinates $A(0,0)$, $B(3,2)$, and $C(0,2)$. What are the new coordinates after it is rotated 90° clockwise about the origin?

18. Triangle *XYZ* has coordinates $X(4,1)$, $Y(7,1)$, and $Z(7,6)$. It is translated 1 unit down and 3 units left. What are the new coordinates?

Solve. (pp. 576–577)

19. The table shows the total cost, c, for the number of tickets bought, t. Write an equation using t that gives the total cost, c.

t	1	2	3	4
c	$4	$8	$12	$16

20. Ella spends $3 on beads for each bracelet that she makes. The table shows her profit, y, when she sells a bracelet for price x. Write an equation using x that gives her profit, y.

x	$3	$4	$5	$6
y	$0	$1	$2	$3

Decide on a plan.

See item **3**.

Use the strategy *draw a diagram* to draw a coordinate plane. Plot the given coordinates and then reflect them across the *x*-axis.

Also see problem **4**, p. H63.

Choose the best answer.

1. Rectangle *ABCD* has coordinates *A*(0,0), *B*(2,0), *C*(2,1), and *D*(0,1). The rectangle is rotated 90° clockwise around the origin. What are the coordinates of point *B′* in the new figure?

   A (0,⁻2)   **C** (1,0)

   B (0,0)   **D** (2,1)

2. Which point represents the ordered pair (⁻1,2) on the coordinate plane below?

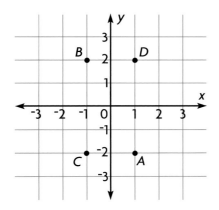

   F *A*   **H** *C*

   G *B*   **J** *D*

3. Rectangle *ABCD* has the coordinates (⁻5,3), (⁻2,3), (⁻5,1), (⁻2,1). What are the coordinates of *A′*, *B′*, *C′*, and *D′* if it is reflected across the *x*-axis?

   A (⁻3,⁻5), (⁻3,⁻2), (⁻1,⁻2), (⁻1,⁻5)

   B (⁻5,⁻3), (⁻2,⁻3), (⁻5,⁻1), (⁻2,⁻1)

   C (2,⁻2), (5,⁻2), (2,⁻4), (5,⁻4)

   D (3,4), (3,1), (1,1), (1,4)

4. What is the least common multiple of 16 and 20?

   F 4

   G 60

   H 80

   J 320

5. Which can be the measure of an acute angle?

   A 45°

   B 90°

   C 95°

   D 180°

6. Bijan rides his bike for 4 hours and travels 30 miles. How many miles per hour is that?

   F 15 miles per hour

   G 10 miles per hour

   H 7.5 miles per hour

   J Not here

Write What You Know

7. When you graph an inequality on a number line, the circle at the end of the graph is sometimes filled in and sometimes not filled in. Explain why that is so, and give an example of an inequality for each case.

8. Triangle *ABC* has coordinates *A*(1,2), *B*(1,0), and *C*(3,0). The triangle is translated 4 units to the left. Find the new coordinates of the triangle. Explain how you found your answers.

MATH DETECTIVE

Analyze This

For each case below, describe at least one way the sequences are the same and at least one way they are different. Then describe a rule and write the next three terms in each sequence.

Case 1

10, 20, 30, 40, . . .

10, 0, ⁻10, ⁻20, . . .

Case 2

4, 7, 10, 13, . . .

4, 12, 36, 108, . . .

Case 3

500, 50, 5, 0.5, . . .

300, 250, 200, 150, . . .

Case 4

M, A, T, H, M, A, T, H, M, A, T, H, . . .

1, 2, 3, 4, 1, 2, 3, 4, 1, 2, 3, 4, . . .

Think It Over!

- **Write About It** Think of a number sequence that starts with 100 and involves division. Then describe the steps you would use to find the next term.

- **Stretch Your Thinking** If the sequence ↑, →, ↓, ←, ↑, →, ↓, ←, . . . is continued, what term will be in the 24th position? Explain.

Stretching Figures

Melissa is making a small quilt. She found a pattern for a star in a quilting book. She will cut out the stars she makes and sew them onto the quilt.

Here are the coordinates of the pattern:
(0,4), (1,1), (3,1), (1,0), (2,⁻2), (0,⁻1), (⁻2,⁻2), (⁻1,0), (⁻3,1), (⁻1,1)

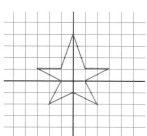

Melissa would like the sides of the pattern to be longer, so she decides to enlarge, or stretch, them by doubling the coordinates.

When she doubles the coordinates, she gets: (0,8), (2,2), (6,2), (2,0), (4,⁻4), (0,⁻2), (⁻4,⁻4), (⁻2,0), (⁻6,2), (⁻2,2)

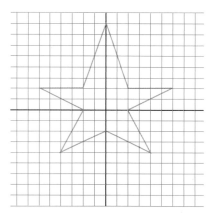

TALK ABOUT IT

What if Melissa wants to make stars of different sizes on her quilt? What coordinates should she use to draw a star if she wants the sides of the star to be five times as long as the original pattern?

TRY IT

Use the figure shown at the right. Give the new coordinates and draw the stretched figure.

1. Stretch the figure so the sides are twice as long.

2. Stretch the figure so the sides are five times as long.

3. Stretch the figure by doubling only the *x*-coordinates.

4. Stretch the figure by doubling only the *y*-coordinates.

VOCABULARY

1. Turning a figure about a point is called a __?__. (p. 550)

2. A figure has __?__ if it can be folded or reflected so that the two parts of the figure match. (p. 560)

3. The horizontal number line in the coordinate plane is called the __?__. (p. 570)

| EXAMPLES | EXERCISES |

Chapter 28

• **Write an equation to represent a function.** (pp. 539–542)

x	15	18	21	24
y	12	15	18	21

The x-values are 3 more than the y-values, so the equation is $y = x - 3$.

Write an equation to represent the function.

4.

x	15	16	17	18
y	10	11	12	13

5.

x	72	64	56	48
y	9	8	7	6

• **Extend geometric patterns.** (pp. 543–545)

What is the next figure in the pattern?

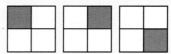

The shaded square moves clockwise 90° in these figures. So, the next figure has the shaded square in the bottom left corner.

Draw the next two figures in the pattern.

6.

7.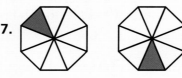

Chapter 29

• **Use translations, rotations, and reflections to transform geometric shapes.** (pp. 550–552)

A *translation* moves a figure along a straight line.
A *rotation* turns a figure around a point.

A *reflection* flips a figure over a line.

Tell which type of transformation the second figure is of the first figure. Write *translation*, *rotation*, or *reflection*.

8.

9.

10.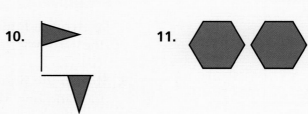

11.

- **Identify rotational symmetry.** (pp. 560–563)

  How much would you have to turn the flower until it matches the original?

  $\frac{1}{4}$ turn, or 90°

Tell whether each figure has rotational symmetry. If it does, identify the symmetry as a fraction of a turn and in degrees.

12.

13.

14.

15.

Chapter 30

- **Solve algebraic inequalities.** (pp. 568–569)

$$a + 7 > 12$$
$$a + 7 - 7 > 12 - 7$$
$$a > 5$$

Subtract 7 from both sides.

Solve the inequality.

16. $x - 9 > 14$

17. $y + 21 < 50$

18. $7z < 63$

19. $\frac{a}{5} > 14$

- **Use an ordered pair to describe a location on the coordinate plane.**

  (pp. 570–573)

  The first number of an ordered pair tells you how far right or left to move from the origin.

  The second number of an ordered pair tells you how far to move up or down. ($^-3,5$) Move 3 units left and 5 units up.

Write the ordered pair for each point on the coordinate plane.

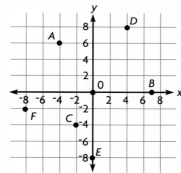

20. *A*

21. *B*

22. *C*

23. *D*

24. *E*

25. *F*

PROBLEM SOLVING APPLICATIONS

26. Nikhil wants to buy a computer that costs $850. His savings grew from $100 in January to $175 in February to $250 in March. If this pattern continues, in what month will he be able to buy the computer? (pp. 534–535)

27. A garden is being enclosed with forty-eight 1-ft sections of fencing. What is the largest possible area of the garden? (pp. 556–557)

Performance Assessment

TASK A • What's the Pattern?

Materials: centimeter cubes

Janine and Bill set up this geometric pattern.

a. Model, describe, and then draw the next three figures in the pattern.

b. Write a numerical sequence to show the number of cubes in the figures in the geometric pattern.

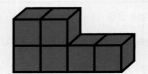

c. What are the next two numbers in the numerical pattern?

d. Describe how you got the numbers in the pattern.

e. Look at the factors below. Write the factors of the next three numbers in the pattern.

$$2 = 1 \times 2 \qquad 6 = 2 \times 3 \qquad 12 = 3 \times 4$$

f. Copy and complete the input/output table for the numerical pattern.

Term (Input)	4	5	6	7	8	9	10
Number (Output)	20	30	■	■	■	■	■

TASK B • Transformations

Materials: graph paper

a. Graph the *x*- and *y*-axes on the graph paper. Then graph the points (0,2), (8,0), (0,⁻4), and (⁻4,0) and connect them in order so that the result is a quadrilateral.

b. Draw the quadrilateral with vertices (⁻4,0), (⁻8,2) (⁻16,0), and (⁻8,⁻4). Is this transformation a rotation, translation, or reflection of the first quadrilateral?

c. Draw a translation of the first quadrilateral. Give the new coordinates of the vertices.

d. Draw a transformation that is a combination of rotation and translation, rotation and reflection, or translation and reflection. Exchange drawings with a classmate and determine which two transformations he or she has used.

Technology Linkup

Calculator • Explore Graphing Equations

Chad works in a cafeteria. To make lemonade, he combines each scoop of lemonade mix with 8 ounces of water.

SCOOPS OF LEMONADE MIX	1	2	3	4	5	6
OUNCES OF WATER	8	16	24	32	40	48

If x equals the scoops of lemonade mix and y equals the ounces of water, then $y = 8x$.

You can use the ordered pairs to graph the equation on graph paper, or you can use a graphing calculator.

• Use the TI-73 graphing calculator to graph $y = 8x$.

Step 1

Press . Press , 8, , .

Step 2

Press **GRAPH** to show the graph on the coordinate plane.

Practice and Problem Solving

Use a graphing calculator to graph each equation. Then sketch the graph on a piece of paper.

1. $y = 6x$

2. $y = 32x$

3. $y = 57x$

4. $y = 0.14.x$

5. $y = x - 6$

6. $y = 18x$

7. $y = 41x$

8. $y = 0.6x$

9. $y = y + 3$

10. $y = x + 8$

11. **Write a problem** involving a function. Write the equation and graph it. Draw the graph on your paper.

Multimedia Math Glossary www.harcourtschool.com/mathglossary

12. **Vocabulary** Look up *equation* in the Multimedia Math Glossary. Write some other algebraic sentences that have the same solution as each of the two examples shown in the glossary.

Problem Solving On Location
In Maryland

Plant Patterns

Sunflowers, daisies, and pine trees all grow in Maryland. The pattern in which these plants grow is also something they have in common. The pattern is an example of the Fibonacci sequence.

The florets at the center of the daisy above follow a Fibonacci pattern.

1. The Fibonacci sequence is a pattern of numbers that begins 1, 1, 2, 3, 5, 8, 13. How is each new number in the pattern formed?

2. Write the next five numbers in the Fibonacci sequence.

For 3–4, use the drawing at the right.

3. Study the pattern of spirals on the pine cone. Start at the center. How many of the spirals turn clockwise? counterclockwise?

4. How are the spirals on the pine cone related to the Fibonacci sequence?

The scales of a pine cone show a Fibonacci pattern in the way they spiral.

5. Sunflowers also model the Fibonacci sequence the way pine cones do. If a sunflower has 55 clockwise spirals in its florets, how many counterclockwise spirals does it have?

The state flower of Maryland is the black-eyed Susan.

For 6–8, use the photo at the right.

6. Tell the types of symmetry you see in the flower.

7. Count the petals on the flower. Is this black-eyed Susan an example of the Fibonacci sequence?

8. Does the flower have rotational symmetry? If so, identify the symmetry as a fraction of a turn.

Camden Yards

The baseball field where the Baltimore Orioles baseball team plays is officially called Oriole Park at Camden Yards. But everyone calls it Camden Yards. It opened in 1992 and its design immediately changed the way people felt a ballpark should look.

Since the day it opened, Camden Yards has been popular with visitors and baseball fans.

For 1–2, use the diagram of Camden Yards at the right.

1. Does the Camden Yards infield have symmetry? Explain.

2. Does the entire Camden Yards field have symmetry? Explain.

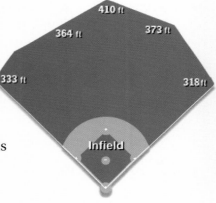

For 3–5, use the table below.

r	0	1	2	3
e	0	2.25	4.5	6.75

3. A pitcher pitched 4 innings on opening day at Camden Yards. His earned run average *(e)* is based upon the number of runs *(r)* he allowed. The function table shows what *e* would be for several different values of *r*. Write an equation you could use to find the pitcher's earned run average.

4. Use the ordered pairs in the table to draw a graph.

5. If a pitcher allows 6 runs in 4 innings, what is his earned run average?

Camden Yards can hold more than 48,000 people. The lower deck has 18,000 seats between the foul poles. During the 2000 season, the total number of fans who watched the Orioles play 81 games at Camden Yards was a little less than 3.3 million. Let *s* represent the number of seats other than the 18,000 and *a* represent the average game attendance.

6. Write an inequality that involves *s*, based on the information in the paragraph above.

7. Solve your inequality in Exercise 6. Explain what your answer means.

8. Write an inequality that involves *a*, based on the information in the paragraph above.

9. Solve your inequality in Exercise 8. Explain what your answer means.

Student Handbook

Troubleshooting H2

Prerequisite Skills Review Do you have the math skills needed to start a new chapter? Use this list of skills to review and remember your skills from last year.

Troubleshooting

Properties

The charts list the basic properties of addition and multiplication.

Addition

PROPERTY	EXAMPLE WITH NUMBERS	EXAMPLE WITH VARIABLES
Commutative	$3 + 7 = 7 + 3$	$a + b = b + a$
Associative	$(4 + 5) + 2 = 4 + (5 + 2)$	$(a + b) + c = a + (b + c)$
Identity Property of Zero	$9 + 0 = 9$ and $0 + 9 = 9$	$a + 0 = a$ and $0 + a = a$

Multiplication

PROPERTY	EXAMPLE WITH NUMBERS	EXAMPLE WITH VARIABLES
Commutative	$8 \times 6 = 6 \times 8$	$a \times b = b \times a$
Associative	$(2 \times 9) \times 5 = 2 \times (9 \times 5)$	$(a \times b) \times c = a \times (b \times c)$
Identity Property of One	$6 \times 1 = 6$ and $1 \times 6 = 6$	$a \times 1 = a$ and $1 \times a = a$
Property of Zero	$7 \times 0 = 0$ and $0 \times 7 = 0$	$a \times 0 = 0$ and $0 \times a = 0$
Distributive	$3 \times (5 + 7) = (3 \times 5) + (3 \times 7)$	$a \times (b + c) = (a \times b) + (a \times c)$

Practice

Name the property shown.

1. $0 \times 12 = 0$

2. $(8 \times 21) \times 5 = 8 \times (21 \times 5)$

3. $72 \times 5 = 5 \times 72$

4. $4 \times (3 + 2) = (4 \times 3) + (4 \times 2)$

5. $9 + 0 = 9$

6. $k + 7 = 7 + k$

7. $(m + 4) + h = m + (4 + h)$

8. $c \times (2 + 3) = (c \times 2) + (c \times 3)$

9. $1 \times x = x$

Represent Decimals

You can use decimal squares to represent decimals. A decimal square is divided into 100 parts. Each part represents 1 hundredth of the whole, or 0.01. Count the number of shaded parts. Write this number as hundredths.

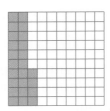

24 out of 100 small squares are shaded.
0.24 of the whole is shaded.

Practice

Write the decimal that is represented.

1.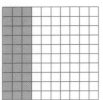

2.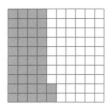

3.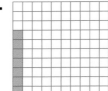

4.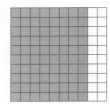

Write and Read Decimals

A place-value chart can help you write and read numbers. Each three-digit group, such as ones or millions, is called a **period**.

Read: "thirty million, one hundred twenty-eight thousand, five hundred ninety-seven and forty-six thousandths."

MILLIONS			THOUSANDS			ONES					
Hundreds	Tens	Ones	Hundreds	Tens	Ones	Hundreds	Tens	Ones	Tenths	Hundredths	Thousandths
	3	0	1	2	8	5	9	7	0	4	6

Example

Name the place value of the digit, 8, in the chart.

To write the value of the 8 in the chart, multiply the digit times the value of the place-value position.

Think: $8 \times 1,000$, or 8,000.

Practice

Name the place value of the digit 4.

1. 327,489,223.78 **2.** 198,238,042.08 **3.** 149,678,935.91 **4.** 728,035.84

Give the value of the blue digit.

5. 538.92 **6.** 82,901,733.006 **7.** 29.35 **8.** 125,674.173

Compare and Order Whole Numbers

You can compare and order numbers by comparing the digits in each place-value position.

Example

Write $<$, $>$, or $=$ to compare the numbers.   14,675 ● 14,228

Step 1

Compare the ten thousands.

14,675
↓   same number of
14,228   ten thousands

Step 2

Compare the thousands.

14,675
↓   same number of
14,228   thousands

Step 3

Compare the hundreds.

14,675
↓ $6 > 2$
14,228   So, $14,675 > 14,228$.

Practice

Write $<$, $>$, or $=$ to compare the numbers.

1. 3,919 ● 3,991 **2.** 188,937 ● 189,066 **3.** 70,001 ● 70,001

Order the numbers from greatest to least.

4. 9,774; 9,718; 9,762 **5.** 82,056; 82,856; 81,978 **6.** 23,091; 23,910; 23,109

Troubleshooting

Round Whole Numbers and Decimals

Follow these steps to round a number to a given place.

Example

Round 168,279 to the nearest ten thousand.

Step 1	**Step 2**	**Step 3**
Find the digit in the place being rounded, the ten thousands place.	Look at the next digit to the right. If it is 5 or greater, round up. Otherwise, round down.	Change to zero each digit to the right of the place being rounded.
1<u>6</u>8,279	16<u>8</u>,279 8 > 5, so round up.	168,279 ⟶ 170,000

Round to the nearest whole number.

1. 7.97   **2.** 15.58

Round to the nearest ten thousand.

5. 530,410   **6.** 12,677

Round to the nearest thousand.

3. 3,378   **4.** 7,607

Round to the nearest tenth.

7. 16.53   **8.** 2.96

Whole-Number Operations

When adding, subtracting, multiplying, and dividing whole numbers, align digits that have the same place value.

Examples

A. Find the sum. 247 + 1,496 + 89

$$\begin{array}{r} \overset{2\,2}{247} \\ 1{,}496 \\ +\ \ 89 \\ \hline 1{,}832 \end{array}$$

Think: 7 + 6 + 9 = 22
Regroup 22 ones as 2 tens 2 ones.
Add the other columns in a similar way.

B. Find the difference. 31− 12

$$\begin{array}{r} \overset{2\,11}{3\,\cancel{1}} \\ -\ 1\,2 \\ \hline 1\,9 \end{array}$$

Regroup when necessary.

C. Find the product. 27 × 86

$$\begin{array}{r} 27 \\ \times\ 86 \\ \hline 162 \\ +\ 2\,160 \\ \hline 2{,}322 \end{array}$$

Multiply by the 6 ones.
Multiply by the 8 tens.
Add the products.

D. Find the quotient. 146 ÷ 9

$$\begin{array}{r} 16\ \text{r}2 \\ 9\overline{)146} \\ -\ 9 \\ \hline 56 \\ -\ 54 \\ \hline 2 \end{array}$$

Divide the 14 tens.
Multiply and subtract.
Bring down the 6 ones.
Divide the 56 ones.
Multiply and subtract.
Write the remainder.

Practice

Add, subtract, multiply, or divide.

1. 79 + 56 + 99   **2.** 345 − 26   **3.** 67 × 76   **4.** 376 ÷ 15

5. 700 − 388   **6.** 19 × 203   **7.** 155 + 9 + 4,823   **8.** 524 ÷ 31

Remainders

When dividing whole numbers, you can write a remainder as a whole number, as a decimal, or as a fraction.

Example

Divide: $15 \div 4$. Write the remainder as a decimal and as a fraction.

Step 1

Write the dividend as a two-place decimal.

$$4\overline{)15.00}$$

Step 2

Divide.

$$\begin{array}{r} 3.75 \\ 4\overline{)15.00} \end{array}$$

Step 3

Write the decimal as a fraction in simplest form.

$$0.75 = \frac{75}{100} = \frac{75 \div 25}{100 \div 25} = \frac{3}{4}$$

$$15 \div 4 = 3\frac{3}{4}$$

Practice

Divide. Write the remainder as a decimal and as a fraction.

1. $4\overline{)30}$ **2.** $8\overline{)50}$ **3.** $10\overline{)3,315}$ **4.** $12\overline{)2,442}$

5. $6\overline{)21}$ **6.** $8\overline{)94}$ **7.** $20\overline{)74}$ **8.** $25\overline{)210}$

Read a Table

Use the title of a table to understand what the data represent. Use the labels to understand what the items represent.

Example

How far did Angie jog on Tuesday?

DAILY JOGGING RECORD (IN MILES)							
Day	Sun	Mon	Tue	Wed	Thu	Fri	Sat
Milo	4	3	5	5	0	7	6
Angie	6	3	4	0	5	8	0

The number in the row marked "Angie" and the column marked "Tue" is 4. So, Angie jogged 4 miles on Tuesday.

Practice

Use the data in the table above to answer the questions.

1. How far did Milo jog on Friday?

2. Which person jogged 5 miles on Wednesday?

3. On which day did Angie jog 6 miles?

4. On which day did Milo and Angie jog the same distance?

5. How far did Angie jog on the two days that she jogged the same distance?

6. Who ran the greater total distance during the week?

Troubleshooting

Mean, Median, Mode, and Range

Example

Find the mean, median, mode, and range for this set of data.
6, 19, 6, 9, 11, 15

Mean Find the sum of the data items. Divide the sum by the number of items.	$6 + 19 + 6 + 9 + 11 + 15 = 66$ mean $= 66 \div 6 = 11$
Median Arrange the items from least to greatest. The median is the middle value. If there are two middle values, the median is the average of the two values.	6, 6, **9**, **11**, 15, 19 9 and 11 are the middle numbers. median $= (9 + 11) \div 2 = 10$
Mode Arrange the items from least to greatest. The mode is the value or values that repeat most often. If no value repeats, there is no mode.	**6, 6**, 9, 11, 15, 19 mode $= 6$
Range Arrange the items from least to greatest. The range is the difference of the greatest and least values.	**6**, 6, 9, 11, 15, **19** range $= 19 - 6 = 13$

Practice

Find the mean, median, mode, and range for each set of data.

1. 2, 1, 3, 5, 9   **2.** 6, 52, 41, 21, 35   **3.** 11, 15, 6, 11, 22   **4.** 9, 5, 2, 5, 6, 1, 14

Read Bar Graphs

A **bar graph** uses bars of different lengths to show and compare data.

Example

How many books did Anita read last month?

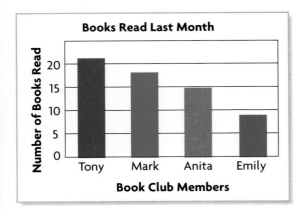

Find the bar representing Anita.

Look at the top of the bar. Read the scale at the left for the number the bar represents.

Anita read 15 books last month.

Practice

Use the bar graph to answer the questions.

1. How many books did Mark read?

2. Who read 22 books?

3. Who read the least number of books?

4. How many more books did Tony read than Anita?

Read Stem-and-Leaf Plots

In a **stem-and-leaf plot**, data are organized by "stems," or the tens digits, and "leaves," or the ones digits. The stems and leaves are listed in order.

Example

Each entry in the stem-and-leaf plot gives the number of cards that were stacked in a house of cards when it collapsed. How many times were there from 40 to 49 cards when the collapse occurred?

Stem	Leaves
1	2 5 7 8 8 9
2	0 0 1 3 5 8
3	1 2 4 7
4	2 8
5	1 6

The numbers from 40 to 49 have the stem of 4, because the tens digit is 4. There are two such numbers in the table:

Stem 4, leaf 2 shows **42**.
Stem 4, leaf 8 shows **48**.

So, twice there were from 40 to 49 cards. Once there were 42 cards, and once there were 48.

Practice

Use the stem-and-leaf plot to answer the questions.

1. How many times were there from 10 to 19 cards when the collapse occurred?

2. What was the greatest number of cards stacked? the least number?

3. How many houses of cards were built in this competition?

4. What are the mode, median, and range of the data in the table?

Prime and Composite Numbers

A **prime number** is a whole number that has exactly two factors, itself and 1. A **composite number** has more than two factors.

Examples

Decide whether the number is *prime* or *composite*.

A. 13 13 has exactly two factors, 13 and 1. These are the only whole numbers that divide 13 evenly. So, 13 is prime.

B. 12 12 has the factors 1, 2, 3, 4, 6, and 12. Each of these whole numbers divides 12 evenly. Since 12 has more than two factors, it is composite.

Practice

Decide whether the number is *prime* or *composite*.

1. 19   **2.** 6   **3.** 10   **4.** 2   **5.** 15

6. 28   **7.** 31   **8.** 99   **9.** 71   **10.** 456

Troubleshooting

Factors and Multiples

Multiples of a number are the products that result when the number is multiplied by 0, 1, 2, 3, 4, and so on.

27 is a multiple of 9 because $27 = 9 \times 3$.

Factors are numbers that divide a whole number evenly.

8 is a factor of 32 because 8 divides 32 evenly: $32 \div 8 = 4$.

Examples

A. List the next three multiples of 7: 7, 14, 21.

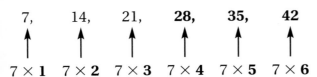

7×1   7×2   7×3   7×4   7×5   7×6

So, the next three multiples are 28, 35, and 42.

B. Find all the factors of 20.

Think: $1 \times 20 = 20$, so 1 and 20 are factors of 20. $2 \times 10 = 20$, so 2 and 10 are factors of 20. $4 \times 5 = 20$, so 4 and 5 are factors of 20.

So, the factors of 20 are 1, 2, 4, 5, 10, and 20.

Practice

List the next three multiples of the number.

1. 5: 5, 10, 15   **2.** 9: 9, 18, 27   **3.** 6: 12, 18, 24   **4.** 20: 40, 60, 80

Find all of the factors of the number.

5. 13   **6.** 12   **7.** 30   **8.** 24

9. 15   **10.** 22   **11.** 28   **12.** 36

Model Fractions

A **fraction** is a number that names a part of a whole or a part of a group.

Example

Write the fraction represented by the shaded part.

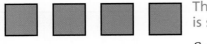

The group has 7 squares. The part that is shaded is 4 squares.

So, the fraction shaded is $\frac{4}{7}$.

Practice

Write the fraction represented by the shaded part.

1. 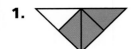   **2.**    **3.**    **4.**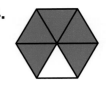

Model Percents

Percent (%) means "per hundred." You can use a hundred square to model percents.

Example

Write the percent for the shaded part of the square.

There are 100 parts, and 75 of the parts are shaded. So, 75% of the parts are shaded.

Practice

Write the percent for the shaded part of the square.

1.   **2.** 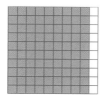   **3.** 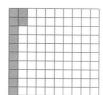   **4.**

Simplify Fractions

A fraction is in **simplest form** when the numerator and denominator have no common factors other than 1.

$\frac{10}{15}$ is not in simplest form, because 10 and 15 have the common factor 5.

$\frac{7}{12}$ is in simplest form, because 7 and 12 have no common factors other than 1.

You can write a fraction in simplest form by dividing the numerator and denominator by the greatest common factor (GCF).

Example

Write the fraction in simplest form. $\frac{16}{28}$

Step 1

Find the greatest common factor of 16 and 28.

factors of 16: 1, 2, 4, 8, 16
factors of 28: 1, 2, 4, 7, 14, 28

The *greatest* common factor is 4.

Step 2

Divide the numerator and the denominator by the GCF.

$\frac{16 \div 4}{28 \div 4} = \frac{4}{7}$ Since the GCF of 4 and 7 is 1, $\frac{4}{7}$ is in simplest form.

Practice

Write the fraction in simplest form.

1. $\frac{9}{15}$ **2.** $\frac{8}{12}$ **3.** $\frac{5}{20}$ **4.** $\frac{18}{6}$ **5.** $\frac{24}{32}$ **6.** $\frac{24}{30}$

Troubleshooting

Add and Subtract Like Fractions

Like fractions are fractions that have the same denominator. To add or subtract like fractions, add or subtract the numerators. Keep the same denominator.

Example

Find the sum. $\frac{4}{9} + \frac{2}{9}$

Step 1

Add the numerators. Write the sum over the like denominator.

$\frac{4}{9} + \frac{2}{9} = \frac{4+2}{9} = \frac{6}{9}$

Step 2

If necessary, rewrite the fraction in simplest form.

$\frac{6}{9} = \frac{6 \div 3}{9 \div 3} = \frac{2}{3}$

Practice

Find the sum or difference. Write the answer in simplest form.

1. $\frac{1}{7} + \frac{4}{7}$
2. $\frac{4}{5} - \frac{2}{5}$
3. $\frac{3}{8} + \frac{3}{8}$
4. $\frac{3}{4} + \frac{1}{4}$
5. $\frac{7}{10} - \frac{5}{10}$

6. $\frac{8}{12} - \frac{4}{12}$
7. $\frac{15}{11} - \frac{9}{11}$
8. $\frac{5}{9} + \frac{8}{9}$
9. $\frac{2}{10} + \frac{6}{10}$
10. $\frac{12}{8} - \frac{4}{8}$

11. $\frac{3}{6} + \frac{1}{6}$
12. $\frac{5}{8} + \frac{1}{8}$
13. $\frac{7}{16} + \frac{5}{16}$
14. $\frac{17}{20} - \frac{9}{20}$
15. $\frac{11}{25} + \frac{4}{25}$

Round Fractions

To round a fraction to 0, $\frac{1}{2}$, or 1, compare the numerator with the denominator.

Examples

Round each fraction to 0, $\frac{1}{2}$, or 1.

A. $\frac{2}{11}$

Compared to 11, 2 is close to 0.

Round the fraction to 0.

B. $\frac{7}{15}$

7 is about half of 15.

Round the fraction to $\frac{1}{2}$.

C. $\frac{13}{16}$

13 is close to 16.

Round the fraction to 1.

Practice

Round each fraction to 0, $\frac{1}{2}$, or 1.

1. $\frac{3}{4}$
2. $\frac{1}{5}$
3. $\frac{2}{7}$
4. $\frac{6}{13}$
5. $\frac{5}{12}$

6. $\frac{14}{15}$
7. $\frac{4}{18}$
8. $\frac{3}{7}$
9. $\frac{4}{5}$
10. $\frac{12}{20}$

11. $\frac{11}{24}$
12. $\frac{17}{20}$
13. $\frac{11}{18}$
14. $\frac{3}{25}$
15. $\frac{15}{16}$

Mental Math and Equations

You can use mental math to solve equations. Try using a related equation to find the value of the variable.

Examples

Use mental math to solve the equation.

A. $k + 6 = 10$

Think: When 6 is added to k, the sum is 10. That means that $k = 10 - 6$.

So, $k = 4$. Check: **4 + 6 = 10** ✔

B. $m - 4 = 11$

Think: When 4 is subtracted from m, the difference is 11. That means that $m = 11 + 4$.

So, $m = 15$. Check: **15 − 4 = 11** ✔

C. $c \div 5 = 7$

Think: When c is divided by 5, the quotient is 7. That means that $c = 7 \times 5$.

So, $c = 35$. Check: **35 ÷ 5 = 7** ✔

D. $h \times 4 = 36$

Think: When h is multiplied by 4, the product is 36. That means that $h = 36 \div 4$.

So, $h = 9$. Check: **9 × 4 = 36** ✔

Practice

Use mental math to solve the equation.

1. $g - 5 = 6$ **2.** $r \times 2 = 12$ **3.** $d \div 3 = 6$ **4.** $x + 9 = 17$

5. $v \div 6 = 5$ **6.** $y - 6 = 9$ **7.** $a + 3 = 15$ **8.** $n \times 3 = 21$

Fractions and Mixed Numbers

A **mixed number** is made up of a whole number and a fraction. You can write a mixed number as a fraction greater than 1, and a fraction greater than 1 as a mixed number.

Examples

A. Write $\frac{14}{3}$ as a mixed number.

$$3 \overline{)14} \quad \text{4 r2}$$

Divide the numerator by the denominator.

$\frac{14}{3} = 4\frac{2}{3}$ Write the remainder as the numerator of a fraction.

B. Write $3\frac{4}{5}$ as a fraction.

$3 \times 5 = 15$ Multiply the denominator by the whole number.

$15 + 4 = 19$ Add the numerator.

$3\frac{4}{5} = \frac{19}{5}$ Write the sum over the denominator.

Practice

Write the fraction as a mixed number.

1. $\frac{13}{7}$ **2.** $\frac{18}{5}$ **3.** $\frac{10}{3}$ **4.** $\frac{22}{9}$ **5.** $\frac{17}{6}$

Write the mixed number as a fraction.

6. $2\frac{4}{5}$ **7.** $1\frac{3}{8}$ **8.** $3\frac{3}{4}$ **9.** $2\frac{22}{25}$ **10.** $2\frac{11}{16}$

Troubleshooting

Compare Fractions

To compare fractions with unlike denominators, rename the fractions so that they have like denominators. Then compare the numerators.

Examples

Compare. Use <, >, or = for ●.

A. $\frac{5}{8} ● \frac{7}{8}$

The denominators are like. Compare the numerators.

$5 < 7$, so $\frac{5}{8} < \frac{7}{8}$.

B. $\frac{7}{10} ● \frac{3}{5}$

The denominators are unlike. Rename one or both fractions.

$\frac{3 \times 2}{5 \times 2} = \frac{6}{10}$

$7 > 6$, so $\frac{7}{10} > \frac{3}{5}$.

C. $2\frac{5}{12} ● 2\frac{1}{3}$

The whole numbers are equal. Rename one or both fractions.

$\frac{1 \times 4}{3 \times 4} = \frac{4}{12}$

$5 > 4$, so $2\frac{5}{12} > 2\frac{1}{3}$.

Practice

Compare. Use <, >, or = for ●.

1. $\frac{4}{5} ● \frac{3}{5}$

2. $\frac{11}{15} ● \frac{13}{15}$

3. $1\frac{4}{6} ● 1\frac{2}{3}$

4. $2\frac{5}{8} ● 2\frac{1}{2}$

5. $\frac{2}{3} ● \frac{5}{12}$

6. $\frac{7}{8} ● \frac{3}{4}$

7. $\frac{5}{15} ● \frac{2}{6}$

8. $\frac{8}{9} ● \frac{2}{3}$

9. $\frac{13}{16} ● \frac{7}{8}$

10. $\frac{5}{9} ● \frac{14}{27}$

Understand Integers

The **integers** are the whole numbers and their opposites. Integers greater than 0 are **positive integers** and are found to the right of 0 on a number line. Integers less than 0 are **negative integers** and are found to the left of 0.

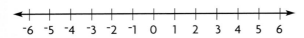

Example

Name the temperatures indicated by point A and point B.

The temperature at point A is $^-35°$. The temperature at point B is $^+68°$.

Practice

Give four examples of each set of numbers.

1. whole numbers

2. negative integers

Write a positive or negative integer for each situation.

3. thermometer point C

4. thermometer point D

5. thermometer point E

6. thermometer point F

7. 35 ft above sea level

8. a loss of $5

Number Lines

You can use a number line to graph numbers and to compare and order numbers. If two numbers are graphed on a number line, the number to the right is greater.

Example

Graph ⁻3, 2, and ⁻1.5 on the number line. Then order the numbers from least to greatest.

Graph each number by placing a dot on the number line.

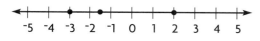

Notice that ⁻1.5 appears halfway between ⁻1 and ⁻2. Since values increase as you move to the right on a number line, the order of the numbers from least to greatest is ⁻3, ⁻1.5, 2.

Practice

Name the number graphed by each point.

```
   A   E      C  B         D
 ◄─●─┼─●─┼─┼─●─●─┼─┼─┼─●─┼─┼►
  ⁻5  ⁻4  ⁻3  ⁻2  ⁻1  0   1   2   3   4   5
```

1. A   **2.** B   **3.** C   **4.** D   **5.** E

Graph the numbers on a number line. Then order the numbers from least to greatest.

6. 4, 2, ⁻1   **7.** ⁻1, ⁻3, 0   **8.** 1.5, 2, ⁻1.5   **9.** 3, ⁻3, ⁻4   **10.** ⁻3.5, ⁻3, ⁻5

Multiplication and Division Facts

To multiply a number by a power of 10, move the decimal point one place to the right for each zero. To divide a number by a power of 10, move the decimal point one place to the left for each zero.

If you forget a multiplication fact that has 11 or 12 as a factor, think of 11 as 10 + 1, and think of 12 as 10 + 2.

Example

Find the product. 12×8

$12 \times 8 = (\mathbf{10 + 2}) \times 8 = (\mathbf{10 \times 8}) + (\mathbf{2 \times 8}) = 80 + 16 = 96$

Practice

Find the product or quotient.

1. 56×10 **2.** $780 \div 10$ **3.** 4.3×100 **4.** $5{,}280 \div 1{,}000$ **5.** $2.61 \times 1{,}000$

6. 0.48×100 **7.** $124 \div 10$ **8.** $5.77 \div 100$ **9.** 11×10 **10.** 12×9

Troubleshooting

Classify Lines

Parallel lines are lines in a plane that are always the same distance apart. **Intersecting lines** cross at exactly one point. **Perpendicular lines** intersect to form 90° angles, or right angles.

Examples

Classify the lines.

A.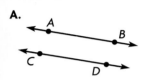

B.

C.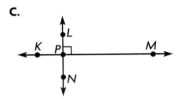

Line *AB* and line *CD* are in the same plane and are always the same distance apart. The lines are *parallel*.

$\overleftrightarrow{AB} \parallel \overleftrightarrow{CD}$ ($\parallel$ means "is parallel to.")

Line *EH* and line *GF* cross at exactly one point, *J*. The lines are *intersecting*.

Lines *KM* and *LN* intersect at *P* to form right angles. The lines are *perpendicular* and *intersecting*.

$\overleftrightarrow{KM} \perp \overleftrightarrow{LN}$ ($\perp$ means "is perpendicular to.")

Practice

Classify the lines.

1.

2.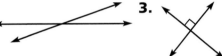

3.

4.

Patterns

Look for a rule for the pattern. Then use it to extend the pattern.

Example

Find the next three possible numbers in the pattern.   1, 2, 4, 8

Step 1   Find a rule.

Each number in the series is 2 times the number before it.
$1 \times 2 = 2$; $2 \times 2 = 4$; $4 \times 2 = 8$
The rule is "multiply by 2."

Step 2   Use the rule.

Multiply each number by 2 to find the next number:
$8 \times 2 = 16$, $16 \times 2 = 32$, $32 \times 2 = 64$
So, the next three numbers are 16, 32, and 64.

Practice

Find the next three possible numbers in the pattern. Write the rule.

1. 10, 14, 18, 22, 26

2. 729, 243, 81, 27

3. 1, 4, 16

4. 63, 55, 47, 39

5. 49, 37, 25, 13

6. $2\frac{1}{2}$, $1\frac{3}{4}$, 1, $\frac{1}{4}$

H14 Prerequisite Skills Review

Exponents

You can use exponents to express powers of numbers. An **exponent** tells how many times the **base** is used as a factor.

$$6 \times 6 \times 6 \times 6 = 6^4 \quad \leftarrow \text{exponent}$$

$\uparrow$ base

The exponent 4 shows that the base 6 is used 4 times as a factor.

Examples

Find the value of 5^3.

A. $5^3 = 5 \times 5 \times 5 = 125$

Write 49 by using an exponent.

B. $49 = 7 \times 7 = 7^2$

Practice

Find the value.

1. 2^4 **2.** 4^3 **3.** 5^2 **4.** 10^3 **5.** 3^4

Write by using an exponent.

6. 9   **7.** 32   **8.** 36   **9.** 100   **10.** 27

Order of Operations

When more than one operation is used in an expression, follow these rules to evaluate the expression.

RULES FOR ORDER OF OPERATIONS	
1. First, do the operations **in parentheses**.	3. Next, **multiply and divide** from left to right.
2. Next, evaluate **exponents**.	4. Finally, **add and subtract** from left to right.

Examples

Evaluate each expression.

A. $3^2 + 5 \times 2$

1. There are no parentheses.

2. Evaluate exponents. $3^2 + 5 \times 2 = \mathbf{9} + 5 \times 2$

3. Multiply. $9 + \mathbf{5 \times 2} = 9 + \mathbf{10}$

4. Add. $\mathbf{9 + 10 = 19}$

B. $(21 - 6) \div 3$

1. Evaluate parentheses. $\mathbf{(21 - 6)} \div 3 = \mathbf{15} \div 3$

2. There are no exponents.

3. Divide. $15 \div 3 = 5$

4. There is no addition or subtraction.

Practice

Evaluate each expression.

1. $4 + 6 \times 9$ **2.** $(5 + 4) \times 3$ **3.** $25 - 12 \times 3$ **4.** $2^2 + 3^2$

5. $(2 + 3)^2$ **6.** $\dfrac{10 - 4}{2} \times 4^2$ **7.** $5^2 \div (1^5 + 4)$ **8.** $2 + 3 \times 3 \times 3$

Troubleshooting

Add and Subtract

To add or subtract unlike fractions, write equivalent fractions with common denominators. Then add or subtract the like fractions.

To add or subtract decimals, align the decimal points. Write equivalent decimals. Then add or subtract as you would with whole numbers.

Examples

Find the sum. $\frac{1}{6} + \frac{1}{2}$

A. $\frac{1}{6} + \frac{1}{2}$ Write equivalent fractions with a denominator of 6.

$$\frac{1 \times 3}{2 \times 3} = \frac{3}{6}$$

$\frac{1}{6} + \frac{3}{6} = \frac{4}{6}$, or $\frac{2}{3}$ Add the numerators and simplify.

Find the difference. $34.7 - 3.651$

B.
$$\begin{array}{r} 34.700 \\ - \ 3.651 \\ \hline 31.049 \end{array}$$
Align the decimal points.
Write an equivalent decimal.
Subtract as for whole numbers.

Practice

Find the sum or difference.

1. $3.12 + 2.7$

2. $1.197 - 1.09$

3. $6 - 3.4$

4. $2.17 + 141.8$

5. $\frac{3}{4} + \frac{1}{6}$

6. $\frac{5}{8} - \frac{3}{16}$

7. $\frac{3}{5} + \frac{1}{4}$

8. $\frac{1}{3} + \frac{5}{12}$

Inverse Operations

Addition and subtraction are **inverse** operations. This means that you can check a sum or difference by using the inverse operation. You can also check a product or quotient by using the inverse operation.

Examples

Use the inverse operation to check the answer.

A.
$$\begin{array}{r} 12 \\ + \ 15 \\ \hline 27 \end{array} \qquad \begin{array}{r} 27 \\ - \ 15 \\ \hline 12 \end{array}$$

B.
$$\begin{array}{r} 322 \\ - \ 180 \\ \hline 142 \end{array} \qquad \begin{array}{r} 142 \\ + \ 180 \\ \hline 322 \end{array}$$

C.
$$\begin{array}{r} 12 \\ \times \ 9 \\ \hline 108 \end{array} \qquad 9)\overline{108}^{\,12}$$

D.
$$14)\overline{56}^{\,4} \qquad \begin{array}{r} 14 \\ \times \ 4 \\ \hline 56 \end{array}$$

Practice

Use the inverse operation to check the answer.

1. $23 + 15 = 38$

2. $57 - 31 = 26$

3. $8 \times 7 = 56$

4. $144 \div 6 = 24$

5. $14 \times 3 = 42$

6. $366 - 218 = 148$

7. $586 + 255 = 841$

8. $250 \div 5 = 50$

Words for Operations

Many different words and phrases can be used in numerical and algebraic expressions to represent the operations of addition, subtraction, multiplication, and division.

Examples

Write the operation described by the phrase.

A. the sum of 6 and 7

To find a sum means to add. The operation is addition.

B. the product of k and 15

To find a product means to multiply. The operation is multiplication.

Practice

Write the operation described by the phrase.

1. k greater than 75

2. 42 less than m

3. the sum of h and 6

4. the quotient of 11 and b

5. c times w

6. the difference of a and 45

7. 9 increased by 7

8. the product of 12 and p

9. 2 reduced by m

Words and Equations

You can write equations to represent some sentences. Use a variable to represent what is unknown in the sentence. Use operation signs or other mathematical signs to represent words or phrases in the sentence.

Example

Write an equation for this sentence: A number increased by 7 is 12.

Represent "a number" by a variable such as x.
Represent "increased by" with a plus sign.
Represent "is" with an equal sign.

So, the equation is $x + 7 = 12$.

Practice

Write an equation for the sentence.

1. The product of a number and 7 is 49.

2. 8 less than a number is 50.

3. 12 and a number have a quotient of 2.

4. 13 times a number is 91.

5. 7 more than a number is 12.

6. 16 divided by a number is $\frac{4}{5}$.

7. A number reduced by 13 is 9.

8. 45 greater than a number is 26.

Troubleshooting

Evaluate Expressions

An expression that includes one or more variables is called an **algebraic expression**. To evaluate algebraic expressions, replace the variables with the given numbers. Then evaluate as you would with a numerical expression. Follow the rules for the order of operations.

Example

Evaluate $3 + m \times 2$ for $m = 6$.

$3 + m \times 2 = 3 + 6 \times 2$ Replace m with 6.

$\qquad\qquad = 3 + 12$ Multiply before adding.

$\qquad\qquad = 15$ Add.

Practice

Evaluate the expressions for the given value of the variables.

1. $4m$ for $m = 9$

2. $\frac{3}{4}p$ for $p = 8$

3. $\frac{z}{24}$ for $z = 96$

4. $t - 5$ for $t = 19$

5. $6h$ for $h = 9$

6. $5 + c$ for $c = {}^-13$

7. $72 \div k$ for $k = 6$

8. $p - 14$ for $p = 27$

9. ^-8f for $f = {}^-7$

Classify Angles

You can classify angles by their sizes.

An *acute* angle has a measure greater than 0° and less than 90°.

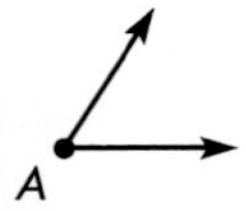

A *right* angle forms a square corner. It measures 90°.

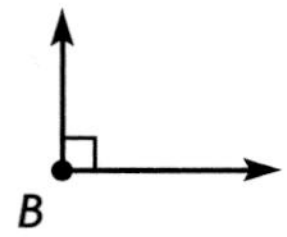

An *obtuse* angle has a measure greater than 90° and less than 180°.

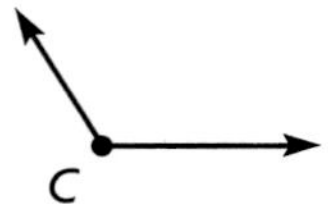

A *straight* angle has a measure of 180°.

A

B

C

D

Practice

Classify each angle by stating whether it is *acute*, *obtuse*, *right*, or *straight*.

1.
K

2.
P

3.
Z

4.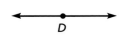
N

5.
V

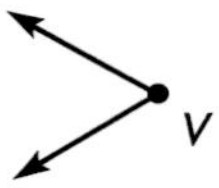

6.
M

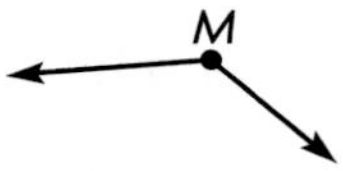

7.
Q

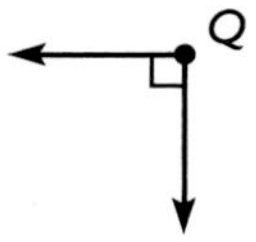

8.
T

H18 Prerequisite Skills Review

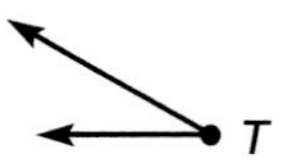

Name Angles

You can name an angle by using one letter, three letters, or a number.

Examples

Name the angle formed by the blue rays.

Use the vertex letter to name the angle. $\angle G$

There are two angles with vertices at *N*. So, use three letters. $\angle ANV$ or $\angle VNA$

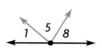

Use a number. $\angle 5$

Practice

Name the angle formed by the blue rays.

1.

2.

3.

4.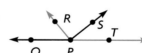

5.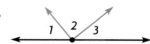

6.

Identify Polygons

A **polygon** is a closed plane figure formed by three or more line segments. A polygon is classified by its number of sides.

Examples

State whether the figure is a polygon or not. If it is, classify it.

A. The figure is not closed, so it is not a polygon.

B. The figure is a 5-sided polygon called a pentagon.

Practice

State whether the figure is a polygon or not. If it is, classify it.

1.

2.

3.

4.

Troubleshooting

Identify Solid Figures

A **polyhedron** is a solid figure with flat faces that are polygons. A **prism** is a polyhedron with two congruent, parallel **bases**. Its **lateral faces** are rectangles. A **pyramid** has a polygon for its base and triangles for its lateral faces.

A **cone** has a circular base. A **cylinder** has two congruent parallel circular bases. Cones and cylinders have curved lateral surfaces.

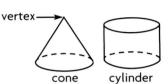

Practice

Identify each solid figure.

1.

2.

3.

4.

5.

6.

7.

8.

Faces, Edges, and Vertices

Each polygon that forms a solid figure is a **face** of the figure. A line segment where two faces meet is an **edge**. A point where three or more edges meet is a **vertex**.

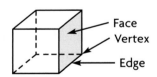

Example

Tell the number of faces, vertices, and edges of the cube.

The cube has 6 faces, 8 vertices, and 12 edges.

Practice

Give the number of faces, edges, and vertices of each solid figure.

1.

2.

3.

4.

Write Equivalent Fractions

To write a fraction equivalent to a given fraction, multiply or divide the numerator and denominator by the same number.

Examples

Complete the number sentence to find an equivalent fraction.

A. $\frac{18}{24} = \frac{3}{\blacksquare}$

Think: To change 18 to 3, divide the numerator of $\frac{18}{24}$ by 6. Then divide the denominator by 6.

$\frac{18}{24} = \frac{18 \div 6}{24 \div 6} = \frac{3}{4}$

So, the equivalent fraction is $\frac{3}{4}$.

B. $\frac{3}{5} = \frac{\blacksquare}{20}$

Think: To change 5 to 20, multiply the denominator of $\frac{3}{5}$ by 4. Then multiply the numerator by 4.

$\frac{3}{5} = \frac{3 \times 4}{5 \times 4} = \frac{12}{20}$

So, the equivalent fraction is $\frac{12}{20}$.

Practice

Complete each number sentence to find an equivalent fraction.

1. $\frac{1}{2} = \frac{\blacksquare}{8}$

2. $\frac{12}{18} = \frac{2}{\blacksquare}$

3. $\frac{3}{4} = \frac{15}{\blacksquare}$

4. $\frac{\blacksquare}{6} = \frac{25}{30}$

5. $\frac{8}{24} = \frac{\blacksquare}{12}$

6. $\frac{\blacksquare}{15} = \frac{30}{45}$

7. $\frac{3}{5} = \frac{48}{\blacksquare}$

8. $\frac{\blacksquare}{40} = \frac{7}{8}$

9. $\frac{27}{36} = \frac{\blacksquare}{12}$

10. $\frac{\blacksquare}{16} = \frac{8}{32}$

11. $\frac{12}{\blacksquare} = \frac{6}{10}$

12. $\frac{\blacksquare}{100} = \frac{11}{20}$

Solve Multiplication Equations

To solve a multiplication equation, use an inverse operation. Divide both sides of the equation by the number that multiplies the variable.

Example

Solve and check. $8n = 40$

$8n = 40$

$\frac{8n}{8} = \frac{40}{8}$ Divide both sides by 8.

$n = 5$ Simplify.

Check: $8n \overset{?}{=} 40$

$8 \times 5 \overset{?}{=} 40$

$40 = 40$ ✔

Practice

Solve and check.

1. $6k = 18$

2. $3h = 27$

3. $5e = 75$

4. $10d = 700$

5. $120 = 8m$

6. $24y = 108$

7. $40 = 160b$

8. $^-1p = 12$

9. $15n = 135$

10. $4k = 30$

11. $1.2\,t = 96$

12. $^-5b = 35$

13. $198 = 18s$

14. $9 = \frac{1}{2}p$

15. $^-7f = ^-98$

16. $^-84 = 14q$

Troubleshooting

Congruent and Similar Figures

Two figures are **similar** if they have the same shape. One may be an enlargement or a reduction of the other. Two figures are **congruent** if they have the same size and shape.

Examples

Tell if the figures in each pair appear to be congruent, similar, both, or neither.

A.

The figures are neither congruent or similar.

B.

The figures are similar.

C.

Both; the figures are congruent and similar.

Practice

Tell if the figures in each pair appear to be congruent, similar, both, or neither.

1.

2.

3.

Write Fractions as Decimals

To write a fraction as a decimal, divide the numerator by the denominator. Or, write an equivalent fraction with a denominator of 10, 100, or 1,000. Then rewrite the fraction as a decimal.

Example

Write $\frac{3}{4}$ as a decimal.

One Way

$$\begin{array}{r} 0.75 \\ 4\overline{)3.00} \\ -\,28 \\ \hline 20 \\ -\,20 \\ \hline 0 \end{array}$$ Divide 3 by 4. So, $\frac{3}{4} = 0.75$.

Another Way

$\frac{3}{4} = \frac{3 \times 25}{4 \times 25} = \frac{75}{100}$ Write an equivalent fraction with a denominator of 100.

$\frac{75}{100} = 0.75$ Write the equivalent fraction as a decimal.

So, $\frac{3}{4} = 0.75$.

Practice

Write the fraction as a decimal.

1. $\frac{1}{2}$ **2.** $\frac{7}{10}$ **3.** $\frac{1}{4}$ **4.** $\frac{3}{8}$ **5.** $\frac{7}{20}$

6. $\frac{2}{5}$ **7.** $\frac{17}{25}$ **8.** $\frac{43}{50}$ **9.** $\frac{48}{250}$ **10.** $\frac{19}{50}$

Multiply with Fractions and Decimals

When multiplying decimals, the number of decimal places in the product is the total of decimal places in the two factors. When multiplying fractions, write the product of the numerators over the product of the denominators.

Examples

Find the product.

A.
$$\begin{array}{r} 2.57 \\ \times \quad 1.2 \\ \hline 514 \\ + \quad 2570 \\ \hline 3.084 \end{array}$$

2 decimal places
1 decimal place

$2 + 1 = 3$ decimal places

B. $\frac{3}{4} \times \frac{8}{9} = \frac{3 \times 8}{4 \times 9} = \frac{24}{36}$ Multiply numerators.
Multiply denominators.

$\frac{24 \div 12}{36 \div 12} = \frac{2}{3}$ Simplify.

Practice

Find the product. Write the answer in simplest form.

1. 7.2×0.9 **2.** 8.4×1.1 **3.** 112.3×0.15 **4.** 17.55×1.05 **5.** 15.82×25

6. $\frac{1}{3} \times \frac{2}{3}$ **7.** $\frac{5}{6} \times \frac{3}{4}$ **8.** $\frac{2}{3} \times \frac{5}{12}$ **9.** $\frac{5}{8} \times \frac{3}{5}$ **10.** $\frac{1}{2} \times \frac{1}{4}$

Fractions, Decimals, and Percents

You can write a percent as a decimal and a decimal as a percent.
You can write a percent as a fraction and a fraction as a percent.

Examples

A. Write 42% as a decimal.

$42\% = 42 \div 100$ Divide by 100.

So, $42\% = 0.42$.

B. Write 0.781 as a percent.

$0.781 = 0.781 \times 100$ Multiply by 100.

So, $0.781 = 78.1\%$.

C. Write 24% as a fraction.

$24\% = \frac{24}{100} = \frac{6}{25}$ Write the percent over 100. Simplify.

So, $24\% = \frac{6}{25}$.

D. Write $\frac{11}{25}$ as a percent.

$\frac{11}{25} = 11 \div 25 = 0.44$ Divide numerator by denominator.

So, $\frac{11}{25} = 44\%$. Write the percent.

Practice

Write each decimal and each fraction as a percent.
Write each percent as a decimal and as a fraction.

1. $\frac{2}{5}$ **2.** 10%   **3.** 0.34   **4.** 25%   **5.** $\frac{1}{2}$

6. 0.81   **7.** $\frac{7}{25}$ **8.** 60%   **9.** $\frac{18}{50}$ **10.** 0.02

Troubleshooting

Certain, Impossible, Likely, Unlikely

An event is **certain** if it is sure to happen. It is **impossible** if it can never happen. It is **likely** if there is a strong chance that it will happen. It is **unlikely** if there is a strong chance that it will not happen.

Examples

Tell if the event is *certain*, *impossible*, *likely*, or *unlikely*.

A. 2 + 2 will equal 5 next Tuesday.

 2 + 2 can never equal 5.

 So, the event is *impossible*.

B. You toss a penny 10 times and get at least 1 head.

 The probability of getting a head is $\frac{1}{2}$ each time you toss a coin.

 So, the event is *likely*.

Practice

Tell if the event is *certain*, *impossible*, *likely*, or *unlikely*.

1. Rain will fall at least once this year.

2. There will be no June next year.

3. You toss a quarter and get either a head or a tail.

4. A person who does not know you will guess your phone number.

Analyze Data

A **line graph** shows changes over time. How many people attended Game 2?

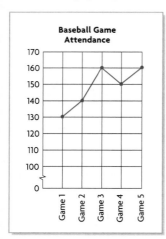

The dot on the vertical "Game 2" line is beside 140. So, 140 people attended the game.

A **circle graph** shows data as parts of a whole. How many people bought peanuts at the fair?

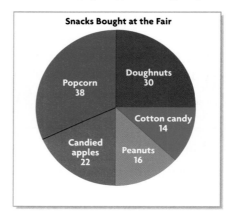

The section of the circle labeled "Peanuts" shows 16. So, 16 people bought peanuts.

Practice

For 1–2, use the line graph.

1. How many people attended Game 4?

2. What was the total attendance at all the games?

For 3–4, use the circle graph.

3. What was the most popular snack?

4. Popcorn and one other snack were half the sales. What was the other snack?

Customary Units

The table shows equivalents in the customary system of measurement. To change from a larger unit to a smaller one, multiply. To change from a smaller unit to a larger one, divide.

UNITS OF LENGTH
12 inches (in.) = 1 foot (ft)
3 feet = 1 yard (yd)
5,280 feet = 1 mile (mi)
1,760 yards = 1 mile

UNITS OF CAPACITY
2 cups (c) = 1 pint (pt)
2 pints = 1 quart (qt)
4 quarts = 1 gallon (gal)

UNITS OF WEIGHT
16 ounces (oz) = 1 pound (lb)
2,000 pounds = 1 ton (T)

Examples

Change to the given unit.

A. 24 in. = ■ ft **Think:** An inch is smaller than a foot. To change from inches to feet, divide by 12.

$24 \div 12 = 2$, so 24 in. = 2 ft.

B. 4 gal = ■ qt **Think:** A gallon is larger than a quart. To change from gallons to quarts, multiply by 4.

$4 \times 4 = 16$, so 4 gal = 16 qt.

Practice

Change to the given unit.

1. 2 yd = ■ ft   **2.** 4 c = ■ pt   **3.** 4 lb = ■ oz   **4.** 2.5 gal = ■ qt

Metric Units

The table shows how prefixes used with the basic units **meter (m)**, **liter (L)**, and **gram (g)** are related in the metric system. To change units, multiply by 10 for each place you move to the right. Divide by 10 for each place you move to the left.

kilo (k)	hecto (h)	deka (da)	BASIC UNIT	deci (d)	centi (c)	milli (m)

Examples

Change to the given unit.

A. 8 m = ■ cm **Think:** From meter to centimeter, the move is 2 places to the right. Multiply by 10 × 10, or 100.

$8 \times 100 = 800$, so 8 m = 800 cm.

B. 4,000 mg = ■ g **Think:** From milligrams to grams, the move is 3 places to the left. Divide by 10 × 10 × 10, or 1,000.

$4,000 \div 1,000 = 4$, so 4,000 mg = 4 g.

Practice

Change to the given unit.

1. 2 m = ■ dm   **2.** 6 L = ■ mL   **3.** 3,000 g = ■ kg   **4.** 5,000 mg = ■ cg

Troubleshooting

Solve Proportions

A **proportion** is a number sentence which states that two ratios are equal. You can solve a proportion by using the fact that in a proportion, the **cross products** are equal.

$$\frac{2}{3} = \frac{4}{6}$$ cross products

$$2 \times 6 = 3 \times 4$$

Example

Solve for n. $\frac{2}{3} = \frac{n}{12}$

$\frac{2}{3} = \frac{n}{12}$ Find the cross products.

$2 \times 12 = 3 \times n$ Use the cross products to write an equation.

$24 = 3 \times n$ Simplify.

$\frac{24}{3} = \frac{3 \times n}{3}$ Divide both sides of the equation by 3.

$8 = n$ Write the solution.

Practice

Solve for *n*.

1. $\frac{n}{3} = \frac{6}{9}$

2. $\frac{4}{n} = \frac{5}{10}$

3. $\frac{8}{10} = \frac{n}{15}$

4. $\frac{20}{15} = \frac{12}{n}$

5. $\frac{n}{6} = \frac{25}{30}$

6. $\frac{n}{5} = \frac{9}{15}$

7. $\frac{8}{n} = \frac{32}{16}$

8. $\frac{48}{36} = \frac{n}{9}$

Perimeter

Perimeter is the distance around a figure. To find the perimeter of a polygon, find the sum of the lengths of the sides.

Example

The opposite sides of a rectangle are congruent, so the sides of the figure that are not labeled measure 4 cm and 2 cm.

So, the perimeter is 12 cm. $P = 4\ \text{cm} + 2\ \text{cm} + 4\ \text{cm} + 2\ \text{cm}$

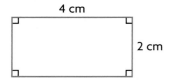

Practice

Find the perimeter of the figure.

1.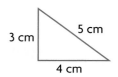

2.

3.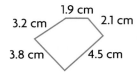

4.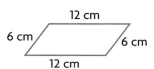

5.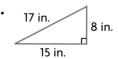

6.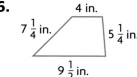

Change Units

When you change from one unit to another, you must decide whether there are more or fewer of the new unit. If there are more units, multiply. If there are fewer units, divide.

Examples

A rug is 4 yd long. How many feet long is this?

| 1 yd | 1 yd | 1 yd | 1 yd |

ft: 1 1 1 1 1 1 1 1 1 1 1 1

There will be *more* feet in the measurement than there are yards. Since 1 yd = 3 ft, multiply to find the answer.

$3 \times 4 = 12$. So, the rug is 12 ft long.

Practice

Change the measurement to the given unit.

1. a 3-ft table, to in.

2. a 5-gal tank, to qt

3. a 15-ft room, to yd

4. a 1,000-m race, to cm

Find the Square and Cube of a Number

The **square** of a number is the product of the number used twice as a factor. The **cube** of a number is the product of the number used three times as a factor.

Examples

A. Find the square of 5.

$5 \times 5 = 25$ Find the product, using 5 as a factor twice.

So, 5 squared is 25, or $5^2 = 25$.

B. Find the cube of 4.

$4 \times 4 \times 4 = 64$ Find the product, using 4 as a factor three times.

So, 4 cubed is 64, or $4^3 = 64$.

Practice

Find the square of each number.

1. 7   **2.** 14   **3.** 25   **4.** 80   **5.** 5.6

Find the cube of each number.

6. 2   **7.** 3   **8.** 5   **9.** 10   **10.** 0

11. 6   **12.** 11   **13.** $\frac{1}{2}$ **14.** $\frac{3}{5}$ **15.** 0.3

Troubleshooting

Areas of Squares, Rectangles, and Triangles

The area of a rectangle or square is the product of the length and the width. The area of a triangle is *half* the product of the base and the height.

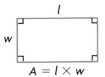

$A = l \times w$

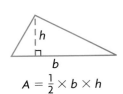

$A = \frac{1}{2} \times b \times h$

Examples

Find the area of the figure.

A.

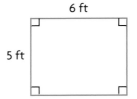

6 ft

5 ft

$A = 6 \times 5 = 30$, or 30 ft^2

B.

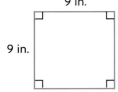

9 in.

9 in.

$A = 9 \times 9 = 9^2 = 81$, or 81 in.^{2}

C.

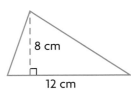

8 cm

12 cm

$A = \frac{1}{2} \times 12 \times 8 = 48$, or 48 cm^2

Practice

Find the area of the figure.

1.

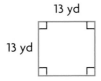

13 yd

13 yd

2.

12 ft

7 ft

3.

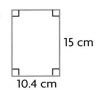

15 cm

10.4 cm

4.

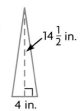

$14\frac{1}{2}$ in.

4 in.

Areas of Circles

The area of a circle is the product of π and the square of the radius. Use 3.14 for the value of π.

$A = \pi r^2$

Examples

Find the area of the circle.

A. radius = 3 cm

3 cm

$A \approx 3.14 \times 3^2$

$\approx 3.14 \times 9$

≈ 28.26 cm^2

B. The radius is half of the diameter, or 8 in.

16 in.

$A \approx 3.14 \times 8^2$

$\approx 3.14 \times 64$

≈ 200.96 in.^{2}

Practice

Find the area of the circle. Use 3.14 for π.

1.

10 cm

2.

12 in.

3.

18 cm

4.

42 in.

Compare Numbers

To compare integers, use a number line. The number to the right on the line is greater.

To compare decimals, align the decimal points. Then compare the digits from left to right.

To compare fractions, rename unlike fractions, then compare the numerators.

Examples

Compare. Write $<$ or $>$ for each ●.

A. 2 ● ⁻3

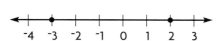

2 is to the right of ⁻3, so 2 $>$ ⁻3.

B. 6.25 ● 6.179

6.2̲5

6.1̲79

0.**2** $>$ 0.**1**, so 6.25 $>$ 6.179.

C. $\frac{1}{2}$ ● $\frac{2}{3}$

$\frac{1 \times 3}{2 \times 3} = \frac{3}{6}$ $\frac{2 \times 2}{3 \times 2} = \frac{4}{6}$

$3 < 4$, so $\frac{1}{2} < \frac{2}{3}$.

Practice

Compare. Write $<$ or $>$ for each ●.

1. 485 ● 579

2. 3.03 ● 3.3

3. $\frac{1}{2}$ ● $\frac{1}{3}$

4. ⁻11 ● ⁻3

5. 5 ● ⁻4

6. 14.97 ● 14.9

7. 523.6 ● 532.8

8. 0.007 ● 0.01

9. ⁻0.5 ● ⁻0.4

10. $\frac{3}{4}$ ● $\frac{5}{8}$

Function Tables

To find an output value in a function table, replace the variable with an input value. Then evaluate the algebraic expression.

Example

Complete the function table.

k	$k + 4$
7	
9	
11	
16	

Replace k in $k + 4$ with 7: $k + 4 = 7 + 4 = 11$.

Replace k in $k + 4$ with 9: $k + 4 = 9 + 4 = 13$.

Replace k in $k + 4$ with 11: $k + 4 = 11 + 4 = 15$.

Replace k in $k + 4$ with 16: $k + 4 = 16 + 4 = 20$.

Input   Output

k	$k + 4$
7	11
9	13
11	15
16	20

Practice

Copy and complete the function table.

1.

x	$x + 7$
9	16
13	
18	
24	

2.

b	$b \times 16$
2	
5	
8	
13	

3.

m	$m \div 6$
192	
150	
90	
51	

4.

r	$r - 7.4$
25.8	
19.2	
13.05	
9.1	

Troubleshooting

Slides, Flips, and Turns

There are three ways that you can **transform** a figure. You can **slide,** or **translate,** the figure along a straight line. You can **flip,** or **reflect,** the figure over a line. Or you can **turn,** or **rotate,** the figure around a point.

Examples

Identify the transformation as a *slide, flip,* or *turn.*

A.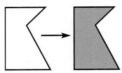

The figure has been translated along a straight line to the right. This is a *slide.*

B.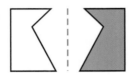

The figure has been reflected across a vertical line. This is a *flip.*

C.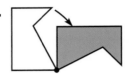

The figure has been rotated clockwise around a point. This is a *turn.*

Practice

Identify the transformation as a *slide, flip,* or *turn.*

1.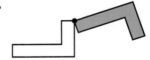
2.
3.
4.

Line Symmetry

A figure has **line symmetry** if it can be folded or reflected so that the two parts of the figure match, or are congruent. A figure can have more than one line of symmetry.

Examples

Is the dashed line a line of symmetry? Write *yes* or *no.*

A. 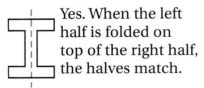 Yes. When the left half is folded on top of the right half, the halves match.

B. No. When the lower left is folded on top of the upper right, the halves do not match.

Practice

Is the dashed line a line of symmetry? Write *yes* or *no.*

1.
2.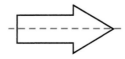
3.
4.

Measure Angles

To measure $\angle ABC$ with a protractor, place the center of the protractor at B.

Align the 0° mark on the protractor with one ray of the angle.

Read the angle measure where the other ray passes through the scale on the protractor.

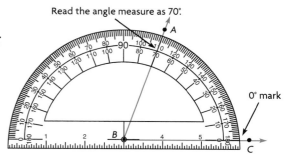

Read the angle measure as 70°.

0° mark

Practice

Use a protractor to measure the angle.

1.

2.

3.

4.

Ordered Pairs

You can use two numbers called an **ordered pair** to locate a point on a grid. The first number tells you how far to move horizontally from (0,0). The second number tells you how far to move vertically from (0,0).

Example

Write the ordered pair for point A.

To reach point A from (0,0), go
$^+3$ units right (horizontally) and
$^-2$ units down (vertically).

So, the ordered pair is (3,$^-$2).

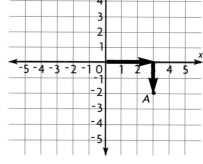

Practice

Write the ordered pair for each point.

1. point A **2.** point C

3. point D **4.** point E

5. point F **6.** point G

7. point I **8.** point J

9. point B **10.** point H

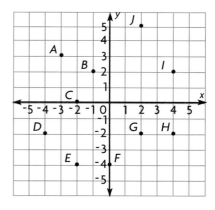

Extra Practice

Set A (pp. 16–19)

Estimate.

1. 589 $+342$	**2.** 865 -383	**3.** 8,926 $+1,674$	**4.** 8,960 9,043 $+8,756$	**5.** 3,406 $- \ 792$

6. $4,496 \div 51$ **7.** 47×98 **8.** 321×43 **9.** $3,552 \div 58$ **10.** $6,251 \div 12$

Tell whether the estimate is an overestimate or underestimate.

11. $848 + 692 \approx 1,550$ **12.** $398 \times 66 \approx 28,000$ **13.** $317 - 67 \approx 230$

Set B (pp. 20–21)

Find the sum or difference.

1. $6,343 - 1,145$ **2.** $19,826 - 3,517$ **3.** $22,912 + 84,718$ **4.** $532,047 + 36,603$

5. $6,052 + 2,791 + 3,047 + 654$ **6.** $7,380 - 1,890 - 408 - 792$

Set C (pp. 22–25)

Multiply or divide, writing any remainder as a fraction.

1. 83 $\times 62$ **2.** $16\overline{)869}$ **3.** $45\overline{)1,675}$ **4.** 458 $\times 194$

Set D (pp. 28–29)

Write a numerical or algebraic expression for the word expression.

1. nine dollars less than twelve dollars **2.** thirteen more than a number, x **3.** twenty-two dogs times two bowls **4.** fifty-two divided by two

Evaluate each expression.

5. $8 \times a$, for $a = 35$ **6.** $p - 38$, for $p = 97$ **7.** $175 \div k$, for $k = 5$ **8.** $96 + m$, for $m = 48$

Set E (pp. 30–31)

Determine which of the given values is the solution of the equation.

1. $5w = 70$; $w = 12, 13$, or 14 **2.** $16 - a = 7$; $a = 8, 9$, or 10 **3.** $48 \div x = 8$; $x = 6, 7$, or 8

Solve each equation by using mental math.

4. $23 + c = 28$ **5.** $k \div 9 = 5$ **6.** $q - 25 = 25$ **7.** $c \times 6 = 426$

Set A (pp. 36–39)

Use mental math to find the value.

1. $26 + (4 + 18)$ 2. $6 \times 4 \times 5$ 3. $28 + 9 + 41$ 4. $83 - 36$

5. $4 + 19 + 26$ 6. 38×6 7. $47 - 29$ 8. $7 + 22 + 13$

9. 51×8 10. $(6 \times 8) \times 5$ 11. $19 + 31$ 12. $92 - 29$

13. $65 + 17$ 14. $10,000 \div 25$ 15. $3 \times 10 \times 7$

16. $51 - 29$ 17. $4 \times 8 \times 2$ 18. $19 + 52 + 21$

19. Ben has three cousins under the age of 13. The product of their ages is 336. How old are they?

Set B (pp. 40–41)

Write the equal factors. Then find the value.

1. 10^4 2. 5^6 3. 2^3 4. 3^2 5. 7^4

6. 1^8 7. 27^2 8. 15^1 9. 12^3 10. 9^4

Write in exponent form.

11. $3 \times 3 \times 3 \times 3$ 12. 10×10 13. $5 \times 5 \times 5 \times 5 \times 5 \times 5$

14. $12 \times 12 \times 12$ 15. $1 \times 1 \times 1 \times 1$ 16. $21 \times 21 \times 21 \times 21 \times 21$

17. Harold was hired to take a survey of 625 people. He wants to write this number in exponent form. If the base is 5, what is the exponent? How did you arrive at your answer?

Set C (pp. 44–45)

Give the correct order of operations.

1. $5 \times 7 + 3 \div 2$ 2. $(6 + 15) \div 3 \times 1 - 6$

3. $12^2 \times 10 - 10^3 - 100$ 4. $10^2 + 5^2 \div 5^2 - 5$

Evaluate the expression.

5. $7 + (2 \times 2)^4 - 9 \times 9$ 6. $90 \times 5 - 4 \times (18 \div 6)$

7. $3^2 \times (4 + 5)^2 - 36$ 8. $15^2 \div (4^2 + 9) + 8^1$

9. $(8^2 + 3^3) \times (9 - 5)^3$ 10. $10^3 \div (10^2 \div 10^1) + 10^2$

11. Scott bought 5 pounds of nails that cost $2.75 per pound and 3 pounds of screws that cost $3.50 per pound. How much did Scott spend?

Set A (pp. 52–55)

Write the value of the blue digit.

1. 6.12053   **2.** 0.0231   **3.** 8.7   **4.** 0.849

Write the number in expanded form.

5. 0.00309   **6.** 5.015   **7.** 3.032

8. 20.0518   **9.** 200.05   **10.** 5.16

Compare the numbers. Write <, >, or = for ●.

11. 5.099 ● 5.999   **12.** 226.5 ● 226.4   **13.** 251.36 ● 241.36

14. 18.3 ● 18.30   **15.** 4.18 ● 4.28   **16.** 49.089 ● 49.098

Write the numbers in order from least to greatest.

17. 82.16, 82, 82.15   **18.** 141.14, 114.41, 141.41   **19.** 5.09, 5.49, 5.23

Set B (pp. 58–59)

Estimate.

1. 3.7
  3.15
+2.98

2. 62.8
× 6

3. 109.7
− 53.622

4. 788.3
× 92

5. 5.92
  3.15
+4.07

6. 21.513
× 9.8

7. 5.816
  3.215
+1.6

8. 465.09
− 73.46

9. 728 ÷ 8.1   **10.** 8.1 − 2.456   **11.** 20.8 ÷ 7   **12.** 123.95 ÷ 61

Set C (pp. 60–61)

Write the decimal and percent for the shaded part.

1. 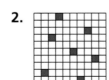   **2.**   **3.**

Write the corresponding decimal or percent.

4. 60%   **5.** 0.9   **6.** 39%   **7.** 0.04   **8.** 0.46

9. 18%   **10.** 0.41   **11.** 0.38   **12.** 7%   **13.** 90%

Set A (pp. 66–69)

Add or subtract. Estimate to check.

1. $12.8 - 4.1$
2. $\$21.85 + \$17.48$
3. $17.3 - 16.5$
4. $8.36 + 5.216 + 0.09$
5. $8 + 7.317 + 3.06$
6. $5.08 - 2.261$

Copy the problem. Place the decimal point correctly in the answer.

7. $13.601 - 10.311 = 329$
8. $18.56 + 3.12 = 2168$
9. $5 - 3.021 = 1979$

Set B (pp. 70–73)

Tell the number of decimal places there will be in the product.

1. 21.3×18.4
2. 7.03×7.05
3. 9.2×2.13

Copy the problem. Place the decimal point in the product.

4. $360.05 \times 12.6 = 4536630$
5. $762 \times 3.285 = 2503170$
6. $10.11 \times 1.02 = 103122$

Multiply. Estimate to check.

7. 37.5×10.26
8. 6.42×9.1
9. 0.05×2.9
10. 19.3×2.41
11. 2.39×7.6
12. 2×6.005

Set C (pp. 76–79)

Rewrite the problem so that the divisor is a whole number.

1. $16.92 \div 0.12$
2. $661.44 \div 31.2$
3. $20.2 \div 0.53$

Copy the problem. Place the decimal point in the quotient.

4. $4.48 \div 2.8 = 16$
5. $28.68 \div 1.2 = 239$
6. $9.87 \div 2 = 4935$

Divide. Estimate to check.

7. $9.72 \div 1.2$
8. $25.0224 \div 3.12$
9. $20.801 \div 6.1$
10. $80.4 \div 4.8$

Set D (pp. 82–83)

Evaluate each expression.

1. $8m$ for $m = 4.2$
2. $6.3 + 5.04 + k$ for $k = 8.4$
3. $(3.4 - c) + 63$ for $c = 2.47$

Solve each equation by using mental math.

4. $p + 2.8 = 7.9$
5. $\frac{k}{3} = 5.6$
6. $a - 17.1 = 9.5$
7. $6s = 7.2$
8. $x + 22.6 = 30.8$
9. $15n = 4.5$

Set A (pp. 94–97)

Determine the type of sample. Write *convenience, random,* or *systematic*.

1. The teacher randomly selected a student and then surveyed every third student on his attendance list.

2. The teacher asked students in English class about their favorite books.

Set B (pp. 98–99)

A survey is to be conducted about the favorite foods of middle-school students. Tell whether the sampling method is *biased* or *unbiased*.

1. Randomly survey 10 people who get off the school bus one morning.

2. Randomly survey 1 of every 10 students in sixth, seventh, and eighth grades.

Set C (pp. 102–105)

Use the table at the right.

1. Copy and complete the table.

2. How large was the sample size?

Favorite Color	Tally	Frequency	Cumulative Frequency
Blue	⊞⊞ ⊞⊞ I	■	■
Green	⊞⊞ IIII	■	■
Red	⊞⊞ III	■	■
Yellow	⊞⊞	■	■
Orange	II	■	■

Set D (pp. 106–108)

Find the mean, median, and mode.

1. 4, 5, 7, 8, 8

2. 6.1, 8.1, 7.4, 7.2

3. 12, 14, 18, 12, 22

4. 1, 1, 7, 9, 3, 4, 2, 3

Set E (pp. 109–111)

Use the data in the table.

NUMBER OF HOURS SPENT ON HOMEWORK										
Week	1	2	3	4	5	6	7	8	9	10
Hours	8	9	10	9	7	9	8	0	7	2

1. Find the mean, median, and mode with the outliers and then without the outliers.

2. How does including the outliers affect the mean? the median? the mode?

Set F (pp. 112–115)

Write *yes* or *no* to tell whether the conclusion is valid. Explain.

1. The first ten females to enter a diner say their favorite drink is orange juice. You conclude that orange juice is the favorite drink of all of the customers.

2. Two hundred students are randomly selected and asked what their favorite kind of music is. Seventy percent say pop music. You conclude that most students prefer pop music.

Set A (pp. 120–123)

1. Make a multiple-bar graph using the data in the table below.

CLUB MEETING ATTENDANCE				
	Sept	Oct	Nov	Dec
Male	42	31	65	61
Female	56	47	51	60

2. Make a multiple-line graph using the data in the table below.

AVERAGE RAINFALL (IN INCHES)				
	Jan	Feb	Mar	Apr
City A	4	5	3	5
City B	8	12	4	7

Set B (pp. 124–125)

1. Make a line graph using the data in the table at the right.

2. If the trend continues, how many miles will be traveled on the sixth day?

DISTANCE TRAVELED					
Day	1	2	3	4	5
Miles	45	85	129	175	220

Set C (pp. 126–128)

1. Make a stem-and-leaf plot of the data below. Find the mode and median.

14	22	37	41	13	18	22	29	33
36	25	21	30	48	44	19	17	28

2. Use the data in the chart below to make a histogram.

NUMBER OF HOURS OF MONTHLY EXERCISE					
0–2	3–5	6–8	9–11	12–15	16–18
12	15	19	22	21	6

Set D (pp. 130–131)

For 1–3, use the box-and-whisker graph. The graph shows the number of tickets sold.

1. What was the greatest number of tickets sold?

2. What is the median?

3. What are the lower and upper quartiles?

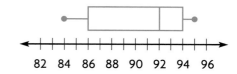

82 84 86 88 90 92 94 96

Set E (pp. 132–135)

For 1–3, use the bar graph at the right.

1. About how many times as high is the bar for Central than the bar for Lee?

2. Has Central won four times as many championships as Lee? Explain.

3. How can you change the graph so it is not misleading?

Set A (pp. 146–147)

Tell whether each number is divisible by 2, 3, 4, 5, 6, 8, 9, or 10.

1. 80   **2.** 99   **3.** 105   **4.** 126   **5.** 234

6. 370   **7.** 591   **8.** 1,620   **9.** 3,048   **10.** 8,020

Set B (pp. 148–149)

Use division or a factor tree to find the prime factorization.

1. 18   **2.** 40   **3.** 36   **4.** 27   **5.** 100

6. 150   **7.** 280   **8.** 12   **9.** 108   **10.** 420

Write the prime factorization of each number in exponent form.

11. 28   **12.** 420   **13.** 48   **14.** 81   **15.** 72

16. 80   **17.** 144   **18.** 221   **19.** 168   **20.** 1,475

Solve for n to complete the prime factorization.

21. $2 \times 5 \times n = 50$ **22.** $2 \times n \times 5 = 30$ **23.** $2 \times 3 \times n = 18$

24. $n \times 3 \times 3 = 27$ **25.** $2 \times 2 \times 2 \times n = 16$ **26.** $n \times 2 \times 3 = 42$

27. $2 \times 3 \times 3 \times n = 54$ **28.** $3 \times n \times 5 = 75$ **29.** $n \times n \times 3 = 12$

Set C (pp. 150–153)

List the first five multiples of each number.

1. 4   **2.** 9   **3.** 14   **4.** 5

5. 30   **6.** 8   **7.** 12   **8.** 21

Find the LCM of each set of numbers.

9. 45, 54   **10.** 10, 35   **11.** 12, 20   **12.** 18, 27

13. 150, 60   **14.** 54, 18   **15.** 55, 33, 44   **16.** 35, 25, 49

17. Jean needs eggs and muffins for the class breakfast. Eggs come in cartons of 12 and muffins come in packages of 8. What is the least number of eggs and muffins she can buy to have an equal number of each?

Find the GCF of each set of numbers.

18. 12, 108   **19.** 148, 84   **20.** 132, 108   **21.** 75, 105

22. 45, 108   **23.** 32, 128   **24.** 252, 336   **25.** 56, 280, 400

Set A (pp. 160–163)

Write the fraction in simplest form.

1. $\frac{24}{80}$ 2. $\frac{14}{63}$ 3. $\frac{24}{56}$ 4. $\frac{15}{60}$ 5. $\frac{16}{40}$ 6. $\frac{50}{90}$

Complete.

7. $\frac{9}{12} = \frac{\blacksquare}{4}$ 8. $\frac{7}{42} = \frac{\blacksquare}{6}$ 9. $\frac{25}{\blacksquare} = \frac{5}{10}$ 10. $\frac{3}{7} = \frac{21}{\blacksquare}$

11. $\frac{1}{12} = \frac{\blacksquare}{144}$ 12. $\frac{9}{20} = \frac{\blacksquare}{60}$ 13. $\frac{5}{\blacksquare} = \frac{15}{27}$ 14. $\frac{12}{\blacksquare} = \frac{3}{8}$

Set B (pp. 164–165)

Write the fraction as a mixed number or a whole number.

1. $\frac{7}{3}$ 2. $\frac{9}{2}$ 3. $\frac{36}{5}$ 4. $\frac{72}{8}$

5. $\frac{10}{2}$ 6. $\frac{13}{4}$ 7. $\frac{36}{6}$ 8. $\frac{12}{9}$

Write the mixed number as a fraction.

9. $4\frac{1}{4}$ 10. $5\frac{2}{3}$ 11. $2\frac{1}{8}$ 12. $7\frac{1}{9}$

13. $3\frac{4}{7}$ 14. $1\frac{6}{11}$ 15. $4\frac{3}{8}$ 16. $15\frac{3}{4}$

Set C (pp. 166–167)

Compare the fractions. Write <, >, or = for each ●.

1. $\frac{5}{9}$ ● $\frac{1}{2}$ 2. $\frac{3}{8}$ ● $\frac{1}{5}$ 3. $\frac{7}{25}$ ● $\frac{3}{5}$ 4. $\frac{3}{5}$ ● $\frac{12}{20}$

Order the fractions from least to greatest.

5. $\frac{1}{3}, \frac{1}{6}, \frac{1}{10}$ 6. $\frac{4}{9}, \frac{2}{3}, \frac{3}{8}$ 7. $\frac{1}{9}, \frac{3}{4}, \frac{5}{12}$

Set D (pp. 169–171)

Write the decimal as a fraction.

1. 0.9   2. 0.081   3. 0.29

Write as a decimal. Tell whether the decimal terminates or repeats.

4. $\frac{3}{8}$ 5. $\frac{5}{16}$ 6. $\frac{1}{3}$ 7. $\frac{7}{9}$

Compare. Write <, >, or = for each ●.

8. $\frac{3}{10}$ ● 0.03   9. $\frac{2}{3}$ ● 0.7   10. 0.79 ● $\frac{5}{8}$ 11. 0.15 ● $\frac{3}{20}$

Write each fraction as a percent.

12. $\frac{6}{25}$ 13. $\frac{7}{10}$ 14. $\frac{9}{20}$ 15. $\frac{4}{5}$

Set A (pp. 176–179)

Estimate the sum or difference.

1. $\frac{4}{7} - \frac{3}{8}$

2. $3\frac{7}{9} + 1\frac{10}{11}$

3. $5\frac{4}{5} - 3\frac{1}{2}$

4. $\frac{8}{9} + \frac{3}{5}$

5. $\frac{4}{9} + \frac{2}{3}$

6. $\frac{8}{9} - \frac{5}{7}$

7. $\frac{4}{9} - \frac{1}{6}$

8. $3\frac{10}{13} + 4\frac{1}{9}$

9. $\frac{3}{7} + \frac{1}{9}$

10. $5\frac{3}{4} - 1\frac{1}{8}$

11. $\frac{5}{9} - \frac{3}{7}$

12. $8\frac{7}{8} + 5\frac{1}{2}$

13. Shawna practiced the tuba for $\frac{5}{6}$ hr on Monday, $\frac{1}{3}$ hr on Wednesday, and $\frac{1}{4}$ hr on Friday. About how many hours did Shawna practice last week?

Set B (pp. 182–185)

Write the sum or difference in simplest form. Estimate to check.

1. $\frac{3}{4} + \frac{1}{2}$

2. $\frac{4}{7} - \frac{1}{3}$

3. $\frac{1}{6} + \frac{5}{18}$

4. $\frac{9}{10} - \frac{2}{5}$

5. $\frac{5}{6} + \frac{1}{4}$

6. $\frac{6}{20} + \frac{3}{5}$

7. $\frac{5}{9} - \frac{1}{2}$

8. $\frac{7}{8} - \frac{1}{16}$

9. $\frac{3}{5} - \frac{1}{9}$

10. $\frac{5}{20} + \frac{11}{20}$

11. $\frac{11}{12} - \frac{1}{6}$

12. $\frac{1}{10} + \frac{3}{8}$

13. Rico read $\frac{1}{5}$ of a book on Saturday and $\frac{1}{3}$ of the same book on Sunday. What portion of the book does he have left to read?

Set C (pp. 186–189)

Draw a diagram to find each sum or difference. Write the answer in simplest form.

1. $2\frac{3}{8} + 1\frac{1}{4}$

2. $2\frac{1}{3} - 1\frac{1}{6}$

3. $1\frac{1}{5} + 1\frac{7}{10}$

4. $3\frac{1}{3} - 1\frac{3}{4}$

Write the sum or difference in simplest form. Estimate to check.

5. $4\frac{5}{12} - 2\frac{1}{6}$

6. $5\frac{2}{5} - 3\frac{3}{10}$

7. $2\frac{1}{9} + 1\frac{2}{5}$

8. $4\frac{2}{3} + 6\frac{1}{6}$

9. Stefan bought $1\frac{1}{2}$ lb of potato salad, $2\frac{1}{4}$ lb of coleslaw, and $4\frac{3}{4}$ lb of chicken at a deli. What is the total weight of his purchase?

Set D (pp. 192–193)

Write the difference in simplest form. Estimate to check.

1. $4\frac{5}{9} - 3\frac{5}{6}$

2. $5\frac{1}{3} - 4\frac{1}{9}$

3. $6\frac{1}{4} - 3\frac{2}{3}$

4. $6\frac{1}{3} - 4\frac{7}{9}$

5. $4 - 1\frac{3}{7}$

6. $5\frac{3}{10} - \frac{2}{5}$

7. $7\frac{1}{6} - \frac{1}{2}$

8. $4\frac{1}{4} - 1\frac{5}{8}$

9. Mr. Norman had $51\frac{1}{3}$ ft of plastic pipe. He installed $25\frac{3}{4}$ ft in the bathroom. How much pipe does he have left?

Set A (pp. 200–201)

Estimate each product or quotient.

1. $\frac{8}{9} \times \frac{3}{5}$ **2.** $\frac{1}{6} \times \frac{4}{9}$ **3.** $5\frac{1}{2} \div \frac{5}{6}$ **4.** $\frac{7}{9} \div \frac{3}{4}$

5. $3\frac{1}{3} \times 5\frac{1}{5}$ **6.** $8\frac{3}{5} \div 2\frac{3}{4}$ **7.** $9\frac{7}{8} \times 10\frac{1}{2}$ **8.** $18\frac{3}{4} \div 5\frac{1}{8}$

9. Liz is making cookies to give to some new neighbors. The recipe calls for $1\frac{3}{4}$ cups of flour. About how much flour does Liz need if she makes $2\frac{1}{2}$ times the recipe?

Set B (pp. 202–205)

Multiply. Write the answer in simplest form.

1. $\frac{2}{4} \times \frac{3}{4}$ **2.** $\frac{2}{5} \times \frac{1}{4}$ **3.** $\frac{3}{8} \times \frac{3}{5}$ **4.** $3 \times \frac{7}{8}$

5. $\frac{2}{7} \times 8$ **6.** $\frac{5}{6} \times \frac{3}{8}$ **7.** $\frac{5}{9} \times \frac{1}{5}$ **8.** $\frac{1}{9} \times \frac{3}{4}$

9. To make a dye for art class, Jan needs $\frac{1}{4}$ tsp of red coloring and $\frac{1}{2}$ that amount of green coloring. How much green coloring does Jan need to make dye?

Set C (pp. 206–207)

Multiply. Write the answer in simplest form.

1. $2\frac{2}{3} \times 2\frac{1}{4}$ **2.** $2\frac{1}{2} \times \frac{1}{5}$ **3.** $\frac{1}{5} \times 2\frac{2}{8}$ **4.** $3\frac{1}{6} \times 5\frac{1}{4}$

5. $2\frac{3}{4} \times \frac{5}{6}$ **6.** $2\frac{2}{7} \times \frac{2}{5}$ **7.** $3\frac{2}{3} \times 2\frac{3}{5}$ **8.** $6\frac{8}{9} \times 2\frac{3}{7}$

Set D (pp. 210–213)

Find the quotient. Write the answer in simplest form.

1. $\frac{2}{9} \div \frac{8}{18}$ **2.** $\frac{3}{8} \div 2\frac{1}{4}$ **3.** $2 \div 2\frac{2}{3}$ **4.** $1\frac{11}{12} \div 1\frac{5}{6}$

5. $\frac{3}{4} \div \frac{1}{8}$ **6.** $\frac{5}{6} \div 1\frac{2}{3}$ **7.** $3\frac{3}{4} \div 2\frac{7}{12}$ **8.** $\frac{5}{7} \div \frac{10}{11}$

Set E (pp. 216–217)

Evaluate the expression.

1. $3r$ for $r = \frac{1}{2}$ **2.** $\frac{5}{6} + d$ for $d = \frac{3}{10}$ **3.** $3\frac{3}{5} \div a$ for $a = \frac{1}{2}$

4. $y + 3\frac{5}{8}$ for $y = 1\frac{1}{4}$ **5.** $2\frac{5}{6} k$ for $k = 1\frac{1}{3}$ **6.** $a \div 4\frac{1}{2}$ for $a = 3\frac{3}{5}$

Solve the equation.

7. $c + \frac{3}{4} = 1\frac{1}{4}$ **8.** $\frac{5}{9}s = \frac{5}{27}$ **9.** $x - 6\frac{1}{8} = 12\frac{1}{4}$

Set A (pp. 228–229)

Write an integer to represent each situation.

1. a temperature increase of 5°
2. the wind speed decreases by 12 mph
3. depositing $510 into a savings account
4. a temperature of 7° below zero

Write the opposite integer.

5. $^-3$
6. $^-12$
7. $^+360$
8. $^-1$
9. $^+160$
10. $^-3,047$
11. $^-1,119$
12. $^-942$

Set B (pp. 230–233)

Write each rational number in the form $\frac{a}{b}$.

1. $3\frac{1}{6}$
2. 0.5
3. 0.27
4. 13.4
5. $2\frac{2}{5}$
6. 3.18
7. 10.02
8. 300
9. 0.36
10. $5\frac{1}{3}$
11. 312
12. $4\frac{1}{4}$

Find a rational number between the two given numbers.

13. $\frac{3}{8}$ and $\frac{5}{6}$
14. $\frac{1}{8}$ and $\frac{1}{4}$
15. $\frac{5}{9}$ and $\frac{11}{15}$
16. 1.8 and 1.9
17. $^-1.5$ and $^-1.3$
18. 4.23 and 4.235
19. 3.8 and 3.82
20. $^-5$ and $^-4.9$
21. $^-12\frac{1}{4}$ and $^-12\frac{1}{3}$

Set C (pp. 234–235)

Compare. Write <, >, or = for each ●.

1. 0.4 ● 0.38
2. $\frac{2}{7}$ ● 0.25
3. $^-0.6$ ● $\frac{-2}{5}$
4. $\frac{7}{9}$ ● 0.8
5. 0.28 ● $\frac{2}{7}$
6. $\frac{3}{13}$ ● 0.13
7. $\frac{-4}{9}$ ● $^-3.1$
8. $^-3$ ● 0.31
9. $2\frac{1}{5}$ ● $2\frac{4}{13}$
10. 0.87 ● 0.868
11. $\frac{-5}{8}$ ● $\frac{5}{8}$
12. $2\frac{1}{16}$ ● $2\frac{1}{10}$

Compare the rational numbers and order them from least to greatest.

13. $\frac{1}{3}, \frac{2}{5}, 0.38, \frac{1}{2}$
14. $\frac{2}{7}, \frac{1}{3}, \frac{1}{4}, 0.26$
15. $0.92, \frac{9}{8}, \frac{8}{9}, 0.924$
16. $0.23, \frac{2}{9}, \frac{1}{4}, \frac{1}{5}$
17. $\frac{-2}{5}, \frac{-1}{3}, \frac{-1}{2}, \frac{-3}{8}$
18. $\frac{1}{10}, ^-3, ^-0.3$
19. $^-7, 6, 1, ^-5$
20. $^-0.1, 0.01, 10, ^-10$
21. $\frac{1}{2}, \frac{1}{6}, \frac{1}{9}, 0.1$

Set A (pp. 244–247)

Find the sum.

1. ⁻3 + 9
2. 1 + ⁻5
3. 5 + 4
4. ⁻2 + 8
5. ⁻9 + 17
6. ⁻14 + ⁻8
7. ⁻19 + 8
8. 22 + ⁻36
9. 31 + 19
10. 32 + ⁻45
11. 63 + ⁻47
12. ⁻71 + 32

13. At 9:00 A.M., the temperature outside was ⁻9°C. By 3:00 P.M., the temperature had risen 5°C. What was the temperature at 3:00 P.M.?

Set B (pp. 250–251)

Find the difference.

1. 2 − 9
2. ⁻8 − ⁻3
3. 4 − ⁻2
4. 4 − 9
5. 8 − 19
6. 14 − ⁻2
7. ⁻12 − ⁻12
8. 3 − 7
9. 23 − 32
10. 14 − 29
11. ⁻48 − 17
12. ⁻24 − ⁻39

13. The water level in the city water tower was 8 ft above normal. Three weeks later, the level was 4 ft below normal. Find the range of the water levels.

Set C (pp. 252–255)

Find the product or quotient.

1. ⁻7 × 6
2. ⁻6 × ⁻8
3. ⁻20 × 4
4. 50 × ⁻8
5. 32 × 10
6. 40 × ⁻9
7. ⁻27 ÷ 9
8. ⁻64 ÷ ⁻8
9. ⁻140 ÷ 10
10. 225 ÷ ⁻25
11. 360 ÷ 15
12. 216 ÷ ⁻12

13. The depth of a diver changed ⁻5 ft every 30 sec. How much of a depth change did the diver have after 2 min?

Set D (pp. 256–259)

Find the sum or difference.

1. $3.8 + 0.7$
2. $8\frac{1}{2} - {}^-17\frac{1}{2}$
3. ⁻4.5 − ⁻2.5
4. ⁻8.5 + 1.8
5. $^-7\frac{1}{2} - 4\frac{3}{4}$
6. $2\frac{1}{2} + {}^-8$
7. ⁻4.9 − 4.1
8. $^-17\frac{1}{2} + {}^-34$

Find the product or quotient.

9. $^-8 \div {}^-2\frac{1}{2}$
10. ⁻2.6 × ⁻20
11. $7.7 \div 0.1$
12. $\frac{4}{7} \times \frac{7}{8}$
13. $0.5 \times {}^-2.6$
14. 85 ÷ ⁻17
15. $3\frac{1}{4} \times {}^-1\frac{3}{5}$
16. $\frac{^-3}{4} \div {}^-4$

Set A (pp. 270–271)

Write an algebraic expression for the word expression.

1. 9 less than y

2. 13 more than a number, x

3. $\frac{3}{4}$ increased by y

4. 7 more than the quotient of 52 and k

5. Fred has 32 more than twice the number of baseball cards that José has. Write an algebraic expression for the number of baseball cards Fred has.

Write a word expression for each.

6. $9^2 + a$

7. $7x + 12$

8. $24 - \frac{1}{5}c$

9. $y + \frac{1}{2}x$

Set B (pp. 272–275)

Evaluate the algebraic expression for the given value of the variable.

1. $x - 6$ for $x = {}^-12$

2. $y + 13$ for $y = {}^-25$

3. $42 - k$ for $k = {}^-30$

4. ${}^-38 + p$ for $p = 22$

5. $a^2 - 25$ for $a = 3$

6. $x^3 - 19$ for $x = 2$

7. $3(a + b)^2 - c$ for $a = {}^-3$, $b = 5$, and $c = {}^-8$

8. $6xyz$ for $x = {}^-4$, $y = {}^-3$, and $z = {}^-6$

9. $3pq + r$ for $p = 2$, $q = {}^-8$, and $r = {}^-15$

Evaluate the algebraic expression for $x = 2, 3, 4,$ and 5.

10. $7x + 12$

11. ${}^-4x + 10$

12. $3x - 26$

13. ${}^-9x - 4$

14. $\frac{60}{x} - 14$

15. $\frac{{}^-60}{x} - 14$

Simplify the expression. Then evaluate the expression for the given value of the variable.

16. $5x + 3x + 12$ for $x = {}^-4$

17. $2y + 7y - 25$ for $y = 2$

18. $42z - 30z - 20$ for $z = {}^-2$

Set C (pp. 278–279)

Evaluate the expression.

1. $\sqrt{100} - 2 + 6$

2. $3 \times (9^2 - 41)$

3. $4 \times \sqrt{25} - 3 \times 2$

4. $\sqrt{36} \cdot \sqrt{36}$

5. $530 \times (\sqrt{4} - 2)^2$

6. $8 + \sqrt{49} - 3^2$

Evaluate the expression for $a = 16, b = 3,$ and $c = 6$.

7. ${}^-a + (b + c)^2$

8. $\frac{c}{b} - \sqrt{a}$

9. $\sqrt{a + b + c}$

Set A (pp. 284–285)

Write an equation for the word sentence.

1. $\frac{2}{3}$ of a number is 12.

2. 1.6 more than a number is 5.

3. $1\frac{1}{2}$ less than a number is 6.

4. A number divided by $^-6$ is 30.

5. 9 times a number is 53.

6. 1.5 increased by a number is 4.2.

7. $3\frac{1}{2}$ decreased by a number is $\frac{1}{2}$.

8. The quotient of a number and 3.3 is 99.

9. The sum of 11.2 and a number is 65.07.

10. 64 divided by a number is 3.2.

11. Leroy sold 250 boxes of apples during a fund-raiser. This was 5 times as many apples as Mary sold. Write an equation that represents the situation.

Set B (pp. 287–289)

Solve and check.

1. $x + 6 = 15$

2. $15 = a + 2$

3. $11 + k = 25$

4. $z + 2.7 = 19.6$

5. $5.7 = b + 8.6$

6. $24.8 = 17.2 + c$

7. $y + 8\frac{2}{3} = 16$

8. $13\frac{1}{4} = 4\frac{4}{5} + s$

9. $13\frac{3}{8} = t + 7\frac{5}{6}$

10. $13.2 = x + 7.12$

11. $t + 7\frac{1}{3} = 10\frac{1}{12}$

12. $4.06 + r = 13.56$

13. A carpenter cut a 72-in. board into two pieces. One of the pieces is 24 in. long. How long is the second piece?

Set C (pp. 290–291)

Solve and check.

1. $x - 7 = 11$

2. $31 = a - 7$

3. $k - 10 = 42$

4. $z - 3.9 = 15.8$

5. $6.5 = b - 21.3$

6. $56.7 = c - 19.8$

7. $y - 5\frac{5}{6} = 13$

8. $22\frac{1}{3} = s - 4\frac{3}{4}$

9. $8\frac{3}{5} = t - 4\frac{5}{7}$

10. $a - 27 = 18$

11. $60.3 = b - 8.07$

12. $4\frac{7}{9} = x + 1\frac{2}{3}$

13. Reggie withdrew $175 from his checking account so he could go shopping for the new school year. His new balance was $234. How much was in the account before the withdrawal?

Set A (pp. 297–299)

Solve and check.

1. $3x = 12$

2. $7k = 56$

3. $\frac{p}{16} = {}^{-}3$

4. $\frac{a}{14} = 2$

5. $27 = {}^{-}3x$

6. $8.8 = 2.2n$

7. $4x = 24$

8. $8x = {}^{-}32$

9. $9 = \frac{p}{3}$

10. $56 = {}^{-}7p$

11. $21 = \frac{s}{3}$

12. $45 = 9n$

13. $180 = 3d$

14. $46.2 = \frac{a}{3}$

15. $1,486 = \frac{a}{2}$

16. $\frac{c}{1.2} = 5.6$

17. $^{-}12a = {}^{-}216$

18. $3.8m = 57$

For 19–20, write and solve an equation to answer the question.

19. Celia divided her baseball cards equally among 4 friends. Each friend got 23 baseball cards. How many baseball cards did Celia have originally?

20. Julie earns $6.75 per hour at her job. She wants to save $324.00. How many hours does she have to work to earn $324.00?

Set B (pp. 300–303)

Use the formula *d* = *rt* to complete.

1. $d = \blacksquare$ mi
   $r = 35$ mi per hr
   $t = 4$ hr

2. $d = 1,600$ km
   $r = \blacksquare$ km per min
   $t = 400$ min

3. $d = 2,100$ km
   $r = 70$ km per sec
   $t = \blacksquare$ sec

4. $d = \blacksquare$ ft
   $r = 90.7$ ft per sec
   $t = 31$ sec

5. $d = 567$ mi
   $r = \blacksquare$ mi per hr
   $t = 17.5$ hr

6. $d = 4,850$ m
   $r = 250$ m per sec
   $t = \blacksquare$ sec

Convert the temperature to degrees Fahrenheit. Write your answer as a decimal.

7. 40°C

8. 2.3°C

9. 14°C

10. 20°C

Convert the temperature to degrees Celsius. Write your answer as a decimal and round to the nearest tenth of a degree.

11. 42°F

12. 47°F

13. 79°F

14. 100°F

15. The Concorde jet has a cruising speed of 1,354 mi per hr. Suppose the Concorde maintained this speed for $3\frac{1}{2}$ hr. How far would the Concorde travel?

16. The air conditioner is on only when the room temperature is greater than 75°F. Room temperature is 22°C. Is the air conditioner on? Explain.

Set A (pp. 318–319)

For 1–4, use the figure at the right. Tell how many of each you can name. Then name them.

1. points
2. line segments
3. rays
4. lines

Name the geometric figure.

5.
6.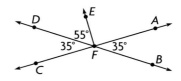
7. 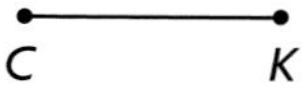 C ——— K
8. •F

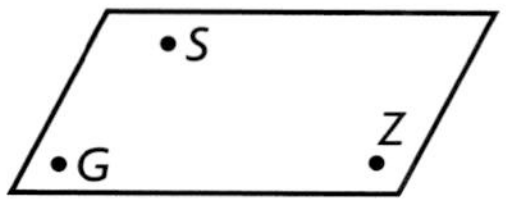

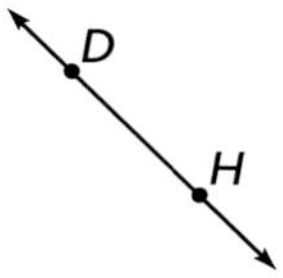

Set B (pp. 322–325)

Tell if the angles are *vertical, adjacent, complementary, supplementary,* or *none of these.*

1. ∠CFD and ∠DFE
2. ∠AFB and ∠CFD
3. ∠EFA and ∠CFD
4. AFB and BFC

Find the unknown angle measure. The type of angle pair is given.

5. complementary

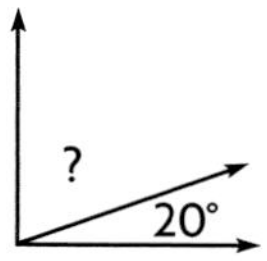

6. complementary

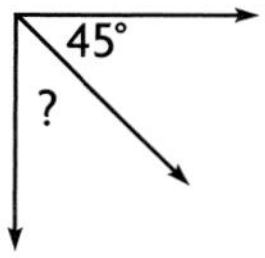

7. supplementary

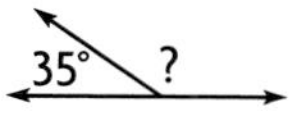

8. supplementary

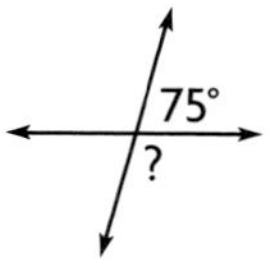

Find the measure of each angle.

9. ∠QSP
10. ∠LSR
11. ∠QSN
12. ∠RSP
13. ∠MSN
14. ∠RSN

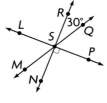

Set C (pp. 326–327)

Use the figure.

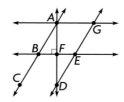

1. Name all the lines that are parallel to $\overleftrightarrow{BE}$.
2. Name a line that is perpendicular to and intersects $\overleftrightarrow{BE}$.
3. Name all the lines that are perpendicular to $\overleftrightarrow{AD}$.
4. Name a line that is parallel to $\overleftrightarrow{AC}$.
5. Name all the lines that intersect $\overleftrightarrow{GE}$.

Student Handbook   H47

Set A (pp. 332–335)

Find the measure of the unknown angle and classify the triangle.

1.

2.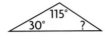

3.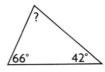

4.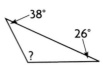

Set B (pp. 338–341)

Give the most exact name for the figure.

1.

2.

3.

4.

Complete the statement, giving the most exact name for the figure.

5. If a rhombus has four right angles, then it is a __?__ .

6. If a parallelogram has four congruent sides, then it is a __?__ .

7. If a polygon has four sides, then it is a __?__ .

8. If a parallelogram has four right angles, then it is a __?__ .

Set C (pp. 342–343)

Draw the figure. Use square dot paper or isometric dot paper.

1. a rectangle

2. a trapezoid

3. a right triangle with all sides different lengths

4. a pentagon with no congruent sides

5. a triangle with all sides the same length

6. a rhombus

Set D (pp. 344–345)

Name the given parts of the circle.

1. center

2. chords

3. radii

4. diameters

5. three arcs

6. Name a sector.

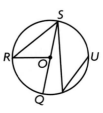

Set A (pp. 350–352)

Name the figure. Is it a polyhedron?

1.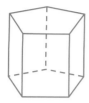

2.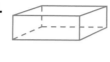

3.

4.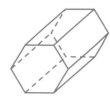

5.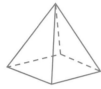

6.

7.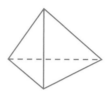

8.

Write *true* or *false* for each statement. Rewrite each false statement as a true statement.

9. In a pentagonal prism, the bases are congruent.

10. A cylinder may have a polygon as its base.

11. All pyramids have rectangular bases.

12. The lateral surface of a cylinder is curved.

13. A prism with square bases may be a cube.

14. An octagonal prism has six faces.

Set B (pp. 353–355)

Name the solid figure that has the given views.

1.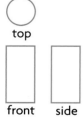

2.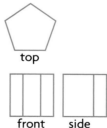

3.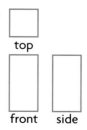

Draw the top, front, and side views for each solid.

4.

5.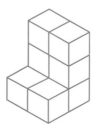

6.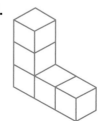

Set A (pp. 364–367)

Use a compass to decide which two line segments in each group are congruent.

1.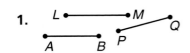

2.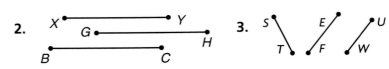

3.

Use a protractor to find the measure of each angle. Then tell whether the angles in each pair are congruent. Write *yes* or *no*.

4.

5.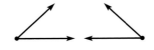

6.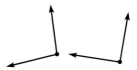

Use a compass and a straightedge to construct a figure congruent to the given figure.

7.

8.

9.

10.

11.

12.

Set B (pp. 368–370)

Trace the figure. Then bisect it.

1.

2.

3.

4.

5.

6.

Set C (pp. 372–373)

Tell whether the figures in each pair appear to be *similar*, *congruent*, *both*, or *neither*.

1.

2.

3.

4.

5.

6.

7.

8.

Set A (pp. 384–386)

Write each ratio in fraction form. Then find the unit rate.

1. $2.50 for 10

2. $2.76 for 12

3. 360 people in 3 sq mi

4. 240 miles in 6 hr

5. $2.75 for 25

6. 500 miles on 20 gallons

Set B (pp. 390–393)

The figures are similar. Write the ratio of the corresponding sides in simplest form.

1.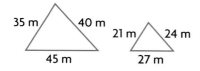

2.

Tell whether the figures in each pair are similar. Write *yes* or *no*. If you write *no*, explain.

3.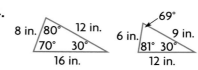

4.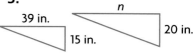

Set C (pp. 394–396)

The figures are similar. Use a proportion to find the unknown length.

1.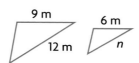

2.

3.

Set D (pp. 397–399)

Find the unknown dimension.

1. scale: 1 in.:5 ft
drawing length: 7 in.
actual length: ■ ft

2. scale: 1 in.:5 ft
drawing length: 4 in.
actual length: ■ ft

3. scale: 1 in.:5 ft
drawing length: ■ in.
actual length: 15 ft

Set E (pp. 400–401)

The map distance is given. Find the actual distance. The scale is 1 in. = 25 mi.

1. $1\frac{1}{2}$ in.

2. 3 in.

3. $5\frac{1}{2}$ in.

4. 7 in.

Set A (pp. 406–407)

Write the percent that is shaded.

1.
2.
3.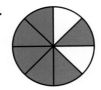
4.

Set B (pp. 408–411)

Write as a percent.

1. 0.3   **2.** 0.09   **3.** 0.43   **4.** $\frac{7}{20}$ **5.** $\frac{3}{8}$

Set C (pp. 412–415)

Find the percent.

1. 20% of 8   **2.** 30% of 90   **3.** 45% of 75   **4.** 50% of 58

5. Of all the cookies Wendy baked, 40% were chocolate chip. If she baked 200 cookies, how many were chocolate chip?

Set D (pp. 418–421)

Find the sale price.

1. regular price: $31.00
25% off

2. regular price: $65.00
50% off

3. regular price: $42.00
75% off

Find the regular price.

4. sale price: $45.00
25% off

5. sale price: $24.95
50% off

6. sale price: $14.40
20% off

Set E (pp. 422–423)

Find the simple interest.

	Principal	Rate	Interest for 1 Year	Interest for 5 Years
1.	$65,000.00	4%	■	■
2.	$735.00	7%	■	■
3.	$1,300.00	3.9%	■	■
4.	$2,250.00	2.3%	■	■

5. Nancy put $2,500 in a savings account for 3 years at a simple interest rate of 8%. How much interest did she earn?

Set A (pp. 428–431)

For 1–5, use the spinner at the right. Find each probability. Write each answer as a fraction, a decimal, and a percent.

1. P(Dee)
2. P(Miles or Lili)
3. P(not Cara)
4. P(Marta)
5. P(Hugo, Miwa, or Chen)

A number cube is labeled 5, 10, 25, 50, 100, and 2,000. Find each probability. Write each answer as a fraction.

6. P(25)
7. P(5 or 25)
8. P(a number ending in zero)
9. P(1,000)
10. P(1,000 or 2,000)
11. P(not 500)

Cards showing pictures of team mascots are placed in a hat. There are 3 lions, 5 bears, 4 cheetahs, and 8 tigers. You choose one card without looking. Find each probability.

12. P(bear)
13. P(tiger or lion)
14. P(member of cat family)

A bag contains some buttons: 8 blue, 12 brown, 10 red, 4 green, and 6 yellow. You choose one button without looking. Find each probability.

15. P(yellow)
16. P(black)
17. P(not brown) –

Set B (pp. 436–437)

A spinner is divided into 5 equal sections. Each section is labeled with one of the letters *A, E, I, O,* and *U*. Anna spins the pointer 100 times and records her results in the table below.

Letter	A	E	I	O	U
Times Landed On	15	5	30	40	10

1. P(*A*)
2. P(*E*)
3. P(*I*)
4. P(*O*)
5. P(*U*)
6. P(*A* or *U*)
7. P(*E, I* or *U*)
8. P(not *A*)

9. Based on her experimental results, how many times can Anna expect the pointer to land on *O* in the next 20 spins?

10. Based on her experimental results, how many times can Anna expect the pointer to land on *E* if she spins 2,000 times?

Set A (pp. 444–446)

Draw a tree diagram or make a table to find the number of possible outcomes for each situation.

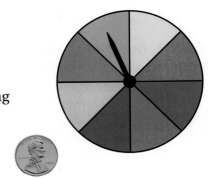

1. choosing vanilla, chocolate, or strawberry yogurt, with cherry or chocolate sauce, and sprinkles or nuts

2. tossing a penny and spinning the pointer on the spinner

Use the Fundamental Counting Principle to find the number of outcomes for each situation.

3. a choice of pancakes, french toast, or waffles, and juice, milk, or tea

4. a choice of 6 salads and 10 dressings

5. rolling 2 number cubes labeled *A* to *F*

Set B (pp. 447–449)

Write *independent* or *dependent* to describe the events.

1. You have a bag of 6 red marbles and 4 green marbles. You draw one marble, record the color, place the marble back in the bag, and draw again.

2. Ana draws one name from a box to select the winner of a movie pass and then draws another name from the same box for the winner of a CD.

A box contains five cards labeled C, L, A, S, S. Without looking in the box, you select a card, replace it, and then select again. For 3–8, find the probability of each event. Then find the probability assuming the first card is not replaced.

| C | L | A | S | S |

3. P(C, L)

4. P(L, C)

5. P(A, S)

6. P(S, L)

7. P(C, L or S)

8. P(C or L, S)

Set C (pp. 450–451)

Seventy-five students from Park Middle School were randomly surveyed about favorite pizza toppings. The results are shown in the table.

1. Suppose there are 225 students who attend Park Middle School. Predict the number of students who prefer black olives as a pizza topping.

2. Suppose there are 314 students at Park Middle School. Predict the number of students who prefer sausage as a pizza topping.

FAVORITE PIZZA TOPPINGS	
Topping	**Number of Students**
Pepperoni	30
Black Olives	15
Sausage	12
Mushrooms	10
Other	8

Set A (pp. 462–463)

Use a proportion to convert to the given unit.

1. 15 ft = m yd
2. 12 c = r pt
3. 8 gal = h qt
4. 120 oz = x lb
5. 6 days = b hr
6. 12 in. = p yd

Set B (pp. 464–465)

Use a proportion to convert to the given unit.

1. 0.22 m = c km
2. 450 mm = h cm
3. 0.0030 kL = p L
4. 1,800 g = ■ kg
5. 10,000 cm = ■ m
6. 0.35 L = ■ mL

7. Paul buys 2 L of orange juice. He drinks 250 mL with breakfast. How many milliliters are left?

8. Tiffany runs 5,000 m. Ashley runs 3.5 km. Who runs farther and by how many meters?

Set C (pp. 466–467)

Use a proportion to convert to the given unit. Use the table on page 456. Round to the nearest hundredth if necessary.

1. 9 in. ≈ ■ cm
2. 11 yd ≈ ■ m
3. 3 mi ≈ ■ km
4. 10 L ≈ ■ gal
5. 55 cm ≈ ■ ft
6. 3.5 kg ≈ ■ lb

7. Eve weighs 51 kg. Beth weighs 116 lb. Who weighs more and by about how many pounds?

8. Zack has a board that is 6 ft long. He wants to cut a length that is 150 cm. About how many centimeters will be left?

Set D (pp. 468–471)

Measure the line segment to the given length.

1. nearest half inch; nearest inch

2. nearest centimeter; nearest millimeter

Tell which measurement is more precise.

3. 7 fl oz or 1 cup
4. 2 qt or 9 c
5. 1,245 m or 1 km

Name an appropriate customary or metric unit of measure for each item.

6. weight or mass of a bag of sugar
7. length of a shoelace
8. water in a fishbowl

Set A (pp. 479–481)

Find the perimeter.

1.
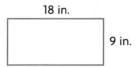
18 in.
9 in.

2.
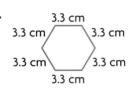
3.3 cm
3.3 cm
3.3 cm
3.3 cm
3.3 cm
3.3 cm

3.
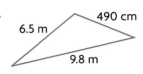
490 cm
6.5 m
9.8 m

The perimeter is given. Find the unknown length.

4.
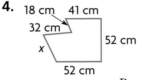
18 cm 41 cm
32 cm
52 cm
x
52 cm
$P = 230$ cm

5.
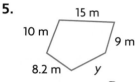
15 m
10 m
9 m
8.2 m y
$P = 54.5$ m

6.
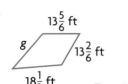
$13\frac{5}{6}$ ft
g
$13\frac{2}{6}$ ft
$18\frac{1}{2}$ ft
$P = 54$ ft

7. Tanya wants to put a string of lights around a rectangular window that is 40 in. wide and 48 in. high. How long will the string of lights need to be to go around the window one time?

8. Linda is building a raised flowerbed $10\frac{1}{2}$ ft long and 4 ft wide. How many feet of lumber will she need to go around the flowerbed?

Set B (pp. 484–487)

Find the circumference. Use 3.14 or $\frac{22}{7}$ for π. Round to the nearest whole number.

1.

7 in.

2.

12 ft

3.

49 cm

4.

$4\frac{1}{2}$ in.

5.

2.7 cm

6.

154 ft

7. $r = 16$ ft

8. $d = 23.5$ cm

9. $d = 28$ ft

10. $r = 25.9$ m

11. $d = 82.1$ mm

12. $d = 3.21$ m

13. A blue circular rug in Dominique's room has a diameter of $6\frac{1}{2}$ ft. To the nearest foot, what is the circumference of the rug?

14. A child's hat has a radius of 8 cm. To the nearest tenth of a centimeter, what is the circumference of the hat?

Set A (pp. 494–497)

Estimate the area. Each square on the grid represents 1 cm².

1.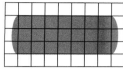

2.

3.

Find the area.

4.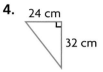

5.

6.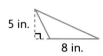

7.

Set B (pp. 498–500)

Find the area of each figure.

1.

2.

3.

4.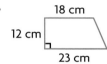

5. A wall plaque is shaped like a trapezoid with a height of 6 in. and bases that measure 3.5 in. and 9.5 in. Find the area.

Set C (pp. 502–503)

Find the area of each circle to the nearest whole number.

1.

2.

3.

4.

5. $d = 24$ cm

6. $r = 9.2$ mm

7. $d = 8$ yd

8. $d = 7\frac{1}{2}$ ft

Set D (pp. 504–507)

Find the surface area.

1.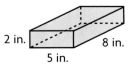

2.

3.

4.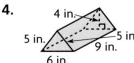

Set A (pp. 512–515)

Find the volume.

1.
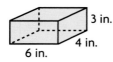
3 in.
4 in.
6 in.

2.
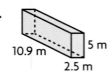
5 m
10.9 m
2.5 m

3
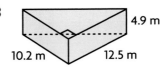
4.9 m
10.2 m
12.5 m

Find the unknown length.

4.
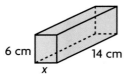
6 cm
14 cm
x
$V = 420 \text{ cm}^3$

5.
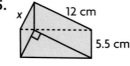
x
12 cm
5.5 cm
$V = 247.5 \text{ cm}^3$

6.

17 in.
x
6 in.
$V = 153 \text{ in.}^3$

Set B (pp. 518–519)

Find the volume.

1.

17 ft
8 ft
8 ft

2.

9.5 m
5 m
7 m

3.
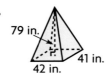
79 in.
41 in.
42 in.

4. Find the volume of a rectangular pyramid with a length of 18 cm, a width of 16 cm, and a height of 12 cm.

5. Find the volume of a rectangular pyramid with a base of 300 ft² and a height of 50 ft.

Set C (pp. 521–523)

Find the volume. Round to the nearest whole number.

1.

17 ft
5 ft

2.
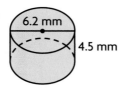
6.2 mm
4.5 mm

3.
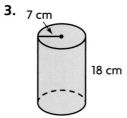
7 cm
18 cm

Find the volume of the inside cylinder to the nearest whole number.

4.
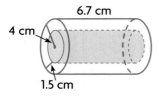
6.7 cm
4 cm
1.5 cm

5.
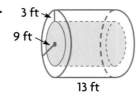
3 ft
9 ft
13 ft

6.
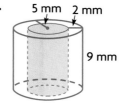
5 mm 2 mm
9 mm

Set A (pp. 537–538)

Write a rule for each sequence.

1. 2, 6, 18, 54, . . .

2. $\frac{1}{10}, \frac{1}{5}, \frac{3}{10}, \frac{2}{5},$ . . .

3. ⁻35, ⁻20, ⁻5, 10, . . .

4. 440, 44, 4.4, 0.44, . . .

Find the next three possible terms in each sequence.

5. $\frac{2}{5}, \frac{4}{5}, 1\frac{1}{5},$ . . .

6. 2.3, 3.9, 5.5, . . .

7. 7, 17, 37, 67, . . .

8. 27, ⁻9, 3, ⁻1, . . .

9. Vittorio practices 6 weeks for the skateboard championship. The first 4 weeks he practices $9\frac{1}{2}$ hr, $11\frac{1}{4}$ hr, 13 hr, and $14\frac{3}{4}$ hr. Following this pattern, how many hours will Vittorio practice the sixth week?

Set B (pp. 539–542)

Write an equation to represent the function.

1.

x	0	1	2	3	4
y	3	4	5	6	7

2.

x	10	9	8	7	6
y	6	5	4	3	2

3.

x	0	1	2	3	4
y	0	8	16	24	32

4.

x	30	27	24	21	18
y	10	9	8	7	6

Set C (pp. 543–545)

Draw the next two figures in the pattern.

1.

2.

Draw the next two solids in the pattern.

3.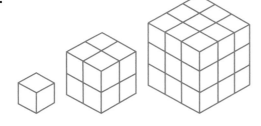

4.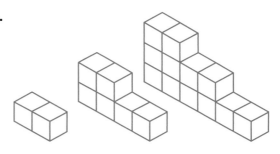

Set A (pp. 550–552)

Tell which type or types of transformation the second figure is of the first figure. Write *translation*, *rotation*, or *reflection*.

1.
2.
3.
4.

Set B (pp. 553–555)

Trace and cut out several of each shape. Tell whether the shape can be used repeatedly to form a tessellation. Write *yes* or *no*.

1.
2.
3.
4.

Set C (pp: 558–559)

Tell how many ways you can place the solid figure on the outline.

1.
2.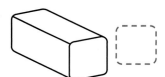
3.

Set D (pp. 560–563)

Trace the figure. Draw the lines of symmetry.

1.
2.
3.
4.

Tell whether each figure has rotational symmetry, and, if so, identify the symmetry as a fraction of a turn and in degrees.

5.
6.
7.
8.

Set A (pp. 568–569)

Graph the solutions of the inequality.

1. $x < 4$ **2.** $x > 9$ **3.** $x \geq 3$ **4.** $x < {}^-5$

Solve the inequality and graph the solutions on a number line.

5. $7a \leq 14$ **6.** $n + 5 > 8$ **7.** $m - 2 < 0$ **8.** $5c \geq 5$

Set B (pp. 570–573)

Write the ordered pair for each point on the coordinate plane.

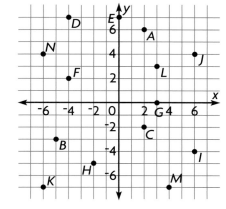

1. point A **2.** point B

3. point C **4.** point D

5. point E **6.** point F

7. point G **8.** point H

9. point I **10.** point J

Set C (pp. 574–575)

Complete the table. Then write an equation relating *y* to *x*.

1.

x	1	2	3	4	5
y	3	4	5	■	■

2.

x	1	2	3	4	5
y	6	12	18	■	■

3.

x	8	9	10	11	12
y	1	2	3	■	■

4.

x	8	4	0	⁻4	⁻8
y	⁻2	⁻1	0	■	■

Set D (pp. 580–583)

Copy the figure onto a coordinate plane. Transform the figure according to the directions given. Name the new coordinates.

1.

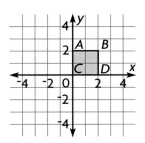

2 units to the right

2.

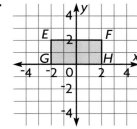

rotate 90° clockwise about (⁻2,0)

3.

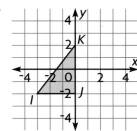

reflect across the *y*-axis

TIPS FOR TAKING MATH TESTS

Being a good test-taker is like being a good problem solver. When you answer test questions, you are solving problems. Remember to **ANALYZE**, **CHOOSE**, **SOLVE**, and **CHECK**.

Analyze
Choose
Solve
Check

Analyze

Read the problem.

- Look for math terms and recall their meanings.
- Reread the problem and think about the question.
- Use the details in the problem and the question.

1. The difference between two numbers is 37. Their sum is 215. What are the numbers?

   A 37 and 178   **C** 89 and 126

   B 83 and 120   **D** 107 and 108

TIP! Understand the problem. The problem requires you to find two numbers for which the difference and sum are given. Reread the problem to compare the details to the answer choices. You can use estimation instead of calculating the sum and difference of each pair of numbers. The answer is **C**.

- Each word is important. Missing a word or reading it incorrectly could cause you to get the wrong answer.
- Pay attention to words that are in **bold** type, all CAPITAL letters, or *italics* and words like *round, best,* or *least to greatest.*

2. Florinda took $\frac{1}{2}$ hour to complete a race. Doris took $\frac{3}{8}$ hour and Violet took $\frac{2}{3}$ hour to complete the same race. Which lists the three runners from fastest to slowest?

   F Florinda, Doris, Violet

   G Doris, Violet, Florinda

   H Violet, Florinda, Doris

   J Doris, Florinda, Violet

TIP! Look for important words. The words *fastest to slowest* are important. The fastest runner takes the least amount of time. Think about where each fraction would be placed on a number line between 0 and 1. Then put the times in order from least to greatest. The answer is **J**.

Think about how you can solve the problem.

- See if you can solve the problem with the information given.
- Pictures, charts, tables, and graphs may have the information you need.
- You may need to recall information not given.
- Sometimes the answer choices have information to help solve the problem.

3. Jeremy wants to make a graph to see the trend of the profit in his lawn-mowing business over the past 10 months. What type of graph would best show this?

A circle graph   **C** line graph

B histogram   **D** stem-and-leaf plot

 Get the information you need.

The answer choices give four different types of graphs or plots. Think about each one and the kind of data that is appropriate for it. The problem states that Jeremy wants to see the trend in his profit over the past 10 months, and line graphs show trends. The answer is **C**.

- You may need to write a number sentence and solve it to answer the question.
- Some problems have two steps or more.
- In some problems you need to look at relationships instead of computing an answer.
- If the path to the solution isn't clear, choose a problem solving strategy and use it to solve the problem.

4. A square tile with 4-inch sides is rotated along the line below. If the tile stopped so that the letter **P** appears in an upright position, which of these distances could it have been rotated?

P

F 40 in.   **H** 48 in.

G 44 in.   **J** 52 in.

Decide on a plan.
From the distances given, you must find the one that could allow for complete rotations of the square tile. Since each side of the square is 4 inches long, it moves 16 inches in one complete rotation. If you *use logical reasoning*, you see that only one of the choices is a multiple of 16. The answer is **H**.

Follow your plan, working logically and carefully.

- Estimate your answer. Compare it to the answer choices.

- Use reasoning to find the most likely choices.

- Make sure you solved all the steps needed to answer the problem.

- If your answer does not match any of the answer choices, check the numbers you used. Then check your computation.

5. Glen and Doug painted a fence. It took Glen three times as long to paint one side of the fence as it took Doug to paint the other side. If h represents the number of hours Doug painted, which expression shows the hours Glen painted?

A $3 \div h$ **C** $3 + h$

B $3 - h$ **D** $3 \times h$

TIP! Eliminate choices.
It is important to understand that h represents the hours it took Doug to paint one side of the fence. Since Glen painted three times as long as Doug, you can eliminate answers **A** and **B** because it does not make sense to divide by or subtract the hours Doug painted. Answer **C** means three more than h, so it can be eliminated. The answer is **D**.

- If your answer still does not match one of the choices, look for another form of the number, such as a decimal instead of a fraction.

- If answer choices are given as pictures, look at each one by itself while you cover the other three.

- If you do not see your answer and the answer choices include Not here, make sure your work is correct and then mark Not here.

- Read answer choices that are statements and relate them to the information in the problem one by one.

6. Tanya and her class helped plant 120 tulip and daffodil bulbs in front of the school. If 30 of the bulbs were tulips, what percent of the bulbs were daffodils?

F 90%   **H** 25%

G $66\frac{2}{3}\%$ **J** Not here

TIP! Choose the answer.
Since 30 of the bulbs were tulips, 90 were daffodils. The question asks what percent are daffodils, so you need to find what percent of 120 is 90 ($90 \div 120$). If your answer doesn't match one of the answer choices, check your computation. If you know your work is correct, mark the letter for Not here. The answer is **J**.

Take time to catch your mistakes.

- Be sure you answered the question asked.
- Check for important words you might have missed.
- Be sure you used all the information you needed.
- Check your computation by using a different method.

7. The temperature was 6°F today. A forecaster predicted that the temperature would drop by 5° each day for the next four days. Which sequence could be used to find the predicted temperatures?

A 6°F, 11°F, 16°F, 21°F, 26°F

B 6°F, 1°F, 6°F, 11°F, 16°F

C 6°F, 1°F, ⁻6°F, ⁻11°F, ⁻16°F

D 6°F, 1°F, ⁻4°F, ⁻9°F, ⁻14°F

TIP! **Check your work.**
You need to find the sequence that shows a drop of 5° in temperature for the next four days. Draw a thermometer or a number line to check your computation. The answer is **D**.

Don't Forget!

Before the test...

- Listen to the teacher's directions and read the instructions.
- Write down the ending time if the test is timed.
- Know where and how to mark your answers.
- Know whether you should write on the test page or use scratch paper.
- Ask any questions you have before the test begins.

During the test...

- Work quickly but carefully. If you are unsure how to answer a question, leave it blank and return to it later.
- If you cannot finish on time, look over the questions that are left. Answer the easiest ones first. Then go back to answer the others.
- Fill in each answer space carefully. Erase completely if you change an answer. Erase any stray marks.
- Check that the answer number matches the question number, especially if you skip a question.

Adding and Subtracting

1. 70
    +8

2. 43
    +5

3. 58
    +7

4. 67
    +18

5. 24
    +36

6. 264
    +58

7. 836
    +54

8. 1,641
    +385

9. 3,231
    +578

10. 4,605
    +2,493

11. 5,619
    +2,537

12. 84,520
    +3,864

13. 274, 051
    +40,318

14. 68
    −5

15. 74
    −43

16. 50
    −38

17. 149
    −58

18. 394
    −145

19. 560
    −217

20. 750
    −192

21. 1,460
    − 316

22. 4,960
    − 681

23. 8,000
    − 562

24. 35,120
    −6,299

25. 36,009
    −19,148

26. 0.4
    +0.9

27. 2.6
    +8.1

28. 0.31
    +2.04

29. 5.92
    +3.15

30. 0.218
    +2.143

31. 2.14
    +6.08

32. 7.04
    +0.13

33. 0.126
    +0.408

34. 8.360
    +5.216

35. 5.816
    +3.215

36. 17.31
    +12.06

37. 8.26
    +26.15

38. 31.18
    +125.50

39. 11.4
    −6.2

40. 12.8
    −4.1

41. 17.3
    −16.5

42. 21.8
    −17.4

43. 13.2
    −8.17

44. 35.51
    −23.17

45. 8.026
    −7.317

46. 19.408
    −1.582

47. 5.848
    −5.261

48. 13.601
    −10.311

49. 5
    −3.21

50. 630.71
    −527.21

Multiplying

1. 20
   ×4

2. 30
   ×2

3. 30
   ×3

4. 24
   ×2

5. 31
   ×3

6. 46
   ×2

7. 37
   ×4

8. 74
   ×6

9. 63
   ×8

10. 58
    ×9

11. 61
    ×20

12. 84
    ×40

13. 76
    ×30

14. 68
    ×50

15. 92
    ×60

16. 26
    ×14

17. 93
    ×15

18. 50
    ×38

19. 81
    ×54

20. 79
    ×43

21. 680
    ×35

22. 710
    ×52

23. 825
    ×173

24. 522
    ×286

25. 463
    ×836

26. 0.5
    ×0.9

27. 3.6
    ×7.1

28. 1.3
    ×2.5

29. 5.9
    ×3.1

30. 4.2
    ×8.7

31. 3.14
    ×6.8

32. 7.24
    ×0.8

33. 9.65
    ×0.4

34. 7.25
    ×1.8

35. 9.97
    ×0.9

36. 17.31
    ×2.06

37. 8.26
    ×0.87

38. 31.82
    ×15.5

39. 15.4
    ×6.27

40. 12.8
    ×4.16

41. 17.3
    ×16.5

42. 165.4
    ×7.4

43. 139.2
    ×8.17

44. 935.5
    ×3.25

45. 987.2
    ×7.8

46. 246.28
    ×25.5

47. 950.8
    ×52.4

48. 825.2
    ×9.63

49. 440.7
    ×425.2

50. 2,145.2
    ×527.5

1. $8\overline{)72}$
2. $27\overline{)108}$
3. $6\overline{)138}$
4. $4\overline{)816}$
5. $5\overline{)525}$

6. $6\overline{)624}$
7. $7\overline{)763}$
8. $4\overline{)836}$
9. $5\overline{)205}$
10. $2\overline{)164}$

11. $7\overline{)642}$
12. $9\overline{)700}$
13. $2\overline{)785}$
14. $5\overline{)277}$
15. $3\overline{)343}$

16. $90 \div 30$
17. $40 \div 20$
18. $160 \div 40$
19. $360 \div 40$
20. $540 \div 90$

21. $630 \div 70$
22. $3{,}000 \div 60$
23. $100 \div 20$
24. $560 \div 70$
25. $3{,}500 \div 50$

26. $630 \div 58$
27. $4{,}801 \div 37$
28. $100 \div 21$
29. $560 \div 82$
30. $1{,}875 \div 19$

31. $900 \div 300$
32. $480 \div 240$
33. $840 \div 105$
34. $1{,}500 \div 300$
35. $9{,}800 \div 800$

36. $1.2 \div 4$
37. $0.12 \div 4$
38. $3.5 \div 5$
39. $6.4 \div 8$
40. $0.18 \div 9$

41. $3.69 \div 3$
42. $83.7 \div 9$
43. $44.8 \div 4$
44. $56.8 \div 8$
45. $19.75 \div 5$

46. $2.24 \div 4$
47. $4.48 \div 2.8$
48. $3.78 \div 3$
49. $12.1 \div 1.1$
50. $229.6 \div 8.2$

51. $0.38\overline{)13.3}$
52. $0.55\overline{)2.42}$
53. $2.48\overline{)1.3392}$
54. $6.41\overline{)135.892}$
55. $15\overline{)10.8}$

56. $9\overline{)43.65}$
57. $18.2\overline{)378.56}$
58. $49.3\overline{)201.144}$
59. $186.24 \div 29.1$
60. $378.56 \div 18.2$

Pronunciation Key

a	add, map	f	fit, half	n	nice, tin	p	pit, stop	yōō	fuse, few
ā	ace, rate	g	go, log	ng	ring, song	r	run, poor	v	vain, eve
â(r)	care, air	h	hope, hate	o	odd, hot	s	see, pass	w	win, away
ä	palm, father	i	it, give	ō	open, so	sh	sure, rush	y	yet, yearn
b	bat, rub	ī	ice, write	ô	order, jaw	t	talk, sit	z	zest, muse
ch	check, catch	j	joy, ledge	oi	oil, boy	th	thin, both	zh	vision,
d	dog, rod	k	cool, take	ou	pout, now	th	this, bathe		pleasure
e	end, pet	l	look, rule	ŏŏ	took, full	u	up, done		
ē	equal, tree	m	move, seem	ōō	pool, food	û(r)	burn, term		

ə   the schwa, an unstressed vowel representing the sound spelled *a* in above, *e* in sicken, *i* in possible, *o* in melon, *u* in circus

Other symbols:
• separates words into syllables
′ indicates stress on a syllable

A

absolute value [ab′sə•lōōt val′yōō] The distance of an integer from zero (p. 228)

acute angle [ə•kyōōt′ an′gəl] an angle whose measure is greater than 0° and less than 90° (p. 320)

acute triangle [ə•kyōōt′ trī′an•gəl] A triangle with all angles less than 90° (p. 332)
Example:

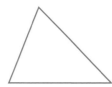

Addition Property of Equality [ə•dish′ən prä′pər•tē əv i•kwol′ə•tē] The property that states that if you add the same number to both sides of an equation, the sides remain equal (p. 290)

additive inverse [ad′ə•tiv in′vûrs] The opposite of a given number (p. 243)

adjacent angles [ə•jā′sənt an′gəlz] Angles that are side by side and have a common vertex and ray (p. 322)
Example:

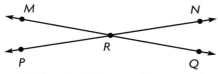

∠MRN and ∠NRQ are adjacent angles.

algebraic expression [al•jə•brā′ik ik•spre′shən] An expression that includes at least one variable (p. 28)
Examples: x + 5, 3*a* − 4

algebraic operating system [al•jə•brā′ik ä′pə•rā•ting sis′təm] A way for calculators to follow the order of operations when evaluating expressions (p. 43)

angle [an′gəl] A figure formed by two rays with a common endpoint (p. 320)
Example:

arc [ärk] A part of a circle, named by its endpoints (p. 344)
Example:

arc *AB* or $\overset{\frown}{AB}$

area [âr′ē•ə] The number of square units needed to cover a given surface (p. 494)

Associative Property [ə•sō′shē•ā•tiv prä′pər•tē] The property that states that the way addends are grouped or factors are grouped does not change the sum or the product (p. 36)
Examples: 12 + (5 + 9) = (12 + 5) + 9
(9 × 8) × 3 = 9 × (8 × 3)

axes [ak′sēz] The horizontal number line (*x*-axis) and the vertical number line (*y*-axis) on the coordinate plane (p. 570)

bar graph [bär′graf] A graph that displays countable data with horizontal or vertical bars (p. 120)

base [bās] A number used as a repeated factor (p. 40)
Example: $8^3 = 8 \times 8 \times 8$; 8 is the base.

base [bās] A side of a polygon or a face of a solid figure by which the figure is measured or named (pp. 350, 495)
Examples:

biased question [bī′əst kwes′chən] A question that leads to a specific response or excludes a certain group (p. 98)

biased sample [bī′əst sam′pəl] A sample is biased if individuals or groups from the population are not represented in the sample. (p. 98)

bisect [bī•sekt′] To divide into two congruent parts (p. 368)

box-and-whisker graph [bäks•ənd•hwis′kər graf] A graph that shows how far apart and how evenly data are distributed (p. 129)

Celsius [səl′sē•əs] A metric scale for measuring temperature (p. 301)

certain [sûr′tən] Sure to happen (p. 429)

chord [kôrd] A line segment with its endpoints on a circle (p. 344)
Example:

chord: $\overline{AB}$

circle [sûr′kəl] A closed plane figure with all points of the figure the same distance from the center (p. 344)
Example:

circle graph [sûr′kəl graf] A graph that lets you compare parts to the whole and to other parts (p. 122)
Example:

FAVORITE HOBBIES

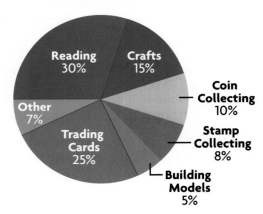

circumference [sûr•kum′fər•əns] The distance around a circle (p. 484)

clustering [klus′tər•ing] A method used to estimate a sum when all addends are about the same (p. 16)

Commutative Property [kə•myoo′tə•tiv prä′pər•tē] The property that states that if the order of addends or factors is changed, the sum or product stays the same (p. 36)
Examples: $6 + 5 + 7 = 5 + 6 + 7$
$8 \times 9 \times 3 = 3 \times 8 \times 9$

compensation [kom•pən•sā′shən] A mental math strategy for some addition and subtraction problems (p.37)

complementary angles [kom•plə•men′tər•ē an′gəlz] Two angles whose measures have a sum of 90° (p. 323)
Example:

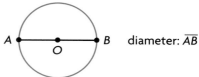

composite number [käm•pä′zət num′bər] A whole number greater than 1 that has more than two whole-number factors (p. 148)

compound event [käm′pound i•vent′] An event made of two or more simple events (p. 444)

congruent [kən•grōō′ənt] Having the same size and shape (p. 390)

convenience sample [kən•vēn′yənts sam′pəl] Sampling the most available subjects in the population to obtain quick results (p. 95)

coordinate plane [kō•ôr′də•nit plān] A plane formed by two perpendicular number lines called axes; every point on the plane can be named by an ordered pair of numbers. (p. 570)

corresponding angles [kôr•ə•spän′ding an′gəlz] Angles that are in the same position in different plane figures (p. 390)
Example:

∠A and ∠D are corresponding angles.

corresponding sides [kôr•ə•spän′ding sīdz] Sides that are in the same position in different plane figures (p. 390)
Example:

 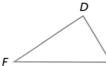

$\overline{CA}$ and $\overline{FD}$ are corresponding sides.

cube [kyōōb] A rectangular solid with six congruent faces (p. 350)
Example:

cumulative frequency [kyōō′myə•lə•tiv frē′kwən•sē] A running total of the number of subjects surveyed (p. 103)

decimal [de′sə•məl] A number with one or more digits to the right of the decimal point (p. 52)

denominator [di•nä′mə•nā•tər] The part of a fraction that tells how many equal parts are in the whole (p. 159)
Example: $\frac{3}{4}$ ←denominator

dependent events [di•pen′dənt i•vənts′] Events for which the outcome of the second event depends on the outcome of the first event (p. 448)

diameter [dī•am′ə•tər] A line segment that passes through the center of a circle and has its endpoints on the circle (p. 344)
Example:

A •————O————• B   diameter: $\overline{AB}$

dimension [di•men′shən] The length, width, or height of a figure (p. 511)

discount [dis′kount] An amount that is subtracted from the regular price of an item (p. 418)

Distributive Property of Multiplication [di•strib′yə•tiv prä′pər•tē əv mul•tə•plə•kā′shən] The property that states that multiplying a sum by a number is the same as multiplying each addend by the number and then adding the products (p. 36)
Example: 14 × 21 = 14 × (20 + 1) = (14 × 20) + (14 × 1)

dividend [di′və•dend] The number that is to be divided in a division problem
Example: In 56 ÷ 8, 56 is the dividend.

Division Property of Equality [di•vi′zhən prä′pər•tē əv i•kwol′ə•tē] The property that states that if you divide both sides of an equation by the same nonzero number, the sides remain equal (p. 297)

divisor [di•vī′zər] The number that divides the dividend
Example: In 45 ÷ 9, 9 is the divisor.

equally likely [ē′kwə•lē li′klē] Having the same chance of occurring (p. 428)

equation [i·kwā′zhən] A statement that shows that two quantities are equal (p. 30)

equilateral triangle [ē·kwə·la′tə·rəl trī′an·gəl] A triangle with three congruent sides (p. 332)
Example:

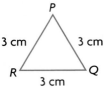

equivalent fractions [ē·kwiv′ə·lənt frak′shənz] Fractions that name the same amount or part (p. 160)

equivalent ratios [ē·kwiv′ə·lənt rā′shē·ōz] Ratios that name the same comparisons (p. 384)

estimate [es′tə·mit] An answer that is close to the exact answer and that is found by rounding, by clustering, or by using compatible numbers (p. 16)

evaluate [i·val′yoo·āt] Find the value of a numerical or algebraic expression (p. 28)

event [i·vent′] A set of outcomes (p. 427)

experimental probability [ik·sper·ə·men′təl prä·bə·bil′ə·tē] The ratio of the number of times an event occurs to the total number of trials or times the activity is performed (p. 436)

exponent [ik·spō′nənt] A number that tells how many times a base is used as a factor (p. 40)
Example: $2^3 = 2 \times 2 \times 2 = 8$;
3 is the exponent.

face [fās] One of the polygons of a solid figure (p. 493)

Example:

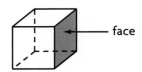

factor [fak′tər] A number that is multiplied by another number to find a product

Fahrenheit [fâr′ən·hīt] A customary scale for measuring temperature (p. 301)

formula [fôr′myə·lə] A rule that is expressed with symbols (p. 300)
Example: $A = lw$

fractal [frak′təl] A figure with repeating patterns containing shapes that are like the whole but of different sizes throughout (p. 543)

frequency table [frē′kwən·sē tā′bəl] A table representing totals for individual categories or groups (p. 103)

function [funk′shən] A relationship between two quantities in which one quantity depends on the other (p. 539)

Fundamental Counting Principle [fun·də·men′təl koun′ting prin′sə·pəl] If one event has m possible outcomes and a second independent event has n possible outcomes, then there are $m \times n$ total possible outcomes. (p. 444)

greatest common factor (GCF) [grā′təst kä′mən fak′tər] The greatest factor that two or more numbers have in common (p. 151)

height [hīt] A measure of a polygon or solid figure, taken as the length of a perpendicular from the base of the figure (p. 495)
Example:

hexagon [heks′ə·gon] A six-sided polygon

histogram [his′tə·gram] A bar graph that shows the number of times data occur in certain ranges or intervals (p. 127)

hypotenuse [hī·pot′ə·n(y)oos′] In a right triangle, the side opposite the right angle (p. 488

Example:

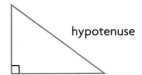

Identity Property of Zero [i·den′tə·tē prä′pər·tē əv zir′ō] The property that states that the sum of zero and any number is that number (p. H2)
Example: $25 + 0 = 25$

Identity Property of One [ĭ·den'tə·tē prä'pər·tē əv wun] The property that states that the product of any number and 1 is that number (p. H2)
Example: 12 × 1 = 12

impossible [im·pos'ə·bəl] Never able to happen (p. 429)

independent events [in·di·pen'dənt i·vents'] Events for which the outcome of the second event does not depend on the outcome of the first event (p. 447)

indirect measurement [in·di·rekt' mezh'ər·mənt] The technique of using similar figures and proportions to find a measure (p. 394)

inequality [in·i·kwäl'ə·tē] An algebraic or numerical sentence that contains the symbol <, >, ≤, ≥, or ≠ (p. 568)
Example: x + 3 > 5

integers [in'ti·jərz] The set of whole numbers and their opposites (p. 228)

isosceles triangle [ī·sä'sə·lēz tri'an·gəl] A triangle with exactly two congruent sides (p. 331)
Example:

7 in. 7 in.
5 in.

lateral faces [lat'ər·əl fās'əz] The faces in a prism or pyramid that are not bases (p. 350)

least common denominator (LCD) [lēst kä'mən di·nä'mə·nā·tər] The least common multiple of two or more denominators (p. 182)

least common multiple (LCM) [lēst kä'mən mul'tə·pəl] The smallest number, other than zero, that is a common multiple of two or more numbers (p. 150)

leg [leg] In a right triangle, either of the two sides that form the right angle (p. 488)

Example:

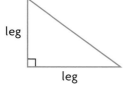
leg
leg

like terms [līk turmz] Expressions that have the same variable with the same exponent (p. 273)

line [līn] A straight path that extends without end in opposite directions (p. 318)
Example: ⟵——————⟶

line graph [līn graf] A graph that uses a line to show how data change over time (p. 121)

line of symmetry [līn əv si'mə·trē] A line that divides a figure into two congruent parts (p. 560)

line plot [līn plät] A graph that shows frequency of data along a number line (p. 121)
Example:

X
X   X
X   X   X   X   X
+—+—+—+—+—+—+—+
1   2   3   4   5   6   7
Miles Jogged

line segment [līn seg'mənt] A part of a line with two endpoints (p. 318)
Example: •——————•

line symmetry [līn si'mə·trē] A figure has line symmetry if a line can separate the figure into two congruent parts. (p. 560)

lower extreme [lō'ər ik·strēm'] The least number in a set of data (p. 129)

lower quartile [lō'ər kwôr'tīl] The median of the lower half of a set of data (p. 129)

mean [mēn] The sum of a group of numbers divided by the number of addends (p. 106)

median [mē'dē·ən] The middle value in a group of numbers arranged in order (p. 106)

midpoint [mid'point] The point that divides a line segment into two congruent line segments (p. 368)

mixed number [mikst num'bər] A number represented by a whole number and a fraction (p. 164)

mode [mōd] The number or item that occurs most often in a set of data (p. 106)

multiple-bar graph [mul′tə•pəl bär′graf] A bar graph that represents two or more sets of data (p. 120)

multiple-line graph [mul′tə•pəl lin′graf] A line graph that represents two or more sets of data (p. 121)

multiple [mul′tə•pəl] The product of a given whole number and another whole number (p. 150)

Multiplication Property of Equality [mul•tə•plə•kā′shən prä′pər•tē əv i•kwol′ə•tē] The property that states that if you multiply both sides of an equation by the same number, the sides remain equal (p. 298)

negative integers [ne′gə•tiv in′ti•jərz] Integers to the left of zero on the number line (p. 228)

net [net] An arrangement of two-dimensional figures that folds to form a polyhedron (p. 356)
Example:

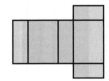

numerator [noo′mə•rā•tər] The part of a fraction that tells how many parts are being used (p. 159)
Example: $\frac{3}{4}$ ←numerator

numerical expression [noo•mâr′i•kəl ik•spre′shən] A mathematical phrase that uses only numbers and operation symbols (p. 28)

obtuse angle [äb•toos′ an′gəl] An angle whose measure is greater than 90° and less than 180° (p. 320)
Example:

obtuse triangle [äb•toos′ tri′an•gəl] A triangle with one angle greater than 90° (p. 332)
Example:

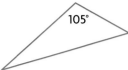

opposites [ä′pə•zəts] Two numbers that are an equal distance from zero on the number line (p. 228)

order of operations [ôr′dər əv ä•pə•rā′shənz] The process for evaluating expressions: first perform the operations in parentheses, clear the exponents, perform all multiplication and division, and then perform all addition and subtraction (p. 42)

ordered pair [ôr′dərd pâr] A pair of numbers that can be used to locate a point on the coordinate plane (p. 570)
Examples: (0,2), (3,4), (⁻4,5)

origin [ôr′ə•jən] The point where the *x*-axis and the *y*-axis in the coordinate plane intersect, (0,0) (p. 570)

outcome [out′kəm] A possible result of a probability experiment (p. 428)

outlier [out′li•ər] A data value that stands out from others in a set; outliers can significantly affect measures of central tendency. (p. 110)

overestimate [ō•vər•es′tə•mət] An estimate that is greater than the exact answer (p. 17)

parallel lines [pâr′ə•lel linz] Lines in a plane that are always the same distance apart (p. 326)
Example:

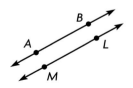

percent (%) [pər•sent′] The ratio of a number to 100; *percent* means "per hundred." (p. 60)

perimeter [pə•ri′mə•tər] The distance around a figure (p. 477)

perpendicular lines [pər•pen•dik′yə•lər linz] Two lines that intersect to form right, or 90°, angles (p. 326)
Example:

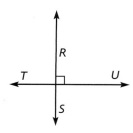

pi (π) [pī] The ratio of the circumference of a circle to its diameter; π ≈ 3.14 or $\frac{22}{7}$ (p. 484)

plane [plān] A flat surface that extends without end in all directions (p. 318)

point [point] An exact location in space, usually represented by a dot (p. 318)

point of rotation [point əv rō•tā′shən] The central point around which a figure is rotated (p. 561)

polygon [pä′lē•gän] A closed plane figure formed by three or more line segments (p. 331)

polyhedron [pä•lē•hē′drən] A solid figure with flat faces that are polygons (p. 350)
Example:

Hexagonal Prism

population [pä•pyə•lā′shən] The entire group of objects or individuals considered for a survey (p. 94)

positive integers [pä′zə•tiv in′ti•jərz] Integers to the right of zero on the number line (p. 228)

prime factorization [prīm fak•tə•ri•zā′shən] A number written as the product of all of its prime factors (p. 148)
Example: 24 = 2³ × 3

prime number [prīm num′bər] A whole number greater than 1 whose only factors are 1 and itself (p. 148)

principal [prin′sə•pəl] The amount of money borrowed or saved (p. 422)

prism [priz′əm] A solid figure that has two congruent, polygon-shaped bases, and other faces that are all rectangles (p. 350)
Example:

probability [prä•bə•bil′ə•tē] See *theoretical probability* and *experimental probability*

product [prä′dəkt] The answer in a multiplication problem (p. 15)

Property of Zero [prä′pər•tē əv zē′rō] The property that states that the product of any number and zero is zero (p. 35)

proportion [prə•pôr′shən] An equation that shows that two ratios are equal (p. 387)
Example: $\frac{1}{3} = \frac{3}{9}$

pyramid [pir′ə•mid] A solid figure with a polygon base and triangular sides that all meet at a common vertex (p. 351)
Example:

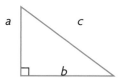

Pythagorean Theorem [pə•thag•ə•rē′ənthē′ə•rem] In any right triangle, if *a* and *b* are the lengths of the legs and *c* is the length of the hypotenuse, then $a^2 + b^2 = c^2$ (p. 488)

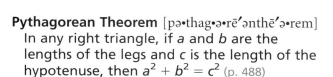

quadrants [kwäd′rənts] The four regions of the coordinate plane (p. 570)

quadrilateral [kwä•drə•lat′ə•rəl] A polygon with four sides and four angles (p. 331)

quotient [kwō′shənt] The number, not including the remainder, that results from dividing (p. 23)

radius [rā′dē•əs] A line segment with one endpoint at the center of a circle and the other endpoint on the circle (p. 344)
Example:

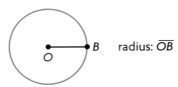

radius: $\overline{OB}$

random sample [ran′dəm sam′pəl] A sample in which each subject in the overall population has an equal chance of being selected (p. 95)

range [rānj] The difference between the greatest and least numbers in a group (p. 103)

rate [rāt] A ratio that compares two quantities having different units of measure (p. 385)

ratio [rā′shē•ō] A comparison of two numbers, *a* and *b*, written as a fraction $\frac{a}{b}$ (p. 230)

rational number [ra′shə•nəl num′bər] Any number that can be written as a ratio $\frac{a}{b}$, where *a* and *b* are integers and $b \neq 0$ (p. 230)

ray [rā] A part of a line with a single endpoint (p. 318)
Example:

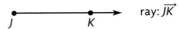

ray: $\overrightarrow{JK}$

reciprocal [ri•sip′rə•kəl] Two numbers are reciprocals of each other if their product equals 1. (p. 209)

reflection [ri•flek′shən] A movement of a figure by flipping it over a line (p. 550)

regular polygon [reg′yə•lər pä′lē•gän] A polygon in which all sides are congruent and all angles are congruent (p. 336)
Example:

repeating decimal [ri•pēt′ing de′sə•məl] A decimal that doesn't end, because it shows a repeating pattern of digits after the decimal point (p. 169)

right angle [rīt an′gəl] An angle that has a measure of 90° (p. 320)
Example:

right triangle [rīt trī′an•gəl] A triangle with one right angle (p. 332)
Example:

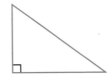

rotation [rō•tā′shən] A movement of a figure by turning it around a fixed point (p. 550)

rotational symmetry [rō•tā′shən•əl si′mə•trē] The property of a figure that can be rotated less than 360° around a central point and still be congruent to the original figure (p. 561)

sales tax [sālz taks] A percent of the cost of an item, added onto the item's cost (p. 420)

sample [sam′pəl] A part of a population (p. 94)

sample space [sam′pəl spās] The set of all possible outcomes (p. 428)

scale [skāl] A ratio between two sets of measurements (p. 398)

scale drawing [skāl drô′ing] A drawing that shows a real object smaller than (a reduction) or larger than (an enlargement) the real object (p. 397)

scalene [skā′lēn] A triangle with no congruent sides (p. 331)

scatterplot [skat′ər•plät] A graph with points plotted to show a relationship between two variables (p. 139)

sector [sek′tər] A region enclosed by two radii and the arc joining their endpoints (p. 344)
Example:

sequence [sē′kwəns] An ordered set of numbers (p. 536)

similar figures [si′mə•lər fig′yərz] Figures with the same shape but not necessarily the same size (p. 372)

simple interest [sim′pəl in′trəst] A fixed percent of the principal, paid yearly (p. 422)

simplest form [sim′pləst fôrm] The form in which the numerator and denominator of a fraction have no common factors other than 1 (p. 161)

solution [sə•loō′shən] A value that, when substituted for a variable in an equation, makes the equation true (p. 30)

square [skwâr] The product of a number and itself; a number with the exponent 2 (p. 276)

square [skwâr] A rectangle with four congruent sides (p. 511)

square root [skwâr roōt] One of two equal factors of a number (p. 277)

stem-and-leaf plot [stem ənd lēf plät] A type of graph that shows groups of data arranged by place value (p. 126)

straight angle [strāt an′gəl] An angle whose measure is 180° (p. 320)
Example:

Subtraction Property of Equality [sub•trak′shən prä′pər•tē əv i•kwol′ə•tē] The property that states that if you subtract the same number from both sides of an equation, the sides remain equal (p. 287)

sum [sum] The answer to an addition problem (p. 15)

supplementary angles [sup•lə•men′tə•rē an′gəlz] Two angles whose measures have a sum of 180° (p. 323)
Example:

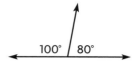

surface area [sûr′fəs âr′ē•ə] The sum of the areas of the faces of a solid figure (p. 504)

survey [sûr′vā] A method of gathering information about a group (p. 94)

systematic sample [sis•tə•ma′tik sam′pəl] A sampling method in which one subject is selected at random and subsequent subjects are selected according to a pattern (p. 95)

T

term [tûrm] Each number in a sequence (p. 536)

terms [tûrmz] The parts of an expression that are separated by an addition or subtraction sign (p. 273)

terminating decimal [tûr′mə•nāt•ing de′sə•məl] A decimal that ends, having a finite number of digits after the decimal point (p. 169)

tessellation [tes•ə•lā′shən] A repeating arrangement of shapes that completely covers a plane, with no gaps and no overlaps (p. 553)

theoretical probability [thē•ə•re′ti•kəl prä•bə•bil′ə•tē] A comparison of the number of favorable outcomes to the number of possible equally likely outcomes (p. 428)

transformation [trans•fər•mā′shən] A rigid transformation is a movement that does not change the size or shape of a figure (p. 550)

translation [trans•lā′shən] A movement of a figure along a straight line (p. 550)

tree diagram [trē dī′ə•gram] A diagram that shows all possible outcomes of an event (p. 444)

triangular number [trī•an′gyə•lər num′bər] A number that can be represented by a triangular array (p. 536)

unbiased sample [un•bī′əst sam′pəl] A sample is unbiased if every individual in the population has an equal chance of being selected. (p. 98)

underestimate [un•dər•es′tə•mət] An estimate that is less than the exact answer (p. 17)

unit rate [yoo′nət rāt] A rate that has 1 unit as its second term (p. 385)
Example: $1.45 per pound

unlike fractions [un′līk frak′shənz] Fractions with different denominators (p. 180)

upper extreme [up′ər ik•strēm′] The greatest number in a set of data (p. 129)

upper quartile [up′ər kwôr′tĭl] The median of the upper half of a set of data (p. 129)

variable [vâr′ē•ə•bəl] A letter or symbol that stands for one or more numbers (p. 28)

Venn diagram [ven dī′ə•gram] A diagram that shows relationships among sets of things (p. 230)

vertex [vûr′teks] The point where two or more rays meet; the point of intersection of two sides of a polygon; the point of intersection of three or more edges of a solid figure; the top point of a cone (pp. 320, 351)
Examples:

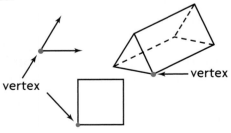

vertex

vertex

vertical angles [vûr′ti•kəl an•gəlz] A pair of opposite congruent angles formed where two lines intersect (p. 322)
Example:

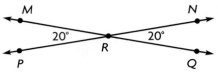

∠*MRP* and ∠*NRQ* are vertical angles.

volume [väl′yəm] The number of cubic units needed to occupy a given space (p. 512)

whole number [hōl num′bər] One of the numbers 0, 1, 2, 3, 4, ... The set of whole numbers goes on without end.

***x*-axis** [eks•ak′səs] The horizontal number line on a coordinate plane (p. 570)

***x*-coordinate** [eks•kō•ôr′də•nət] The first number in an ordered pair; it tells the distance to move right or left from (0,0).

***y*-axis** [wī•ak′səs] The vertical number line on a coordinate plane (p. 570)

***y*-coordinate** [wī•kō•ôr′də•nət] The second number in an ordered pair; it tells the distance to move up or down from (0,0).

Chapter 1

Page 15

1. product   **3.** difference   **5.** sum
7. hundreds   **9.** hundred thousands
11. hundred millions   **13.** ten millions
15. 90,000   **17.** 300,000,000
19. 20,000,000,000   **21.** 2,000   **23.** 29,000
25. 35,000   **27.** 6,000   **29.** 135,000
31. 60,000   **33.** 50,000   **35.** 10,000
37. 640,000   **39.** 90,000

Pages 18–19

1. underestimate since both addends are
rounded down   **3.** 1,500   **5.** 12,000   **7.** 600
9. 24,000   **11.** 360   **13.** 1,500   **15.** 80
17. 70   **19.** 6,000   **21.** 18,000   **23.** 350
25. 14,000   **27.** 16,000   **29.** 4,000
31. 150,000   **33.** 90   **35.** 12   **37.** 3,200,000
39. over; $400 + 700$ **41.** over; 100×20
43. over; 300×30 **45.** >   **47.** <   **49.** <
51. about 4,500,000 books   **53.** 9,936 people
55. 235 and 345   **57.** 36 in.; 54 in.^{2}

Page 21

3. 1,439   **5.** 26,208   **7.** 36,374   **9.** 1,111,384
11. 46,395,619   **13.** 75,125   **17.** 11,306
19. 1, 2, 3, 4, 6, 8, 12, 24

Pages 24–25

1. There were 102 full packages. The 18
newspapers left over were not enough to
make a full package.   **3.** 792,456   **5.**
3,553,275   **7.** 54   **9.** 38,480   **11.** 850,038
13. 56 r4   **15.** 961,184   **17.** 3,496,458
19. 1,745,677   **21.** 26,522   **23.** 14
25. 20,900   **27.** 26 r2   **29.** 89   **31.** 1,312,800
33. 1,855,998   **35.** 84   **37.** $367\frac{9}{20}$ **39.** $846\frac{1}{4}$
41. 415   **43.** $159   **45.** 10   **47.** 64

Page 27

1. 13 apples, 27 oranges   **3.** B   **5.** April 26,
May 2, and May 8   **7.** 48 cards   **9.** 121
stamps   **11.** 5 min

Page 29

1. An algebraic expression has one or more
variables.   **3.** $125 - 46$ **5.** $y \div 15$, or $\frac{y}{15}$
7. 46   **9.** 25×20 **11.** $76 - k$ **13.** 465
15. 9,000   **17.** 140   **19.** 3   **21.** $n + 12$
23. Deidre   **25.** 215   **27.** 7,643

Page 31

1. No, because $4 + 3$ equals 7; $x = 6$
3. $f = 21$ **5.** $x = 4$ **7.** $k = 4$ **9.** $x = 5$
11. $k = 18$ **13.** $m = 9$ **15.** $x = 20$
17. $k = 29$ **19.** $v = 20$ **21.** $d = 9$
23. $p = 6$ **25.** 25 students   **27.** 41
29. 4,990

Chapter 2

Page 35

1. 27   **3.** 64   **5.** 10,000   **7.** 512   **9.** 343
11. 256   **13.** Commutative of Addition
15. Property of Zero   **17.** Associative of
Addition   **19.** Distributive   **21.** Identity
of Multiplication   **23.** Distributive   **25.** $a =$
5   **27.** $r = 3$ **29.** $t = 8$ **31.** 20   **33.** 32
35. 8   **37.** 32   **39.** 18

Pages 38–39

3. 204   **5.** 157   **7.** 56   **9.** 145   **11.** 55
13. 142   **15.** 93   **17.** 43   **19.** 168   **21.** 495
23. 185   **25.** 108   **27.** 212   **29.** 72
31. 244   **33.** 85   **35.** 45   **37.** 1,500
39. 1,028   **41.** 160   **43.** 64   **45.** 52
49. 72 CDs   **51.** 18 CDs each   **53.** 68 more
signatures   **57.** 33,672

Page 41

1. 7 **3.** $2 \times 2 \times 2$; 8 **5.** $3 \times 3 \times 3 \times 3$; 81
7. $1 \times 1 \times 1 \times 1$; 1 **9.** $7 \times 7 \times 7$; 343
11. $5 \times 5 \times 5$; 125 **13.** 34×34; 1,156
15. $10 \times 10 \times 10 \times 10 \times 10 \times 10 \times 10 \times 10$;
100,000,000 **17.** $2 \times 2 \times 2 \times 2 \times 2 \times 2 \times$
$2 \times 2 \times 2 \times 2$; 1,024 **19.** 1 **21.** 25; 25
23. 12^3 **25.** 4^4 **27.** n^2 **29.** 8^2 **31.** 10^4
33. 1940 **35.** How many more games does
Scott have than Aaron? **37.** 46,000
39. 745 r12

Pages 42–43

1. parentheses, exponent, multiplication,
addition; 154 **3.** multiplication, division,
addition; 16 **1.** No; 17 is the correct value,
not 9 **3.** $15 \div 3 + 12$ **5.** 29 **7.** 216
9. 31

Pages 44–45

1. $(420 - 100) \div 40 = 8$ **3.** 25 **5.** 28
7. 25 **9.** 68 **11.** 100 **13.** 1,218
15. 504 **17.** 60 **19.** 77 **21.** 11
23. $\$20 - (\$3.25 + 3 \times \$1.29) = \$12.88$
25. 18,447 people **27.** 204 **29.** 32 r3

Page 47

1. Aquarium Tour, Underwater Acrobats,
Animal Acts, Whale Acts **3.** H **5.** 3 books,
4 magazines **7.** $4,700 **9.** 24 nails

Chapter 3
Page 51

1. 0.2 **3.** 0.09 **5.** 836.23 **7.** 93,450.38
9. 306,007.06 **11.** 7 **13.** 1 tenth **15.** <
17. < **19.** > **21.** > **23.** > **25.** >
27. 1 **29.** 42 **31.** 72 **33.** 26.4 **35.** 8.6
37. 14.5 **39.** 55.6

Pages 54–55

1. 2.0769 **3.** 0.007 **5.** 0.06 **7.** 0.001 +
0.00003 **9.** 300 + 40 + 2 + 0.04 + 0.006
11. = **13.** 1.351, 1.361, 1.363 **15.** 0.0007
17. 0.00006 **19.** 0.03 + 0.006 + 0.0002
21. 2 + 0.4 + 0.05 + 0.006 **23.** < **25.** =
27. > **29.** < **31.** 0.405, 1.05, 1.125, 1.25,
1.45 **33.** 8.91, 9.082, 9.285, 9.82, 9.85
35. 125.4, 125.35, 125.33, 125.3 **37.** 41.01,
$14\frac{1}{10}$, $14\frac{3}{100}$, 14.01 **39.** $94,563,020; ninety-
four million, five hundred sixty-three
thousand, twenty **43.** 228 **45.** 29

Page 57

1. C **3.** air conditioner **5.** calculator,
$10.50; pen, $1.50; notebook, $2.00 **7.** 60
cards **9.** Blaster

Page 59

3. 40 **5.** 90 **7.** 210 **9.** 32 **11.** 9
13. 50 **15.** 140 **17.** 1,800 **19.** 27,000
21. $17,000 **23.** 36 **25.** > **27.** Possible
answer: about 120 lb **31.** 2.235, 2.325, 2.523,
2.532 **33.** 1,121 r18

Page 61

1. 18% means 18 per hundred, so 18 are red.
3. 0.06, 6% **5.** 0.7 or 0.70 **7.** 3% **9.** 0.5,
0.50 **11.** 0.11, 11% **13.** 0.62 **15.** 0.28
17. 0.53 **19.** 85% **21.** 40% **23.** 0.79, 79%
25. Write 0.6 as 0.60 to show hundredths.
Then write 60 hundredths as 60%. **27.** 172
29. 82 r26

Chapter 4
Page 65

1. 47 **3.** 44 **5.** 10 **7.** 222 **9.** 252
11. 1,444 **13.** 4,000 **15.** 3,000 **17.** 24
19. 47 **21.** 14 **23.** 29 **25.** 43 **27.** 840
29. 19,405 **31.** 8.75 **33.** 12.5 **35.** 160.8
37. $9\frac{5}{8}$ **39.** $39\frac{2}{7}$

Pages 68–69

3. $0.63   **5.** 18.585   **7.** 20.923   **9.** 37.31
11. 124.94   **13.** 87.80   **15.** 1,438.66
17. 0.6   **19.** 287.673   **21.** 488.55
23. 529.484   **25.** 257.14   **27.** 9.42   **29.** yes
31. no   **33.** 1.25   **35.** 12.3   **37.** 2.216
39. 0.090   **41.** 210; more than   **43.** 441 lb
45. 0.46

Pages 72–73

3. 0.35   **5.** 2   **7.** 3   **9.** 2.87   **11.** 1.218
13. 25.84   **15.** 0.24   **17.** 3   **19.** 6
21. 194.208   **23.** 0.410   **25.** 13.8   **27.** 36.5
29. 4.92   **31.** 0.08   **33.** 0.441   **35.** 5.5080
37. 26.8467   **39.** 0.39501   **41.** 0.8934
43. $7.31   **47.** 65.65   **49.** 299 r14

Page 74

1. 1.01   **3.** 0.45

Page 75

1. 6   **3.** 3   **5.** 4   **7.** 3   **9.** 8   **11.** 0.57
13. 89 r21

Pages 78–79

1. multiply by 100   **3.** 96 ÷ 16
5. 482.4 ÷ 24 **7.** 14.86   **9.** 1.97   **11.** 4.4
13. 819 ÷ 9   **15.** 239 ÷ 5   **17.** 2,335.8 ÷ 102
19. 6.1   **21.** 6.9   **23.** 9.91   **25.** 0.21
27. 21.2   **29.** 8.2   **31.** 15.05   **33.** 220
35. 2.36   **37.** 3.03   **39.** 5 weeks
41. 30 pages   **43.** greater than

Page 81

1. C   **3.** 4 magnets   **5.** 8 hr 5 min
7. 65 min, or 1 hr 5 min   **9.** 108 pieces

Page 83

1. $a \times 6.45$; 12×6.45; $77.40   **3.** 11.7
5. 5.08   **7.** $m = 19$ **9.** 4.6   **11.** 14   **13.** 3.9
15. $r = 7.7$ **17.** $a = 8.2$ **19.** $p = 17.6$
21. $n \div 6$; 4.8 mi   **23.** 4 mi   **25.** 37
27. 3.08, 3.508, 3.58, 3.85

Chapter 5

Page 93

1. median   **3.** range   **5.** 13 more girls
7. 29 girls   **9.** 8   **11.** 8.475   **13.** 28; 32
15. 11; 10.8   **17.** 6

Pages 96–97

5. population, because there are not too
many students and they are easily available
7. systematic   **9.** convenience
11. population; There are not too many
basketball team members.   **13.** systematic
15. convenience   **21.** 9.3   **23.** 2

Pages 98–99

1. Do you enjoy movies that are longer than
two hours?   **3.** biased; excludes males
5. unbiased; all members have an equal
chance   **7.** unbiased; all shoppers have an
equal chance   **9.** biased; excludes people
who shop in other stores   **11.** biased
13. Yes. It leads others to agree with the
opinion.   **15.** 4   **17.** 9,000

Page 101

1. D   **3.** 12 members   **5.** 1 3-point,
7 2-point; 3 3-point, 4 2-point; 5 3-point,
1 2-point   **7.** $42.50   **9.** 37 birds
11. Saturday

Pages 104–105

1. 1–10, 11–20; Since 2 × 10 = 20, you would
have 2 intervals that include 10 numbers
5. 16   **13.** 9   **15.** No. The data are
categories and not numerical.   **19.** unbiased

Pages 107–108

1. The mean would become 21.2. The mean
or median would be most useful to describe
the data.   **3.** 7 hours   **5.** 24; 23; 22
7. 15.375 points   **9.** 10 and 12   **11.** 336.8;
360; no mode   **15.** Tim should divide by 4,
not 3; 6   **17.** 71 people   **19.** 90%

Pages 110–111

1. The median and mode would not change. The mean would not increase as much.
3. 16; 14; 12   **5a.** Abe: mean 7.5, median 6, mode 7; Bart: mean 7.5, median 8.5, mode 0;
5b. Abe: mean 9.3, median 7, mode 7 and 20; Bart: mean 6.4, median 5, mode 0;
5c. Abe's: all increase; Bart's: mean/median decrease; mode doesn't change   **7a.** 75
7b. No. With a perfect score of 100, the mean would be 65.7.   **9.** 5.75; 5.5; 5   **11.** 9.5; 10; 10 and 12

Pages 113–115

1. No. The sample would be a convenience sample and not a random sample.
3. Yes. The sample is random and representative; the question is unbiased.
5. No. The sample is not representative.
7. No. The sample is not representative of the population.   **9.** Yes. An unbiased question was asked of a random sample from the correct population.   **11.** The interest in crafts decreased.   **13.** The interest in trading cards increased.   **15.** A random sample was chosen from the correct population and the question was unbiased.   **17.** The sample is biased since it only includes middle school athletes.   **19.** 4.25, 4, 4 and 6   **21.** 7, 7, no mode

Chapter 6
Page 119

1. Saturn V, Ariane IV, Titan-Centaur, Titan IIIC   **3.** about 55 m   **7.** about 15,000 sq mi
9. 92   **11.** 22   **13.** 3 or more   **15.** 85

Pages 122–123

1. so you can tell the difference between the sets of data   **3.** line graph   **5.** bar graph
7. circle graph   **11.** Each month, more comedies were rented.   **13.** Comedy video rentals will increase and action video rentals will decrease.   **17.** No. The sample is not representative of the student population.
19. 0.56

Page 125

1. Extend the graph until it intersects with 24 mi and find the corresponding time.
3. 5 hr   **5.** 6 hr   **7.** 8 hr   **9.** about 45 mi
11. 5.6; 5; 4   **13.** 6.5; 7; 9

Pages 127–128

5. bar graph   **7.** bar graph   **9b.** 24.5 in.; 28 in.   **13.** random   **15.** 73%

Pages 130–131

1. A box-and-whisker graph shows the distribution of data, not all of the data.
3. 11; 22; 11   **5.** 58; 67; 9   **7.** 85; 104; 19
9. 72 points; 97 points   **13.** $4.07   **15.** 0.02 or two hundredths

Pages 133–135

3. Yes. The question is biased and could lead people to choose the Mustangs as the best baseball team.   **5.** No. Angel Falls is about 3,200 ft high while Tugela Falls is about 3,000 ft high.   **7.** Yes. The question is biased and could lead people to choose oranges.   **9.** No. Brand B costs $30 and Brand A costs $20.
13. Lin looked only at the graph and did not look at the scales.   **15.** 43 and 55

Chapter 7
Page 145

1. prime number   **3.** yes   **5.** no   **7.** yes
9. yes   **11.** yes   **13.** no   **15.** yes   **17.** no
19. yes   **21.** no   **23.** 16, 20, 24   **25.** 48, 60, 72   **27.** 20, 25, 30   **29.** 6, 12, 18, 24, 30
31. 30, 60, 90, 120, 150   **33.** 9, 18, 27, 36, 45
35. 1, 2, 4, 8   **37.** 1, 11   **39.** 1, 2, 3, 6, 9, 18, 27, 54

Page 147

3. 2, 4, 8   **5.** 2, 4, 8   **7.** 3   **9.** 2, 3, 4, 6, 9
11. 2, 4   **13.** 3, 5   **15.** 2, 3, 4, 6, 8   **17.** 3, 5
19. 2, 3, 6, 9   **21.** 3   **23.** T   **25.** T
29. $28.88   **31.** 5%   **33.** 16,302

Page 149

1. $2^2 \times 3 \times 13$ **3.** $2 \times 2 \times 3$ **5.** $2 \times 2 \times 2 \times 2$ **7.** 3×7 **9.** 2×127 **11.** $2 \times 2 \times 2 \times 2 \times 2 \times 2 \times 2$ **13.** $2 \times 2 \times 19$ **15.** $2 \times 3 \times 3; 2 \times 3^2$ **17.** $7 \times 7; 7^2$ **19.** $2 \times 2 \times 7 \times 19; 2^2 \times 7 \times 19$ **21.** 2×373 **23.** $n = 2$ **25.** $n = 5$ **27.** $c = 2$ or 3 **31.** 60% **33.** 13

Pages 152–153

3. 3, 6, 9, 12, 15 **5.** 11, 22, 33, 44, 55 **7.** 21 **9.** 18 **11.** 3 **13.** 3 **15.** 4, 8, 12, 16, 20 **17.** 16, 32, 48, 64, 80 **19.** 10, 20, 30, 40, 50 **21.** 14, 28, 42, 56, 70 **23.** 24 **25.** 60 **27.** 120 **29.** 108 **31.** 2 **33.** 3 **35.** 1 **37.** 8 **39.** 9 and 12 **41. a.** 60 **b.** 4 packages of cereal samples, 3 packages of pamphlets **43.** She found the GCF instead of LCM. The LCM is 30. **45.** 124 r14

Page 155

1. October 15 **3.** H **5.** 9 in. **7.** 18 blocks

Chapter 8

Page 159

1. > **3.** denominator **5.** > **7.** < **9.** > **11.** 62,000; 61,600; 61,060 **13.** $\frac{5}{6}$ **15.** $\frac{1}{3}$ **17.** 20% **19.** 75%

Pages 162–163

1. Also multiply the denominator by 5; $\frac{10}{15}$ **3.** 12 **5.** 4 **7.** 1 **9.** 16 **11.** 1, 2, 4 **13.** 1, 2 **15.** $\frac{1}{8}$ **17.** $\frac{1}{6}$ **19.** $\frac{11}{4}$ **21.** $\frac{3}{10}$ **23.** 5 **25.** 4 **27.** 6 **29.** 8 **31.** 12 **33.** 7 **35.** 1 **37.** 1, 3 **39.** 1, 5 **41.** 1 **43.** $\frac{1}{6}$ **45.** $\frac{1}{8}$ **47.** $\frac{5}{9}$ **49.** $\frac{1}{5}$ **51.** $\frac{2}{3}$ **53.** $\frac{10}{7}$ **55.** $\frac{1}{4}$ **57.** $\frac{1}{5}$ **59.** $\frac{1}{2}$ **61.** $\frac{2}{3}$ **63.** $\frac{1}{4}$ **65.** What fractions of the muffins are bran? **67.** 24 **69.** 0.035

Page 165

1. It has a whole number part and a fraction part. **3.** $3\frac{1}{2}$ **5.** $3\frac{2}{3}$ **7.** $\frac{5}{4}$ **9.** $\frac{8}{3}$ **11.** $\frac{37}{7}$ **13.** $4\frac{1}{2}$ **15.** $5\frac{3}{4}$ **17.** $5\frac{1}{6}$ **19.** $12\frac{6}{7}$ **21.** $16\frac{2}{3}$ **23.** $3\frac{2}{3}$ **25.** $\frac{13}{2}$ **27.** $\frac{19}{10}$ **29.** $\frac{37}{4}$ **31.** $\frac{53}{11}$

33. $\frac{93}{5}$ **35.** no, only a fraction whose numerator is greater than its denominator **37.** $2^2 \times 3^2$ **39.** 23×1

Pages 166–167

1. The denominators are the same. $4 < 5$, so $\frac{4}{9} < \frac{5}{9}$ **3.** < **5.** = **7.** $\frac{1}{4}, \frac{6}{12}, \frac{2}{3}$ **9.** > **11.** = **13.** = **15.** > **17.** $\frac{2}{6}, \frac{1}{2}, \frac{9}{12}$ **19.** $\frac{1}{2}, \frac{7}{12}, \frac{5}{6}$ **21.** $\frac{3}{10}, \frac{2}{5}, \frac{1}{2}$ **23.** $\frac{1}{6}, \frac{1}{3}, \frac{1}{2}$ **25.** $\frac{1}{12}, \frac{5}{8}, \frac{3}{12}$ **27.** 3 pieces of each pizza remain. That is $\frac{3}{8}$ of the mushroom and $\frac{3}{12}$ cheese pizza. $\frac{3}{12} = \frac{1}{4} = \frac{2}{8} < \frac{3}{8}$, so more of the mushroom pizza is left. **29.** 7 **31.** 24

Page 168

1. 0.3 **3.** 0.7 **5.** 0.90 **7.** 0.4 **9.** 0.85

Pages 170–171

3. $\frac{7}{10}$ **5.** $\frac{105}{1,000}$ **7.** 0.25; T **9.** $0.\overline{6}$; R **11.** > **13.** = **15.** 20% **17.** 40% **19.** $\frac{6}{100}$ **21.** $\frac{61}{100}$ **23.** $\frac{205}{1,000}$ **25.** $\frac{9}{1,000}$ **27.** $0.1\overline{6}$; R **29.** 0.625; T **31.** 0.3; T **33.** 0.1; R **35.** 0.36; T **37.** $0.\overline{15}$; R **39.** = **41.** < **43.** < **45.** 75% **47.** 6% **49.** 50% **51.** 0.5% **53.** 0.72 **55.** Megan **57.** $\frac{9}{2}$ **59.** 76.89

Chapter 9

Page 175

1. mixed number **3.** divide **5.** $\frac{3}{4}$ **7.** $\frac{1}{3}$ **9.** $\frac{4}{3}$, or $1\frac{1}{3}$ **11.** $\frac{3}{2}$, or $1\frac{1}{2}$ **13.** $\frac{1}{2}$ **15.** $\frac{2}{3}$ **17.** $\frac{5}{3}$, or $1\frac{2}{3}$ **19.** $\frac{1}{3}$ **21.** $\frac{1}{2}$ **23.** $\frac{5}{4}$, or $1\frac{1}{4}$ **25.** $\frac{2}{3}$ **27.** $\frac{1}{2}$ **29.** $\frac{1}{2}$ **31.** $\frac{1}{3}$ **33.** $\frac{5}{6}$ **35.** 1 **37.** $\frac{2}{9}$ **39.** $\frac{5}{7}$

Pages 178–179

3. close to 1 **5.** between 0 and $\frac{1}{2}$ **7.** $1\frac{1}{2}$ **9.** $\frac{1}{2}$ **11.** $7\frac{1}{2}$ **13.** 2 **15.** close to 1 **17.** close to 0 **19.** 2 **21.** $\frac{1}{2}$ **23.** 17 **25.** 3 **27.** 0 **29.** $4\frac{1}{2}$ **31.** 9 to $9\frac{1}{2}$; $9\frac{1}{4}$ **33.** $4\frac{1}{2}$ to 5; $4\frac{3}{4}$ **35.** $5\frac{1}{2}$ to 6; $5\frac{3}{4}$ **37.** about 7 **39.** about 6 feet **41.** about 26 **43.** about 30 **47.** about $1\frac{1}{2}$ yd **49.** March 15, 24; April 2, 11, 20, 29 **51.** 45%

Page 180

1. $\frac{3}{4}$ **3.** $\frac{9}{10}$ **5.** $\frac{7}{12}$ **7.** $\frac{2}{3}$ **9.** $\frac{7}{10}$ **11.** $\frac{1}{2}$

Page 181

1. $\frac{5}{12}$ **3.** $\frac{1}{12}$ **5.** $\frac{1}{10}$ **7.** $\frac{1}{12}$ **9.** $\frac{1}{8}$ **11.** $\frac{1}{2}$
13. = **15.** >

Pages 183–185

1. $\frac{5}{12}$ c more **3.** $\frac{7}{10} + \frac{2}{10}$ **5.** $\frac{12}{15} - \frac{5}{15}$ **7.** $\frac{1}{3}$
9. $1\frac{5}{12}$ **11.** $\frac{1}{15}$ **13.** $\frac{7}{9}$ **15.** $\frac{24}{28} - \frac{21}{28}$ **17.** $\frac{5}{6}$
19. $\frac{4}{5}$ **21.** $\frac{3}{14}$ **23.** $\frac{1}{5}$ **25.** $1\frac{3}{8}$ **27.** $\frac{5}{8}$ **29.** $\frac{7}{15}$
31. $\frac{2}{9}$ **33.** $\frac{6}{8}$, or $\frac{3}{4}$ **35.** 2 **37.** $\frac{5}{12}$ mile
39. $r = 1$ **41.** $c = \frac{3}{5}$ **43.** $m = 1$ **45.** $\frac{5}{8}$ tsp
47. $\frac{2}{15}$ is left; add to find the total spent and saved; subtract to find how much is left.
49. 24 and 28 **51.** $\frac{4}{9}$ **53.** 6,132

Pages 187–189

3. $3\frac{4}{5}$ **5.** $1\frac{3}{10}$ **7.** $6\frac{7}{12}$ **9.** $1\frac{1}{6}$ **11.** $1\frac{1}{18}$
13. $2\frac{1}{2}$ **15.** $8\frac{3}{10}$ **17.** $10\frac{1}{18}$ **19.** $4\frac{1}{10}$ **21.** $11\frac{3}{20}$
23. $8\frac{3}{14}$ **25.** $28\frac{1}{8}$ **27.** Commutative, $2\frac{1}{4}$
29. Associative, $1\frac{5}{6}$ **31.** longer; $\frac{1}{6}$ in.; subtraction; to find the difference in length
33. $3\frac{1}{4}$ c **35.** 3rd prize **37.** $\frac{16}{15}$, or $1\frac{1}{15}$
39. $\frac{4}{5}, \frac{2}{3}, \frac{1}{2}$

Page 191

1. $1\frac{2}{3}$ **3.** $1\frac{4}{9}$ **5.** $\frac{7}{10}$ **7.** $\frac{11}{12}$ **9.** $3\frac{3}{4}$ **11.** 21.42

Page 193

1. $3\frac{1}{3} = 2\frac{4}{3}$ **3.** $2\frac{1}{12}$ **5.** $3\frac{9}{10}$ **7.** $3\frac{17}{18}$ **9.** $1\frac{11}{12}$
11. $4\frac{7}{9}$ **13.** $2\frac{7}{12}$ **15.** $1\frac{3}{8}$ **17.** $2\frac{7}{10}$, or 2.7
19. $2\frac{4}{5}$ **21.** $\frac{7}{8}$ mi shorter **25.** $7\frac{1}{3}, \frac{22}{3}$
27. 55 r5

Page 195

1. $18\frac{1}{2}$ mi **3.** C **5.** 12 cans **7.** $4\frac{3}{4}$ mi
9. Gary

Chapter 10

Page 199

1. equation **3.** 0 **5.** $\frac{1}{2}$ **7.** $\frac{1}{2}$ **9.** $\frac{1}{2}$ **11.** $\frac{1}{2}$
13. $\frac{1}{2}$ **15.** 0 **17.** $\frac{1}{2}$ **19.** $c = 32$ **21.** $x = 0.17$ **23.** $t = 14$ **25.** $r = 2.6$ **27.** $1\frac{1}{6}$

29. $1\frac{1}{3}$ **31.** $\frac{15}{2}$ **33.** $\frac{14}{5}$

Page 201

1. about 500 million pounds **3.** $\frac{1}{2}$ **5.** 20
7. 70 **9.** 118 **11.** $\frac{1}{2}$ **13.** 1 **15.** 4 **17.** 3
19. 20 **21.** $\frac{1}{4}$ **23.** > **25.** about 7 min
27. 8 kg **29.** $3\frac{4}{9}$ **31.** 11

Pages 204–205

3. $\frac{3}{8}$ **5.** $\frac{2}{10}$, or $\frac{1}{5}$ **7.** $\frac{3}{10}$ **9.** $\frac{12}{7}$, or $1\frac{5}{7}$ **11.** $\frac{8}{3}$, or $2\frac{2}{3}$ **13.** $\frac{5}{9}$ **15.** $\frac{3}{16}$ **17.** $\frac{1}{16}$ **19.** $\frac{2}{9}$ **21.** $\frac{2}{15}$
23. $\frac{7}{10}$ **25.** $\frac{1}{10}$ **27.** $\frac{4}{15}$ **29.** 2 **31.** <
33. < **35.** 108 voters **37.** 6 **39.** $b = 12.09$

Page 207

1. $3 \times 4\frac{2}{3} = (3 \times 4) + (3 \times \frac{2}{3})$; $12 + 2 = 14$.
3. $1\frac{1}{8}$ **5.** $2\frac{1}{4}$ **7.** $18\frac{3}{8}$ **9.** $2\frac{1}{4}$ **11.** $8\frac{1}{6}$
13. 33 **15.** 15 **17.** 85 **19.** $7\frac{1}{5}$ **21.** $22\frac{1}{2}$
23. < **25.** $2\frac{1}{12}$ mi **29.** $6\frac{7}{20}$ **31.** 352,053

Page 208

1. 9 **3.** 4

Page 209

1. $n = 4$ **3.** $n = \frac{3}{4}$ **5.** 8 **7.** $\frac{3}{2}$, or $1\frac{1}{2}$ **9.** $9\frac{9}{20}$
11. $606.64 > 606.074$

Pages 212–213

3. $\frac{3}{2}$ **5.** $\frac{1}{7}$ **7.** $\frac{3}{13}$ **9.** $\frac{4}{5}$ **11.** $\frac{1}{24}$ **13.** 3
15. $1\frac{7}{32}$ **17.** $\frac{1}{10}$ **19.** $\frac{9}{2}$ **21.** $\frac{7}{15}$ **23.** $\frac{1}{9}$
25. $\frac{5}{13}$ **27.** $1\frac{1}{6}$ **29.** $9\frac{1}{3}$ **31.** $2\frac{1}{4}$ **33.** 5
35. $5\frac{1}{15}$ **37.** $1\frac{8}{13}$ **39.** 18 **41.** 32 **43.** 81
45. 44 **47.** $2\frac{5}{14}$ **49.** $\frac{5}{18}$ **51.** $\frac{4}{5}$ **53.** 48
burgers **55.** $1\frac{1}{4}$ yd **57.** $46.90 **59.** $13\frac{3}{8}$

Page 215

1. $1\frac{5}{6}$ min; subtraction **3a.** B **3b.** H
5. $2\frac{2}{3}$ mi **7.** $10\frac{1}{3}$ mi

Page 217

1. $b \times \frac{3}{4}$; $32 \times \frac{3}{4}$; 24 oz **3.** $2\frac{1}{5}$ **5.** $5\frac{5}{8}$
7. $y = \frac{3}{8}$ **9.** $7\frac{5}{6}$ **11.** $1\frac{1}{4}$ **13.** $\frac{1}{20}$ **15.** $\frac{3}{8}$
17. 1 **19.** $x = 2$ **21.** $x = \frac{1}{4}$ **23.** $x = 1\frac{1}{3}$
27. $\frac{3}{4}, \frac{3}{5}, \frac{3}{8}$ **29.** $a = 4.5$

Chapter 11

Page 227

1. negative numbers **11.** 1, 2, 3, 4
13. 1, 3, 5, 7 **15.** < **17.** = **19.** >
21. 32°F **23.** $^-16$°F **25.** 60°F

Page 229

3. $^-350$ **5.** $^-14$ **7.** $^+25$ **9.** $^-742$ **11.** $^-5$
13. $^+12{,}000$ **15.** $^+2$ **17.** $^+31$ **19.** $^-207$
21. 28 **23.** 660 **25.** $^-1{,}310$ ft **27.** $\frac{1}{5}$
29. $\frac{12}{20}$, or $\frac{3}{5}$ **31.** 13

Pages 232–233

1. All integers can be written in the form $\frac{a}{1}$
3. $^-\frac{37}{100}$ **5.** $\frac{889}{1{,}000}$ **7.** $^-\frac{22}{3}$ **9.** $^-\frac{1}{2}$ **11.** 0
13. $^-\frac{71}{100}$ **15.** $^-\frac{21}{8}$ **17.** $^-0.3$ **19.** $^-0.7$
21. $^-7\frac{1}{4}$ **23.** 16.05 **25.** $^-\frac{5}{8}$ **27.** yes
29. yes **31.** yes **33.** 90 steps **35.** Not all integers are whole numbers; $^-8$ is not a whole number. **37.** 0.34; $\frac{34}{100}$

Page 235

3. < **5.** < **7.** < **9.** > **11.** < **13.** >
15. > **17.** 0.4, 0.46, 0.6 **19.** $^-\frac{1}{4}$, $^-0.2$, 0, $\frac{1}{4}$
21. 1°, $^-3$°, $^-5$° **23.** $^-450$ **25.** $\frac{24}{7}$

Page 237

1. Jeffrey, sixth grade; Victoria, seventh grade; Arthur, eighth grade **3.** C **5.** Arlene, Kathy, Helene **7.** Canadian Dollars **9.** 3:15 P.M.

Chapter 12

Page 241

1. integers **3.** $^+62$ **5.** $^-13$ **7.** $^+12$ **9.** $^-3$
11. $^+11$ **13.** $^-8$ **15.** 63 **17.** 10 **19.** 4
21. 56 **23.** 12 **25.** 5 **27.** 12
29. $n=3$ **31.** $n=8$ **33.** $n=132$

Page 242

1. $^+13$ **3.** $^-10$

Page 243

1. $^-2$ **3.** 0 **5.** $^+7$ **7.** $^-9$ **9.** $^-6.8$, $^-6.4$, $^-6.2$ **11.** $^-3.6$, $^-3\frac{4}{7}$, $^-3\frac{1}{2}$

Pages 246–247

1. Use sign of the addends. **3.** The sum was 12, which indicates a gain of 12 yards, not a loss of 12 yards **5.** $5 + {^-9} = {^-4}$ **7.** $^-7$
9. $^-2$ **11.** $^-10$ **13.** 0 **15.** $^-2 + 6 = 4$
17. $^-1$ **19.** 3 **21.** $^-20$ **23.** 8 **25.** $^-28$
27. $^-12$ **29.** $^-11$ **31.** 100 **33.** $x = {^-7}$
35. $x = 5$ **37.** $^-12$°F **39.** The Wildcats lost 2 yards. **41.** 12 cm **43.** greater than

Page 249

1. $^-11$ **3.** $^-22$ **5.** 2 **7.** $^-5$ **9.** 6 **11.** $^-8$

Page 251

1. $^-9 + 40$ **3.** $3 + 6$ **5.** $^-4 + {^-6}$ **7.** $^-5$
9. $^-7$ **11.** $8 + 11$ **13.** $^-9 + {^-11}$ **15.** $^-4$
17. $^-2$ **19.** 0 **21.** $^-5$ **23.** $^-4$ **25.** $^-5$
27. 5 **29.** $^-14$°F **31.** What was the range of temperatures? **33.** $^-213$ **35.** $2 \times 2 \times 3 \times 7$

Pages 254–255

1. $^-35$; 35 **3.** $^-54$ **5.** $^-16$ **7.** $^-3$ **9.** $^-24$
11. $^-40$ **13.** 14 **15.** 12 **17.** 9 **19.** 288
21. $^-300$ **23.** 36 **25.** 80 **27.** $y=3$
29. $y={^-4}$ **31.** $y={^-35}$ **33.** $^-14$ in. **35.** $^-30$
37. $^-70$ ft per minute **39.** What's the sign of the quotient? **41.** 22 **43.** $38\frac{1}{4}$

Pages 258–259

3. 1 **5.** $^-0.7$ **7.** $13\frac{1}{5}$ **9.** $^-8.0$ **11.** $^-10.8$
13. 19 **15.** 3.5 **17.** $^-2$ **19.** $^-1$ **21.** $^-9\frac{1}{4}$
23. $^-3\frac{1}{2}$ **25.** 15.6 **27.** $^-87.4$ **29.** $^-17$
31. $^-12$ **33.** $\frac{2}{9}$ **35.** $^-48$ **37.** 0.7 **39.** $^-1.5$
41. 24 **43.** 4 **45.** $^-10.5$ **47.** 4.4 **49.** $^-25$
51. 557.4°F **53.** $1\frac{1}{8}$ lb **57.** 8.5%

Chapter 13

Page 269

1. exponent **3.** algebraic expression **5.** 32
7. 125 **9.** 16 **11.** 81 **13.** 49 **15.** 216
17. 25 **19.** 7 **21.** 2 **23.** 23 **25.** 20
27. 4 **29.** 21 **31.** 1, 2, 4, 8, 16 **33.** 1, 3, 11, 33 **35.** 1, 17 **37.** 1, 2, 3, 6, 7, 14, 21, 42
39. 1, 2, 4, 5, 8, 10, 20, 40

Page 271

3. $y \div 1.5$ **5.** 54 less than 19 times x
7. $3c + 2.9$ **9.** $h \times 4j \times k$ **11.** $6.3p - 5m$
19. $14.25 + 3.5h$ **21.** $2 + 2x$ **25.** $^-36$
27. 6 and 8

Pages 274–275

1. You will have to perform fewer computations when evaluating the expression if you simplify it first.
3. $^-16, ^-11, ^-6, ^-1$ **5.** $^-2, 6, 14, 22$ **7.** 1, $^-2, ^-3, ^-2$ **9.** $7x - 8; ^-29$ **11.** $^-16, ^-20,$ $^-24, ^-28$ **13.** $20\frac{1}{2}, 20, 19\frac{1}{2}, 19$ **15.** $^-5, 1, 3, 4$
17. $12x - 41; ^-89$ **19.** $356 + 6a; 236$
21. $^-2$ **23.** Associative; 67   **25.** $x = 2$
27. $x = ^-1$ **31.** $^-5$

Page 276

1. 4   **3.** 49

Page 277

1. 8   **3.** 14   **5.** 22   **7.** 81; 9   **9.** $^-49$
11. $>$

Page 279

1. Operate inside the parentheses, evaluate $\sqrt{36}$, then multiply; 12   **3.** $^-33$ **5.** 24
7. 150   **9.** $^-24$ **11.** $^-10$ **13.** 77   **15.** $>$
17. 14 ft   **19.** $^-2, ^-6, ^-10, ^-14$ **21.** $2, ^-4,$ $^-6, ^-7$

Chapter 14
Page 283

1. equation   **3.** 37   **5.** 23   **7.** 11.37
9. 2.7   **11.** $1\frac{7}{12}$ **13.** $\frac{1}{2}$ **15.** $7\frac{11}{12}$ **17.** $2\frac{9}{28}$
19. $43 - 19 = 24$ **21.** $16 + 3 = 19$
23. $46 + 196 = 242$ **25.** $125 \div 5 = 25$
27. $2 \cdot 6 = 12$ **29.** $21 \cdot 12 = 252$
31. $25 \cdot 16 = 400$ **33.** addition
35. subtraction   **37.** multiplication
39. subtraction

Page 285

1. any quantity that you do not know
3. $x + 7 = 20$ **5.** $5 \cdot m = 35$; or $5m = 35$
7. $14 = n + 12$ **9.** $n \div 2\frac{3}{4} = \frac{5}{6}$
11. $72,000 = 9e$ **15.** $8x - 15; ^-31$
17. $25 + 9z; ^-47$

Page 286

1. $x = 4$ **3.** $x = 7$ **5.** $x = 3$ **7.** $x = 1$

Pages 288–289

3. $x = 7$ **5.** $c = 2.7$ **7.** $m = 15\frac{1}{4}$ **9.** $x = 8$
11. $k = ^-41$ **13.** $b = 2.9$ **15.** $y = 8\frac{3}{4}$
17. $t = ^-19$ **19.** $x =$ unknown length; $x + 12 + 10 = 29; x = 7$, or 7 cm   **21.** $11 = n + 8$
23. 13

Page 291

1. You add the number that is being subtracted from the variable.   **3.** $b = ^-6$
5. $y = 23.2$ **7.** $d = 34\frac{7}{15}$ **9.** $a = 33$
11. $z = 17.0$ **13.** $c = 23.3$ **15.** $s = 22\frac{11}{12}$
17. $m = 11.1$ **19.** $f = \frac{19}{24}$ **21.** $s =$ original amount in savings; $s - (110 + 90 + 40) = 527;$ $s = 767$ **23.** $61 - 12 - 13 - 14 + 5 - 7$
25. $b = 4.4$ **27.** 9, 7, 5, 3

Chapter 15
Page 295

1. Celsius   **3.** $18 = n + 6$ **5.** $\frac{n}{2} = \frac{2}{3}$
7. $2x = 47$ **9.** 14   **11.** $^-73$ **13.** $^-140$
15. 9   **17.** 25   **19.** 99   **21.** 27.3   **23.** $3\frac{4}{5}$
25. $y = 5$ **27.** $a = 9$ **29.** $c = 30$
31. $q = 56$ **33.** $y = 0.04$

Page 296

1. $c = 4$ **3.** $b = 2$

Page 299

1. You use inverse operations.   **3.** $x = ^-7$
5. $y = 7.2$ **7.** $k = 6$ **9.** $a = 18$ **11.** $p = 16$
13. $n = ^-5$ **15.** $a = 16.48$ **17.** $w = 2.73$
19. $a = 10$ **21.** $\frac{m}{3} = 14; m = 42$; 42 marbles
23. $0.45 = 0.9w; w = 0.5$ cm   **25.** $x = 22$
27. $z = ^-12$

Pages 302–303

1. The rate of speed is in feet per minute.
5. 20   **7.** 50°F   **9.** 86°F   **11.** 41°F
13. 13.9°C   **15.** 37.8°C   **17.** 36.8   **19.** 2.5
21. 98.6°F   **23.** 203°F   **25.** 0°C   **27.** 8.3°C
29. 34.4°C   **31.** 87.5 mi per hr   **33.** Yes. The
shuttle would be traveling 17,550 mi per hr.
35. Earth   **39.** $y = {}^-6$

Page 305

1. $4x + 1 = 5; x = 1$ **3.** $y = 2$ **5.** Add 1 to
both sides, then divide by 2; $x = 3$.   **7.** 3
9. 374

Page 307

1. 18 mi   **3.** C   **5.** 4:20 P.M.   **7.** 8 dimes,
4 nickels   **9.** 6 students

Chapter 16

Page 317

1. angle   **3.** line   **5.** obtuse   **7.** acute
9. straight   **11.** acute   **13.** $\angle MNP$ or $\angle PNM$
15. $\angle 3$ **17.** $\angle a$ **19.** $\angle PBF$ or $\angle FBP$

Page 319

1. plane, point, line, line segment, or ray
3. point R **5.** point C **7.** plane DHY
9. $\overrightarrow{XY}$ **11.** $\overleftrightarrow{PQ}, \overleftrightarrow{QR}, \overleftrightarrow{PR}$ **13.** $\overrightarrow{PQ}, \overrightarrow{QP}, \overrightarrow{RQ},$
$\overrightarrow{QR}, \overrightarrow{RP}, \overrightarrow{PR}$ **15.** a point   **17.** a point
23. $n = 96$ **25.** $^-2$

Page 321

1. 135°; obtuse   **3.** 73°; acute   **9.** 45°
11. 90°   **13.** $b = 8$ **15.** $\frac{2}{15}$

Pages 324–325

3. $\angle AED$ and $\angle BEC$, $\angle AEB$ and $\angle DEC$
5. $\angle AED$ and $\angle BEC$ **7.** 51°   **9.** 33°
11. $\angle DOC$ **13.** $\angle AOB, \angle DOC$ **15.** 18°
17. 131°   **19.** 90°   **21.** supplementary,
adjacent   **23.** none of these   **25.** $\angle 1, \angle 2$;
$\angle 2, \angle 3; \angle 3, \angle 4; \angle 4, \angle 5; \angle 5, \angle 1$; They are
side by side and have a common vertex.
27. $\angle 1, \angle 5; \angle 4, \angle 5$; Together they form a

straight line.   **29.** Their measures are equal.
31. $\overrightarrow{PQ}$ or $\overrightarrow{QP}$ **33.** $^-6$

Pages 326–327

1. perpendicular and intersecting
3. intersecting / perpendicular
5. intersecting   **7.** intersecting
9. intersecting   **11.** $\overleftrightarrow{AC}, \overleftrightarrow{BD}, \overleftrightarrow{CG}, \overleftrightarrow{DH}$
13. $\overleftrightarrow{BD}, \overleftrightarrow{FH}, \overleftrightarrow{CD}, \overleftrightarrow{GH}$ **17.** Perpendicular
lines always intersect but intersecting lines
are not necessarily perpendicular.
19. 59.2 mi per hr   **21.** $\frac{5}{12}$

Chapter 17

Page 331

1. isosceles   **3.** pentagon   **5.** hexagon
7. quadrilateral   **9.** triangle   **15.** add 4; 20,
24, 28   **17.** subtract 6; 13, 7, 1   **19.** multiply
by $\frac{1}{2}$; $\frac{1}{32}$; $\frac{1}{64}$; $\frac{1}{128}$

Pages 333–335

3. 49°; acute   **5.** 51°; obtuse   **7.** 61°; acute
9. 20°; obtuse   **11.** 40°; acute   **13.** 45°
15. 75°   **17.** 30°   **19.** 115°   **21.** 120°
23. 55°   **25.** 90°   **27.** 46°   **29.** 18°
31. 153°   **33.** perpendicular   **35.** 21

Page 337

1. 16, 22, and 29 dimes   **3.** C
5. Silvia: blue; Rhoda: green; David: brown
7. about $4 billion   **9.** 24 mi per gal

Pages 340–341

3. trapezoid   **5.** square   **7.** rectangle
or square   **9.** square   **11.** rectangle
13. quadrilateral   **15.** rhombus   **17.** square
19. quadrilateral   **21.** parallelogram,
trapezoid   **23.** 11 in. × 15 in.   **27.** 84°
29. 0.08

13. cannot **15.** can **17.** 9 in. **19.** 55°
21. $\overrightarrow{XY}$, ray XY **23.** 0.9

Page 345

1. Divide the diameter by 2. **3.** O
7. M **9.** $\overline{MY}, \overline{MW}$ **11.** Possible answer:
WY **13.** The length of the radius increases.
15. 25 min **17.** trapezoid **19.** 20

Chapter 18

Page 349

1. cone **3.** rectangular prism
5. triangular prism **7.** rectangular pyramid
9. 4; 4; 6 **11.** 5; 5; 8 **13.** 7; 10; 15

Pages 351–352

1. octagonal prism; pentagonal pyramid;
polyhedron **3.** cone; no **5.** square
pyramid; yes **7.** cylinder; no **9.** pentagonal
pyramid; yes **11.** False; a cone has one flat
surface. **13.** true **15.** False; some
pyramids have triangular bases **17.** hexagon
21. parallelogram, square, or rectangle
23. sector **25.** $3\frac{6}{7}$

Pages 354–355

3. triangular pyramid **5.** rectangular prism
7. rectangular pyramid **19.** a rectangle
21. triangular prism **23.** 0

Page 359

1. 20 balls of clay **3.** D **5.** Monday; 8°C
9. 9 sides

Chapter 19

Page 363

1. congruent **3.** congruent **5.** yes **7.** no
9. no **11.** yes **13.** parallel **15.** parallel
17. intersecting

Pages 366–367

3. $\overline{HJ} \cong \overline{KL}$ **5.** 90°, 48°, no **7.** 48°, 48°, yes
9. $\overline{WX} \cong \overline{YZ}$ **11.** 140°, 135°, no **21.** 66°
23. $1\frac{1}{9}$

Pages 369–370

1. 2 congruent angles; 2 congruent line
segments **3.** 19 in. **5.** 0.15 m **7.** 26°
9. 4.05° **13.** 8.5 ft **15.** 49.675 cm
17. 32.5° **19.** 71.3° **27.** a smaller square
29. 38° **31.** 60° **33.** $\frac{12}{25}$

Page 373

1. yes; no **3.** both **5.** neither **7.** both
9. neither **11.** both **13.** false
15. Yes; they are the same shape but not
necessarily the same size. No; squares come
in different sizes. **17.** 36.5° **19.** 1

Chapter 20

Page 383

1. equivalent **3.** polygon **5.** 27 **7.** 20
9. 6 **11.** 5 **13.** 7 **15.** 6 **17.** $p = 7$
19. $m = 5$ **21.** $d = 26$ **23.** $y = 17$
25. $m = 240$ **27.** $h = 1$ **29.** neither
31. similar **33.** both

Pages 385–386

1. Multiply or divide both terms of a ratio
by the same number. **3.** $\frac{1}{2}; \frac{8}{16}$ **5.** $\frac{3}{4}; \frac{6}{8}$
7. $\frac{150 \text{ points}}{10 \text{ games}}$; 15 points per game **9.** $\frac{90 \text{ words}}{2 \text{ min}}$;
45 words per min **11.** $\frac{\$15}{6 \text{ lb}}$; \$2.50 **13.** $\frac{1}{2}, \frac{3}{6}$
15. $\frac{1}{3}, \frac{2}{6}$ **17.** $\frac{3}{8}, \frac{12}{32}$ **19.** $\frac{10}{18}, \frac{15}{27}$ **21.** $\frac{\$15}{5 \text{ tapes}}$; \$3
per tape **23.** $\frac{60 \text{ mi}}{3 \text{ gal}}$; 20 mi per gal **25.** $\frac{\$2.10}{6 \text{ fish}}$;
\$0.35 per fish **27.** $\frac{4}{2}$ **29.** 15 **31.** 60
33. 4:1 **35.** Since C has 6 boxes and B has 3
boxes, double the price of B and compare it
to the price of C. **37.** \$2.10 **39.** squares of 1,
2, 3, 4…; 25 **41.** $2 \times 3 \times 3 \times 3$

Page 389

1. $27\frac{1}{2}$ lb **3.** 81 mi **5.** J **7.** 600 in.^{2}
9. 98 **11.** 19 salespeople **13.** 60
15. He misinterpreted the labels. Four people
sold between 21–30 cars each.

Pages 392–393

1. Yes, compare the small triangle with the large triangle: $\frac{20}{30} = \frac{2}{3}, \frac{12}{18} = \frac{2}{3}$, and $\frac{22}{33} = \frac{2}{3}$.
3. $\overline{RS}$ and $\overline{MN}$, $\overline{ST}$ and $\overline{NO}$, $\overline{TU}$ and $\overline{OP}$, $\overline{UR}$ and $\overline{PM}$; $\angle R$ and $\angle M$, $\angle S$ and $\angle N$, $\angle T$ and $\angle O$, $\angle U$ and $\angle P$; $\frac{2}{1}$ or $\frac{1}{2}$ **7.** No; ratios are not equivalent.   **9.** Yes   **11.** $\angle T = 70°$; $\angle Y = 110°$; $ST = 3.5$ m   **13.** No; some or all pairs of sides may not have equivalent ratios.
17. 45 ft $\times$ 60 ft   **21.** $y = 3$

Pages 395–396

1. The ratios of corresponding sides of similar figures are equal   **3.** 12 in.
5. $x = 39$ ft   **7.** $\frac{n}{8} = \frac{3}{8}$; $n = 3$ **9.** $\frac{30}{n} = \frac{18}{24}$; $n = 40$ cm   **11.** $\frac{n}{6.2} = \frac{2.5}{5}$; $n = 3.1$ cm
13. 24 in.   **15.** No. Corresponding sides do not have the same ratio.   **17.** 19

Pages 398–399

1. The real bike is 36 times the bike in the drawing.   **5.** 28   **7.** 30   **9.** 20   **11.** 4
13. 96 boxes   **15.** Each length of 1 cm on the drawing represents a length of 2 m on the actual object.   **17.** false   **19.** 30

Pages 400–401

1. 35 mi   **3.** 50 mi   **5.** 425 mi   **7.** 240 mi
9. 420 mi   **11.** 1,440 mi   **13.** 370 mi
15. 6 in.   **17.** $2\frac{1}{2}$ in.   **19.** about $1\frac{1}{4}$ hr
21. the auto club map   **23.** 4 in.   **25.** 9

Chapter 21
Page 405

1. ratio   **3.** 0.25   **5.** 0.1   **7.** 0.8   **9.** 0.2
11. 0.81   **13.** 17.28   **15.** 26.65   **17.** 28.8
19. 34.2   **21.** $\frac{1}{6}$ **23.** $\frac{2}{21}$ **25.** $\frac{3}{5}$ **27.** $\frac{16}{21}$
29. 0.45   **31.** 14%   **33.** 0.4   **35.** 0.53
37. 1   **39.** 0.03

Page 407

1. 63%   **3.** 16%   **5.** 31%   **7.** 60%   **9.** >
11. =   **13.** 8%   **15.** 18%   **17.** 84%   **19.** <
21. >   **23.** 64%   **25.** $2\frac{1}{2}$ in.   **27.** 13.015

Pages 410–411

1. move left 2 places, move right two places
3. 25%   **5.** 60%   **7.** $\frac{1}{2}$ **9.** $\frac{3}{4}$ **11.** $1\frac{4}{5}$
13. 0.08   **15.** 0.002   **17.** 15%   **19.** 9%
21. 0.7%   **23.** 250%   **25.** 12.5%   **27.** $\frac{22}{25}$
29. $\frac{1}{25}$ **31.** $\frac{1}{3}$ **33.** 0.07   **35.** 2.2
37. >   **39.** <   **41.** 20%   **43.** What percent of the instruments are woodwinds?
45. 26 pg per hr

Pages 414–415

1. Simply double the value of 25% of the number.   **3.** 6   **5.** 36   **7.** 7.8   **9.** 21
11. 20   **13.** 44   **15.** 1.08   **17.** 59.64
19. 18.48   **21.** 42.4   **23.** 15.3   **25.** 123
27. 3.85   **29.** 57   **31.** $1.20   **33.** $2.37
35. $4.83   **37.** $5   **39.** $0.80   **41.** 25%; 15
43. 10%; 0.6   **45.** 140,000,000 mi^2 **49.** 0.1%
51. similar

Pages 420–421

3. $60   **5.** $76.80   **7.** $100   **9.** $33.53
11. $46.62   **13.** $38.08   **15.** $30.72
17. $4.65   **19.** $46.20   **21.** $61.51
23. $0.68 per lb   **27.** $b = 17$

Page 423

1. It is a formula to find simple interest. Interest equals principal · rate · time.   **3.** $4
5. $1,728   **7.** $208   **9.** $1.40, $7.00
11. $82.80, $414.00   **13.** Bank B. At Bank A, he will owe $1,255 while at Bank B he will owe $1,200.   **15.** He used 0.6 for 6%. The correct simple interest is $36.   **17.** 5 in.
19. 10

Chapter 22
Page 427

1. impossible   **3.** equally likely   **5.** 0.5, 50%
7. 0.75, 75%   **9.** 0.1, 10%   **11.** 0.6, 60%
13. 0.72, 72%   **15.** 0.12, 12%   **17.** 0.95, 95%
19. $\frac{1}{3}$ **21.** $\frac{1}{2}$ **23.** $\frac{2}{3}$ **25.** $\frac{2}{5}$ **27.** $\frac{7}{8}$ **29.** $\frac{1}{3}$
31. $\frac{7}{12}$ **33.** $\frac{1}{4}$ **35.** certain   **37.** impossible

Pages 430–431

1. 1, 2, 3, 4, 5, 6   **3.** $\frac{1}{4}$, 0.25, 25%   **5.** $\frac{1}{4}$, 0.25, 25%   **7.** $\frac{3}{4}$, 0.75, 75%   **9.** $\frac{1}{4}$, 0.25, 25%   **11.** $\frac{0}{8}$, 0, 0%   **13.** $\frac{1}{2}$, 0.50, 50%   **15.** $\frac{1}{6}$ **17.** $\frac{1}{3}$ **19.** $\frac{1}{2}$ **21.** $\frac{1}{3}$ **23.** 1   **25.** =   **27.** >   **29.** <   **31.** 0.6   **33.** $\frac{5}{12}$ **35.** 77%   **37.** the probability the event will not occur   **41.** 9 blue; 15 are not blue   **43.** $^-$7.75, $7\frac{1}{4}$, $7\frac{3}{8}$, 7.5

Page 433

1. *too much*; $\frac{4}{7}$ **3.** *too little*; need the 1970 land speed record   **5.** D   **7.** 4 hr   **9.** 8,528 steps

Page 435

1. $\frac{5}{8}$ **3.** 89

Page 437

3. $\frac{1}{5}$ **5.** $\frac{1}{10}$ **7.** $\frac{2}{15}$ **9.** $\frac{1}{4}$ **11.** $\frac{2}{5}$ **13.** $\frac{2}{5}$ **15.** 15 times   **17.** number of times 4 lands ÷ total number of rolls   **19.** $\frac{4}{7}$ **21.** 9.75

Chapter 23

Page 441

1. sample space   **3.** outcome   **5.** $\frac{1}{3}$ **7.** $\frac{3}{16}$ **9.** $\frac{2}{5}$ **11.** $\frac{1}{8}$ **13.** $\frac{1}{25}$ **15.** about 26 in.   **17.** 100 students   **19.** 16 students

Page 443

1. 12 choices   **3.** B   **5.** 1,870,737   **7.** 4 groups of 10; 3 groups of 8

Pages 445–446

1. 24   **3.** 9 outcomes   **5.** 36 outcomes   **7.** 16 outcomes   **9.** 18 outcomes   **11.** 216 outcomes   **13.** yes; 12 · 4 · 8 = 384, and 384 > 365   **15.** $\frac{1}{5}$ **17.** 163°

Pages 448–449

1. 16%; 36%; yes, because there are more odd numbers on the spinner than even numbers, and the probability of two odd numbers is greater.   **3.** dependent   **5.** $\frac{1}{36}$ **7.** 0   **9.** $\frac{1}{20}$ **11.** $\frac{1}{5}$ **13.** independent   **15.** $\frac{1}{36}$; $\frac{1}{30}$ **17.** $\frac{1}{6}$; $\frac{1}{5}$ **19.** $\frac{1}{9}$; $\frac{2}{15}$ **21.** $\frac{1}{27}$; 0   **23.** $\frac{1}{108}$; 0   **27.** 15 outcomes   **29.** 18 outcomes

Page 451

1. Write and solve the proportion $\frac{5}{75} = \frac{n}{210}$ **3.** $\frac{7}{20}$, 0.35, or 35%   **5.** about 102 sixth graders   **7.** about 600 cars   **9.** $\frac{1}{3}$ **11.** $\frac{1}{2}$

Chapter 24

Page 461

1. feet   **3.** multiply   **5.** 3   **7.** 56   **9.** 4   **11.** 24   **13.** 12   **15.** 5   **17.** 1,000   **19.** 10   **21.** 1,000   **23.** 2   **25.** 4,000   **27.** 9   **29.** $n = 45$ **31.** $n = 9$ **33.** $n = 10$

Page 463

1. Use the ratio $\frac{4 \text{ qt}}{1 \text{ gal}}$ on one side of the proportion and the number of quarts over x gallons on the other side.   **3.** 4   **5.** 128   **7.** 12   **9.** $\frac{1}{2}$ **11.** $1\frac{1}{2}$ **13.** 60   **15.** $2\frac{1}{4}$ **17.** $3\frac{1}{4}$ **19.** =   **21.** $4\frac{1}{2}$ yd   **25.** $1\frac{13}{55}$ **27.** 2.361

Page 465

1. Use the ratio $\frac{10 \text{ dm}}{1 \text{ m}}$ on one side of the proportion and the number of decimeters over x meters on the other side.   **3.** 0.005   **5.** 9,000   **7.** 200,000   **9.** 0.5   **11.** 1.2   **13.** 440,000   **15.** 18,000   **17.** 425   **19.** >   **21.** =   **25.** $\frac{1}{3}$ **27.** 40%   **29.** $^-$9

Page 467

1. greater; it takes 1.61 km to make 1 mi.   **3.** 25.4   **5.** 427.7   **7.** 10   **9.** 76.2   **11.** 4.75   **13.** 26.37   **15.** <   **17.** >   **19.** 1.38 in.   **21.** 18°C   **23.** $^-$10   **25.** $\frac{7}{6}$

Pages 470–471

1. gram; because the gram is a smaller unit than the kilogram **3.** 2 cm; 23 mm **5.** 85 in. **7.** 65 oz **9.** meter, yard, or foot **11.** gram or ounce **13.** $1\frac{1}{2}$ in.; $1\frac{3}{4}$ in. **15.** $\frac{3}{4}$ in.; $\frac{7}{8}$ in. **17.** 8 ft **19.** 9 oz **21.** millimeter or part of an inch **23.** = **25.** = **27.** To the nearest millimeter because a millimeter is a tenth of a centimeter, which is smaller than a half centimeter. **31.** 17.055 L **33.** $^-96$

Page 473

1. estimate; no **3.** estimate; no **5.** C **7.** $9\frac{3}{20}$ in. **9.** $2\frac{3}{8}$ in. below **11.** 23 years old

Chapter 25
Page 477

1. perimeter **3.** 32 ft **5.** 20 cm **7.** 104 in. **9.** 4 **11.** 36 **13.** 4 **15.** 3,000 **17.** 5,000 **19.** 0.00009 **21.** 88 **23.** 2,625 **25.** 526.5 **27.** 60 **29.** 73.6

Page 481

3. 13.57 m **5.** 9 ft or 108 in. **7.** 43 m **9.** $x = 13.5$ cm **13.** 5.40 **15.** $a = 1,264$

Page 483

1. 66 in. **3.** B **7.** 42.4% **9.** 1 in.

Pages 486–487

1. They both use π; one uses the diameter and the other uses 2 times the radius, which is equal to the diameter. **3.** 16 m **5.** 12 cm **7.** 283 yd **9.** 27 in. **11.** 33 cm **13.** 20 in. **15.** 9 yd **17.** 220 cm **19.** 316 in. **21.** 22 ft **23.** 16 cm **25.** 4.5 ft **27.** 9.3 cm **29.** The circumference is twice as long. **31.** 88 yd **33.** 149 m

Page 489

1. 25 **3.** 41 **5.** no **7.** yes **9.** 109.9 mm **11.** 314 in.

Chapter 26
Page 493

1. square **3.** faces **5.** 144 **7.** 400 **9.** 81 **11.** 5.76 **13.** 2,401 **15.** 16,641 **17.** 25 **19.** 192 **21.** 16 **23.** 5 faces **25.** 4 faces

Pages 496–497

1. The area of the triangle is $\frac{1}{2}$ the area of the rectangle. **3.** about 20 m^2 **5.** 117 in.^{2} **7.** 45.5 in.^{2} **9.** about 6 m^2 **11.** 558.25 mm^2 **13.** 0.2 m^2 **15.** 22.5 ft^2 **17.** 273 yd^2 **19.** 12 **21.** 60%

Pages 499–500

1. $A = \frac{1}{2}h(b_1 + b_2) = \frac{1}{2} \times 4 \times (4.2 + 6.5) = 21.4$; 21.4 m^2 **3.** 10 ft^2 **5.** 102.3 m^2 **7.** 136.95 m^2 **9.** 40 ft^2 **11.** 1,008 cm^2 **15.** 24 m^2 **17.** 0.003

Page 501

1. 3 m^2 **3.** 154 m^2

Page 503

3. 28 cm^2 **5.** 1,809 ft^2 **7.** 50 yd^2 **9.** 7 m^2 **11.** 50 mm^2 **13.** 3,419 in.^{2} **15.** 77 cm^2 **17.** 804 ft^2 **19.** 100 ft^2 **21.** 3

Pages 506–507

1. Find the area of each pentagonal face and the area of the five rectangular faces and add. **3.** 184 ft^2 **5.** 336 cm^2 **7.** 340 m^2 **9.** 73.5 m^2 **11.** 108 cm^2 **13.** 69.36 cm^2 **15a.** $3 \times 6 \times 12$; 252 m^2 **15b.** $3 \times 6 \times 6$; 144 ft^2 **15c.** $10 \times 5 \times 15$; 550 in.^{2} **15d.** $2 \times 8 \times 32$; 672 cm^2 **17.** 2 cans **19.** 5 in. **21.** 113 ft^2 **23.** 0.25

Chapter 27
Page 511

1. dimensions **3.** height **5.** 64 **7.** 27 **9.** 216 **11.** 0.008 **13.** 1,728 **15.** 20 ft^2 **17.** 21 m^2 **19.** $58\frac{7}{12}$ ft^2 **21.** 52.36 m^2 **23.** 141 cm^2 **25.** 346 cm^2

Pages 514–515

1. Find $26 \times 3 \times 18$, or 1,404 cubes.
3. 24 in.^{3} **5.** 48 cm^3 **7.** 324 ft^3
9. 294 ft^3 **11.** 144 in.^{3} **13.** $x = 10$ m
15. 189 ft^3 **17.** 648 ft^3 **19.** 45.63

Page 517

1. twice the volume of the original container
3. B **7.** 10 shirts, 0 shorts; 0 shirts, 8 shorts;
5 shirts, 4 shorts **9.** Mike, Sharon, Jasmine,
Hugh

Page 519

1. Both include the area of the base times the
height; prism: $V = bh$; pyramid: $V = \frac{1}{3}Bh$
3. 48 m^3 **5.** 224 cm^3 **7.** 24 cm^3, or 24,000
mm^3 **9.** 4,500 ft^3 **11.** 24 yd^3, or 648 ft^3
15. 110 **17.** 0.79

Page 520

1. about 75 cm^3

Pages 522–523

1. 6,029 in.^{3} **3.** πr^2 represents the area of
the base and h represents the height.
5. about 346 in.^{3} **7.** about 942 in.^{3}
9. about 1,409 cm^3 **11.** about 311 cm^3
13. about 2,374 cm^3 **15.** about 7,436 ft^3
17. The volume is about eight times as large.
21. about 4 ft^3 **23.** $\frac{1}{2} \div \frac{1}{10}$

Chapter 28
Page 533

1. equation **3.** $<$ **5.** $<$ **7.** $>$ **9.** $>$
11. $<$ **13.** $>$ **15.** $<$ **17.** $>$ **19.** 10; 21
21. 25; 25.5 **23.** 15 **25.** $^-5$ **27.** 63
29. 29 **31.** $\frac{3}{14}$ **33.** 3.1

Page 535

1. the eighth day **3.** A **5.** 7 videos
7. 660 ft **9.** 40 ft^2

Pages 537–538

1. Multiply by 4; 2,048; 8,192; 32,768 **3.** Add
15 to each successive term **5.** 43, 54, 65
7. Divide each term by 3. **9.** Add 0.89 to
each term. **11.** 335, 485, 665 **13.** 81, 76, 70
15. 9, 12.7, 16.4, 20.1, . . . **17.** add \$54; \$462
21. 754 ft^3 **23.** 264

Pages 541–542

3. $b = a - 6$ **5.** $d = c + 1.1$ **7.** $l = 4w$
9. $g = k \div 2$; 27 **11.** $d = c \div \,^-4$; $^-8$
15. in $y^2 = x$, if $x = 1$, $y = 1$ or $y = \,^-1$
17. $t = 0.75n$; \$37.50 **19.** \$209 **21.** $2\frac{5}{104}$

Page 545

9. $5^0, 5^1, 5^2, 5^3, \ldots$ **11.** 16 prisms; 25 prisms
13. $y = \,^-9, ^-3, 3, 9, 15$ **15.** 5×13

Chapter 29
Page 549

1. congruent **3.** slide **5.** slide **7.** turn
17. 25° **19.** 135°

Pages 551–552

1. The figure does not change in size or
shape. **3.** rotation or reflection
5. reflection **7.** rotation, reflection
9. rotation or translation **11.** rotation
13. translation, reflection, reflection
19. horizontal reflection **25.** $\frac{5}{8}$

Pages 554–555

3. yes **5.** no **11.** yes **13.** no
19. 288 sq yd. **21.** rotation **23.** $\frac{11}{12}$

Page 557

1. triangle and square; square and octagon
3. C **5.** 63° **7.** a **9.** Gerta, Bernie, Nash,
Lisa

Page 559

1. Yes, the figure's size and shape do not change. **3.** 8 ways **5.** 6 ways **7.** 8 ways **9.** 12 ways **11.** rotate it 180° **13.** reflection **15.** 24 cm

Pages 562–563

1. Line symmetry means that the figure can be folded into two congruent halves that are mirror images. Rotational symmetry means that the figure matches itself when rotated less than 360°. **11.** yes; $\frac{1}{4}$; 90° **13.** yes; $\frac{1}{2}$; 180° **27.** yes; $\frac{1}{5}$; 72° **29.** no **31.** 8 pieces; yes **33.** To have rotational symmetry, it must match up with a rotation of less than 360° **35.** 125.6 in.^{3}

Chapter 30

Page 567

1. ordered pair **3.** perpendicular **5.** $=$ **7.** $>$ **9.** $<$ **11.** $>$ **13.** ($^-$2,2) **15.** (0,0) **17.** (2,1) **19.** (0,3) **21.** ($^-$2,$^-$2)

Page 569

1. Yes. **19.** $p > ^-11$ **21.** $a < 20{,}320$ **25.** 37.5% **27.** $\frac{3}{5}$

Pages 572–573

1. (5,4) is above the x-axis. ($^-$5,$^-$4) is below the x-axis. **3.** (2,4) **5.** (0,0) **7.** (5,$^-$5) **9.** (0,$^-$5) **11.** (2,$^-$3) **13.** A, H **15.** B **17.** (2,6) **19.** (2,$^-$2) **21.** (0,7) **23.** (3,0) **25.** (6,$^-$4) **27.** ($^-$7,$^-$6) **29.** A, L, J **31.** B, H, K **41.** triangle; 28 sq units **43.** It is 0; it is 0. **45.** $y \geq ^-7$ **47.** 157 cm

Page 575

1. a table, a graph of ordered pairs. **7.** (45,15), (48,16), (51,17), (54,18); $w = \frac{1}{3}c$ **9.** Quadrant I **11.** 40 cm^2

Page 577

1. C **3.** $y = 9x$ **5.** 1,333 mi **7.** 11:00 A.M. **9.** about 800

Pages 582–583

3. $A'(0,3)$, $B'(2,3)$, $C'(2,1)$, $D'(0,1)$ **5.** $A'(0,0)$, $B'(^-6,0)$, $C'(^-4,2)$, $D'(^-2,2)$ **7.** $E'(0,^-1)$, $F'(3,^-1)$, $G'(3,^-4)$, $H'(0,^-4)$ **9.** $E'(1,^-3)$, $F'(4,^-3)$, $G'(4,^-1)$, $H'(1,^-1)$ **11.** No. **13.** $y = 7x$ **15.** $\frac{1}{4}$

Index

Draw a Diagram strategy, 2, 194–195, 482–483
Draw a Picture strategy. *See* Draw a Diagram strategy

Edges, 358–359, H20, H50
E-Lab activities, 43, 75, 129, 143, 162, 168, 181, 191, 208, 249, 277, 286, 296, 305, 353, 369, 387, 417, 435, 459, 484, 501, 520, 534, 544, 581
Endpoints , 318, 320
Equations
    addition, 286–289, H46
    decimals in, 82–83
    division, 298–299
    fractions in, 216–217
    integers in, 284–291, H46
    models of, 286, 296, 304
    multiplication, 296–299, H47
    proportions, 387–389, 394–401, 462–467, H26, H51
    solving, 30–31, H46, 82–83, 216–217, 286–291, 296–299, H11, H21–H47
    subtraction, 286, 290–291, H46
    two-step, 304–305
    writing, 82–83, 284–285, 388–389, 540–542, 574–577, H46, H59
Equilateral triangles, 332
Equivalent decimals, 52–55, H34
Equivalent fractions, 160–163, H21, H39
Equivalent ratios, 384–386, H51
Error analysis. *See* What's the Error?
Estimate or Find Exact Answer, 472–473
Estimation
    of area, 494, 496
    of circumference, 484
    by clustering, 16–19, 58–59
    compatible numbers, 17–19, 58–59
    of decimals, 58–59, 67–69
    of differences, 16–19, 177–178
    of fractions, 176–179, 200–201
    multiples of ten, 16–19
    overestimate, 17–19
    of percent, 412–415
    of perimeter, 478
    of products, 17–19, 200–201
    of quotients, 17–19, 200–201
    reasonableness, 66–69
    and rounding, 16–19, 58–59
    of sums, 16–17, 58–59, 176–179, H32, H34, H40
    underestimate, 17–19
    of unknown quantities, 124–125
    of volume, 512
    when to estimate, 472–473
Events
    compound, 444–446
    dependent, 448–449
    impossible versus certain, 429
    independent, 447–449, H54
Expanded form, 52–55
Experimental probability, 434–437, H53
Exponents, 40–41, 263, 276–279, H15, H33
    negative, 263
Expressions
    algebraic, 28–29, 69, 82–83, 212, 216–217, 246, 270–275, 278–279, H18, H32
    evaluating, 82–83, 216–217, 272–275, 278–279, H15, H18
    numerical, 28–29, 278–279, H15
    writing, 28–29, 82–83, 216–217, 270–271, H45
Extra Practice, H32–H61

Faces, 350–355, 358–359, H20, H49
Factor trees, 148–149
Factors
    common, 151
    finding prime factors of a number, 148–149, H38
    greatest common, 151–153
    missing, 30–31
    of numbers, H8
    prime, 148–149
Fahrenheit temperature, 301–303
Feet, 462–463, H25, H55
Fibonacci sequence, 538
Find a Pattern strategy, 7, 336–337, 534–535
Finding a rule. *See* Algebra *and* Functions
Flips, H30
    See also Reflections
Fluid ounces, 462–463
Focus on Problem Solving, 1–13
Formulas
    for area
        of circles, 501–503
        of parallelograms, 498–500
        of rectangles, 494–497
        of squares, 495–497, H28
        of trapezoids, 499–500
        of triangles, 495–497
    Celsius to Fahrenheit, 301–303
    for circumference of circles, 484–487
    distance, 124–125, 300–303
    Fahrenheit to Celsius, 301–303
    generating, 495, 498–499, 501
    for perimeter of polygons, 479
    for simple interest, 422–423
    for surface area of solids, 504–507
    for volume
        of cylinders, 521–523
        of rectangular prisms, 513–515
        of triangular prisms, 514–515
    See also Table of Measures *on back cover*
Fractals, 544
Fractions
    adding, 180, 182–185, H10, H16, H40
    of circles, 160, 208
    comparing, 166–167, H12, H29
    decimals and, 168–171, H22, H23
    dividing, 208–213, H41
    in equations, 216–217
    equivalent, 160–163, H21, H39
    estimating, 176–179
    finding common denominators, 180–185, H40
    greatest common factor (GCF), 203–205
    improper, 164
    least common denominator (LCD), 182–185
    least common multiple (LCM), 150–153, 180–182
    like, 180, H10
    in measurements, 462–463
    in mixed numbers, 164–165, H11
    models of, 160–161, 180, 202, 208–209, H8
    multiplying, 202–205, H23, H41
    negative, 230–235, H42
    on number lines, 166–167, 176–178
    ordering, 166–167, 231–232
    as parts of a group, H8
    as parts of a whole, H8
    percent and, 170–171, 406–411, H23, H39
    probability and, 428–431, 436–437
    as ratios, 384–386, H51
    rounding, 176–178, 200, H10

two-dimensional figures. *See also* Plane figures

volume
    of cylinders, 520–523, H58
    of prisms, 512–515, H58
    of pyramids, 518–519, H58

Glossary, H73–H82

Grams, 464–465, H25, H55

Graphs
    analyzing, 120–135
    bar, 120–123, H6, H37
    box-and-whisker, 129–131, H37
    choosing, 120–123, 126–128
    circle, 122–123, 416–417, H52
    comparing, 120–123, 132–135
    coordinate plane, 570–575, 578–579, H61
    of equations, 578–579
    of functions, 574–575, H61
    histograms, 127–128, H37
    inequalities, 568–569, H61
    interpreting, 120–135
    intervals of, 127–128
    key of, 120
    labels on, 416–417
    line, 121–125, H37
    line plots, 102–105
    making, 102–105, 120–123, 126–128, 129, H36–H37
    misleading, 132–135, H37
    multiple-bar, 120–123, H37
    multiple-line, 121–123, H37
    ordered pairs, 570–575, 578–579, H61
    predictions from, 124–125
    as problem solving strategy, 8
    scatterplots, 139
    stem-and-leaf plots, 126–128, H7, H37
    transformations, 580–583
    tree diagram, 444–446, H54

"Greater than" symbol, 568–569, H3

Greatest common factor (GCF), 151–153, 161–163, 203–205

Guess and Check strategy. *See* Predict and Test strategy

H

Harcourt Math Newsroom Videos, 18, 234, 273, 338, 390, 484, 539

Health, 50, 174

Height
    of parallelograms, 498–500
    of prisms, 512–515
    of pyramids, 505–506, 518–519
    of trapezoids, 498–500
    of triangles, 495–497

Hexagonal prisms, 350–352

Hexagonal pyramids, 350–352

Hexagons, 336

Histograms, 127–128, H37

History, 154, 166, 440

Hour, 462–463

Hypotenuse, 488–489

I

Identity Property
    of Addition, H2
    of Multiplication, H2

Images
    reflection, 550–552, H60
    translation, 550–552, H60

Impossible events, 429, H24

Improbable events. *See* Unlikely outcomes

Improper fractions, 164

Inches, 462–463, H25, H55

Independent events, 447–449, H54

Indirect measurement, 394–396, H51

Inequalities, 24, 568–569, H61

Input/output tables, 539–542, H29

Integer equations, 284–291, H45

Integers
    adding, 242–247, H14, H43
    comparing and ordering, 228–229, H29
    concept of, 228–229, H12
    dividing, 253–255, H43
    in equations, 284–291, H45
    in expressions, 257
    models of, 242–244, 246, 248–249, 252
    multiplying, 252–255, H43
    negative, 228–229, 263
    on a number line, 228–229, 244, 246, 252, H12, H29
    opposites, 228–229
    subtracting, 248–251, H14, H43

Interest, simple, 422–423, H52

Internet activities. *See* E-Lab activities

Interpret the remainder, 80–81

Intersecting lines, 326–327, H48

Intervals, 127–128

Inverse operations, 306, H16

Irregular figures
    area of, 494, 496
    estimate perimeter of, 478

Isosceles triangles, 332, 342

J

Journal. *See* Write About It *and* Write a Problem

K

Key, of graph, 120

Kilograms, 464–465, H25, H55

Kiloliters, 464–465, H25, H55

Kilometers, 464–465, H25, H55

L

Labels, on graphs, 416–417

Lateral faces, 350–352

Least common denominator (LCD), 182–187

Least common multiple (LCM), 150–153, 166–167, 180–182

Leg, 488–489

Length
    customary system, 462–463, H25
    measuring, 468–471, H55
    metric system, 464–465, H25, H55

"Less than" symbol, 568–569, H3

Like fractions, 180, H10

Likely events, 428–431, H24

Line graphs, 121–125, H37

Line plots, 102–105

Line segments
    bisecting, 368–370, H50
    congruent, 364–367, H50
    defined, 318–319

Line symmetry, 560–563, H30, H60

Linear relationships, 578–579

to millions, 52–55, H3
models of, H2–H3
periods, H3
powers of ten and, 40
stem-and-leaf plots and, 126
to ten-thousandths, 52–55
to thousandths, 52–55, H3

Plane figures
area of, 494–503, H57
congruent, 390–393
drawing, 342–343, H49
perimeter of, 478–481, H56
similar, 390–396, H51
symmetry in, 560–563
transforming, 550–552, 580–583
See also Circles, Geometry, *and* Polygons

Planes, 318–319, H48

Plots
line, 102–105
stem-and-leaf, 126–128, H7, H37

Points, 318–319

Points, with ordered pairs, 570–573

Polygons
classifying, 332–335, 338–341, H19, H49
hexagons, 336
octagons, 336
parallelograms, 338–341, H49
pentagons, 336
quadrilaterals, 336, 338–341, H49
rectangles, 338–341, H49
regular, 336
rhombuses, 338–341, H49
squares, 338–341, H49
trapezoids, 338–341, H49
triangles, 332–335, H49
See also Geometry

Polyhedron, 350–352, H50

Populations, 94–97

Positive integers, 228–229

Possible outcomes, 428, 444–445, H54

Pounds, 462–463, H55

Powers of ten, 40, 87, 263, H13
See also Exponents

Precision, 468–471, H55

Predict and Test strategy, 4, 26–27

Predictions
about populations, 450–451, H54
of certain events, 436–437, H53
of outcomes, 436–437

Prefixes, in metric units, 464–465, H25, H55

Prime factorization, 148–150

Prime numbers, 148, H7

Prisms
bases of, 350–352, H50
classifying, 350–352, H50
heights of, 512–515
models with nets, 356–357, 512
surface area of, 504–507, H57
volume of, 512–515, H58

Probability
calculating for
one event followed by another independent event, 447–449, H54
two disjoint events, 444–446
certain events, 429, H24
compound events, 444–445, H54
dependent events, 448–449, H54
equally likely events, 428–431
estimating future events, 450–451, H54
experimental, 434–437, H53

Fundamental Counting Principle, 444–446, H54
geometric, 431
impossible events, 429, H24
independent events, 447–449, H54
likely outcomes, 429, H24
possible outcomes, 428, 444–445
predictions, 436–437, 450–451, H53–H54
random, 445
sample space, 428, 444–445
simulations, 434–435
that an event will not occur, 429–431, 447–449
theoretical, 428–431, 447–451, H53
tree diagrams, 444–446, H54
unlikely events, 429, H24
verify reasonableness, 430, 448

Problem solving. *See* Problem solving applications, Problem solving Linkup, Problem Solving on Location, Problem solving skills, Problem solving strategies, and Problem solving Thinker's Corner

Problem solving applications
art, 396
astronomy, 399
geography, 415
geometry, 247, 289, 573
graphs, 303
measurement, 569
number sense, 193, 255, 285
reasoning, 29, 99, 111, 123, 149, 205, 259, 279, 286, 302, 325, 335, 342, 345, 352, 386, 396, 415, 421, 423, 431, 465, 487, 497, 500, 507, 515, 541, 552, 563
science, 285, 303, 437
space, 303
technology, 163
use data, 19, 21, 38, 41, 47, 55, 57, 59, 61, 73, 114, 163, 171, 179, 184, 188, 291, 303, 337, 345, 433, 443, 451, 471, 473, 483, 507, 517, 542, 557, 569, 577

Problem Solving on Location 91A–91B, 143A–143B, 225A–225B, 267A–267B, 315A–315B, 381A–381B, 459A–459B, 531A–531B, 591A–591B

Problem solving skills
choose the operation, 214–215
compare strategies, 12
estimate or find exact answer, 472–473
interpret the remainder, 80–81
make generalizations, 576–577
multistep problems, 13
sequence and prioritize information, 46–47
too much or too little information, 432–433

Problem solving strategies
draw a diagram/picture, 2, 194–195, 482–483
find a pattern, 7, 336–337, 534–535
make a model, 3, 516–517, 556–557
make a table or graph, 8, 56–57, 100–101
make an organized list, 6, 154–155, 442–443
predict and test, 4, 26–27
solve a simpler problem, 9, 358–359
use logical reasoning, 11, 236–237
work backward, 5, 306–307
write an equation, 10, 388–389

Products. *See* Multiplication

Properties
Associative, 36–39, H2, H33
Commutative, 36–39, 274–275, H2
Distributive, 36–39, 206–207, 274–275, H2
of Equality, 287–291, 297–299
Identity, of One, H2
Identity, of Zero, H2
Reflexive, 311
Symmetric, 311
Transitive, 311
of Zero in Multiplication, H2

with mixed numbers, 186–191
in subtraction, 190–191
Regular polygons, 336
Remainders, 23–24, 80–81, H5
Repeating decimals, 169–171
Review. *See* Mixed Applications, Mixed Review and Test Prep,
Mixed Strategy Practice, *and* Standardized Test Prep
Rhombuses, 338–341, H49
Right angles, 320–321, 332, H18
Right triangles, 332–335, H49
Pythagorean Theorem, 488–489
Rotational symmetry, 560–563
Rotations, 550–552, 558–559, 580–583
Rounding
decimals, 58, H4
fractions, 176–178, 200, H10
mixed numbers, 186
whole numbers, 16, H4

S

Sales tax, 418–421
Sample space, 428, 431, 443, 444–446
Samples
of a population, 94–99
biased, 98–99, H36
characteristics of, 94–99
convenience, 95–97, H36
different ways of selecting, 94–97
limitations of, 94–99
random, 95–97, H36
responses to survey, 94–97
systematic, 95–97, H36
unbiased, 98–99, H36
when to use, 94–97
Scale, 398–401, H51
Scale drawings, 397–399, H51
Scalene triangles, 332
Scatterplots, 139
Science, 14, 20, 34, 36, 66, 69, 91A, 91B, 102, 107, 118, 120, 130,
143A, 143B, 158, 161, 164, 170, 186, 187, 192, 200, 216, 234,
240, 250, 259, 267A, 267B, 297, 332, 381A, 384, 387–388,
459A, 531A, 531B, 548, 553, 560, 573, 591A
Scientific notation, 87, 263
Sequence and prioritize information, 46–47
Sequences, 536–538, 586, H59
Shapes. *See* Geometry
Sides. *See* Polygons *and* Polyhedron
Similar figures
corresponding parts, 391–396, H51
identifying, 390–393, H22, H51
indirect measurement, 394–396, H51
proportions, 394–396, H51
ratios, 390–393, H51
Simple interest, 422–423, H52
Simplest form of fractions, 161–163, H9
Simulations, 434–435
Skills Review, H66–H68
Slides, H30
See also Translations
Social Studies, 30, 60, 103, 225A, 225B, 300, 315A, 315B, 381A,
394, 410, 459A, 476, 484, 487, 510
Solid figures
cones, 350–352, H50
cubes, 350–352, H50
cylinders, 350–352, H50

drawing, 353–354
edges of, 358–359, H20
faces of, 350–355, 358–359, H20, H50
prisms, 350–352, H50
pyramids, 350–352, H50
surface area of, 504–507, H57
transformations of, 558–559
views of, 353–355
volume of, 512–515, 518–519, H58
Solve a Simpler Problem strategy, 358–359
Solving equations, 30–31, 82–83, 216–217, 286–291, 296–299,
H11, H21, H46–H47
Sports, 16, 198, 282, 426, 460, 532
Square pyramids, 350–352
Square roots, 277–279, H45
Square unit, 494
Squares
area of, 495–497, H28
of numbers, 276, 278–279, H27, H45
Standard form, 52–55
Standardized Test Prep, 33, 49, 63, 85, 117, 137, 157, 173, 197,
219, 239, 261, 281, 293, 309, 329, 347, 361, 375, 403, 425, 439,
453, 475, 491, 509, 525, 547, 565, 585
Statistics
analyzing data, 100–115, 120–135, H5, H24
averages. *See* Measures of central tendency
bar graph, 120–123, H6, H37
circle graph, 122–123, 416–417
collecting data, 94–99, H36
frequency tables, 103–105, H56
histogram, 127–128, H37
line graph, 121–125, H37
measures of central tendency
mean, 106–111, H6, H36
median, 106–111, H6, H36
mode, 106–111, H6, H36
organizing data, 56–57, 100–105, 120–123, 126–131
outliers, 110–111, H36
range, 103, H6
samples
biased, 98–99, H36
convenience, 95–97, H36
of a population, 94–99
random, 95–97, H36
systematic, 95–97, H36
unbiased, 98–99, H36
stem-and-leaf plots, 126–128, H7, H37
surveys, 94–99
tally tables, 100–101
See also Data
Stem-and-leaf plots, 126–128, H7, H37
Straight angles, 320–321, H18
Stretching, 587
Student Handbook, H1–H110
Study Guide and Review, 88–89, 140–141, 222–223, 264–265,
312–313, 378–379, 456–457, 528–529, 588–589
Subtraction
addition as an inverse operation, H16
of decimals, 66–69, H35
equations, 286, 290–291, H46
estimation and, 16–19, 177–179
of fractions, 181–185, H10, H16, H40
of integers, 248–251, H14, H43
mental math and, 37–39
of mixed numbers, 187–193, H40
models of, 180–181, 186–188, 248–249
Property of Equality, 287–288
of rational numbers, 256–259, H43
of whole numbers, 20–21, 37–39, H4
Sums. *See* Addition

estimating, 512
of prisms, 512–515, H58
of pyramids, 518–519, H5

Weight, 462–463, H25, H55
What If? 26, 38, 56, 80, 100, 107, 113, 193, 201, 203, 236, 271, 289, 291, 306, 388, 442, 445, 482, 513, 514, 515, 516, 521, 582
What's the Error? 25, 45, 55, 79, 83, 108, 115, 123, 153, 165, 167, 189, 213, 217, 233, 247, 255, 275, 291, 299, 327, 341, 343, 389, 396, 423, 463, 487, 523, 535, 563, 569
What's the Question? 27, 31, 41, 57, 81, 105, 131, 155, 163, 185, 205, 237, 251, 255, 279, 289, 307, 325, 335, 370, 386, 411, 431, 449, 465, 483, 507, 517, 577, 583
Whole numbers
  comparing, H3
  ordering, H3
  place value of, H3
  rounding, 16, H4
  See also Addition, Division, Multiplication, *and* Subtraction
Word expressions, translating, 270–271, 284–285, H17, H46
Word form, 52–55
Work Backward strategy, 306–307
World Wide Web. *See* E-Lab activities
Write a Problem, 21, 59, 73, 101, 125, 163, 179, 195, 229, 271, 285, 303, 337, 407, 433, 443, 451, 467, 473, 481, 523, 538, 545, 575
Write About It, 19, 29, 39, 47, 61, 69, 86, 97, 99, 128, 138, 149, 165, 171, 193, 201, 215, 220, 235, 262, 310, 319, 327, 355, 367, 373, 376, 393, 399, 401, 437, 446, 454, 471, 519, 526, 555, 559, 573, 586

Write an Equation strategy, 10, 388–389
Write What You Know, 33, 49, 63, 85, 117, 137, 157, 173, 197, 219, 239, 261, 281, 293, 309, 329, 347, 361, 375, 403, 425, 439, 453, 475, 491, 509, 525, 547, 565, 585
Writing equations, 82–83, 284–285, 388–389, 540–542, 574–577, H46, H59
Writing in math. *See* What's the Question?, Write About It, Write a Problem, and Write What You Know

x-axis, 570
x-coordinate, 570

Yards, 462–463, H25, H55
y-axis, 570
y-coordinate, 570
Year, 462–463

Zero properties
  addition, H2
  multiplication, H2

Photo Credits

Table of Measures

METRIC	CUSTOMARY

Length

METRIC	CUSTOMARY
1 millimeter (mm) = 0.001 meter (m)	1 foot (ft) = 12 inches (in.)
1 centimeter (cm) = 0.01 meter	1 yard (yd) = 36 inches
1 decimeter (dm) = 0.1 meter	1 yard = 3 feet
1 kilometer (km) = 1,000 meters	1 mile (mi) = 5,280 feet
	1 mile = 1,760 yards
	1 nautical mile = 6,076.115 feet

Capacity

METRIC	CUSTOMARY
1 milliliter (mL) = 0.001 liter (L)	1 teaspoon (tsp) = $\frac{1}{6}$ fluid ounce (fl oz)
1 centiliter (cL) = 0.01 liter	1 tablespoon (tbsp) = $\frac{1}{2}$ fluid ounce
1 deciliter (dL) = 0.1 liter	1 cup (c) = 8 fluid ounces
1 kiloliter (kL) = 1,000 liters	1 pint (pt) = 2 cups
	1 quart (qt) = 2 pints
	1 quart (qt) = 4 cups
	1 gallon (gal) = 4 quarts

Mass/Weight

METRIC	CUSTOMARY
1 milligram (mg) = 0.001 gram (g)	1 pound (lb) = 16 ounces (oz)
1 centigram (cg) = 0.01 gram	1 ton (T) = 2,000 pounds
1 decigram (dg) = 0.1 gram	
1 kilogram (kg) = 1,000 grams	
1 metric ton (t) = 1,000 kilograms	

Volume/Capacity/Mass for Water

1 cubic centimeter (cm^3) → 1 milliliter → 1 gram

1,000 cubic centimeters → 1 liter → 1 kilogram

TIME

1 minute (min) = 60 seconds (sec)	1 year (yr) = 12 months (mo), or about 52 weeks
1 hour (hr) = 60 minutes	1 year = 365 days
1 day = 24 hours	1 leap year = 366 days
1 week (wk) = 7 days	